Chromosome Techniques
Theory and Practice

Chromosome Techniques
Theory and Practice

ARUN KUMAR SHARMA, D.SC., F.N.A.
Professor and Head,
Department of Botany, University of Calcutta

and

ARCHANA SHARMA, D.PHIL., D.SC.
Reader
Department of Botany, University of Calcutta

BUTTERWORTHS LONDON
UNIVERSITY PARK PRESS BALTIMORE

THE BUTTERWORTH GROUP

ENGLAND
Butterworth & Co (Publishers) Ltd
London: 88 Kingsway, WC2B 6AB

AUSTRALIA
Butterworth & Co (Australia) Ltd
Sydney: 586 Pacific Highway Chatswood, NSW 2067
Melbourne: 343 Little Collins Street, 3000
Brisbane: 240 Queen Street, 4000

CANADA
Butterworth & Co (Canada) Ltd
Toronto: 14 Curity Avenue, 374

NEW ZEALAND
Butterworth & Co (New Zealand) Ltd
Wellington: 26–28 Waring Taylor Street, 1

SOUTH AFRICA
Butterworth & Co (South Africa) (Pty) Ltd
Durban: 152–154 Gale Street

First published 1965
Reprinted 1967
Second edition 1972

ISBN 0 408 70307 5

Suggested UDC number 576.312.32: 578

Published 1972 jointly by
BUTTERWORTH & CO (PUBLISHERS) LTD, LONDON
and
UNIVERSITY PARK PRESS, BALTIMORE

Library of Congress Cataloging in Publication Data

Sharma, Arun Kumar, 1923
Chromosome techniques.

Includes bibliographies.
1. Chromosomes. 2. Cytology–Technique.
I. Sharma, Archana (Mookerjea) joint author.
II. Title.
QH605.S38 1972 547.8'732 71–39189
ISBN 0–8391–0628–9

Filmset by Filmtype Services Limited, Scarborough, Yorkshire
Printed in England by Fletcher & Son Ltd, Norwich

Preface to Second Edition

The last eight years, since the first edition went to press, have witnessed tremendous advances made in all aspects of chromosome methodology. In addition to overall refinements to the methods already in vogue, outstanding progress has been achieved in the fields of human and mammalian chromosome methodology, including the study of malignant cells, mammalian chromosome analysis, and chromosomes in culture, together with somatic cell hybridisation, electron microscopy, high-resolution autoradiography and other modes of quantitation. These advancements in technology, along with a wide appreciation of the first edition—which resulted in its reprinting—added to the criticisms and suggestions received from reviewers, prompted us to venture on a new edition of the book. Each chapter has been rewritten with suitable additions, and deletions of the relatively less important techniques. Several new chapters have been incorporated but limitations imposed by the cost of production did not allow us to deal with the details of instrumentation on which a number of treatises are available.

In addition to those scientists, who had kindly given us the original microphotographs of their preparations for the first edition, the present edition has been further enriched by the inclusion of valuable photographs presented by the following experts, to whom we wish to express our sincere gratitude: Dr. E. G. Barry (Dept. of Botany, University of North Carolina, Chapel Hill); Prof. W. Beermann (Max-Planck-Institute Tübingen); Prof. J. G. Gall and Dr. M. L. Pardue (Dept. of Biology, Yale University, New Haven); Prof. H. Harris (William Dunn School of Pathology, University of Oxford); Dr. T. C. Hsu (M. D. Anderson Hospital, University of Texas); Prof. B. John (Dept. of Genetics, University of Southampton); Prof. B. A. Kihlman (Dept. of Genetics, Royal Agricultural College, Uppsala); Dr. T. N. Khoshoo (National Botanic Gardens, Lucknow); Dr. S. Ohno (Dept. of Biology, City of Hope Nat. Med. Center, Duarte, Calif.); Prof. S. P. Raychaudhuri (Dept. of Zoology, Banaras); Dr. M. Ray (Children's Hospital, Winnipeg); Dr. V. Sorsa (Dept. of Genetics, Univ. of Helsinki); Dr. M. S. Swaminathan (Indian Agricultural Research Institute, New Delhi); Prof. J.Wahrman (Department of Zoology, Hebrew University, Jerusalem); Dr. F. Weiner (Karolinska Institute, Stockholm); and, Prof. M. J. D. White (Department of Genetics, University of Melbourne, Victoria).

This endeavour would be considered a success if it proves useful to all workers interested in any aspect of chromosome study. Lastly, we are grateful to our publishers for their patience and meticulous care in attending to the revised manuscript.

Calcutta

A. K. Sharma
Archana Sharma

Preface to First Edition

The tremendous progress of the discipline of cytology within the last twenty years has been responsible for making the study of chromosomes a science in itself with its own theories and techniques and its own achievements. The continued enthusiasm for refinement in methods owes its impetus to the outstanding discoveries on the finer structure of chromosomes, the chromosomal basis of differentiation and the role and association of chromosomes in human abnormalities and cancer. Technological advances have also led to widening of the outlook on chromosome structure from a purely cytogenetical level towards a cytochemical and cytophysical analysis. This approach has given the cell biologist an insight into the pattern or organisation at the microscopical, submicroscopical, ultrastructural and even molecular levels. Achievements in methodology have further revealed the dynamic nature of chromosomes, in spite of their basic uniformity and their multiplicity in structure and chemistry, together with their development and physiology at different phases of growth.

A serious handicap to a cytologist is the absence of any comprehensive single treatise on the methodology of all aspects of chromosome structure and behaviour in all organisms and at different stages of their differentiation. Chromosome science, responsible for the unification in biology, should be viewed as a whole—as fundamental to all organisms and not as a series of compartmental sciences in separate realms of plants, animals or human beings. Not only does the need for a comprehensive treatise limit the study of chromosome science, but the absence of any book dealing with the chemistry of reactions between reagents and chromosomes is also a serious impediment to advanced research. The idea that the chemistry of the reactions is not known does not represent the true state of affairs. A number of reactions, though studied by experimenters, have not been published; those that are published, have not been compiled and presented for the convenience of the worker. The absence of any information on the chemical principles underlying chromosome techniques has resulted in the general practice of random trials with different fluids in search for a suitable one.

This unscientific approach must necessarily be replaced by scientific and rational treatment. The difficulties that we faced as student, as research worker and as teacher, prompted one of us to start a probe into the technological aspects about twenty years ago. The intervening years have seen considerable advances in this field from different centres, including our own. The need for the presentation of the data obtained, in a critical and comprehensive form, is strongly felt—hence this publication.

In this book, efforts have been directed at presenting the technological aspects of chromosome study with particular emphasis on the principles underlying the different techniques and on the outlining of schedules. Methods have been described along with an account of their advantages, limitations and applicability, to enable the worker to plan a project directed either towards refining the methods given, discovering newer ones or applying them to various objects. As a natural sequence, the achievements too have been pointed out in brief. In order to meet the above ends a considerable amount of theoretical discussion has been incorporated which, it is hoped, would be of use to anyone interested in chromosome study.

This book is designed to meet the requirements of teachers, research workers and students alike, engaged in chromosome studies on plants, animals and human beings. No claim is made however to regard it as an encyclopaedia; for critical discussions on certain aspects of instrumentation, such as in electron microscopy or ultra-violet photometry, the reader is referred to the works of experts in the field. Efforts have been made to provide the widest possible fundamental knowledge for those whose only instruction in chromosome science would be this book, and maximum specialisation for those who intend to pursue it further. Modern developments in chromosome science have not only undergone evolution towards extreme specialisation but simultaneously towards greater simplification of the existing schedules. That is why chromosome science can now be pursued in any laboratory, however ill-equipped it may be, the category of research naturally depending on the availability of equipment. The object of the book will be achieved if it can meet the demands of workers engaged in this field in any type of laboratory.

The task has been made easier by the existence of certain books, invaluable in their own fields, chief of which are: *Histochemistry* by A. G. E. Pearse; *An Encyclopaedia of Microscopic Stains* by G. T. Gurr; *The Microtomists Vade-mecum* edited by B. Lee, and *Cytological Technique* by J. R. Baker. Our thanks are due to our co-workers in the field of cytology from several countries, whose good wishes gave us the impetus to finish this work. Special thanks are due to Prof. P. C. Koller of the Chester Beatty Research Institute, Drs. S. Makino and M. S. Sasaki of Hokkaido University, Drs. A. Levan and W. W. Nichols of South Jersey Medical Research Foundation, Prof. H. G. Callan of the University of St. Andrews, Drs. Wenner Schmid and T. C. Hsu of the University of Texas M. D. Anderson Hospital and Research Institute, Prof. J. H. Taylor of the University of Florida, Drs. H. H. Smith and W. Prensky of Brookhaven National Laboratory, Drs. P. Harris and D. Mazia of the University of California, Dr. J. Mitra of the University of New York and Prof. R. P. Roy of Patna University, for very kindly sending original photographs of their preparations for the book. We would also like to thank our colleagues here, notably, Drs. N. K. Bhattacharyya, U. C. Bhattacharyya, S. Sen (Miss), M. Chaudhuri, A. K. Chatterjee, C. Talukdar and R. Mallick, for their preparations, original photographs of which have been included. Lastly our thanks are due to our publishers, Butterworths, for their patience and co-operation in finishing this work.

Calcutta   A. K. Sharma
                                                                        A. Sharma

Contents

PART 11 CHEMICAL NATURE OF CHROMOSOMES

Part 1
Physical Nature of Chromosomes

1 Introduction

The analysis of chromosome structure starts with the understanding of a clear delimitation between the chromosomes of lower organisms, including viruses and bacteria on the one hand and those of higher organisms on the other (Gay, 1966). In the former, the chromosome is a genophore, being essentially a long chain DNA molecule, not discernible under the light microscope and studied mainly through electron diffraction patterns and mutation and recombination data. In the latter, the chromosome is an extremely complex body, which has other constituents in addition to the DNA genophore, and it may be analysed under a light microscope. The genic effect on metabolism in the former can be immediately studied, whereas in the latter there is a wide gap and a series of reactions intervenes between the initial reaction at gene level and its manifestation in the phenotype. A concept of the evolutionary advance, step by step, from the genophore of the prokaryotes to the complex chromosome structure of eukaryotes, correlated with functional diversities and specialisation of segments, had been proposed (Sharma, 1969).

Research in the nineteenth and twentieth centuries has clearly established that 'chromosomes', which bear the hereditary materials or 'genes' in a linear sequence, are of permanent fibrous constitution. The term 'chromosome', meaning *colour body*, is based on its property to absorb certain dyes, thus showing basophilia. The importance and scope of the different techniques in the study of chromosomes can only be realised as the gradual advancement, aided through technical refinements resulting in the modern concept of chromosome structure, is outlined.

Notwithstanding the fact that the fibrous constitution of chromosomes has been unequivocally accepted, much discussion has been occasioned with regard to the longitudinal constitution of the thread, and the ultimate number of fibrous units composing it in Eucaryota. Leaving aside the *alveolar* theory, which has been completely discarded and in which chromosomes were supposed to be constituted of 'alveoles', reconciliation between the other two, i.e. the *chromomere* and *chromonema* hypotheses, has been

1

effected in later years. The filamentous nature of chromosomes was first visualised in the chromomere hypothesis. This hypothesis assumed that the chromosome is composed of longitudinally aligned chromatic granules, otherwise termed 'chromomeres', joined by an achromatic thread. The chromonema hypothesis, on the other hand, visualised a uniformly thick continuous structure in a chromosome, capable of spiralisation in mitosis.

The amalgamation of the two theories was brought about by Sharp (1943). He assumed that the varying distribution of the extra-chromatin substance, in addition to the one constituting the chromomeres, is responsible for the chromomeric and chromonemic appearance of the chromosome at different stages of the divisional cycle. In the later phases of division, when the deposition of extra-chromatin matter is heavy, the chromomeric appearance becomes pronounced. In telophase and earlier stages of division, the structure appears to be of nearly uniform thickness. It is now accepted by several authors that chromomeres represent specially differentiated portions of the chromonema thread, showing chemical reactions different from the interchromomeric segments. Whether the genes are located on the chromomeres or on the interchromomeric segments has also been debated. Chromomeres have also been assumed to represent tightly coiled regions of the chromosome thread (*see* Ris, 1957).

With regard to the multiplicity of the chromonemata, three different theories have been prosposed in the past, based on the number of threads constituting each chromonema. Darlington (1937, 1965) postulated the single-thread conception, in which the anaphase thread was supposed to be single, becoming double in the subsequent prophase. The double nature of the anaphase thread was maintained by Gates (1938), who also assumed that the chromosomes divided at the prometaphase condition, the metaphase thread being quadripartite in nature. The bipartite nature of the anaphase chromosome has been demonstrated by nitric acid vapour treatment. The other view held is that the anaphase chromosome is quadripartite, becoming 8-partite in the subsequent prometaphase (Nebel and Ruttle, 1936). Irrespective of the different theories held with regard to the unitary or multiple nature of the chromonemata, the generally accepted idea is that the anaphase thread is not unitary in nature. Opinions still differ as to the exact number of fibres present in the anaphase thread.

Later observations (Taylor, 1967), indicating the completion of DNA synthesis before zygotene have raised important issues regarding the initiation of meiotic pairing. The multistranded nature of chromosomes has been confirmed by various workers, both through ultrastructural and other evidence, and the implication of these findings has been discussed in several treatises and excellent reviews (Keyl, 1965; Sparvoli *et al.*, 1966; Gay, 1967; Sueoka *et al.*, 1967). The problem of interpreting the series of 200–250 Å thick fibrils in chromosomes has been outlined by Dupraw (1965), Wolfe (1965), Gall (1966) and Ris (1967).

The synchronous functioning of multi-stranded fibrils with respect to gene action has been explained in terms of a master gene, the rest being derogated to the category of slave genes (Callan, 1967; Whitehouse, 1967; *see* Edström, 1968).

With the aid of electron microscopy, ultrastructural units of chromosomes have been clarified. The minute lamellar constitution of the chromosomes has been claimed by a number of workers dealing with electron micrographs (Kaufmann and De, 1956; Ris, 1957; Kaufmann and colleagues, 1960), interpretation of the orientation of the molecules being different, however (Taylor, 1968). It has been suggested that 32, 64 or even 128 microfibrils may constitute the chromosome thread. This multilamellar concept has serious implications on the existing theories of crossing over and mutation as to how far all the lamellar units can undergo recombination of segments simultaneously, and how far mutation may originate in all the corresponding loci of a segment of chromosome lamellae. To explain mutation, therefore, fortuitous and simultaneous changes in the same locus in all fibrils have to be assumed. Recombination presents a similar problem where crossovers must involve all the fibrils at the same time (Uhl, 1965). To reconcile these discrepancies, the existence of a single long chain DNA molecule is often assumed but the gap between molecular and microscopic dimensions is too wide to be justified.

The theories of chromosome division do not necessarily require any modification in the light of the lamellar concept (Sharma and Sharma, 1958). When refinements in technique make it possible to reach the molecular level, more sub-divisions of lamellae may be clarified. The fundamental issue regarding their behaviour in anaphase is their segregation as binary units during division, and fortunately no case has yet been recorded where chromosomes have not behaved as binary units in anaphase. Therefore, on the basis of circumstantial evidence, a reasonable assumption is that, whatever may be the number of lamellae constituting the chromosome, the two daughter halves at anaphase are composed of an equal number of lamellar units. It may therefore be stated that the functional unit of the chromosome is the chromatid. Lamellar nature can be studied only with the aid of electron microscopy (*see* Chapter on Electron Microscopy), whereas the multiplicity of the chromonemata is revealed by acid or ammonia vapour (*see* Chapter 8). Such treatments not only expose the chromosome thread, but also the relational spiral between the two chromatids and the spirals between the constituent units of chromatids.

The refinements in technique have brought about a complete reorientation of outlook in the study of chromosome morphology. Formerly the conventional method of describing a chromosome was to refer to it as a *J* or *V* shaped structure, or rather as acro- or meta–centric, without any description of its morphology. This was principally due to the inadequacy of the techniques then available, which could not clarify the details of the structure. Gradually, as the need for an intensive research into chromosomal details was realised, techniques were invented from different centres (Flemming, La Cour, Lewitsky, etc.), which allowed further resolution of the details, including the clarification of primary and even secondary constriction regions responsible for the formation of the nucleolus. Further advances led to an understanding of the heterochromatic and euchromatic segments of chromosomes, which could be differentiated through temperature and other treatments. These two types of segments are characterised by different staining cycles in the different phases of division. The presence of different types of heterochromatin, with regard to different functions, has been realised (Brown, 1966; Wolf and Wolf, 1969).

The molecular pattern has yet to be correlated with the observed transverse differentiation of chromosome segments, though several models have been outlined (*see* Hamilton, 1968). The centromere, secondary constriction regions, telomeric segments and other heterochromatic regions must fit in the molecular structure of the genophore, since they are regions which have been differentiated functionally. The occurrence of divalent cations, histones and even lipids are often regarded as transverse incorporations.

A major achievement in the study of chromosome science is the understanding of its dynamicity, replacing the concept of its uniformity and monotonous behaviour in all organs (Sharma, 1968, 1971). The dynamicity in chromosome behaviour in respect to differentiation and development has been seen clearly in *Drosophila* and other Diptera (Beermann, 1967; Pavan and Da Cunha, 1969). The synthesis of ribonucleoprotein for certain segments in the lampbrush chromosome at certain stages in vertebrate oöcytes as well as puffing at different segments of salivary chromosomes in different growth phases in *Drosophila* present direct evidence of chromosomal control of metabolism.

Further examples of this behaviour are the differential replications in meristematic and adult differentiated nuclei. While the behaviour of the chromosome thread follows the usual sequence in the former, it undergoes endomitotic replication in adult nuclei. This interpretation can explain the genic control of differentiation, which is maximum at the adult stage, as well as the apparently non-dividing state of the chromosomes. The cause of chromosome duplication, without separation, in adult nuclei has been attributed to regulated deficiency of the sugar component of DNA (Sharma and Mookerjea, 1954). The polytene state, due to endomitotic replication, has been confirmed by inducing division in these nuclei through IAA treatment by Huskins (1947) and by the precursors of nucleic acids and related chemicals in our laboratory. However, a generalisation from the evidence obtained from lower organisms implies that transcription and translation responsible for gene action and differentiation are not necessarily associated with gene duplication. The endomitotic replication, occurring with diploidy, and the polytenic constitution of some differentiated nuclei, are, however, observational facts, the association of which in differentiation requires clarification. These observations raise the problem of the transcribing limit of a DNA strand and the necessity of fresh strands for transcription. In any case, it is a clear index of the dynamicity of chromosome behaviour in response to the need for differentiation.

In the last several years considerable strides have been made in the study of chromosome structure, due to the invention of a number of pre-treatment chemicals. Special treatments, previous to fixation, have been responsible for a clear understanding of the structure of the different parts of chromosomes, including the centromere—the chromosome segment necessary for attachment and movement along the spindle. Knowledge regarding the structure of the centromere was vague until its quadruple nature in metaphase was observed, which has been further clarified through later works. Refinements in methods have led to a clear understanding of the structure of the spindle and its relationship with the chromosome (Wada, 1966; Sakai, 1968).

Such changes in pattern, associated with phasic development, growth and differentiation of plants, indicate that, in spite of a basic genetic uniformity, the structural pattern of chromosomes is dynamic. Even the study of their chemical nature, discussed later, has shown that the chromosomes which are packaged for transmission are not identical in all respects with those of the somatic cells responsible for differentiation.

Further evidence of dynamicity and its control in the reproduction of species is seen in the chromosome behaviour in asexually reproducing species studied extensively in our laboratory. Chromosome complements of the somatic tissue universally exhibit numerical and structural changes and the constancy of the chromosomes is not maintained as such in different cells of the same tissue, but a chromosome *mosaic* is formed. This *regulated* abnormal behaviour is of great importance in species identification since the changed chromosome complements may enter into the growing apex of the vegetative shoots and form genotypically new individuals without the act of fertilisation. This outstanding example of dynamicity in chromosome behaviour, observed with the aid of refinements in techniques, shows the response to reproductive necessity under abnormal conditions.

The invention of methods for chromosome study in normal and cancerous cells of human beings has yielded further information of this dynamic behaviour suggesting that the chromosomes do not follow a normal pattern of behaviour in malignant cells. Mitotic instability, uncontrolled cell proliferation and mosaics of different chromosome numbers characterise the different tumourous and cancerous cells. The technique for such studies, though simple, was not available for a considerable period and its discovery has led to outstanding achievements in this field. The convenient schedules evolved for culturing leucocytes from blood have opened up new avenues of research and several congenital anomalies have been correlated with definite chromosomal characteristics. The advances in the study of human chromosomes have been outstanding. The significance of the dynamicity in chromosome behaviour, shown through recent developments in methodology, cannot be overrated since it has been instrumental in understanding the overall control of chromosomes in maintaining hereditary stability, as well in species replication and the hitherto unexplained processes of development and differentiation.

Finally, before dealing with the details of techniques, a few words about the modifications of chromosome structure, occasioned by physical and chemical treatments, is necessary.

Since the discovery of polyploidisation through colchicine, and mutation and chromosome breakage through x-rays, vigorous research on these and allied aspects is being carried out throughout the world. The fundamental and utilitarian implications of such findings will be outlined in their respective chapters. This aspect of study involves two lines of investigation, one dealing with the techniques for inducing chromosomal abnormalities, and the other concerning the methods adopted for scoring the results.

For the study of the different aspects of the physical structure of chromosomes, varied techniques have been devised from time to time. The details of these techniques—their principles, applicability, drawbacks and recent developments—are discussed in the subsequent chapters.

REQUISITES

Chromosome study from living cells presents considerable difficulties. The phase difference or, more precisely, the difference in refractive indices between cellular components, being insignificant, it is difficult to resolve the components separately under the light microscope. For the study of chromosomes from living cells, ultrastructural pattern and quantitative estimation of chromosome components, different types of microscopy are employed.

In order to study chromosome structure under an ordinary light microscope, the tissue is killed and the tissue part selectively preserved without causing any appreciable distortion of nuclear matter, through a process known as 'fixation'. During this process, the cells are killed and dividing nuclei are fixed in the particular phase of division in which they were lying at the time of treatment. Chromosome study is usually performed after subsequent stages of washing, processing and staining in suitable dyes.

It has been observed that mere fixation alone may not prove adequate for the analysis of chromosome structure, especially of the somatic cells. This is particularly true for nuclei having a high chromosome number or very small chromosomes. In the former case, all the chromosomes form an aggregated lump, the individual structure thus not being discernible. In the latter case, on the other hand, the chromosomes may undergo too much swelling to allow any analysis of the details. The non-availability of any adequate method of fixation, and the lack of any suitable special schedule prior to fixation, led to the publication of a large number of papers in which the description of chromosome structure was either absent or inadequate. In the study of phylogeny or affinities between species, over-emphasis on the number of chromosomes with practically no reference to their structure, as is done in a number of cases, can also be attributed to these technical limitations. This is principally the reason why the role of structural changes of chromosomes in evolution was not properly assessed in the past.

The invention of a number of pre-treatment chemicals which are applied to fixation has considerably aided the study of chromosome structure in all its details. The principle of pre-treatment is mostly to bring about physical changes in the cytoplasm and nucleus which would help in clarifying the details of chromosome morphology. Thus, for the study of the physical nature of chromosomes, (a) pre-treatment, (b) fixation, (c) processing and (d) staining are the four essential steps which are discussed in the following chapters.

REFERENCES

Beermann, W. (1967). In *Heritage from Mendel*, Madison; Univ. of Wisconsin Press
Brown, S. W. (1966). *Science* **151**, 417
Callan, H. G. (1967). *J. Cell. Sci.* **2**, 1
Darlington, C. D. (1937, 1965). *Recent Advances in Cytology*. London; Churchill
Dupraw, E. J. (1965). *Proc. Nat. Acad. Sci. US* **53**, 161
Edström, J. E. (1967). *Nature* **220**, 1196
Gall, J. G. (1966). *Chromosoma* **20**, 221
Gates, R. R. (1938). *J. R. micr. Soc.* **58**, 97
Gay, H. (1966). *Carnegie Inst. Year Book* **581**, 1587
Gay, H. (1967). *Science* **158**, 528
Hamilton, L. D. (1968). *Nature* **218**, 633
Huskins, C. L. (1947). *Amer. Nat.* **81**, 401
Kaufmann, B. P. and De, D. N. (1956). *J. biophys. biochem. Cytol.* Suppl. 419
Kaufmann, B. P., Gay, H. and McDonald, M. R. (1960). *Int. Rev. Cytol.* **9**, 77
Keyl, H. G. (1965). *Experientia* **21**, 191
Nebel, B. R. and Ruttle, M. L. (1936). *Amer. J. Bot.* **28**, 651
Pavan, C. and Da Cunha, A. B. (1969). *Ann. Rev. Genet.* **3**, 425
Ris, H. (1957). *Symposium on Chemical Basis of Heredity*, p. 23. Baltimore; John Hopkins Press
Ris, H. (1967). In *Regulation of nucleic acid and protein biosynthesis* 11, Amsterdam; Elsevier
Sakai, H. (1968). *Int. Rev. Cytol.* **23**, 89
Sharma, A. K. (1968). *Sci. Cult.* **34** Suppl., 69
Sharma, A. K. (1969). *Nucleus* **12**, 86
Sharma, A. K. (1971). Abstr. *All-India Congr. Cyt. Genet.*
Sharma, A. K. and Mookerjea, A. (1954). *Caryologia* **6**, 52
Sharma, A. K. and Sharma, A. (1958). *Bot. Rev.* **24**, 511
Sharp, L. W. (1943). *Fundamentals of cytology*. New York; McGraw-Hill
Sparvoli, E., Gay, H. and Kaufmann, B. P. (1966). *Abst. Int. Cong. Rad. Res.* III, Cortina
Sueoka, N., Chiang, K. S. and Kates, J. R. (1967). *J. Mol. Biol.* **25**, 47
Taylor, J. H. (1967). *Handbuch Pflanzenphysiologie* **18**, 344
Taylor, J. H. (1968). In *Molecular Genetics*, New York; Academie Press
Uhl, C. H. (1965). *Nature* **206**, 1003
Wadar, B. (1966). *Cytologia Suppl.* **30**
Whitehouse, H. L. K. (1967). *J. Cell. Sci.* **2**, 9
Wolf, B. E. and Wolf, E. (1969). In *Chromosomes today* **II**, 44, Oliver & Boyd; Edinburgh
Wolfe, S. L. (1965). *J. Ultrastruct. Res.* **12**, 104

2 Pre-treatment

Pre-treatment for the study of chromosomes is generally performed for several special reasons. It may be carried out for: (*a*) clearing of the cytoplasm, (*b*) separation of the middle lamella causing softening of the tissue, or (*c*) bringing about scattering of chromosomes with clarification of constriction regions. Pre-treatment may also be needed to achieve rapid penetration of the fixative by removing undesirable deposits on the tissue as well as for the study of the spiral structure of chromosomes (La Cour, 1935). The first two applications involve removal of extranuclear limitations, where as the third and most important one exerts a direct effect on the chromosomes.

PRE-TREATMENT FOR CLEARING THE CYTOPLASM AND SOFTENING THE TISSUE

In order to clear the cytoplasm from its heavy contents, acid treatment has often been found to be very effective. Short treatment in normal hydrochloric acid often brings about transparency of the cytoplasmic background. Such treatments, however, require thorough washing for the removal of excess acid and acid soluble materials. However, acid treatment affects the basophilia of the chromosomes heavily and so mordanting may become necessary after acid treatment and washing.

In addition to acid, several enzyme preparations are applied for clearing the cytoplasm and cell separation through their inherent capacity of digestion. McKay and Clarke (1946) applied pectinase with success, not only achieving tissue softening, but also securing clearing of the cytoplasm. Similarly, cytase extracted from snail stomach at a particular pH was successfully applied by Fabergé (1945). Cytoplasmic clearing was obtained by Brachet (1940) with the application of ribonuclease, which brought about digestion of ribonucleic acid—the principal constituent of the cytoplasm. A complex enzyme preparation 'Clarase' was used by Emsweller and Stuart (1944), the application yielding brilliant preparations following staining. In the authors' laboratory, treatment with cellulase has yielded very satisfactory preparations in difficult dicotyledonous materials. Once a particular difficulty in obtaining successful results in any material is ascertained, the application of special enzymes for its elimination may be regarded as an accurate and scientific approach towards the solution of the problem. But one of the serious limitations of this aspect of pre-treatment is that enzyme preparation, completely free from contaminants, is difficult to obtain and, as such, several nuclear components may be undesirably affected after this procedure. Even in the absence of contaminants, some of the enzymes such as 'ribonuclease' may affect chromosomes in the purest form. This is understandable, as ribonucleic acid has been unequivocally claimed as a constituent of chromosomes, at least of animals, by a number of workers, and therefore, in cases requiring critical observation, enzyme treatment, if necessary, should be cautiously performed. The use of enzymes for cell separation has been dealt with in the chapters on tissue culture and mammalian chromosomes.

Occasionally it also becomes necessary to employ alkali solutions as pre-treatment agents. This step is desirable for materials having heavy oil content in the cytoplasm. Alkalis cause removal of the oil by saponification. Alkali solutions commonly employed are sodium hydroxide or sodium carbonate. As in acid treatment, thorough washing in water is necessary after the tissue has been kept in alkali solution.

In certain cases, pre-treatment may be necessary to remove deposits of secretory and excretory substances from the surface of the tissue which, if not removed, can hinder the access of the fixing fluid. The best example is provided by the application of hydrofluoric acid to remove siliceous deposits prior to fixation in bamboos; similarly, a very short treatment in Carnoy's fluid, containing chloroform, is needed to remove oily or other secretory deposits on the cell walls before fixation in a number of plant materials.

SEPARATION OF CHROMOSOME AND CLARIFICATION OF CONSTRICTIONS

The underlying principle of these important aspects of pre-treatment is the viscosity change in the cytoplasm. As spindle formation is dependent on the viscosity balance between cytoplasmic and spindle constituents, a change in cytoplasmic viscosity brings about a destruction of the spindle mechanism with the chromosomes remaining free or, more precisely, not attached to any binding force within the cell. Pressure applied during smearing results in the scattering of chromosomes throughout the cell surface. Changes in cytoplasmic viscosity simultaneously affect the chromosome too, which undergoes differential hydration in its segments, and due to this differential effect, constriction regions in chromosomes appear well clarified. Several pre-treatment chemicals have so far been applied for the purpose, colchicine being the most important one.

In addition, pre-treatments are often performed on tissues to secure a high frequency of metaphase stages obtained through spindle inhibition. For this purpose colchicine is supposed to be the most active substance, and also compounds having similar properties, such as chloral hydrate, gammexane, acenaphthene, etc.

Colchicine

Colchicine was first isolated from *Colchicum autumnale* by Zeisel in 1883, the alkaloid being isolated from the roots. The empirical formula of colchicine is $C_{22}H_{25}O_6N$, as determined by T. Y. Johnstone in Glasgow. The general method of extraction is through the use of ethyl alcohol and subsequent dilution with water; finally the aqueous solution is extracted with chloroform and crystals of colchicine are obtained along with the solvent. By gradual evaporation and distillation of chloroform, amorphous colchicine can be obtained. The chemical structure of colchicine, as suggested by Windaus in 1924, has been modified by Dewar (1945). All these authors agree that it has a three-ringed structure. Barton, Cook and Loudon (1945) consider that the second ring is 7-membered and Dewar suggests that the third ring is also 7-membered. Steinegger and Levan (1947) have suggested that Dewar's modification is the most reasonable one.

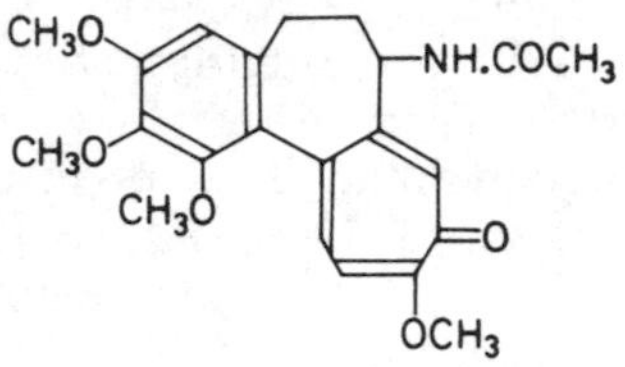

Colchicine

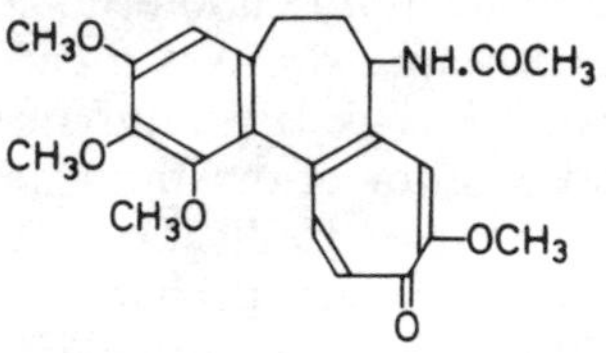

Isocolchicine

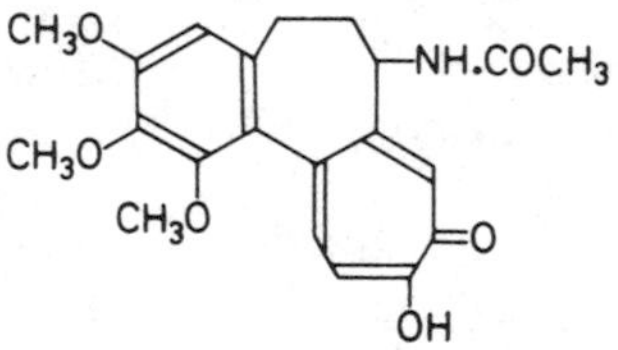

Colchiceine

Eigsti and Dustin (1957) have summed up all the evidence presented with regard to the structure of colchicine and hold that colchicine is the methyl ether of an enolone containing three additional methoxyl groups, an acetylated primary amino group and three non-benzenoid double bonds. Levan (1949) has noted that the threshold regions of colchicine-mitotic activity are identical for both crystalline and amorphous forms; chloroform exerts no appreciable effect. One of Levan's significant findings in *Allium cepa* is that whenever isocolchicine is used instead of colchicine, no c-mitotic action is observed. The latter differs from the former only in minor details of structure.

Colchicine is soluble in water (500·000 in 10^{-6} mol/l in water). Ferguson (1939) worked out the thermodynamic activity of several narcotics and showed that they can be classified under two categories: (a) one in which there is a direct correlation between activity and physical properties, such as water solubility, and (b) one showing no such correlation.

Steinegger and Levan (1947) and Levan (1949) also demonstrated a relationship between water solubility and activity threshold, the two being directly proportional. Colchicine is remarkable in the sense that, though highly water-soluble, it is very active at an extremely low concentration. It falls under Ferguson's second category of compounds and the reaction is chemical. Originally the role of colchicine was considered to be that of a catalyst (Bhaduri, 1939). It brings about a change in the colloidal state of the cytoplasm, causing spindle disturbance by increasing the fluidity of the nuclear substance.

With regard to the exact reactive groups in the colchicine molecule, Eigsti and Dustin (1957) reviewed in detail the observations presented by different workers (Ludford, 1936; Lettré and colleagues, 1947, 1952; Branch and colleagues, 1949; Leiter and colleagues, 1952). These data imply that: (a) at least one methoxyl group in ring A is necessary for colchicine action; (b) ring C must be 7-membered and the hydroxyl group should preferably be replaced by an amino group; (c) esterification of amino group in ring B increases the activity; and (d) isocolchicine and its derivatives are less active. A proper distance should be maintained between esterified side chains of rings B and C. This last statement is based on the fact that in isocolchicine, where the position of the methoxyl group is reversed in ring C, the decrease in efficacy has been considered to be due to the presence of hydrogen bonds between the side chains of rings B and C. In the case of 'colchiceine', the weak action is the result of the iso-form of this molecule.

Though these data indicate the reactive groups responsible for colchicine action, the exact reaction involved in spindle inactivation is not yet fully clear. Pertinent suggestions have been made on the basis of results obtained with spindle poisons of different chemical structure. Östergren (1950), who regards colchicine action as narcosis, suggested a relationship between spindle poisoning and lipoid solubility. Eigsti and Dustin (1957) pointed out that the exact relationship between lipoids and the function of the spindle is yet to be clarified; they emphasised the need of specific biochemical evidence to substantiate this statement. A number of lipoid-soluble hydrocarbons have been found to have the property of narcotising mitosis. Lettré and colleagues (1952) suggested that ATP may be responsible for spindle activity and mitosis, and that colchicine may modify this mechanism. A number of sulphydryl poisons, such as iodoacetamide, dimercaptopropanol (BAL), mercaptoethanol, sodium diethyldithiocarbamate, etc. (Dustin, 1949, 1950), act in the same way as colchicine. These results may not strictly be regarded as suggesting that sulphydryl poisoning is involved in colchicine action, but at least they imply that colchicine enters into a chemical combination with some intracellular receptor. In order to arrive at a definite conclusion about the chemistry of colchicine action, simpler molecules, allied to colchicine, should be tried on a wide variety of plants and animals. A biochemical approach to the problem, involving an analysis of cellular products before and after colchicine treatment, is also necessary.

In the study of chromosomes, without inducing polyploidy, colchicine has to be applied in a low concentration, such as 0·5 per cent for 1 h, whereby straightening of the chromosome arms can be achieved, which allows a thorough study of the constriction regions. It is especially effective for long chromosomes, as otherwise the chromosome arms remain foreshortened. The chromosome arms, though split, remain held together at the constriction region.

Different types of fixatives, both metallic and non-metallic, can be applied after colchicine treatment. One of the most essential requisites in the use of this alkaloid is that the tissue must be thoroughly washed to free it of colchicine before fixation; otherwise the superficial deposition of this alkaloid hampers visibility of the chromosomes and penetration of the fixative. It is always necessary to use a rapidly penetrating fluid, such as acetic–alcohol after colchicine treatment, as otherwise the nuclei may rapidly enter into the metabolic phase at the time of fixation. Though such fixation yields the desired results, colchicine should not be regarded as incompatible with other fixatives. McKay and Clarke (1946) have shown that satisfactory results can be obtained if both enzyme treatment and fixation are done after the application of colchicine.

In the authors' laboratory it has been found that colchicine can be effectively employed in the study of chromosome structure if after treatment, the tissue is studied following smearing instead of through block preparation. The scattering effect, obtained in squash, is difficult to secure in block preparations. In the latter, chromosomes may overlap after colchicine treatment due to lack of orientation. The two effects—straightening of chromosome arms and loosening of spirals—are best observed in squash preparations.

Compared with other pre-treatment chemicals, colchicine has the added advantage of being active within a very wide range of temperature. In the tropics, even during summer when the temperature rises to 43·3 °C, no significant effect on colchicine action has been noted. In this laboratory recently, colchicine has been tried on tissues kept at 8–9 °C and the frequency of metaphase plates is observed to be distinctly higher in these tissues than in those sets kept for a similar period at room temperature (Sharma and Sarkar, 1963). The clarification of chromosome structure too is obtained within a very short period.

Regarding the applicability of colchicine to metaphase arrest, nearly all plant organs and a number of animal organs as well respond to its treatment. In plants, the somatic tissue responds more readily to colchicine action than do the meiotic cells. Effective spindle arrest, resulting in a high frequency of metaphase plates, with the chromosomes well clarified and straightened, has been obtained in meristematic cells of roots, shoot apex and also in both the first and second divisions of the pollen grain nucleus. Germination of seeds in colchicine solution gives a large number of metaphase plates in the root-tips (Wolff and Luippold, 1956). For the study of the haploid mitosis, pollen tube culture, with colchicine added in the medium, is considered to be most satisfactory. In animals, spermatogonial cells as well as the larval tissue respond well to colchicine. However, colchicine action has not been so well studied in animals as in plants, and data so far obtained in the former point to a comparatively poor response. In mammalian tissue, including human tissue, colchicine treatment has yielded a high number of divisional figures, particularly in the endometrium, bone marrow, regenerating liver, normal and neoplastic cells, fibroblast cultures, eggs of sea urchin, etc. Though metaphase has been arrested in these cases, the polyploidising effect of colchicine is not marked in animals; investigation into the cause of this differential susceptibility is necessary. Colcemid, a derivative of colchicine, has been used successfully in mammalian tissue to obtain well-scattered metaphase plates. It is free from the limitations of colchicine and if injected, is absorbed more easily by the tissue.

The period necessary for the manifestation of colchicine effect varies in different plant and animal groups. The range of concentrations is also wide, lying between 0·001 and 1 per cent. In animal cells the method of application is preferably by injection, whereas in plant tissue it is applied through soaking, plugging and injection. It can also be used in lanolin paste and in agar. In artificial culture of both plant and animal tissues, colchicine is added in the medium. To avoid toxicity affecting other metabolic processes, this alkaloid is generally applied in a low concentration to animal cells.

In general, it may be stated that the drug is exceptionally suitable for the study of chromosome structure and metaphase arrest, provided that strict control is maintained over the concentration and period of treatment; any excess of these factors may lead to a heavy contraction of chromosomes and polyploidy, obscuring analysis.

Acenaphthene

In order to secure an effect similar to colchicine, extensive studies have been carried out with different groups of chemicals, and Kostoff (1939) demonstrated that acenaphthene has the same property of arresting metaphase. As its structure is quite different from that of colchicine, several other aromatic compounds were tried and various derivatives of benzene and naphthalene were found to be effective.

Acenaphthene

Several haloid derivatives were also tried by Gavauden and shown to have the same effect (Gavauden and colleagues, 1938). He observed that though benzene itself is ineffective, the presence of side chains is responsible for c-mitotic property; the case with naphthalene derivatives is similar. Acenaphthene too follows the principle of Ferguson's (1939) thermodynamic activity and its mitotic poisoning effect is related to low water solubility.

As acenaphthene is sparingly soluble in water, it should be applied in saturated solution in water for treatment of the plant tissue. From trials with acenaphthene on different types of plant tissue, it is seen to be most effective in clarifying the chromosomes of the pollen tubes (Swanson, 1940). Comparative effects of this chemical have been studied on different organs of the same plant in the authors' laboratory and it has been observed that, with the exception of the pollen grains, the tissues do not respond to the treatment at all, this differential behaviour being attributed to the varied constitution of the plasma in different organs (Sharma and Sarkar, 1957). Being effective principally on the microspores, the actual use of acenaphthene is limited to the culture medium only. The tubes being narrow, the straightened chromosomes are exposed as rods, linearly arranged along the tube. In view of the fact that only a few species of plants produce pollen of favourable constitution, the application of this technique is extremely restricted.

Chloral Hydrate

Another agent used for inducing metaphase arrest is chloral hydrate. It has been used as a narcotic since 1904 (Nemec, 1904), and as a pre-treatment chemical in cytology since 1935 (Peto, 1935). It has the formula $CCl_3 \cdot CH(OH)_2$, a characteristic smell and sharp taste. It has the same effect as colchicine, and is less expensive. It was originally intended as an important hypnotic and stimulant. Curiously, however, the concentration required for chloral hydrate treatment is the same as that of colchicine, though having a different chemical composition. It is quite effective in smears, and both metallic and non-metallic fixatives can be used after the treatment. In the authors' experience, this chemical, though effective in certain species of plants, is ineffective in others. Moreover, its erratic behaviour requires serious consideration. The results obtained are very inconsistent and the response varies not only in the same plant but also in the same tissue. It appears that this erratic behaviour may be attributed to the dependence of its action on several subtle factors, which under normal conditions, though appearing to be constant, may vary in the finer details. Unless a complete understanding about the reaction of chloral hydrate with several intra- and extra-cellular factors is achieved, wide use of this chemical cannot be recommended.

Gammexane

Metaphase arrest has also been secured by the use of γ-hexachlorocyclohexane, commonly known as gammexane, by different authors. Principally, the chemical has been tried on somatic cells of *Allium cepa* and other species.

In addition to metaphase arrest, polyploidy and fragmentation have also been recorded in some cases. Later, its δ-isomer was found to be similarly effective. Meso-inositol, which has the same isomeric structure as gammexane, acts as an antagonist of the latter, possibly due to the substitution of the Cl atoms by OH groups.

Polyploidy induced by gammexane has, however, not been found to be of permanent nature, at least in the plants tried out in this laboratory (Sharma and Chaudhuri, 1959). If the somatic tissue is cultured in gammexane medium, there is a direct duplication of chromosome number up

Gammexane

to a certain threshold point, after which the chromosome number undergoes reduction due to many spindle formations within the cell. Though such multi-spindle and multi-nucleate conditions have been observed in a high frequency, it is not certain whether any regularity exists in the formation of these nuclei. How far the normal number with the normal complement is restored through such reductional separation is not fully known.

A study of the effects of gammexane at different temperatures has shown it to be most active at low temperatures (12–16 °C) (Sharma and Chaudhuri, 1961). Metaphase arrest can be secured following treatment with a saturated solution within a period of 3 h. Compared with colchicine and acenaphthene, gammexane is less active than the former but more active than the latter. Its action has been analysed, at least in the case of plants, both in meiotic and somatic cells. In the meiotic cells, with metaphase arrest, a considerable amount of mitotic disturbance has been recorded, though polyploidy is very rare. The mitotic instability indicates that instead of causing absolute spindle inactivation, gammexane leads to abnormalities in spindle formation.

With regard to the exact metabolic paths affected by gammexane, analysis so far carried out has revealed that it has the capacity of affecting all the metabolic paths except those in which Mg, S and Ca are involved. Further, being effective at a very low concentration, it has been assumed that chemical changes are involved in gammexane-induced reactions. The metabolic disorders caused by colchicine and gammexane have been found to differ significantly. Nybom and Knutsson (1947) inferred that the action of gammexane is dependent on its lipoid-solubility, the evidence being obtained from the highly narcotic reactions in γ-isomers, which have a heavy lipoid-solubility. Levan and Östergren (1943) suggested that possibly certain enzymic functions of the cell are affected, whereas Östergren and Levan (1950) pointed out that fibrous components of the spindle are brought to assume a corpuscular shape (*see* Mazia, 1955, 1959). All this evidence, taken in conjunction with its activity at low concentration and temperature dependence, suggests the chemical nature of the reaction.

Coumarin and its Derivatives

Coumarin—The importance of coumarin in chromosome analysis was pointed out by Sharma and Bal (1953). It was also used for the study of the gross chromosome number in cancer of human cervix by Manna and Raychaudhuri (1953).

Coumarin

Coumarin is the *o*-coumaric acid lactone found mostly in the glycoside form in plants and is responsible for the fragrance of new-mown hay and woodruff. Several natural and synthetic coumarins are available and Quercioli (1955) studied in detail the cytological property of synthetic coumarins. Data on natural coumarins are available from work in this laboratory (Sharma and Bal, 1953; Sharma and Sarkar, 1955; Sharma and Chaudhuri, 1962).

This chemical was tried on a variety of plants, especially those belonging to the monocotyledonous group, and clarification of constriction regions and especially of the centromere has been obtained through this treatment. It is effective even at a temperature of 30 °C, though cold treatment causes rapid action. In plants so far studied it is ineffective in species with a high chromosome number, even after 2–3 h of treatment. With chromosomes of the human cervix, even a short treatment of 5 min results in their perfect scattering. Being sparingly soluble in water—its solubility is approximately 0·02 per cent—coumarin is used as a saturated solution in water.

Coumarin, as an agent for chromosome breakage, was demonstrated by D'Amato and Avanzi (1954). Earlier, Sharma and Bal (1953) had observed that agents inducing chromosome breaks can be effectively employed for karyotype analysis if they are applied at a lower concentration or for small periods of treatment. The same holds good for coumarin.

Aesculine—A derivative of coumarin is aesculine. It is extracted from *Aesculus hippocastanum* Tourn. Its chemical structure is:

Aesculine

It is the direct derivative of aesculetin, in which the H from the —OH group in the 6-position is replaced by $C_6H_{11}O_5$. This particular compound has been found to be suitable for different groups of plants having chromosome numbers ranging from very high to very low (Sharma and Sarkar, 1955). The length of application needed for chromosome analysis also varies, from 30 min in species of *Allium* up to 24 h in different species of palms. In the latter, though the chromosomes are very short, even the constriction regions can be resolved in well scattered plates after aesculine treatment. Aesculine has a wide applicability, yet one of its serious limitations (shared by coumarin) is its capacity of inducing fragmentation of the chromosomes after prolonged treatment. Its approximate solubility is 0·04 per cent in water, and for chromosome analysis it is applied both in saturated and half-saturated solutions. It is effective only at very low temperatures, ranging between 4 and 16 °C.

Isopsoralene—Another natural derivative of coumarin, namely isopsoralene, gives excellent results in different groups of plants (Chaudhuri, Chakravarty and Sharma, 1962), which have so far proved difficult to analyse, such as species of *Sanseviera, Ophiopogon*, etc. Even in species with very high chromosome numbers, such as species of *Pteris*, clarification of the chromosomes is very satisfactory. Isopsoralene is extracted from *Psoralea corylifolia*. Its chemical structure is:

Psoralene   Isopsoralene

Both psoralene and its isomer isopsoralene were tried out, and the former was found to be unsuitable for chromosome study. In view of this differential action Chaudhuri, Chakravarty and Sharma (1962) inferred that the position of the furan ring is an important controlling factor in the manifestation of karyotype clarification.

Umbelliferone—Umbelliferone is another derivative of coumarin shown to have the property of clarifying chromosomes. It is extracted from several species belonging to the family Umbelliferae and has the chemical structure:

Umbelliferone

Its application is very limited. In some of the plant species having a high chromosome number, umbelliferone has been found to yield very good results (Sharma and Chaudhuri, unpublished).

The majority of the coumarin derivatives can be employed for the study of chromosome analysis, but the best results have been obtained with the four mentioned above. Their applicability is restricted mostly to specific groups of plants, with the exception of aesculine which has a very wide application.

As viscosity change constitutes the principal basis of action of all of these compounds, attempts were made to find out the degree of change in viscosity induced by these derivatives (Sharma and Chaudhuri, 1962). The technical principle involved was to study the effect of the chemical on the streaming rotation movement of the protoplasm of *Vallisneria spiralis*—the same study can be performed on the staminal hairs of *Tradescantia paludosa*—the movement of the protoplasm being inversely proportional to the degree of viscosity change induced. One of the advantages of this technique is that the effect can be studied on a portion of the leaf attached to the plant. If the leaf had been detached from the mother plant, the general effect of the shock caused would have masked the effect of the chemical.

Another method, which has also be successfully employed, is to centrifuge the treated tissue. The time taken for the total displacement of the nuclei to the corner of the cell in tissues with various chemicals and at a constant revolution per minute gives an idea of the differential change in viscosity. The more viscous the cytoplasm, the longer is the time taken for nuclear displacement.

Following the above two methods, Sharma and Chaudhuri (1962) observed that, of coumarin and three of its derivatives (namely, aesculine, daphnetin, and aesculetin), daphnetin causes an increase in viscosity of the plasma within a very short period, the other three differing with respect to this property, the effect of coumarin being least. These results indicate the differential efficacy of different chemicals as pre-treatment agents.

The chemical structures of these compounds are:

Aesculetin   Daphnetin

A noteworthy feature is that daphnetin, which is extracted from *Daphne mezereum* Linn., and aesculetin, derived from *Euphorbia lathyrus* Linn., are isomers of each other. The solubility of aesculetin is 0·007 per cent while that of daphnetin is 0·005 per cent. The comparatively rapid action, with increased solubility, deviates slightly from Ferguson's law of thermodynamic activity as specially elaborated by Gavauden and Gavauden (1939). Coumarin and aesculine, however, follow Ferguson's principle, as coumarin is the most water soluble member (0·02 per cent) but also the least active one.

Oxyquinoline (OQ)

One of the most important groups of compounds investigated for their action in studying chromosome structure, is the quinoline complex. Tjio and Levan (1950) first worked out the importance of 8-hydroxyquinoline in chromosome analysis. They showed that it not only causes mitotic arrest, thus demonstrating its c-mitotic property, but at the same time it is also endowed with certain characteristics not shared by colchicine. The spindle is inactivated and, as such, does not cause any hindrance to the chromosomes being spread out during squashing; the chromosome arms contract equally. Unlike colchicine, OQ allows the metaphase chromosomes to maintain their relative arrangements at the equatorial plane. Both primary and secondary constrictions become conspicuous and the satellite gap appears to be greatly exaggerated. The centromeric structure can easily be analysed.

The c-mitotic property of oxyquinoline is not unusual as it is highly lipoid-soluble, but the very fact that the arrangement of the chromosomes is maintained intact on the spindle suggests that they are still attached to the spindle (the increased viscosity of the plasma does not allow the movement of the chromosomes). Further evidence of the attachment of the chromosomes to the spindle is provided by the 'centromeric reaction': this region becomes exaggerated due to a heavy contraction of chromosome arms against a rigid plasma which keeps the arms in their original position. According to Tjio and Levan (1950), the contraction operates towards the middle of each chromosome arm from the ends and the centromere, the centromeric gaps becoming pronounced due to the fact that the primary constriction region acts as a fixed point.

With regard to the mode of action of OQ, Stålfelt (1950) studied in detail the possibility of change in plasma viscosity. He observed that OQ, in concentrations of the same strength as used by Tjio and Levan, increased the viscosity in leaf cells of *Elodea* sp. No dislocation of plastids, after centrifuging, was observed, due to the rigidity attained by the plasma in strong concentrations.

Though a similar increase in rigidity of the plasma has been obtained with a number of lethal chemicals, yet their strong pyknotic reaction on chromosomes makes them unsuitable agents for karyotype. With colchicine as well, the increase in viscosity comes gradually and the chromosomes undergo movement during treatment, so that metaphase arrangement is disturbed. The influence of OQ on plasma viscosity is of such a nature that it does not alter the metaphase arrangement.

Oxyquinoline has been tried on different plant materials and it has been found to be specially suitable for species having long chromosomes. The study of human chromosomes too, following OQ treatment, as demonstrated by Tjio and Levan (1956), has been made possible. One of the limitations of pre-treatment with OQ in plants is that it requires a long period of treatment. Other chemicals, worked out later, have given equally satisfactory results in much less time. Radiomimetic action of OQ too has been recorded in some cases (Tjio and Levan, 1950), which should be considered during pre-treatment. However, such effects are very rare and if properly controlled OQ can be safely recommended as a pre-treatment agent for chromosome analysis.

Phenols

The radiomimetic action of phenols was observed by Levan and Tjio (1948). In view of the fact that a number of radiomimetic chemicals can be applied for the analysis of chromosomes, different types of mono-, di- and trihydric phenols were investigated in the authors' laboratory (Sharma and Bhattacharyya, 1956a). Of the dihydric types, resorcinol and hydroquinone were tested, whereas trihydric ones were represented by phloroglucinol and pyrogallol. Of the monohydric forms, several cresols have been tried. Guaiacol, in which there is an —OCH$_3$ group in addition to an —OH, was included to study the effect of the addition of methoxy group. A derivative of pyrogallol—gallic acid—was also tried out (Sharma and De, 1954).

It has been observed that phenols can be effectively employed for the study of chromosome morphology if applied at concentrations below the one causing chromosomal abnormalities. The most interesting feature is the absence of any effect in guaiacol. The role of the —OCH$_3$ group in suppressing the property of viscosity change is evident.

The importance of the presence and location of hydroxyl (—OH) groups is seen from the data so far obtained. Though all the three types of

Gallic acid   Phloroglucinol   Resorcinol

Hydroquinone   Pyrogallol   Guaiacol

phenols, mono-, di- and tri-hydric, can help in the clarification of chromosome morphology, yet dihydric forms are the most effective ones, that is, they are active at a very low concentration. Even amongst the dihydric types, such as resorcinol and hydroquinone, the location of the —OH groups is possibly responsible for their differential activity.

Low temperature appears to be essential for the activity of phenols. The roots have to be treated at a temperature ranging between 10° and 16 °C approximately; moreover the extremely low concentration, such as 10^{-3}M, which is sufficient for the activity, indicates that the reaction involved is chemical in nature. As with other groups of chemicals, not only is there spindle inactivation, but the differential hydration of chromosome segments also leads to the clarification of chromosome morphology.

The time required for treatment in phenols varies in different materials, although in no case has less than 3 h of treatment been found to be effective. They are applicable to almost all groups of plants, with modifications in the period of treatment and concentration, and are compatible with both metallic and non-metallic fluids, which can be used as fixatives after treatment in phenol. Following appropriate schedules for squashes and paraffin blocks, these compounds can be used in various procedures without in any way hindering the staining of chromosomes with any dye.

Gallic acid is useful in a number of species of water plants. The effective concentration range is not very wide (0·0005M) and the period of treatment is quite long, at least 3 h. There are no true sub-narcotic and narcotic zones in the action of this chemical, and with higher concentrations lethality ensues; therefore it cannot be safely recommended for karyotype analysis.

p-Dichlorobenzene (pDB)

Of the benzene derivatives, the most useful one in chromosome work is *p*-dichlorobenzene (*p*DB). It is sparingly soluble in water and has the following chemical structure:

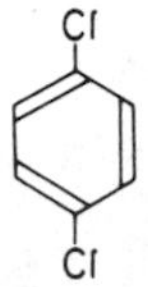

Meyer (1945) first demonstrated the importance of *p*DB as a pre-treatment agent. It is also employed as a reagent for the production of polyploidy in plants, and because of its very low solubility is used as saturated solution in water. Later workers (Dermen and Scott, 1950; Conagin, 1951) employed it effectively for chromosome counts. A detailed use of this compound has been made by Sharma and Mookerjea (1955) in plants. In their technique the entire schedule has been shortened by the omission of fixation after pre-treatment so that the tissue can be directly transferred to the staining mixture. The staining with acid–dye mixture, applied after pre-treatment, serves the purpose of fixation to some extent, but it has been noted that the direct transfer of the tissue to the acid–dye mixture, without pre-treatment, results in inadequate fixation with badly preserved chromosome structure. In different organs of the body the chemical was applied for the clarification of chromosomes and in all cases, even including leaves, successful preparations have been secured. In the case of leaves, the compound is specially helpful since, being a benzene derivative, it aids in clearing the cytoplasm.

Similar to aesculine and other coumarin derivatives, *p*DB not only causes spindle inhibition but also leads to clarification of chromosome constrictions due to the contraction and differential hydration of chromosome segments. Of all the chemicals tried so far, it has a comparatively much wider application, and complements with both long and short chromosomes appear to respond equally, even though the period of treatment may require modification. It is especially advantageous in the case of complements with many chromosomes as well as for tissues from different parts of the plant body which make it more useful in cytological research, and it has been recommended for difficult materials having chromosome complements with high number and small size.

One of the serious drawbacks of *p*DB treatment is that in the majority of cases the period of treatment necessary is at least 3 h. In the case of aesculine, this period may even be as little as 30 min. The temperature required too is rather specific and cold treatment at 10–16 °C is essential. Another limitation of *p*DB technique has recently been brought out by Sharma and

Bhattacharyya (1956b). They observed that after only $3\frac{1}{2}$ h of treatment, profuse chromosome fragments appeared in the preparations. As the period needed for optimum clarification of the chromosome morphology, that is 3 h, and the period necessary for inducing radiomimetic activity are fairly close, a cautious use of this agent is necessary to secure an accurate picture of the chromosomes. In order to minimise this error, dilution of the saturated solution is recommended. If utilised under properly controlled conditions, it can be applied in almost all groups of plants and animals.

Monobromonaphthalene

This naphthalene derivative has the structural formula given below and is sparingly soluble in water.

The property of spindle inhibition of 5-bromoacenaphthene and α-bromonaphthalene was first demonstrated in rye and wheat by Schmuck and Kostoff (1939). Östergren (1944) worked out the mitotic property of α-bromonaphthalene. O'Mara (1948) utilised this chemical as a preparation agent meant for contraction and clarification of constriction regions of chromosomes. Bhaduri and Ghosh (1954) established its efficacy in the study of wheat chromosomes, and claimed that transparency of the chromosomes is not lost even after Feulgen staining. O'Mara (1948) suggested non-metallic fixation after a minimum of 2 h of treatment in α-bromonaphthalene. For the study of prophase chromosomes, methanol-added fixative is chosen. Sharma (1956) suggested that the purpose of such special treatment after the application of monobromonaphthalene is possibly twofold. It is possible that the chromosome, at the time of pre-treatment, undergoes certain chemical changes so that it cannot respond to any of the fixing agents, except the one recommended, or the chemical nature of the cytoplasm undergoes such an alteration that it becomes impermeable to any other fixative. In any case, a chemical change in the chromosome can be visualised, as even the effect on the cytoplasm would consequently affect the chromosomes.

Although in O'Mara's schedule prolonged pre-treatment for 2–4 h was suggested, it has been observed that even 10–15 min treatment at low temperature in saturated aqueous solution of monobromonaphthalene may bring about the desired representation of chromosome morphology in certain groups of plants. For aquatic angiosperms especially, pre-treatment with this chemical is usually effective. The principal cause is the rapid rate of penetration of this chemical into the tissues. In hydrophytes, both submerged and aquatic, the adaptations of their tissues to their habit, such as air spaces, etc., necessitate a rapid penetration of the chemical for exerting any effect. Like p-dichlorobenzene, it has a clearing effect on the cytoplasm but it is not as pronounced as that of the former.

Veratrine

Another important alkaloid, shown to possess the property of clarifying chromosome structure, is veratrine (Sharma and Sarkar, 1956). Cold treatment is found to be essential for its activity. Radiomimetic action of this chemical has also been demonstrated. At lower concentrations, it effectively clarifies chromosome morphology. Both dicotyledonous and monocotyledonous plants respond well to its application, the former group being more susceptible than the latter. An advantage of its application is its rapid action on somatic chromosomes, which can be observed within 30–40 min of treatment. Its effective concentration range is wide, between 0·05 and 4 per cent. It is sparingly soluble in water and its molecular formula is $C_{32}H_{49}NO_9$.

Hormones, Heterocyclic Bases, Vitamins and Vegetable Oils

The importance of hormones in chromosome analysis has been worked out (Sharma and Mookerjea, 1954). β-Naphthoxyacetic acid, α-naphthylacetic acid, phenylacetic acid, indolylacetic acid, indolylpropionic acid and indolylbutyric acid have been shown to induce viscosity changes in the plasma at 10–16°C, so that scattered metaphase plates, with clarified chromosome structure, are obtained. No correlation between water solubility and the effect of the hormones has been observed. It has only been noted that the higher the maximum point or upper threshold of the hormone, the lower is its minimum point or lower threshold of activity. The concentration needed for clearing chromosomal details is always higher than that needed for the promotion of growth. As there are a variety of hormones, they can be used in a variety of organisms.

Some of the heterocyclic bases, such as guanine, uracil and maleic hydrazide, have also been seen to induce change in plasma viscosity leading to spindle arrest and chromosome clarification (Sharma and De, 1956). The chemical structures of these compounds are:

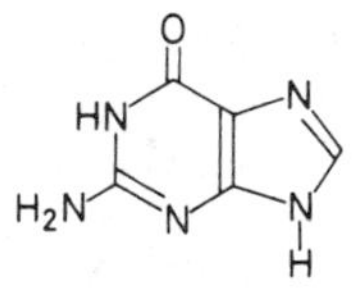

Guanine   Uracil

(2 – amino – 6 – oxypurine)   (2,4 – dioxypyrimidine)

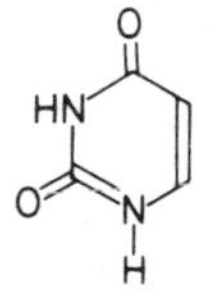

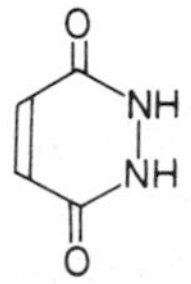

Maleic hydrazide

Their solubility is low and low temperature is essential for their effect. In all cases, the concentration to bring out the details of chromosome structure is always below the one causing sub-narcotic effect. Evidently, such pre-treatment agents should be applied under controlled conditions, as prolonged treatment or higher concentrations will lead to chromosome breaks.

The capacity of some vitamins in clearing chromosome morphology was also explored. Of those studied, only ascorbic acid (Sharma and Datta, 1956) is effective. In some species of Liliaceae it acts at a concentration of 0·05M. Low temperature is an essential prerequisite for its action. The application of this group of chemicals is, however, very limited.

Swaminathan and Natarajan (1957) have demonstrated that scattered chromosome in metaphase plates can be secured through treatment in a number of vegetable oils, specially the oil of *Sesamum*. Evidently, molecules of these compounds react with their solvent in the cytoplasm and thus alter the viscosity of the plasma. As sub-narcotic effects have also been observed with these oils, investigations are needed to explore the property of a number of essential oils in chromosome clarification such as eugenol, lavender, etc., the radiomimetic actions of which have been reported by D'Amato and Avanzi (1949). A homogenised mixture of 1 drop of castor oil and 5 cm^3 of 0·003M aq. OQ solution has been used for chromosomes of *Sequoia sempervirens* (Fozdar and Libby, 1968).

O-Isopropyl-n-phenylcarbamate (IPC)

Pure IPC is a white crystalline powder, highly soluble in ethanol, but only soluble in water to the extent of 250 p.p.m. at 25 °C. It is decomposed in soil by micro-organisms and also in unrefrigerated aqueous solutions. In mammals, its toxicity is low, the LD$_{50}$ being 3000 mg/kg body weight. Concentrations effective on plants lie between 2·5 and 20 p.p.m., for periods of treatment ranging between 1–4 h. The solvent ethanol must be removed for pre-treatment schedules. It is preferable to use solutions in water. The main effects of this chemical are, immediate cessation of mitotic activity, followed by contraction of chromosomes in prophase, metaphase and anaphase. In some cases, the chromosomes may be reduced to about one-fourth of their original length (Mann *et al.*, 1965; Storey and Mann, 1967). Sawamura (1965) noted that CIPC, a meta-chloroanalogue of IPC, causes inhibition of phragmoplast and spindle development.

Water in Pre-treatment

Several of the pre-treatment chemicals tested so far are used in aqueous solutions and are sparingly soluble in water. As many are almost insoluble in distilled water, tap water is often used as a solvent. The use of isotonic solution for swelling and chromosome scattering, has been described in the chapter on mammalian chromosomes. The effects of water distilled under different conditions—tank water, tap water (Sharma and Sen, 1954) and isotonic solution of $CaCl_2$ (Sharma and Mookerjea, 1955)—have been studied in the clarification of chromosome morphology. In order to check the degree of their effect, controlled experiments were performed in which fixation and staining were carried out without any pre-treatment. Three types of distilled water were used, namely: (*a*) distilled in metallic retort, (*b*) distilled once in glass distillator and (*c*) distilled several times in glass distillator. The purpose of using glass-distilled and re-distilled water was to eliminate any error arising through the presence of any metal in the water.

Root tips, directly treated in the fixing-*cum*-staining mixture, revealed clumped chromosomes. Satisfactory contraction of chromosomes and straightening of arms were obtained with tap water, tank water and with water distilled in a metallic retort, though the effect of the latter was less than that of the former. The high frequency of metaphase plates, as compared to control, suggests spindle abnormality. Pronounced scattering of chromosomes could be secured in some of the water plants, such as species of *Vallisneria* and *Hydrilla*, with water distilled twice in a glass distillator. These data suggest that the chemicals need not be present in significant amount to change plasma viscosity, necessary for chromosome clarification. Not only tap water, which is rather alkaline, and tank water, yielded satisfactory results, but even an isotonic solution of calcium chloride in water yielded well scattered chromosome plates in *Dieffenbachia picta*, a member of the Araceae having as high a chromosome number as $2n = 126$. It is remarkable that in this species this pre-treatment yielded better results than *p*-dichlorobenzene and coumarin. Of course, as all these pre-treatments were performed under low temperature, it is obvious that cooling accelerated the reaction. That the entire effect on this species is not due to cooling is proved by the complete absence of any reaction in roots of this species which are merely cooled without being kept in the pre-treatment medium. However, Lesins (1954) recorded that keeping plants on snow overnight prior to direct fixation gave good results for certain species. Any nucleic acid starvation effect due to cold was counteracted by a treatment in 3 per cent ferric ammonium sulphate solution for 3 h or overnight before treatment. For animal chromosomes, specially of cancerous tissue, water pre-treatment is very effective. In certain species, if the water treatment is prolonged, chromosome breakage may result, which has also been noted by Macfarlane (1951) and others. Cold treatment alone at 0–2 °C, and 8–10°C, for 4–24 h in water, has been used for screening wheat aneuploids (Tsunewaki and Jenkins, 1960).

The distinct response of nucleus and cytoplasm to such dilute solutions as isotonic calcium chloride, tap water or even re-distilled water suggests the existence of an extremely subtle metabolic equilibrium necessary for the functioning of the spindle and maintenance of chromosome structure. Any change in the medium, however minimal the chemical concentration may be, may upset the balance, resulting in spindle disturbance and contraction of the chromosome and its segments. The differences in water treatment can be explained by taking into consideration the fact that re-distilled water does not constitute the normal medium for growth of roots in nature. As

Table 1 SOME COMMON PRE-TREATMENT CHEMICALS

Chemical	Effective concentration of aqueous solution	Period of treatment needed for karotype clarification	Temperature °C	Remarks
Acenaphthene	Saturated	1–4 h	Room temperature	Application is limited to specific organs, particularly to pollen tubes
Aesculine	Saturated or half saturated	5 min–24 h	4–16	Widely applicable in both plants and animals
Aesculetin	Saturated	Variable	10–16	Limited to only a few plants
α-Bromonaphthalene	Saturated	10 min–4 h	10–16	Very effective particularly for aquatic plants and wheat chromosomes
Chloral hydrate	0·5–1%	30 min–2 h	10–16	Effective in a number of plants but the results are rather inconsistent
Colchicine	0·5–1%	30 min–1 h for plants but 5–30 min for animals	Preferably between 8 and 16	Very wide range of effects. Equally suitable for plants and animals
Coumarin	Saturated	3–6 h for plants, 5–60 min for animals	Both cold and room temperature	Effective for long chromosomes and for animals
Daphnetin	Saturated	Variable	10–16	Limited application

Substance	Concentration	Time		Effect
p-Dichlorobenzene	Saturated	3–5 h	12–16	Very effective, particularly on plants with a high number of chromosomes
Gammexane	Saturated	1–3 h	12–16	Effective mainly on plant tissue
Hormones	0·0002–0·02%	3–3½ h	10–16	Effective on plants with both low and high number of chromosomes
Isopsoralene	Saturated	1–2 h	12–14	Very effective on mitotic chromosomes of a number of plants
Oxyquinoline	0·002M	3–4 h	12–16	Very effective on plants, particularly with medium sized to long chromosomes
Phenols	Variable, usually low doses	3–6 h	10–16	Applicable to almost all groups of plants with changes in concentration
Umbelliferone	Saturated	1–3 h	12–16	Applicable to some plant groups having high chromosome number
Veratrine	0·05–4%	30–40 min	12–16	Applicable to several groups of plants

Miscellaneous fluids

(a) Hungerford's fluid (1955)	(i) 0·6% NaCl in double-distilled water: 6 parts (ii) Colchicine in modified Ringer's solution (0·82% NaCl, 0·02% KCl, 0·02% $CaCl_2 \cdot 2H_2O$ in double-distilled water) in proportion $1/10^6$:1 part	Used for minced mouse foetus. Incubated at 37°C in covered depression slide for 30–60 min with agitation after treatment for 5–10 min

(b) Dilute isotonic salt solution can also be used

evidently the cytoplasmic constitution of the different groups of plants is different, the degree of viscosity change required to cause spindle disturbance is not identical in different species. Because of this difference in requirement all the species do not respond in a similar way to a pre-treatment chemical. It has already been mentioned that pre-treatment chemicals differ with respect to their capacity for causing viscosity change in the cytoplasm. This is the reason why different groups of organisms may require different pre-treatment chemicals for spindle arrest and clarification of chromosome morphology. In the absence of inventions whereby cytoplasmic constitution of a species can be detected without detailed analysis, successive trials are necessary to find out the pre-treatment chemical necessary for a species. In general, it may only be stated that related species with similar genotypes have preference for a common pre-treatment agent.

REFERENCES

Barton, N., Cook, J. W. and Loudon, J. D. (1945).   *J. Chem. Soc.* 176
Bhaduri, P. N. (1939).   *J. R. micr. Soc.* **59,** 245
Bhaduri, P. N. and Ghosh, P. N. (1954).   *Nature* **174,** 934
Brachet, J. (1940).   *Arch. Biol.* **51,** 151
Branch, C. F., Fogg, F. C. and Ullyot, G. E. (1949).   *Acta Un. int. Cancr.* **6,** 439
Chaudhuri, M., Chakravarty, D. P. and Sharma, A. K. (1962).   *Stain Tech.* **37,** 95
Conagin, C. H. T. (1951).   *Stain Tech.* **26,** 274
D'Amato, F. and Avanzi, M. G. (1949).   *Caryologia* **1,** 175
D'Amato, F. and Avanzi, M. G. (1954).   *Caryologia* **6,** 131, 150
Dermen, H. and Scott, D. H. (1950).   *Proc. Amer. Soc. hort. Sci.* **56,** 145
Dewar, M. J. S. (1945). *Nature* **155,** 141
Dustin, P. Jr. (1949).   *Exp. Cell Res.* Suppl. **1,** 153
Dustin, P. Jr. (1950).   *C. R. Soc. Biol., Paris* **144,** 1297
Eigsti, O. J. and Dustin, P. Jr. (1957).   *Colchicine in agriculture, medicine, biology & chemistry.* Iowa; Iowa State College Press
Emsweller, S. L. and Stuart, N. W. (1944).   *Stain Tech.* **19,** 109
Fabergé, A. C. (1945).   *Stain Tech.* **20,** 1
Ferguson, J. (1939).   *Proc. Roy. Soc. B.* **127,** 387
Fozdar, B. S. and Libby, W. J. (1968).   *Stain Tech.* **43,** 97
Gavauden, P. (1938).   *J. R. micr. Soc.* **58,** 97
Gavauden, P. and Gavauden, N. (1939). *C. R. Acad. Sci., Paris* **209,** 805
Hungerford, D. A. (1955).   *J. Morph.* **97,** 497
Kostoff, D. (1939).   *Cellule* **48,** 179
La Cour, L. F. (1935).   *Stain Tech.* **10,** 57
Leiter, J., Hartwell, J. L., Kline, I., Nadkarni, M. V. and Shear, M. J. (1952).   *J. nat. Cancer Inst.* **13,** 1201
Lesins, K. (1954).   *Stain Tech.* **29,** 261
Lettré, H. (1947).   *Angew. Chem. A.* **59,** 218
Lettré, H., Fernholz, H. and Harwig, E. (1952).   *Liebigs Ann.* **576,** 147
Levan, A. (1949).   *Proc. 8th Int. Congr. Genet., Hereditas,* Suppl. 325
Levan, A. and Östergren, G. (1943).   *Hereditas, Lund.* **29,** 381
Levan, A. and Tjio, J. H. (1948).   *Hereditas, Lund.* **34,** 453, 250
Lewitsky, G. A. (1931).   *Bull. appl. Bot., St. Petersb.* **27,** 19
Ludford, R. J. (1936).   *Arch. exp. Zellforsch.* **18,** 411
Macfarlane, E. W. E. (1951).   *Growth* **15,** 241
Macfarlane, E. W. E., Messing, A. M. and Ryan, M. H. (1951).   *J. Hered.* **42,** 95
Mann, J. D., Jordan, L. S. and Day, B. E. (1965).   *Weeds* **13,** 63
McKay, H. H. and Clarke, A. E. (1946).   *Stain Tech.* **21,** 111

Manna, G. K. and Raychaudhuri, S. P. (1953). *Proc. 40th Ind. Sci. Congr.* **3,** 181

Mazia, D. (1955). *Symp. Soc. exp. Biol.* **9,** 335

Mazia, D. (1959). 'Cell division'. *The Harvey Lectures,* 1957–58. New York; Academic Press

Meyer, J. R. (1945). *Stain Tech.* **20,** 121

Nemec, B. (1904). *Jahr. wiss. Bot.* **35**

Nybom, N. and Knutsson, B. (1947). *Hereditas, Lund.* **33,** 220

O'Mara, J. G. (1948). *Stain Tech.* **23,** 201

Östergren, G. (1944). *Hereditas, Lund.* **30,** 429

Östergren, G. (1950). *Hereditas, Lund.* **36,** 371

Östergren, G. and Levan, A. (1950). *Hereditas, Lund.* **36,** 371

Peto, H. H. (1935). *Canad. J. Res. C.* **13,** 301

Quercioli, E. (1955). *Caryologia* **7,** 350

Sawamura, S. (1965). *Cytologia* **30,** 325

Schmuck, A. and Kostoff, D. (1939). *C. R. Acad. Sci. U.R.S.S.* **23,** 263

Sharma, A. K. (1956). *Bot. Rev.* **22,** 665

Sharma, A. K. and Bal, A. K. (1953). *Stain Tech.* **28,** 255

Sharma A. K. and Bhattacharyya, N. K. (1956*a*). *Genetica* **28,** 121

Sharma, A. K. and Bhattacharyya, N. K. (1956*b*). *Cytologia* **21,** 353

Sharma, A. K. and Chaudhuri, M. (1959). *Curr. Sci.* **28,** 498

Sharma, A. K. and Chaudhuri, M. (1961). *Nucleus* **4,** 157

Sharma, A. K. and Chaudhuri, M. (1962). *Nucleus* **5,** 137

Sharma, A. K. and Datta, A. (1956). *Φyton* **6,** 71

Sharma, A. K. and De, D. (1954). *Caryologia* **6,** 180

Sharma, A. K. and De, D. (1956). *Φyton* **6,** 23

Sharma, A. K. and Mookerjea, A. (1954). *Caryologia* **6,** 52

Sharma, A. K. and Mookerjea, A. (1955). *Stain Tech.* **30,** 1

Sharma, A. K. and Sarkar, A. (1963). *Revta port. Zool. Biol. ger.* **4,** 29

Sharma, A. K. and Sarkar, S. (1955). *Nature* **176,** 261

Sharma, A. K. and Sarkar, S. (1956). *Caryologia* **8,** 240

Sharma, A. K. and Sarkar, S. (1957). *Proc. Indian Acad. Sci.* **45,** 288

Sharma, A. K. and Sen, S. (1954). *Genet. iber.* **6,** 19

Stålfelt, M. G. (1950). Appendix to the paper by Tjio, J. H. and Levan, A. in *An. Estac. exp. Aula Dei* **2,** 21

Steinegger, E. and Levan, A. (1947). *Hereditas, Lund.* **33,** 515

Storey, W. B. and Mann, J. D. (1967). *Stain Tech.* **42,** 15

Swaminathan, M. S. and Natarajan, A. T. (1957). *Stain Tech.* **32,** 43

Swanson, C. P. (1940). *Stain Tech.* **15,** 49

Tjio, J. H. and Levan, A. (1950). *An. Estac. exp. Aula Dei* **2,** 21

Tjio, J. H. and Levan, A. (1956). *Hereditas, Lund.* **42,** 1

Tsunewaki, K. and Jenkins, B. C. (1960). *Cytologia* **25,** 373

Windaus, A. (1924). *Liebigs Ann.* **439,** 59

Wolff, S. and Luippold, H. E. (1956). *Stain Tech.* **31,** 201

Zeisel, S. (1883). *Mh. Chem.* **4,** 162

3 Fixation

Fixation may be defined as the process by which tissues or their components are fixed selectively to a desired extent. The purpose of fixation is to kill the tissue without causing any distortion of the components to be studied, as far as is practicable. In chromosome study, the purpose may vary to a significant extent. In certain cases, it may be necessary to study the phospholipid component of the chromosome, and for this purpose a specialised schedule is used. On the other hand, for nucleoprotein precipitation the method is different. In order to study the enzyme activity in the chromosome, freeze–drying methods are often applied. These instances indicate the extent to which the purpose of study varies even when studying the chromosome alone.

Suitable fixation, which is absolutely essential for the study of chromosome· structure, has always intrigued cytologists. Recently the scope of cytological study has been extended to include different disciplines of biological science and the need of critical fixation is being more and more realised. Even now, very little is known about the exact changes undergone at an intracellular level following fixation, and as such the cytologist is handicapped in having to rely partly on chance.

The term fixation was previously used in a wide sense involving, principally, the mere visibility of chromosome structure. With the extension of the scope of cytological study, the necessity for a differential analysis of chromosome segments was felt. Fixation, as is now realised, must be *critical*. It should not only increase visibility of the chromosome structure but should also clarify the details of chromosome morphology, such as the chromatic and heterochromatic regions and the primary and secondary constrictions. In view of the limited number and inadequacy of techniques that were available in the past, somatic chromosomes could not be studied to the desired extent in the majority of species. The inadequacy of the techniques was chiefly due to the fact that the earlier workers did not realise the importance of critical fixation in the analysis of chromosome structure.

The study of fixation and its principles is in itself a specialised line of investigation. Species differ greatly in their response to a fixing chemical. In the case of plants, monocotyledonous and dicotyledonous groups show different responses. Similarly, a fixative proved to be suitable for a plant species may be inadequate for animals. Mammalian chromosomes require special methods of fixation. All these facts imply that the cytoplasmic constitution of each individual is one of the principal characteristics of the individual and is therefore one of the principal factors controlling its response to a chemical fixative. The main limitation in the analysis of the exact effect of fixation at an intracellular level is the inherent complexity of all biological objects. In every multicellular living object, so many intra-cellular variable factors are present that it is very difficult to analyse the effect of the fixing chemical on any one of them.

Although a number of fixing reagents have been devised by various workers, all of them possess certain common characteristics which are essential for a fixative. Each fixing chemical is lethal in its action. In fact, Levan (1949) has classified the lethal chemicals into two categories.

Under one category fall those compounds which cause pyknosis of chromosomes or detachment of nucleic acid from the protein thread, and in the other are included compounds which maintain the chromosome structure intact. In a fixing chemical, therefore, lethality must not be associated with pyknosis of chromosomes or the dissolution of the nucleic acid from the protein thread. The structural integrity of the chromosome must be maintained intact. Precipitation of the chromatin matter is essential to render the chromosome visible and to increase its basophilic nature in staining. Under living conditions, the phase difference between different components of the cell is not enough to permit them to be observed as distinct entities. Coagulation of protein and consequent precipitation cause a marked change in the refractive index of the chromosomes, helping them to appear as differentiated bodies within the cell. All fixatives, so far used, have the property of crosslinking proteins. The primary requisite of a fixative for the study of chromosomes is therefore the possession of the *property of precipitating chromatin*. A fixative for chromosomes is, generally, a mixture of several compounds and at least one of its constituents must have this property.

Another important requisite of a fixative is that it should have the property of rapid penetration so that the tissue is killed instantaneously, the divisional figures being arrested at their respective phases. *Immediate killing* is essential as otherwise the nuclear division may proceed further and attain the so-called 'resting' or metabolic phase. Divisional configurations are necessary for visualising the chromosomes correctly. Conversion to the resting stage makes it useless for chromosome study.

With the death of the cells, certain consequent changes occur which are detrimental to the preservation of chromosome structure. The most important change is the autolysis of protein. Under normal conditions for the tissue, enzymes are present which help to build up the proteins. With lethality, as the medium becomes acidic, these enzymes act in the reverse direction and cause a breakdown of the complex protein into simpler amino acids. As polypeptide forms one of the principal constituents of the chromosomes, the effect of denaturation will cause ultimate disorganisation of the chromosome structure. Therefore, a fixative should also be able to *check autolysis of proteins.*

With the onset of lethality, bacterial action causes the tissue to decompose. Another prerequisite for a good fixative is to *prevent this decomposition* by maintaining an aseptic condition in which bacterial decay cannot take place.

As the purpose of chromosome study is to observe the minute details of chromosome morphology, the staining should be perfect. A number of chemicals, e.g. formaldehyde, though otherwise possessing all the qualifications of good fixatives, may often affect the basophilia of chromosomes. Such chemicals, when used alone, cannot be recommended for chromosome fixation. A proper fixative should, in general, *enhance the basophilia* of the chromosome.

A fixing mixture which fulfils all the conditions detailed above can be considered to be a truly effective fixative for chromosome study, but since all these properties are rarely to be found within a single chemical, a fixative is generally a combination of several compatible fluids which jointly satisfy all the above requirements.

Before dealing with the different fixatives and their properties, it should be stated that, even with the best fixatives, the chemical changes undergone by the nuclear bodies cannot be ignored as there are certain inherent disadvantages in the entire process. Firstly, at least some of the post-mortem changes are difficult to overcome. Secondly, there is a possibility of the extraction of diffusible components. Lastly, even with all precautions, it is very difficult to keep the enzyme activity unaffected, and in addition the tissue shows a tendency to shrink on coming into contact with the chemicals.

All these disadvantages can be eliminated by fixing through freezing at low temperature, followed by drying the tissue—the principle involving rapid cooling of the tissue to a low temperature, followed by extraction of water in vacuum. This method allows a life-like preservation of the tissue. The cooling process must be so rapid that the water cannot crystallise during freezing, as this results in the distortion of cellular components, and consequent misinterpretation of chromosome structure. The initial water contents of the tissue, the shape and size of the material, the temperature of the tissue and of the cooling bath are all factors contributing to proper fixation by freezing without forming ice crystals. Water can be frozen into amorphous ice in a cooling bath of $-175\,°C$, secured by condensation and liquid nitrogen (Stephenson, 1956; Gersh, 1959; Burstone, 1969).

After fixation by freezing, water must be removed. The process is usually carried out in vacuum at a low temperature. The material can also be dried by passing it through a stream of dry cool gas (Treffenberg, 1953; Jensen and Kavaljian, 1957). It has been argued that the water molecules may form crystals during the process of drying (De Nordwall and Staveley, 1956), but up to the present time no evidence in favour of this contention has been obtained. The material can be cut directly or before sectioning, and may be infiltrated with paraffin or some other medium. Post-fixation too, depending on the purpose for which it is needed, may be employed

(Bell, 1956; Clements, 1962; McClintock, 1964; Rutherford *et al.*, 1964; Lotke and Dolan, 1965; Pearse, 1968; Burstone, 1969). Several freeze-drying equipments and cryostats have now been developed.

The freezing method of fixation has a number of advantages, such as: (*a*) minimum distortion of the tissue after its death, (*b*) least possibility of diffusion, and (*c*) no significant effect on the enzyme system. Further, the tissue can be directly embedded in paraffin without dehydration or clearing. Even then, however, its inherent drawbacks cannot be ignored. For example, all the above advantages can be nullified by the distortion of cellular components during embedding or sectioning. Interference may occur with very small materials by crystallisation of water while preparing the material in bulk, and disintegration of the tissue between developing ice crystals, which cannot be checked. The extreme cost involved in setting up the apparatus hinders its use in all laboratories.

Some of these limitations, however, have been eliminated in the process of dehydration by freezing–substitution. In this technique, the specimen is rapidly frozen, followed by dehydration at a very low temperature ($-20°$ to $-78\,°C$) through any one of the following reagents: *n*-propanol, *n*-butanol, methanol, ethanol, methyl cellosolve, or the chemical fixatives, as would be discussed later. After complete dehydration, the material is brought back to room temperature slowly (Bullivant, 1965; Rebhun, 1965; Pease, 1966; Malhotra, 1968; Cope, 1968). During this process there is a possibility of the extraction of diffusible substances and lipids but the proteins are kept intact. The extraction of lipids prevents a proper preservation of chromosome structure as phospholipid is one of the likely components of the chromosome.

The freeze–drying method of fixation no doubt has numerous advantages over the chemical method of fixation. It is particularly accurate for the study of the effect of chemical and physical agents on the chromosome, where the immediate effect has to be analysed, but it is not very useful for the study of the structure and behaviour of chromosomes during the process of division under normal conditions. Other advantages of chemical fixation, such as increase in the basophilia of chromosomes, differential precipitation of chromatin matter in its different segments and so on, cannot be obtained by means of the freezing–substitution technique. These factors, taken together, have contributed largely to the wide use of chemical fixatives in routine work on chromosome studies, and only under special circumstances, as mentioned above, is freezing–substitution applied.

The fixing chemicals, in general, may be classified into two categories, precipitants and non-precipitants. The classification is based on their property of precipitating proteins within the cell. The best examples of precipitant fixatives are chromic acid, mercuric chloride, ethyl alcohol, etc. Among the non-precipitant fixatives can be included osmium tetroxide, potassium dichromate, etc. There are certain fixatives which undergo chemical combination with proteins, as Baker (1966) has pointed out, some of which precipitate out proteins, while others do not. A fixative may also alter the nature of proteins without the necessary addition of new atoms to them, thus causing denaturation, and subsequently the solubility of the protein is lost (*see* Wolman, 1955). That denaturation does not involve any addition of atoms is proved by the observation of a similar effect after heat treatment. It is possible that denaturation causes the molecules to straighten out into a fibrous form, the chemical change being the appearance of reactive sulphydryl groups.

Whenever a fixative has a strong precipitating action, this is usually counteracted to a certain extent by the addition of other reagents. To deal with any possible error in the interpretation of their combined action at an intracellular level, Baker (1944) suggested that the best fixation would be possible with a simple non-precipitant (unmixed) fixative, to which non-fixative salts have been added. No doubt the action of such a fixative can easily be interpreted, but for chromosome study, where a variety of reactions is needed for differential clarification of segments, it is nearly impossible to secure a single chemical combining all the desirable qualities.

Fixatives having the property of precipitating proteins cause the nuclei to have a spongy appearance. Telleyesniczky (1905) first suggested that such sponge work is absent in the living nucleus and it appears only after fixation with a protein precipitant. Even then, precipitation has an added advantage in paraffin embedding, as the latter infiltrates readily into the meshes of the sponge. Non-precipitant fixatives, on the other hand, preserve the structure very badly after embedding. For the proper observation of chromosomes, some amount of precipitation is necessary and therefore in a fixative a slight precipitating action is preferred.

The swelling of the cells and shrinkage of chromosomes are the two principal factors controlling the merit of a fixative. It has always been difficult to check the initial changes in volume occurring during fixation. Undoubtedly, with larger molecules of fixatives, the initial swelling is retarded (Bähr, 1957; Bloom and Friberg, 1957). A heavy shrinkage was observed in xylol and paraffin infiltration when lower alcohols were used. An overall alteration in the volume of the cell or a contraction or swelling of chromosome structure is not harmful, but a differential shrinkage of the constituents of the cell results in distortion of chromosome structure, which must be avoided. Osmotic concentration was previously considered to be one of the principal factors affecting the swelling or distortion of the cellular structures. Therefore isotonic solution was often used for fixation. Telleyesniczsky (1905) claimed that osmotic concentration is not the controlling factor because the tissue, at the time of fixation, loses its capacity to respond to any change in osmotic pressure. Baker (1966) also confirmed the validity of this statement. Dilute acetic acid, having a pressure of 20 atm, causes the tissue to swell, whereas picric acid, with a low pressure of $2\frac{1}{2}$ atm, induces shrinkage to a significant extent. All this evidence suggests that osmotic concentration or pressure is of little significance in controlling the efficacy of a fixative. On the other hand, picric acid is a strong protein precipitant and acetic acid is not. Therefore Baker (1950) suggested that this property is possibly the controlling factor in causing shrinkage. Hence, all the above factors are taken into consideration when a fixative is selected.

The main chemicals which have been used as fixatives, or more precisely, as ingredients of a fixing mixture, are: (*a*) non-metallic—ethyl alcohol, methyl alcohol, acetic acid, formaldehyde, propionic acid, picric acid, chloroform; (*b*) metallic—chromic acid, osmic acid, platinic chloride, mercuric chloride, uranium nitrate, lanthanum acetate, etc. Several of these compounds are also used as vapour fixatives, with the sole object of converting the soluble substances into insoluble ones, before coming into contact with water or other solvents, so that *in situ* preservation is maintained.

Most non-metallic fixatives, except formalin in cytochemical work, have one advantage over the metallic ones—that no washing in water is required after fixation.

Of all the fixatives employed so far, Law's method (1943) is supposed to be the most simple, in which the fixation was performed by boiling water. This brings about coagulation of protein and specially uncoiling of chromosome spirals. This method is only useful for temporary study of merely the divisional figures. Chromosome structure cannot be well preserved following this technique of fixation.

The properties of the different fixing chemicals, the merits and demerits of their uses and their applicability are discussed below.

FIXING FLUIDS

NON-METALLIC FIXATIVES

Ethyl Alcohol or ethanol (C_2H_5OH)

Ethyl alcohol is used extensively as a constituent of chromosome fixatives. The suitable percentage of alcohol for fixation varies from 70 to 100 per cent. One of the most important advantages of the use of ethyl alcohol is its capacity for immediate penetration. Fischer (1899) noted that alcohol precipitates nucleic acid. The dehydrating property of alcohol is well known and it causes an irreversible denaturation of proteins. It also has an undesirable hardening effect on the tissue. The denaturating action on proteins due to precipitation may convert the molecule into a form impermeable to reagents or approximating closely to the separated groups in protein chains (Pearse, 1968). It has been claimed to break the hydrogen bonds and salt links in protein chains, thus revealing several side groups (Kauzmann, 1959; Okunuki, 1961). As the stereochemical pattern is often altered, its effect in misrepresenting the structure is to be considered.

Being a reducing agent, it undergoes immediate oxidation to acetaldehyde and then to acetic acid in the presence of an oxidiser, and so it cannot be used in combination with many good metallic fixatives, such as chromic acid or osmium tetroxide. Baker (1950) further suggested that it is not an effective fixative for chromatin. Therefore it can be used principally in combination with acetic acid, formaldehyde or chloroform. Chilled ethanol fixation is often used for the preservation of certain enzymes, as the reactive groups of enzymes generally remain undisturbed. It also does not affect the isoelectric point appreciably, which is an added advantage for cytochemical purposes (Lojda, 1965). For enzyme studies on chromosomes, chilled 80 per cent ethanol fixation for 1 h or more is recommended and for monolayer cultures, absolute or 96 per cent ethanol fixation for 1–15 min is often applied (Smetana, 1967).

Acetic Acid (CH_3COOH)

One of the primary advantages of using acetic acid in a fixing mixture is that it can be combined with any of the other fixatives so far studied. It can be mixed in all proportions with water and alcohol. It is the chief constituent of vinegar and has a pungent smell. It can be used from very low concentration, i.e. 1 per cent, to even glacial (100 per cent) form. The term 'glacial' is derived from the fact that it freezes to a form resembling ice at very cold temperatures. This acid has a remarkable penetrating property, even higher than alcohol; possibly its smaller ions are responsible for this property.

Pischinger (1937) noted that acetic acid can precipitate nucleic acid and dissolve the histones, but it is incapable of fixing cytoplasmic protein. It cannot be recommended for the observation of phospholipids in the chromosomes. It may be noted that potassium dichromate, which is a good fixative for lipids, loses this property if used in combination with acetic acid. Baker (1966) suggested that it can precipitate nucleoprotein but not albumin. Wolman (1955) pointed out that acetic acid affects fixation of proteins like nucleoproteins and mucoproteins, the isoelectric point of which is near the pH value of the acid solution employed. The bound water layer, which is normally present around the ionised groups of the protein molecules, disappears with the loss of the electric charge, and the protein molecules are then free to form as the reactive points come closer together. The main use of acetic acid in the study of chromosome structure is to check shrinkage and to preserve the chromosome structure without distortion. The materials fixed in acetic acid can resist hardening in alcohol.

Acetic acid is, in general, an ideal fixative for chromosomes, and in spite of Pischinger's finding that it dissolves histones, it has been observed to maintain the chromosome structure intact, presumably not causing any distortion of the nucleoprotein. One of the limitations of this fixative is the excessive swelling of the chromosome segments induced by it. Therefore where a study of detailed structure of chromosomes is needed, this point must be borne in mind, and acetic acid should be used in combination with alcohol, or a similar chemical, which shrinks and hardens the tissue. For the study of meiotic chromosomes, where the purpose is to study the behaviour instead of the structural details, acetic acid is quite suitable. In pachytene chromosome analysis, where clear chromosomal details are required, acetic acid serves as an ideal fixative.

Acetic acid is also a good solvent for aniline dyes, and due to this property it is a necessary component of staining-*cum*-fixing mixtures, like acetic-carmine, acetic-orcein, acetic-lacmoid, etc.

Formaldehyde (H.CHO)

Formaldehyde is a gas and its commercial form, known as 'formalin', contains a 40 per cent solution of formaldehyde in water. It is a bifunctional compound capable of forming crosslinks between protein end groups (Pearse, 1968). For the fixation of chromosomes 10–40 per cent solution of commercial formalin in water is used. It has a very low precipitating action on protein in high concentrations.

The action of formaldehyde is chiefly on proteins. It reacts with the amino groups of proteins, with the production of water, or simply attaches itself to the amino acid without the liberation of water (*see* Woodroffe, 1941; Baker, 1966). The reactions can be interpreted as:

$$\text{H-}\underset{\underset{\text{H}}{|}}{\overset{\overset{\text{NH}_2}{|}}{\text{C}}}\text{-C}\underset{\text{OH}}{\overset{\text{O}}{\diagdown}} + \text{HCHO} \longrightarrow \text{H-}\underset{\underset{\text{H}}{|}}{\overset{\overset{\text{N=CH}_2}{|}}{\text{C}}}\text{-C}\underset{\text{OH}}{\overset{\text{O}}{\diagdown}} + \text{H}_2\text{O}$$

$$\text{H-}\underset{\underset{\text{H}}{|}}{\overset{\overset{\text{NH}_2}{|}}{\text{C}}}\text{-C}\underset{\text{OH}}{\overset{\text{O}}{\diagdown}} + \text{HCHO} \longrightarrow \text{H-}\underset{\underset{\text{H}}{|}}{\overset{\overset{\text{HN-C-OH}}{|}}{\text{C}}}\text{-C}\underset{\text{OH}}{\overset{\text{O}}{\diagdown}}$$

Levy (1933) suggested that two molecules of formaldehyde may react with one of amino acid:

$$\text{H-}\underset{\underset{\text{H}}{|}}{\overset{\overset{\text{NH}_2}{|}}{\text{C}}}\text{-C}\underset{\text{OH}}{\overset{\text{O}}{\diagdown}} + 2\ \text{HCHO} \longrightarrow \text{H-}\underset{\underset{\text{H}}{|}}{\overset{\overset{\text{N}\diagup^{\text{CH}_2\text{OH}}_{\diagdown\text{CH}_2\text{OH}}}{|}}{\text{C}}}\text{-C}\underset{\text{OH}}{\overset{\text{O}}{\diagdown}}$$

The reaction does not necessarily mean that formaldehyde can react only with one amino group at the end of a polypeptide chain, because in a number of amino acids, like lysine, there are two amino groups, one being engaged in the polypeptide chain and the other remaining free and reacting with formaldehyde. Moreover, not only $-\text{NH}_2$ groups, but also $-\text{NH}$ groups can react with formaldehyde. The methyl compounds formed may further react with amines of other groupings containing active H in another protein molecule, forming methylene bridges. Such bridges may be formed between two $-\text{NH}_2$, or $-\text{NH}_2$ and $-\text{NH}$, or $-\text{NH}_2$ and $-\text{CONH}$ groups (Pearse, 1968). In strong alkaline solutions, with more than pH 8·0, it reduces S—S to $-\text{SH}$ groups and then reacts to form a methylene bridge (S—CH_2—S) (Middlebrook and Phillips, 1942). These cross-links may be responsible for the hardening effect of formaldehyde. Several other chemical groups of the protein molecules can react with H.CHO (*see* French and Edsall, 1945; Walker, 1953; Lojda, 1965). Further, as concentrated formalin exists principally in a polymerised form and depolymerisation is not instantaneous; the possibility of a reaction between polymerised formaldehyde and tissue proteins has been hinted at by Wolman (1955).

Formaldehyde hardens the tissue to a remarkable extent and it should not be considered, by itself, as a good fixative for chromosomes. The vapour has, however, the capacity of preserving the cell constituents in an excellent way so that their life-like structure is maintained (Falck and Owman, 1965). This reagent does not afford any protection to the chromosomes against damage caused by ethanol, benzene or paraffin dehydration and block preparation.

Formaldehyde easily oxidises to formic acid and should never be used in conjunction with oxidising agents like chromic acid or osmium tetroxide. However, it is often used even in such combinations in classical fixatives because, the speed of oxidation being very low, fixation can be accomplished by the time the oxidation has been completed. Often neutral formalin is used (Lillie, 1954). It can be prepared by adding basic magnesium or calcium carbonate to the solution. For cytochemical work, fixation at 4 °C for 30 min to 2 h in 10 per cent neutral formalin, after adjusting the pH by calcium carbonate or buffer is recommended (Marinozzi, 1963; Smetana and Busch, 1964; Smetana, 1967). In short, the use of formaldehyde in fixation lies in its property of combining with proteins and forming bridges between adjacent molecules.

Tissues, which are placed in fixatives containing formaldehyde, often show well scattered chromosomes, especially after sectioning from paraffin blocks. The cell volume increases considerably, resulting in spreading of the chromosomes over a larger area. The constriction regions appear slightly exaggerated due to contraction of the euchromatic segments. This effect is possibly due to the action of formaldehyde on the chromosome proteins—doubtless one of the advantages of fixation in formaldehyde. In cytochemical work, washing after formalin fixation is essential so that the reactive groups of proteins remain unmasked to combine with the reagents, e.g., Sakaguchi reaction for arginine cannot be carried out satisfactorily without washing off excess formalin as otherwise guanidyl groups remain blocked (Pearse, 1968).

Another serious disadvantage in using formaldehyde as a fixing agent is possibly the fact that the tissue treated with this reagent is difficult to smear; the exceptional hardening, which is a result of its action on protein, being responsible for this. Further, even in block preparations, if formaldehyde is not used in suitable proportions, extreme granulation of chromosomes can be observed, which implies that the linkage between nucleic acid and protein may be affected by formaldehyde at certain susceptible segments, so that these segments become slightly depleted of nucleic acids. The differential distribution of nucleic acids, induced thereby, has an adverse effect on the basophilia of chromosomes and culminates in the appearance of granulated segments. This appearance closely resembles the nucleic acid starved areas of Darlington, observed after cold treatment. The granular representation makes study of the constriction regions difficult, as they cannot be distinguished from the areas showing low basophilia. Therefore formaldehyde should always be used as a component in a fixing mixture in suitable concentrations for chromosome analysis. It is widely used for both plant and animal materials.

Methyl Alcohol or Methanol (CH₃OH)

Methyl alcohol, or methanol, is occasionally used in chromosome studies of plants but is extensively employed in fixing mammalian chromosomes. It is obtained from destructive distillation of wood, as a component of pyroligneous acid, but does not possess the essential requirements of a typical alcoholic fixative. While ethyl alcohol causes heavy shrinkage of chromosomes, methyl alcohol causes swelling and this property has been used advantageously in the preparation of fixatives where a swelling agent is often needed to compensate for the shrinking effect of other chemicals. Its effective concentrations are the same as ethyl alcohol. It is a colourless poisonous liquid and is miscible with water in all proportions. It should not be used with an oxidising agent, as it is then immediately oxidised to formic acid. In properties, methyl alcohol closely resembles ethyl alcohol, except for the fact that the oxidation products of the two chemicals are different and methyl alcohol does not give the haloform reaction given by ethyl alcohol.

Acrolein and glutaraldehyde

These two aldehydes, namely, acrylic aldehyde, $H_2C{=}CH.CHO$ and glutaric dialdehyde $(CH_2)_3CHO.CHO$, have the property of crosslinking protein molecules more actively than formaldehyde (Bowes, 1963; Sabatini *et al.*, 1963). Though their use in ultrastructural studies has been established (*see* chapter on electron microscopy), yet they cannot be recommended for wide use in chromosome fixation.

Propionic acid (C₂H₅COOH)

Lately, propionic acid has been used extensively in the fixation of chromosomes. It is a fatty acid like acetic acid, and is a colourless liquid with an acrid odour; it is also miscible with water, ethyl alcohol and ether in all proportions, and is a good solvent for aniline dyes. It is present in pyroligneous acid though it can be commercially obtained from the oxidation of *n*-propyl alcohol.

In view of the above properties, it is generally used as a good substitute for acetic acid. Its penetration is not as rapid as that of acetic acid but it causes much less swelling of the chromosomes. It can be used, similar to acetic acid, in staining-*cum*-fixing mixtures, like propionic–carmine, propionic–orcein, etc.

Chloroform (CHCl₃)

Chloroform is a trihalogen derivative of methane; it is a colourless liquid with a sickly sweet smell and is sparingly soluble in water. It is miscible in all proportions with alcohols, ether and acetone. It has a narcotising action and is used as an anaesthetic. It is slowly converted into highly poisonous carbonyl chloride in the presence of air and light. It is a good solvent for fatty substances, a property which has been advantageously used in fixation. In the study of plant chromosomes, chloroform is generally used in the fixative to dissolve the fatty and waxy secretions from the upper surface, facilitating the penetration of the fixative. In the study of chromosomes from animal tissues, chloroform is frequently used to dissolve the fats which are present as accessories in the desired tissue.

A judicious use of chloroform in fixing mixtures is recommended, as an excessive dose or long period of treatment may be toxic. Alcohol should be an essential component of a fixative containing chloroform as the former checks the decomposition of the latter into carbonyl chloride. As the tissue becomes extremely brittle after treatment in chloroform, it has a limited use in smear preparations.

Picric Acid

Picric acid is 2,4,6-trinitrophenol having the chemical structure of:

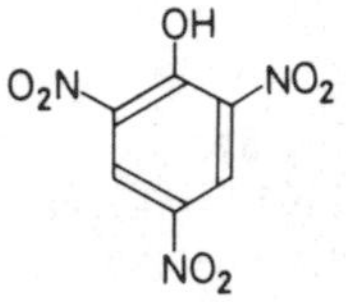

It is a yellow crystalline solid with a bitter taste; insoluble in cold water but soluble in hot water, ether, alcohol, benzene and xylol. The NO_2 groups in *o*- and *p*- positions are responsible for the maximum amount of resonance of this acid and, being a nitrophenol, it is strongly acidic. Its yellow colour is due to NO_2 chromophoric groups.

It is used in saturated or nearly saturated solutions for cytological work. Jones (1920) first observed its protein–precipitating action on nuclein. Pischinger (1937) observed that it can precipitate both histone and nucleic acid. It is remarkable that neither *p*-nitrophenol nor 2,4–dinitrophenol is capable of precipitating protein (Holmes, 1944). Evidently, all three —NO_2 groups are important factors controlling the activity of picric acid. Baker (1950) suggested that the action of picric acid on proteins is different from other protein precipitants in the sense that, as complex anions precipitate proteins, an actual chemical compound, protein picrate, is formed in this case. Badder and Mikhail (1949), and Anderson and Hammick (1950) observed that 2,4,6-nitro groups behave as inducing dipoles and activity may increase further, due to several resonance states of the picric acid molecules. Protein molecules may be bridged by picric acid due to weak electrostatic forces, linkage being due partly to ionic forces and partly to polarity induced in many polarisable moieties of the protein molecule. The protein precipitating action has also been worked out by Lison (1960) and Lojda (1965).

The penetration of picric acid is slow and a significant amount of shrinkage of chromosomes is observed. Being an acid, it should cause swelling of the cellular components but it is possible that shrinkage due to protein precipitation over-balances this effect. Due to its acidity, however, the hardening of the tissue, which is inherent in protein shrinkage, is checked. Picric acid, therefore, may be considered as a good ingredient in a fixative for protein precipitation without causing undesirable hardening.

The usual practice of washing after picric acid treatment is carried out by passing through different grades of ethyl alcohol. This is done during the usual dehydration procedure followed for block preparation.

For plant tissues, picric acid is not generally recommended in chromosome study because of heavy protein precipitation and shrinkage. These effects often result in a distortion of the nuclear and cytoplasmic components. In animal material, however, where the main object is to study the gross behaviour of the chromosomes, it is used always as an ingredient of a fixing mixture, such as Bouin's fluid, but never alone.

Dioxane

Dioxane is diethylene dioxide with a structural formula:

$$\begin{array}{c} CH_2-CH_2 \\ O \diagup \qquad \diagdown O \\ CH_2-CH_2 \end{array}$$

It is a colourless liquid, mixing in all proportions with water and most organic solvents, is fairly toxic in action and isomeric with ethyl acetate. It dissolves resins, fatty oils, etc., and because of this property, which assists the rapid penetration of the fluid, and also because of its ready miscibility with most solvents, it forms a good ingredient of fixing mixtures.

Ether

The ether of commercial use is diethyl ether with a formula C_2H_5—O— C_2H_5. It is a colourless volatile liquid with a characteristic smell, fairly soluble in water and mixes well with alcohols or liquid hydrocarbons in all proportions, being a solvent for fats and oils. Its use in local anaesthesia can be attributed to intense cooling, resulting from rapid evaporation. Ethyl alcohol is generally present as an impurity in ether. It should always be kept in airtight containers in cold dark places.

Formerly, ether was not used in fixing mixtures because chloroform, which is less toxic and easier to handle, could be used for the same function. It has been used in a few fixing mixtures, like Newcomer's fluid (1952), to clear the cytoplasm and the cell of the fatty and oily substances, so that the other components can reach the substrate easily.

Isopropyl Alcohol

Isopropyl alcohol is a colourless liquid with a formula $CH_3.CH(OH)CH_3$, being soluble in water, ethyl alcohol and ether. Although having the same effect as ethyl alcohol on chromatin, it has been preferred by some authors due to its comparatively less drastic action.

Acetone

Acetone is dimethyl ketone with a formula, $CH_3—CO—CH_3$, and is a colourless liquid with a pleasant smell, being miscible with water, ethyl alcohol and ether in all proportions. It is a good solvent for cellulose acetate and for many organic compounds, and when acting in a fixing mixture serves the same purpose as chloroform or ether in clearing the cytoplasm by dissolving the organic matter.

Similar to ethanol, it is likely (Wolman, 1955) that lipids, which are bound to protein molecules by their hydrophilic groups, neutralise some of the actively charged molecular groups of proteins. As soon as the lipids are dissolved by ethanol or acetone, the charged groups of protein are unmasked and made more reactive; consequently the protein molecules attract each other and form crosslinks (cf. Kauzmann, 1959; Okunuki, 1961).

METALLIC FIXATIVES

Osmium Tetroxide (OsO_4)

Osmium tetroxide, an expensive chemical, was first used in cytology by Schultze and Rudneff in 1865. The metal osmium is closely related to platinum and occurs in platinum ore. It is shining white in texture and at high temperatures undergoes oxidation in the presence of air to OsO_4.

The metal itself has the remarkable property of remaining unaffected by any acid acting alone. It is a strong oxidising agent and should never be mixed with formaldehyde or alcohol. In solution, it oxidises aliphatic and aromatic double bonds, alcoholic (OH) groups, amines, —SH groups and other nitrogenous groups as well, but generally the carboxyl and carbonyl groups are not affected. The details of its chemical action were worked out by Wolman (1955). Berg (1927) first studied the chemistry of its action and divided the effect into two categories, primary and secondary. During the primary effect, the entire molecule combines with the amino groups of proteins. In the secondary phase, the compound formed undergoes oxidation, during which the residual part of osmium tetroxide is reduced to a lower oxide or hydroxide. Due to this, the tissue fixed in osmium tetroxide turns black. Criegee (1936) attributed blackening to the oxidation of double bonds between adjacent carbon atoms (cf. Wigglesworth, 1957). Bahr (1955) obtained positive reaction of osmium tetroxide with several amino acids and noted that the reaction of amines is directly proportional to chain length, whereas aldehydes and ketones are reactive only in long chains. Adams (1960) did not get confirmatory proof of reduction of osmium tetroxide by proteins or polysaccharides, but his observations were refuted by Wigglesworth (1964) and Hake (1965).

Stoeckenius and Mahr (1965) concluded that, excepting phosphatidyl serine, carbon double bonds are the primary site of reaction of osmium tetroxide. Details of its reaction with lipoproteins have been worked out by Finean (1954), Khan *et al.* (1961), Riemersma and Booij (1962), Riemersma (1963), Salem (1962), Hayes *et al.* (1963) and *see* Pearse (1968). Bahr (1954), however, suggested that nucleic acids do not react with osmium. Porter and Kallman (1953) claimed that intermolecular linkages are formed at points containing double bonds in osmic tetroxide fixation. Wolman (1955) suggested that it is quite likely that ethylene bonds are not the only points where the links are formed.

Osmium tetroxide fixes both fats and lipids and the blackening is due to the fact that in all animal cells, an unsaturated substance, olein, is present which reduces the compound to lower oxide or hydroxide. In general, it fixes homogeneously, maintaining a life-like preservation of the tissue. It penetrates very slowly and a number of post-mortem changes may occur if the tissue is fixed in this fluid alone. In fact, Sjöstrand (1956) suggested that only the outer layer of tissue, up to 40 µm in depth, is fixed properly. In the case of thick materials, the fixation is very uneven, the outermost layers being overfixed, the middle region properly fixed and the innermost layers not fixed at all. Gibbons and Bradfield (1956) considered that there is at least partial preservation of the distribution of chromatin.

An excellent advantage of osmium fixation is that it does not cause much shrinkage of the tissue, but on the other hand, there is a slight swelling. The texture of the osmium-fixed tissue, though not very soft, allows good material for cutting for observation under light microscopes. The low degree of shrinkage is often compensated for during dehydration in ethanol (Fernandez-Moran and Finean, 1957). Osmium tetroxide does not allow ethanol to cause precipitation during dehydration.

The effect of osmium fixation depends to a significant extent on the pH, toxicity and temperature of the fixing mixture (Baker, 1966; Sjöstrand, 1969). Hairston (1956) suggested isosmotic fixation where vacuoles show osmotic effect. For the fixation of bacteria, too, the ionic environment is a significant factor (Maaløe and Birch-Andersen, 1956). Sjöstrand (1956, 1969) also observed that reduction of temperature aids fixation in osmium tetroxide. It is often called 'osmic acid', though actually it is not. It is neutral to indicators and is a non-electrolyte. Although osmium fixation preserves chromosomes during the divisional cycle, it cannot be recommended for the study of the interkinetic nuclei. Moreover, as it often results in protein loss (Dallam, 1957), it is to be used with caution (Pearse, 1968).

The method of preparation of osmium tetroxide solution requires special mention. It is generally sold in sealed glass containers in measured quantities. The container is cut open by a glass scratcher and, with its contents, is put in a glass bottle and a required quantity of water is added. The solution is shaken and the glass bottle stoppered tightly to prevent evaporation. It is wrapped in dark paper and kept in a cool place, as light causes its reduction to lower hydroxides. Care should be taken while handling this chemical as it has a damaging effect on eyes and mucous membranes. 1 to 2 per cent solution in distilled water or buffer is used. Fixation in the cold is necessary if collidine buffer is employed (Marinozzi, 1963; Smetana, 1967).

A serious limitation of osmium fixation is the blackening of the tissue. Bleaching with hydrogen peroxide forms an essential step in fixation by osmium tetroxide, but on the other hand, affects chromosome stainability. Washing overnight in water is essential after fixation. With these limitations, it is often preferred as a fixative due to the excellent preservation of the nuclear and cytoplasmic structures. In animals it is often applied even when the blood circulation is continuing in order to minimise the time interval between fixation and interruption of circulation.

The best result with osmium tetroxide is obtained if the fixative is applied in the form of vapour (Gibson, 1885, referred to in Gatenby and Beams, 1950, and in Darlington and La Cour, 1968). This can, however, be applied only on small materials, such as prothalli of ferns, unicellular objects or materials having no cellulose wall, such as smeared animal tissues. Rapid penetration without any deformation of the tissue is the special advantage of this method. In electron microscopy, osmium fixation is widely utilised.

Materials, after osmium fixation, require bleaching to remove the black precipitate produced by fats. Usually, bleaching is performed with hydrogen peroxide when the slide is brought down to water prior to staining. The slide is transferred from 80 per cent ethanol to a jar containing H_2O_2 and 80 per cent ethanol in equal proportions and kept from 1 to 12 h. It is examined to see if the black precipitate has been bleached and is then processed and stained as usual. For bleached tissues, pre-mordanting in 1 per cent chromic acid solution is necessary.

Platinum Chloride (H_2PtCl_6)

Platinum chloride is hydrochloroplatinic acid, the deliquescent brown red crystals being water soluble (Gatenby and Beams, 1950). Platinum chloride solution in water is often applied in place of osmium tetroxide, especially in the somatic tissue of plants. It does not have the same capacity as osmium tetroxide of preserving the life-like structures of the cell, but its capacity for penetration is decidedly greater. It is sold as shining yellow crystals in glass capsules and its preparation is similar to that of osmium tetroxide. It is compatible with formalin and can be used as a substitute for chromic acid in chromic–formalin mixed fixing fluids, but its application is rather limited. It is very effective for solanaceous groups of plants. Fixation in platinum chloride requires bleaching of the fixed tissue.

Chromic Acid (H_2CrO_4)

Chromic acid is formed when chromium trioxide reacts with water. Chromium trioxide is crystalline, light red or brown in colour and is deliquescent and fully soluble in water. Chromic acid is a very weak acid and its salts can be dissociated even by acetic acid. It has a strong oxidising action and is itself reduced to CrO_3; because of this, it should never be used in combination with alcohol or formalin. In a number of fixing fluids, however, chromic acid is used together with formalin—the reducing action being slow, the fixation is completed before the acid is fully reduced. It is a strong precipitant of protein (Baker, 1966) but Berg (1927) found it to be a very weak precipitant of nuclein. The dissociation of chromic acid in water may result in H^+ and $HCrO_4^-$ or $2H^+$ and CrO_4^- ions. According to Berg (1927), protein undergoes denaturation and precipitation by the primary action of chromic acid, and the secondary action results in hardening. He holds that the ion $HCrO_4^-$ is responsible for the secondary action. Chemical reaction probably occurs between protein and chromic acid, but the exact steps are not precisely known. However, the principal affinity of chromium is for the carboxyl and hydroxyl groups (Bowes and Kenten, 1949; Strakov, 1951). Green (1953) suggested that co-ordinates with —OH and —NH_2 are formed after reaction with carboxyl groups. Proteins, acted upon by chromic acid, are resistant to the action of pepsin and trypsin. Chromic acid penetrates the tissues slowly and the hardening induced by this acid makes the tissue resistant to hardening by ethanol in subsequent processing. It does not cause excessive shrinkage of the tissue.

Materials fixed in this acid require thorough washing in water, at least overnight, otherwise the deposition of chromic crystals not only hinders staining but also hampers the observation of chromosomes. Because of its slight hardening action it is difficult to use this fluid as a fixative for squash preparations, unless softened by some strong acid, which may hamper staining. It should never be used alone, as then heavy precipitates are formed causing shrinkage of nucleus and cytoplasm. Materials treated in chromic acid should not be kept in strong sunlight due to the chance of break-down of proteins. Basic dyes adhere closely to tissue fixed in chromic acid.

In general, chromic acid is considered an essential ingredient of several fixing mixtures. It imparts a better consistency to the tissue and aids staining better than osmium tetroxide.

Potassium Dichromate ($K_2Cr_2O_7$)

The crystals of potassium dichromate are orange or orange red in colour, and it was first used in biological studies by Müller in 1859. It is prepared from chromium ore through heating with potassium carbonate and afterwards treating in H_2SO_4. It is not as soluble in water as chromic acid, its solubility being 9 per cent. It is a strong oxidising agent, though not so effective as chromic acid and evidently should not be mixed with formalin or alcohol. Baker (1950, 1966) claimed that unacidified $K_2Cr_2O_7$ cannot fix proteins effectively, though it renders the latter insoluble in water. The life-like appearance of the cells is maintained and, if acidified, chromatin can also be preserved as such.

Zirkle (1928) demonstrated that over pH 4·6, $K_2Cr_2O_7$ can maintain the structure of chromosomes, whereas, with less acidity, only the cytoplasmic structures are preserved. Any of the ions, anions or cations, obtained from $K_2Cr_2O_7$, can react with proteins, depending on the acidity of the fixing fluid. In addition to chromium combining with proteins, the reduction of the chromate must also be involved, resulting in change of colour to green. Green (1953) suggested that carboxyl groups must combine with proteins. For the stabilisation of the structure he further suggested that this reaction is followed by co-ordination between amino and hydroxyl groups. Wolman (1955) regarded the primary reaction as that of chromium bridges between protein molecules.

Potassium dichromate, on the whole, is a good fixative for lipids. It has a rapid rate of penetration and shrinkage is not very marked. It does not harden the tissue significantly and, as such, shrinkage in alcohol during subsequent treatment is considerable—this being one of the disadvantages of potassium dichromate fixation. After dichromate fixation cellular constituents respond well to acid dyes, and the response of chromatin to basic dyes can be maintained if the fixation is performed at an acidic level. Because of its rapid rate of penetration, it is often preferred to chromic acid in mixtures in which a nuclear precipitant is added. It is widely used for the fixation of both plant and animal chromosomes, and can be washed away with dilute solution of chloral hydrate in water.

Mercuric Chloride ($HgCl_2$)

Mercuric chloride is available in the form of white powder and was first introduced in cytological practices by Lang in 1878. It has a moderate solubility (7 per cent) in water, and dissolves very well in alcohol. One of the advantages of mercuric chloride as a fixative is its compatibility with the majority of the fixing fluids. Slightly acidified mercuric chloride precipitates protein very stongly.

In water, mercuric chloride undergoes partial hydrolysis into hydrogen and chlorine ions, and acidity is due to hydrogen ions. Mercury, like several cations, has the property of precipitating protein in an insoluble form (Baker, 1950), this reaction taking place at a comparatively moderate temperature. Calvery (1938) suggested that $-NH_2$ groups of amino acids react with mercuric chloride. Hughes (1950) showed that the mercuric ion may be bound by the SH group of proteins and it can form a bridge between two molecules of protein. This suggestion is quite reasonable, as mercury is a bivalent heavy metal. Mercury ions are capable of forming inter-molecular links through their reaction with —SH, —COOH and —NH$_2$ groups, a property which is principally responsible for the powerful protein-precipitating action of mercuric chloride.

Because of its strong action, it is preferable not to use mercuric chloride alone as a fixative, especially for delicate materials. Its rapid penetration and strong protein reaction can be advantageously employed in mixtures with compatible fluids. Its effect on staining, too, is remarkable, and chromosomes respond well to most of the dyes after mercuric fixation. In spite of its beneficial effect, one of the serious limitations of mercuric fixation is that a needle-like precipitate of mercurous chloride is often formed in the tissue following this fixation. No doubt metallic mercury is often removed by alcoholic iodine, but under certain conditions it may remain, necessitating further washing. In view of all these limitations and its strong action on protein, it is not widely used.

Lanthanum Acetate

Lanthanum salts can also precipitate nucleic acid. In order to secure a good preservation of chromatin, lanthanum salts are added either in the fixative or after treatment with a recognised fixing fluid. For cytochemical studies, its use is rather limited because it often affects enzyme activity (Opie and Lavin, 1946).

Uranium Nitrate

Often in fixing mixtures potassium dichromate is replaced by uranium salts, but it is of extremely limited application. It is preferable to use it as an additional component of the fixing fluid rather than as a substituted metallic compound.

Iridium Chloride

If acidified with dilute acetic acid, iridium chloride has been found to be very effective in the fixation of *Triton* sp. However, it is not effective elsewhere, due to its low capacity of nucleo-protein precipitation.

FIXING MIXTURES

Of the different types of mixtures employed for chromosome studies two categories, at least, can be formulated. In one of these, both metallic and non-metallic fluids have been included, whereas the other is constituted purely of non-metallic fixing fluids. The majority of the fixatives, the principal ones of which are listed below, fall under the first category.

Flemming's Mixture and Other Related Fluids

The science of chromosome study owes a good deal to Walther Flemming who introduced the term 'chromatin' in 1879. The formula of his fixing fluid was published in 1882, and even now his recipe is considered as one of the few excellent ones so far proposed for chromosome study.

The principal constituents of Flemming's fluid are osmic acid, chromic acid and acetic acid. Use of the first two fluids for fixation was first made by Flesch in 1879. It remained for Flemming to point out that the use of acetic acid helps in bringing out nuclear details clearly and also accelerates chromosome staining. In 1882, Flemming published a weak formula, using an aqueous solution of chromic acid (0·25 per cent), osmic acid (0·1 per cent) and glacial acetic acid (0·1 per cent). The proportion of the acetic acid was, however, variable. In fact, in this mixture, the two constituents, acetic acid and chromic acid, are good precipitants for chromatin, whereas osmic acid helps to maintain life-like preservation of the cell. Protein precipitation is also aided by chromic acid. The presence of acetic acid inhibits chromic acid from causing heavy shrinkage. The cutting quality of this fixed material is also very satisfactory. The fixation period generally varies between 4 and 24 h. It is always preferable to keep the three constituents separate, and mix them just before use. If necessary, chromic acid and osmic acid can be mixed in required proportions and stored, and acetic acid should be added at the time of fixation.

The most serious drawback of this fluid lies in the unequal fixation of the tissue. It is not uniform in all cell layers. In multicellular, many-layered objects, such as in root tips, the outer peripheral layers are over-fixed due to rapid action of osmic acid and there is an intense blackening of the tissue. The middle layers show excellent fixation in which precipitation is very homogeneous, interkinetic nuclei being excellently preserved due to osmic acid. The other two constituents help in the good preservation of chromosomes in divisional figures. Due to satisfactory precipitation and the presence of heavy metals, the nuclear staining is intense in this region. The innermost segments, on the other hand, show only the effects of chromic acid and acetic acid due to their comparatively rapid entry, whereas osmic acid, because of its slow rate of penetration, does not show its effects at all except causing a blackening of this segment. In spite of this unevenness of fixation, it is widely recommended for chromosome preservation and staining of chromosomes and interkinetic nuclear details which are properly fixed. Weak Flemming is suitable for smaller objects, whereas for larger objects strong Flemming is desirable to secure better penetration. Fixation at a very low temperature often causes better preservation of nuclear details due to lessening of the effect of osmium tetroxide, but this practice has been discouraged by certain authors (Baker, 1950) who consider it improper to use a costly chemical like osmic acid in order to inactivate it at a particular temperature.

A number of fixing fluids was later proposed, based principally on the use of constituents of Flemming's fluid. Benda in 1902 used a modification of Flemming's fluid in which the proportion of acetic acid was much minimised, being reduced to 2–3 drops. In the experience of the authors, Flemming's mixture in plants holds good for meiotic chromosomes whereas for the study of chromosome morphology in mitosis, Benda's modification gives very satisfactory details. Strong acetic acid causes heavy swelling, obscuring chromosome details, an effect which is much minimised in the formulae recommended by Benda.

Young (1935) mentioned that fixation can be spread over a wider area if a little sodium chloride solution, which helps in solubility and penetration, is added in the Flemming's fluid for fixation in cold- and warm-blooded animals. Meves (1908) observed that the action of chromic acid is accelerated by the addition of sodium chloride. Gatenby and Beams (1950) observed that sodium chloride causes disintegration of the solution and so it should be kept separately and mixed with osmic acid if its use is necessary.

La Cour (1931) suggested a series of formulae, in all of which the principal constituents are chromic acid, potassium dichromate, osmic acid, acetic acid and saponin. In the three mixtures, namely, 2BD, 2BE and 2BX, the relative proportions of the constituents vary in different fluids: 2BD is recommended for all organisms in general, 2BE for plants only, whereas 2BX is meant for bulk fixation. In the case of tissues providing penetration difficulties, a pre-treatment in Carnoy's fluid has been recommended prior to fixation. The use of saponin is meant to increase surface activity, and Baker (1950) has further claimed that it prevents the fluid from becoming stable. The use of acetic acid shifts the pH of the medium to the acidic side, thus helping in the precipitating effect of potassium dichromate, which then behaves like chromic acid. The purpose of using both chromic acid and potassium dichromate seems to be obscure, and Baker has rightly pointed out that a strong solution of chromic acid would have served the purpose equally well. In the experience of the authors, La Cour's fluids are excellently suited for the study of meiotic chromosomes, but their efficacy is dubious for the study of chromosome morphology from somatic cells.

Champy (1913) also used chromic acid, potassium dichromate and osmic acid in his recommended fluid which is widely used for the study of animal chromosomes, but the reason for using both potassium dichromate and chromic acid is not clear. It is possible that the dichromate's property of rendering the protein insoluble is taken advantage of in this fluid. Minouchi and Koller (referred to in Darlington and La Cour, 1960) have suggested modifications of Champy's fluid. In Minouchi's modification (1928), which is meant for animal chromosomes in general, all the constituents of Champy's fluid have been maintained and only the proportions have been altered. Koller's modification, meant for mammalian chromosomes in particular, does not involve the use of potassium dichromate at all; it therefore appears to be quite rational, especially in view of the properties of the fixing fluids so far known.

Other important modifications of Flemming's fluid were suggested by Johnson (referred to in Gatenby and Beams, 1950) and Hermann (1899). Johnson's modification involved the use of potassium dichromate, osmic acid, platinum chloride and acetic or formic acid. He also considered that acetic acid should be added just before use, the reduction of platinum and osmium being very rapid. The presence of three metallic compounds aids remarkably in the staining of chromatin.

In Hermann's modification (1899), on the other hand, chromic acid has been completely substituted by platinum chloride. Platinum also helps in the staining of chromatin with basic dyes, but is not suitable when staining in both cytoplasm and nucleus is desired, as it prevents cytoplasmic coloration and so acidic dyes do not adhere to the tissue: this modification is thus useful in order to obtain bright staining of chromosomes against comparatively colourless cytoplasm. Because of the presence of platinum, the background generally takes on a light yellowish orange colour and, against this, the chromosome staining with any basic dye appears sharp.

Taylor (1924) and Catcheside (1934) published, for plant chromosomes, different modifications of Flemming's fluid in which maltose was added to prevent clumping and to help spreading of the chromosomes. Carpenter and Nebel (1931) and Catcheside (1935) substituted osmium tetroxide by ruthenium tetroxide and uranium trioxide respectively.

Zenker's Fluid

In the remainder of the fluids listed under the first category, formalin forms an invariable constituent except in the mixture recommended by Zenker as early as 1894. The principle of action was clarified later in the works of Zirkle (1928).

Navashin's Fluid

Of all the other mixtures having both metallic and non-metallic constituents for chromosome work, Navashin's fixative is the most important.

The original formula of S. Navashin, the Russian worker, was published in 1912 (*Mem. Soc. Nat. Kiev.* as referred to in Gatenby and Beams, 1950) in which the principal reagents used were chromic acid, acetic acid and formalin. A number of modifications were later suggested by different authors for different types of materials (San Felice, 1918, modified by White, 1940, referred to in Darlington and La Cour, 1960; Navashin, M., 1925; Karpechenko, 1924, 1927; Webber, 1930; Randolph, 1935; Hill and Myers, 1945; Belling, as referred to in Gatenby and Beams, 1950). Of these, Navashin's modification is meant specially for root tips, Karpechenko's for pollen mother cells, Randolph's for plants in general and San Felice and White's for animal tissues. In all these modifications, the constituents remain the same but their proportions vary in different recommended fluids. The only precaution that is needed in fixation with Navashin's fluid is that chromic acid should never be kept mixed with formalin, for the simple reason that oxidisers are not to be kept with reducers. Even fixation with chromic acid and formalin together in a fluid is theoretically unsound, but as the penetration of these fluids is quite rapid, the fixation is nearly complete before the oxidation. Though recommended, it is therefore useless to keep the tissue in Navashin's fluid for more than 3–4 h. For smears, 1 h only is sufficient, whereas for blocks of tissue, as much as 3–4 h fixation may be needed. In the experience of the authors, Navashin's fluid, especially the modification of Belling, is excellently suited for the study of meiotic chromosomes of plants in general. San Felice's fluid holds good for animal chromosomes; for mitotic analysis, it is just adequate for chromosome counts, but not critical enough for the study of chromosome morphology in detail. Possibly the presence of strong acetic acid results in chromosomes being so much swollen that all the chromosomal details are obscured. Navashin's fluid can be considered as a modification of Flemming's, in which osmium tetroxide has been replaced by formalin. For both plants and animals Navashin's fluid or its modifications can be strongly recommended, especially where the purpose is to study the behaviour and gross morphology of chromosomes.

Bouin–Allen's Fluid

The use of picric acid in cytological fixation was made by Bouin in 1897 (as referred to in Gatenby and Beams, 1950). In the original formula as presented by Bouin, picric acid, formalin and acetic acid were the constituents, and the fixative was principally used for animal materials, but later this original fluid did not find much application in chromosome work, because of the strong precipitating and shrinking action of picric acid. A number of modifications of this mixture were proposed, in all of which the effect of picric acid was much minimised through the use of other reagents. Of these, the most important mixtures are those of Allen (1916), Painter (1924, referred to in Darlington and La Cour, 1960, 1968) and other authors. A modification by Carothers (referred to in Gatenby and Beams, 1950) is specially meant for Orthopteran chromosomes. In both Allen and Painter's modifications, chromic acid and urea were added as special components. The former aided nucleoprotein precipitation and the latter was possibly supposed to secure the effect of osmotic pressure of fluid. Penetration in these fluids is very rapid, but the extent to which urea can help in adjusting the osmotic pressure, which is principally controlled by smaller ions, is not yet known.

Carnoy's Fixative and Other Related Mixtures

Makino (1932) recorded that for Amphibian chromosomes in general very good results can be secured if acetic acid is omitted from Flemming's fluid. The swelling due to acetic acid may prevent proper representation of chromosomes. He has further observed that the meiosis in eggs can, however, best be studied if a fixing mixture containing only a strong solution of mercuric chloride and very little acetic acid is used. Combinations with mercuric chloride have secured wider use in animals as compared to those in plants.

Carnoy and Lebrun (1887), as well as Sansom (referred to in Gatenby and Beams, 1950), have employed mercuric chloride even in combination with chloroform, acetic acid and ethyl alcohol. Satisfactory results have been secured with ova of *Ascaris* within 30 s, but the combination of reagents is not rational and compatible, as precipitates of ethyl acetate appear within a short period. It is of very limited application. In the Susa fluid and its modification as made by Ludford (referred to in Gatenby and Beams, 1950), not only is mercuric chloride added to acetic acid and formalin, but trichloracetic acid also forms one of its constituents. Even though it has found application in vertebrate materials, it cannot be recommended for critical study as TCA under certain conditions may remove nucleic acid from chromosomes (Schneider, 1945). In the second category of fixatives come those which are constituted of entirely non-metallic fluids, and in all such instances acetic acid and ethyl alcohol are invariable constituents. In fact, acetic–alcohol mixture, the use of which was first intrduced by Carnoy in 1886, is considered one of the most rapidly penetrating and quickly acting fixatives for chromosome study. In the same year, Carnoy introduced another formula in which chloroform was added to aid in rapid penetration of the fluid and to remove fatty substances from chromosomes, which helps to secure a clear background for study. For the study of animal chromosomes, where fats remain associated with several organs, such as testes, etc., Carnoy's fluid is exceptionally useful. It has also been recommended for the study of nucleic acids in cytochemical procedures (Lillie, 1954; McManus and Mowry, 1960; Amano, 1962). In plants, the general practice is to use it as a pre-treatment agent in the fixation of flower buds (Kihara, 1927) though it has also been applied in root-tip fixation. Dipping for just one or two seconds in Carnoy's fluid followed by washing prior to fixation removes the secretory or excretory products from the surface, allowing an easy passage of the fixative to the chromosomes.

One limitation of Carnoy fixation, either with or without chloroform, is that the tissues fixed as such do not respond to many of the aqueous solutions of the basic dye unless some special mordanting for staining is employed. This difficulty in staining is principally due to the absence in the fluid of any metal which aids in basophilia of chromosomes. As such it may be necessary to mordant in chromic acid prior to staining the material. In any case, especially because of the rapid penetration, Carnoy's fluids are widely recommended for bulk fixation for animal tissue. However, prolonged fixation in Carnoy's fluid may cause extraction of DNA (Pearse, 1968), and as such, long duration treatment should be avoided.

Smith (1943) and La Cour (1944) not only modified the proportions of acetic acid and alcohol in their fluids but also added formalin as one of the constituents, Smith's fixative being specially meant for insects and La Cour's for blood smears. The addition of formalin helps to increase cell volume and

Table 2 SOME COMMON FIXATIVES

Chemical	Molecular weight	pH in fresh condition (Lassek and Lunetta, 1950)	Used alone or in mixture	Concentration commonly used
Acetic acid . . .	60·05	2·3 (5% aq.)	Both	100% and 45%. When used alone 10% soln. is effective
Acetone . . .	58	7·0 (abs.)	Mixture	100%
Chloroform . . .	119·59	7·2 (abs.)	Mixture	100%
Chromic acid . . .	100·01	1·2 (1% aq.)	Mixture	1 and 2%
Dioxane . . .	88·11		Mixture	100%
Ethyl alcohol . . .	46	8·4 (abs.)	Both	50–100%
Ether	74		Mixture	Absolute
Formaldehyde . .	30	3·4 (comm.)	Mixture	40%, 5%
Formic acid . . .	46	2·0 (1% aq.)	Mixture	1%
Hydrochloric acid . .	36·5		Both	Normal
Isopropyl alcohol . .	60		Mixture	100%
Methyl alcohol . .	32		Mixture	100 and 95%
Mercuric chloride . .	272	3·0 (6·9%)	Mixture	—
N-butyl alcohol . .	74·12	6·2 (abs.)	Mixture	100%
Nitric acid . . .	63		Mixture	10%
N-propyl alcohol . .	60	6·7 (abs.)	Mixture	100%
Osmic acid . . .	255	6·1 (1% aq.)	Both	0·5–2%
Picric acid . . .	229	1·3 (sat.)	Mixture	Sat.
Platinum chloride . .	518·08	2·5 (1% aq.)	Mixture	0·5–2%
Potassium dichromate .	294	3·9 (3% aq.)	Mixture	$2-7\frac{1}{4}\%$
Propionic acid . .	74·08		Mixture	45 and 100%
Trichloracetic acid . .	163·40	1·2 (1%)	Both	—

secures chromosome spreading, but the hardening caused by formalin makes the tissue unsuitable for preparing smears. Moreover, the absence of any metallic fluid, in conjunction with the absence of formalin, renders staining difficult, thus making the process limited in application.

Another very widely used fixative, especially employed for somatic chromosomes of plants, is a mixture of 1 per cent chromic acid and 10 per cent formalin, devised by Lewitsky (1931). A number of modifications (Prokofieva, 1935) of this fluid was later published, depending on the type of tissue investigated. In general, the increase in the proportion of formalin helps to cause wide scattering of chromosomes, which is essential for the study of their morphology, and the constriction regions come out clearly. However, too much formalin may lead to a granulated appearance of chromosomes and a reduction of basophilia.

Chromic acid and formalin, being oxidiser and reducer respectively, are incompatible, but even then satisfactory results have been obtained in chromosome fixation because the period required for oxidation is more than the period needed for fixation. Massive objects or bulk materials cannot be fixed in this fluid as the period of fixation needed will be too long. Mitotic chromosomes of soft and small materials, such as thin root tips, respond well to chromic–formalin fixation.

Kahle (1908) as well as Telleyesniczky (1905) used all three constituents, namely, formalin, acetic acid and alcohol, for cytological studies. The former is widely recommended for animal material, the latter for the ovular tissues of higher plants. Though they may serve the purpose of chromosome counting, the absence of a good shrinking agent makes critical analysis of chromosomes difficult.

Immersion of plant tissue for fixation is facilitated by the addition of a drop or two of liquid detergent to each 10 cm^3 of fixing solution (Miller and Colaiace, 1968). The use of a commercial bleaching agent, like Chlorox, Purex or Hy-Pro, mixed with conc. HCl and 50 per cent ethanol (2:2:1), after soaking root tips in water, has been found to improve separation of orchid chromosomes (Freytag, 1964).

ALCOHOLIC

1. *Carnoy's Fluid (1886)*

| Glacial acetic acid | 1 part |
| Absolute ethyl alcohol | 3 parts |

It is effective for all plant, animal and human materials both for squash and block preparations, the period of fixation varying from 15 min to 24 h in cold or at room temperature. The fluid should be washed out with 70 or 90 per cent alcohol. Difficult materials can be mordanted in ethyl alcohol–3 per cent ferric ammonium sulphate (7:2·5) mixture for 3–12 h after fixation (Lesins, 1954). Certain Rhodophyceae are mordanted in aqueous 0·5–5 per cent ferric ammonium sulphate solution after fixation (Austin, 1959).

Modifications include mixtures in the proportions 1:1 (Von Beneden and Neyt, 1887) and 1:2. Zacharias (1888) added a few drops of osmic acid solution to acetic acid–alcohol (1:4) mixture. Burns and Yang (1961) fixed *Nicotiana* microspores in acetic acid–95 per cent alcohol mixture (3:2) for 15 min after dissolving the pollen grain walls. For electron microscopic studies, anthers can be fixed in a mixture of cadmium chloride and absolute alcohol for 30 min or in only 0·5M $CdCl_2$ for 10 min followed by refixation in acetic–alcohol (1:3). A modification containing acetic acid, 96 per cent ethanol, concentrated HCl and distilled water (1:3:2:2) has been used for mosquito chromosomes after 30 s in Carnoy's fixative (Amir-khanian, 1968).

2. *Carnoy's Fluid II (1886)*

Glacial acetic acid	1 part
Chloroform	3 parts
Absolute ethyl alcohol	6 parts

It is widely used for animal and human tissues and for flower buds. The period of treatment is from 15 min to 24 h in cold or in room temperature. Metzger and Leng (1955) modified it by saturating it with $HgCl_4$ for certain Leguminous plants. Another modification, with ingredients in the proportion of 1:3:4 and 1:1:3 (Semmen's fluid), has been used in Compositae and other families (Turner, 1956). The fluids, mixed in equal proportions (1:1:1) have been used on certain Heteroptera.

3. *Carnoy and Lebrun's Fluid (1887)*

Glacial acetic acid	1 part
Chloroform	1 part
Absolute ethyl alcohol	1 part
Corrosive sublimate to saturation	

A modification by G. S. Sansom contains

Glacial acetic acid	1 part
Chloroform	6 parts
Absolute ethyl alcohol	13 parts
Corrosive sublimate to saturation	

The mixture is prepared just before use and is successful for insect ovum with shell and for certain vertebrate materials. It penetrates very rapidly, the period of fixation being from 10–30 min and is washed out with absolute alcohol.

4. *Schaudinn's Fluid*

Absolute ethyl alcohol	1 part
Sublimate soln.	2 parts

If necessary 1–5 per cent of glacial acetic acid may be added. It is a common fixative for Protozoa and is also used for higher organisms.

5. *Chromo-nitric Acid*
(a) *Original formula* (Perenyi, 1882)

10% aq. nitric acid soln.	4 parts
Absolute ethyl alcohol	3 parts
0·5% aq. chromic acid soln.	3 parts

The period of fixation is 4–5 h, followed by washing in 70 per cent and absolute ethyl alcohol for 2–3 days.

(b) *Modified formula* (Perenyi, 1888)

20% aq. nitric acid soln.	3 parts
Absolute ethyl alcohol	4 parts
1% aq. chromic acid soln	3 parts

The period of fixation is 20–30 min, followed by washing for 1 h each in 70 per cent and absolute alcohol. These fixatives are used in embryological studies and for segmenting eggs and their nuclei.

6. *Propionic Acid Modification*
Propionic acid substitutes acetic acid effectively in several mixtures:

(a) Propionic acid	1 part
95% ethyl alcohol	3 parts

The mixture has been used effectively on perithecia of Ascomycetes. The period of fixation is 24–36 h (Cutter, 1946). It can also be used on other plant tissues and should be washed out with 70 per cent alcohol.

(b) Propionic acid	100 cm^3
95% ethyl alcohol	100 cm^3
Ferric hydroxide	0·4 g

To each 10 cm^3, a few drops of carmine are added. This fixative is very effective for plants with small chromosomes (Hyde and Gardella, 1953) and for potato cultivars (Marks, 1960).

(c) Propionic acid	1 part
Chloroform	1 part
Absolute alcohol	2 parts

Newcomer (1952) found 12–24 h fixation in this fluid to be suitable for avian chromosomes. This mixture, in the proportion 1:4:3, has been used on different plants, e.g. species of *Plantago*.

7. *Iron Acetate Modification*

(a) Acetic acid	1 part
Absolute ehtyl alcohol	3 parts
Iron acetate	A small quantity

It is suitable for anthers with small chromosomes. Period of fixation is 12 h followed by keeping 5–15 min in:

Saturated soln. of iron acetate in 45% acetic acid	3 parts
45% acetic acid	5 parts
1% aq. formaldehyde soln.	2 parts

The tissue is rinsed in 45 per cent acetic acid (Marks, 1952)

(b) 95% ethyl alcohol	3 parts
Acetic-carmine soln. with added iron acetate	1 part
A flake of rusted iron	

The period of fixation is 12–24 h followed by washing and storage in 95 per cent ethyl alcohol with iron flake for 5–10 days. The mixture is useful for flower buds of species of *Cucurbita* (McGoldrick and colleagues, 1954).

8. *Alcohol-Ether Mixture*

Absolute ethyl alcohol	1 part
Pure ether	1 part

Fresh liquefied semen is fixed for 3 min in the mixture and air dried (Casarett, 1953).

9. *Newcomer's Fluid (1953)*

Isopropyl alcohol	6 parts
Propionic acid	3 parts
Petroleum ether	1 part
Acetone	1 part
Dioxane	1 part

It can be applied to both plants and animals for smears or sections. It is a very stable fixative and can be used in combination with pre-treatment chemicals.

10. *Lactic Acid Modification (Julien, 1958)*

Glacial acetic acid	1 part
Absolute ethyl alcohol	6 parts
Lactic acid	1 part

It is effective in fixing perithecia of species of *Venturia*.

FORMALIN

1. *Original Navashin's Fluid (M. Navashin's Modification of S. Navashin's Fluid, 1925)*

Solution A

Chromic anhydride	1·5 g
Glacial acetic acid	10 cm^3
Distilled water	90 cm^3

Solution B

40% aq. formaldehyde soln.	40 cm^3
Distilled water	60 cm^3

The two solutions, previously prepared, are mixed together in equal proportions immediately before use. The fixative is suitable for both block and squash preparations of flower buds and root tips. The period of fixation is 24 h, followed by washing in running water for 3 h. Diluted Navashin's (50 per cent) fluid has been used in some plants like *Gentiana* species.

2. *The Svalöv Modification*

Solution A

Chromic anhydride	1 g
Glacial acetic acid	10 cm^3
Distilled water	85 cm^3

Solution B

40% aq. formaldehyde soln.	30 cm^3
95% ethyl alcohol	10 cm^3
Distilled water	55 cm^3

The application is the same as Navashin's fluid.

3. *Randolph's (CRAF) Modification (1935)*

Solution A

Chromic anhydride	1 g
Glacial acetic acid	7 cm^3
Distilled water	92 cm^3

Solution B

40% aq. formaldehyde soln.	30 cm^3
Distilled water	70 cm^3

Its use is similar to Navashin's fluid; 70 per cent alcohol can be used in washing. The period of fixation varies from 10 min in partial vacuum (Bowden, 1949) to overnight.

4. *Karpechenko's Fluid (1924)*

1% aq. chromic acid soln.	15 parts
Glacial acetic acid	1 part
16% aq. formaldehyde soln.	3 parts
Distilled water	17 parts

A later modification (1927) contains:

2% aq. chromic acid soln.	100 cm^3
20% acetic acid	67 cm^3
40% aq. formaldehyde soln.	11 cm^3
Distilled water	300 cm^3

The ingredients are mixed just before use. Its use is similar to Navashin's fluid.

5. *Webber's Fluid (1930)*

Solution A

Chromic anhydride	1 g
Glacial acetic acid	10 cm^3
Distilled water	65 cm^3

Solution B

40% aq. formaldehyde soln.	40 cm^3
Distilled water	35 cm^3

6. *Belling's Modification*

Solution A

Chromic anhydride	5 g
Glacial acetic acid	50 cm^3
Distilled water	320 cm^3

Solution B_1 (for prophase)

40% aq. formaldehyde soln.	20 cm^3
Distilled water	175 cm^3

Solution B_2 (for metaphase)

40% aq. formaldehyde soln.	100 cm^3
Distilled water	275 cm^3

Solution A is mixed with an equal quantity of B_1 or B_2 just before use. The fixation period is 3–12 h. The fixative is effective for studying meiosis in flower buds.

7. Hill and Myer's Fluid (1945)

Solution A

Chromic anhydride	1 g
Propionic acid	15 cm^3
Distilled water	85 cm^3

Solution B

40% aq. formaldehyde soln.	30 cm^3
95% ethyl alcohol	10 cm^3
Distilled water	60 cm^3

Its use is similar to the use of original Navashin's fluid. It is effective in studying chromosomes of grasses.

8. Langlet's Fluid (1948)

Solution A

Chromic anhydride	1 g
Glacial acetic acid	10 cm^3
Distilled water	8 cm^3

Solution B

40% aq. formaldehyde soln.	30 cm^3
95% ethyl alcohol	10 cm^3
Distilled water	130 cm^3

The solutions are mixed just before use in the proportion of 1A:9B.

9. San Felice's Fluid (1918)

1% aq. chromic acid soln.	16 cm^3
40% aq. formaldehyde soln.	8 cm^3
Glacial acetic acid	1 cm^3

The ingredients should be freshly mixed before use. The period of fixation varies from 3 h to overnight. Washing in running water is necessary after fixation. The fixative is very effective for block preparation of animal tissues.

10. Bouin's Fixatives

	Original	Allen's B.15	Painter's modification	Allen's P.F.A.3	B.3	Carother's fluid
Sat. aq. picric acid	75 cm^3	75 cm^3	75 cm^3	75 cm^3	75 cm^3	75 cm^3
40% aq. formaldehyde soln.	25 cm^3	25 cm^3	25 cm^3	15 cm^3	15 cm^3	15 cm^3
Glacial acetic acid	5 cm^3	5 cm^3	10 cm^3	10 cm^3	10 cm^3	10 cm^3
Urea	—	2 g	2 g	1 g	1 g	1·5 g
Chromium trioxide anhydride	—	1·5 g	1·5 g	—	1 g	—

Picric acid is dissolved in distilled water and kept as a mixture with formalin and acetic acid. Just before use, the mixture is heated to 37 °C and chromic anhydride crystals added, stirred, and followed by urea. The tissue is now kept in the fluid at 37–39 °C and allowed to cool gradually. The period of fixation is 4–12 h. The material is then washed in repeated changes of 70 per cent alcohol until no more yellow colour is extracted. The fixative is chiefly applied to animal tissues.

Cleland added 1 g of chromic anhydride to freshly prepared Bouin's solution and substituted 1 g of maltose or lactose for the urea. This modification, followed by iron-haematoxylin staining, gave good results in Onagraceae. Carother's fluid is used specially for Orthopteran chromosomes. Painter's modification is effective on mammalian chromosomes, the period of fixation being $1\frac{1}{2}$–3 h. Washing is carried out successively through different grades of alcohol and aniline oil. Bouin–Allen's modification is useful for amphibian chromosomes.

11. *Levitsky's Fixatives*
Ingredients:

1% aq. chromic acid soln.	(A)
10% aq. formaldehyde soln.	(B)

Preparation—They are mixed in different proportions just before use for different materials, such as, 3A:2B, 4A:1B, 1A:1B, 1A:2B, 1A:3B. They are effective in fixing root tip chromosomes of plants. The period of fixation is 12–24 h in cold or at room temperature for block preparations and squashes. The root tips require at least 3 h washing in running water after fixation. Brain and ganglion tissues can also be fixed in these fluids (Prokofieva, 1935).

12. *Prokofieva's Fluid (1934)*

5% aq. chromic acid soln.	1 part
50% formalin soln.	1 part

It has been used for studying Teleost chromosomes.

13. *Kahle's Fluid (1908)*

Glacial acetic acid	2 cm^3
40% aq. formaldehyde soln.	12 cm^3
95% ethyl alcohol	30 cm^3
Distilled water	60 cm^3

It is effective for animal and human tissues, particularly eggs of insects.

Kahle–Smith's modification contains 95 per cent alcohol (15 parts), 40 per cent aqueous formaldehyde (6 parts) and glacial acetic acid (2 parts).

14. La Cour's Fluid (1944)

Glacial acetic acid	2 parts
40% aq. formaldehyde	
soln.	7 parts
Absolute methyl alcohol	100 parts
Distilled water	70 parts

The mixture is suitable for blood smears.

15. Formaldehyde–Acetic Alcohol and Formaldehyde–Propionic Alcohol

Glacial acetic acid	1 part
40% aq. formaldehyde	
soln.	6 parts
Absolute ethyl alcohol	14 parts

It is applied especially to insect gonads before squashing in acetic–carmine or Feulgen solution. Propionic acid can be used to replace acetic acid in bulk fixation of ovules (Paolillo, 1960).

A propionic acid modification is:

40% aq. formaldehyde	
soln.	1 part
95% ethyl alcohol	15 parts
Propionic acid	2 parts

The period of fixation is 1–2 h and it can be applied to both plant and animal tissues (Morrison and colleagues, 1959).

Lillie's modification (1954):

Formalin	10 parts
Glacial acetic acid	5 parts
Absolute ethanol	85 parts

16. Susa Mixtures

(a) Original Susa mixture of Heidenhain

Mercuric chloride	4·5 g
Sodium chloride	0·5 g
Trichloracetic acid	2 g
Glacial acetic acid	4 cm^3
40% aq. formaldehyde	
soln.	20 cm^3
Distilled water	80 cm^3

The period of fixation is 1–24 h followed by washing in 90 per cent alcohol. For preparing the mixture, dissolve mercuric chloride and sodium chloride in water and, just before use, add the remaining ingredients.

(b) Romeis 'Susa' fluid

5% trichloracetic acid	20 cm^3
40% aq. formaldehyde	
soln.	5 cm^3
Distilled water	25 cm^3
Saturate with mercuric chloride	

Period of fixation is 1–2 h for small tissues and up to 24 h for larger ones. The tissues are washed in 80 or 90 per cent alcohol. The fixative is very effective for amphibian larvae.

17. *Smith's Fluid (Referred to in Gatenby and Beams, 1950)*

40% aq. formaldehyde soln.	5 cm^3
Glacial acetic acid	2·5 cm^3
7$\frac{1}{4}$% aq. potassium dichromate soln.	13 cm^3
Distilled water	80 cm^3

18. *Chrome alum Fixative (Ammerman, 1950)*

C.P. chrome alum	3 g
40% aq. formaldehyde soln.	30 cm^3
Glacial acetic acid	2 cm^3
Distilled water	238 cm^3

This fixative is suitable for yolk-rich amphibian eggs, *Euglena* and insect larvae.

19. *Battaglia's 5111 Mixture (1957)*

95% ethyl alcohol	5 parts
Chloroform	1 part
Glacial acetic acid	1 part
40% aq. formaldehyde soln.	1 part

The mixture has been used by the inventor on both plant and animal tissues. For most plant materials, fixation for 5 min is sufficient.

20. *Baker's mixture (1958, 1966)*

Formalin	10 to 15 per cent
Calcium chloride	1 per cent

OSMIC ACID

1. *Flemming's Fixatives (1882 and 1884)*
(a) *Strong Flemming*

1% aq. chromic acid soln.	15 cm^3
Glacial acetic acid	1 cm^3 or less
2% aq. osmic acid soln.	4 cm^3 according to Baker (1950)

or

2% aq. chromic acid soln.	100 cm^3
2% aq. osmium tetroxide soln.	53 cm^3
10% aq. acetic acid soln.	133 cm^3 according to Darlington and La Cour (1960)

(b) *Medium Flemming*

1% aq. chromic acid soln.	30 cm^3
5% aq. acetic acid soln.	25 cm^3
2% aq. osmic acid soln.	10 cm^3

(c) *Weak Flemming*

2% aq. chromic acid soln.	12·5 cm^3
Glacial acetic acid	0·1 cm^3
2% aq. osmic acid soln.	5 cm^3
Distilled water	83 cm^3

(d) *Benda's modification (1902)*

1% aq. chromic acid soln.	15 cm^3
2% aq. osmic acid soln.	4 cm^3
Glacial acetic acid	3 drops

Benda's fluid, diluted with an equal part of water, is useful for spermatogonial divisions of Teleosts (Makino, 1934).

(e) *Hermann's modification*

1% aq. platinic chloride soln.	15 cm^3
Glacial acetic acid	1 cm^3
2% aq. osmic acid soln.	2–4 cm^3

It is used also for embryonic gonads with reduced acetic acid.

Flemming's fixatives should always be mixed just before use. They are suitable for both plant and animal tissues. Strong Flemming is used for bulk fixation. The tissues are fixed from 1 h to overnight, depending on the size, and require washing in running water from 1 h to overnight after fixation. Fixation is more effectively carried out in cold (McClung, 1918).

2. *Fixatives Allied to Flemming's Fluid*
(a) *Modified formula with NaCl*

Solution A

1% aq. chromic acid soln.

Solution B

2% osmium tetroxide in
4% aq. sodium chloride soln.

The two solutions are mixed before use in the proportion 15A:4B. For preparing B, osmium tetroxide crystals are first dissolved in water and then sodium chloride is added (Gatenby and Beams, 1950). Meves (1908) and Young (1935) had worked out this modification earlier, but as the solution with NaCl disintegrated, the present formula is more satisfactory.

(b) *Hance's fluid (1917)*

> Flemming's strong fixative 100 cm^3
> 0·5% urea soln.

The tissue is fixed at 4–5 °C for 24 h. It is suitable for mammalian chromosomes and also for embryonic divisions in birds.

(c) *Oguma and Kihara's fluid (1923)*
This is effective for human material. Thin slices of testes are fixed for 1 min in Carnoy's fluid II and then fixed in strong Flemming's solution for 24 h.

(d) *Bonn's fluid*

> 10% aq. chromic acid soln. 0·33 cm^3
> 10% aq. acetic acid soln.   3 cm^3
> 2% aq. osmic acid soln. in
>   2% chromic acid   0·62 cm^3
> Distilled water   6·27 cm^3

(e) *Newton and Darlington's fluid*

> 1% aq. chromic acid soln.   60 cm^3
> 2% aq. osmic acid soln.   20 cm^3
> 10% aq. acetic acid soln.   25 cm^3

It is suitable for smears of plant anthers.

(f) *Taylor's fluid (1924)*

> 10% aq. chromic acid soln. 0·2 cm^3
> 10% aq. acetic acid soln.   2 cm^3
> 2% osmic acid in 2% chromic
>   acid   1·5 cm^3
> Distilled water   8·3 cm^3
> Maltose   0·15 g

(g) *Catcheside's fluid*

> 10% aq. chromic acid soln.   3 cm^3
> 10% aq. acetic acid soln.   2 cm^3
> 2% aq. osmic acid soln.   1·5 cm^3
> Maltose   0·2 g
> Distilled water   10 cm^3

It is used for p.m.c. smears with small chromosomes.

(h) *Belar's modification (1929)*

> 2% aq. osmic acid soln.   4 cm^3
> 1% aq. chromic acid soln.   15 cm^3
> Glacial acetic acid   1–2 drops

It is found to be satisfactory for smears.

3. *Champy's Fluid and Allied Fixatives*
   (a) *Original Champy's fluid (1913)*

2% aq. chromic acid soln.	20 cm^3
7¼% aq. potassium dichromate soln.	16 cm^3
2% aq. osmic acid soln.	22 cm^3
Distilled water	33 cm^3

(b) *Minouchi's modification (1928)*

2% aq. chromic acid soln.	5 cm^3
2% aq. potassium dichromate soln.	15 cm^3
2% aq. osmic acid soln.	6 cm^3

(c) *Koller's modification (Referred to in Darlington and La Cour, 1960)*

2% aq. chromic acid soln.	20 cm^3
2% aq. osmic acid soln.	10 cm^3
Distilled water	25 cm^3

In all these fluids, the ingredients are mixed freshly before use though the mixture has been found to keep for a few weeks. These mixtures have been used successfully on animal tissue, the last one on mammals. The periods of treatment and washing in water are 6–24 h. Original Champy's fluid works well on spermatocyte stages of Teleosts.

(d) *Nakamura's mixture (1928)*

2% aq. osmic acid soln.	3 parts
1·6% aq. chromic acid soln.	6 parts
6% aq. potassium dichromate soln.	4 parts

It is useful for reptilian chromosomes. The period of fixation is 24 h, followed by washing for the same period.

(e) *Makino's fluid (1956)*

0·75% aq. chromic acid soln.	8 cm^3
2% aq. osmic acid soln.	4 cm^3
3·5% aq. potassium dichromate soln.	8 cm^3
Urea	1 g

It is effective on Avian testes.

4. *Mann's Fluid (1894)*

7% aq. mercuric chloride soln.	50 cm^3
2% aq. osmic acid soln.	25 cm^3
Distilled water	25 cm^3
Sodium chloride	0·75 g

The fluid is prepared freshly before use. It is effective in fixing animal tissues, the period of fixation varying from 15 min to 3 h depending on the size of the tissue. Washing in water is necessary after fixation.

5. *La Cour's Fixatives (1931)*

	2BD (cm³)	2BE (cm³)	2BX (cm³)
2% aq. chromic acid soln.	100	100	100
2% aq. potassium dichromate soln.	100	100	100
2% aq. osmic acid soln.	60	32	120
10% aq. acetic acid soln.	30	12	60
1% aq. saponin soln.	20	10	10
Distilled water	210	90	50

2 BD is a good general fixative, 2 BE is recommended for plant tissues and 2 BX for bulk fixation. The usual period of fixation is 24 h in cold or in room temperature followed by washing in running water for 3–12 h. The ingredients are mixed just before use. For studying *Hordeum* chromosomes, Morrison and colleagues (1959) kept roots in a mixture of 1 per cent chromic acid and 2 BD (1:1) for 12 h after fixation in 2 BD.

6. *Osmic, Dichromate and Platinic Mixture*

2% aq. potassium dichromate soln.	70 cm³
2% aq. osmic acid soln.	10 cm³
1% aq. platinum chloride soln.	15 cm³
Acetic or formic acid	5 cm³

The acetic or formic acid should be added just before use. Period of fixation is 12 h followed by washing in running water.

7. *Smith's Fluid (1935)*

	S_1	S_2
1% aq. chromic acid soln.	100 cm³	75 cm³
2% aq. osmic acid soln.	35 cm³	25 cm³
5% aq. acetic acid soln.	25 cm³	12·5 cm³
Potassium bichromate	0·5 g	1 g
Saponin	0·05 g	0·05 g
Distilled water	50 cm³	46 cm³

S_1 is suitable for early prophase and S_2 for diakinesis and metaphase stages in p.m.c.s. Substitutes for osmic acid, such as ruthenium tetroxide or uranium trioxide, have also been used.

MISCELLANEOUS FIXING MIXTURES

1. *Chromo-acetic Acid (Flemming, 1882)*

Anhydrous chromium trioxide	0·25 g
Glacial acetic acid	0·1 cm³
Distilled water	100 cm³

The period of fixation is 12 h followed by washing in running water.

Though originally recommended for achromatic elements of karyokinesis by Flemming, this mixture is also useful for studying nuclei.

2. *Chromo-formic Acid (Rabl, 1884)*

0·33% aq. chromic acid soln.	200 cm^3
Conc. formic acid	4–5 drops

The mixture should be freshly prepared. Period of fixation is 12–24 h, followed by washing in water.

3. *Copper Chloride and Acetate Mixture*

Camphor water (not saturated)	75 cm^3
Distilled water	75 cm^3
Glacial acetic acid	1 cm^3
Copper acetate	0·30 g
Copper chloride	0·30 g
Optional: Osmium tetroxide	

The fixative is very moderate and useful for objects studied in as fresh a state as possible. Objects fixed stain easily in methyl green.

4. *Davidson's Fixatives (1949)*

(a)	Tertiary butyl alcohol	3 parts
	Absolute ethyl alcohol	1 part

and

(b)	Tertiary butyl alcohol	3 parts
	Ethyl phosphate	1 part

They are found to be suitable for vaginal smears.

5. *Kollman's Fixative (1885)*

Potassium dichromate	5 g
Chromium trioxide	2 g
Conc. nitric acid	2 cm^3
Distilled water	100 cm^3

The period of fixation is 12 h followed by washing in water for an equal period. The fixative has been used on animal ova.

6. *Makino's Fluid (1934)*

Glacial acetic acid	1 cm^3
Distilled water	100 cm^3
Mercuric chloride	saturation

This mixture is used for meiotic divisions in mature eggs. The gelatinous envelope is removed. The eggs are fixed for 10–15 min and washed in 70 per cent alcohol.

7. *Picric Acid–Sulphosalicylic Acid Mixture Soln.*

5% aq. sulphosalicylic acid soln.	1 part
Sat. aq. picric acid soln.	1 part

This fixative is used for fixing mammalian tissues for block preparation. The period of fixation is 48 h at 25 °C followed by washing in distilled water for 30 min. The tissues are then stained in Feulgen's solution before dehydration (Lhotka and Davenport, 1947).

8. *Telleyesniczky's Fluid (1889)*

Glacial acetic acid	5 cm^3
Distilled water	100 cm^3
Potassium dichromate	3 g

Period of fixation is 12–24 h followed by washing in running water.

9. *Zenker's Fluid (1894) and Lillie (1954)*

Glacial acetic acid	5 cm^3
Distilled water	100 cm^3
Mercuric chloride	5 g
Potassium dichromate	2·5 g
Optional: Sodium sulphate	1 g

To the rest of the fluid, glacial acetic acid is added immediately before use. The pH is 2·3. Fixation for several hours to overnight is followed by washing for 24 h in running water and treatment with 0·5 per cent iodine in 70 per cent alcohol to remove the mercury precipitate.

In a modification of Zenker's fluid for vertebrate material acetic acid can be replaced by 5 cm^3 of 40 per cent aqueous formaldehyde solution (Helly's fluid).

10. *Picric Acid—Parker's Medium TC199* for 24 h for human spermatocytes at pachytene (Annéren *et al.*, 1969).

DRY PRESERVATION AND DRIED OUT FIXATIONS

1. Occasionally the fixative evaporates from the phials and leaves the fixed tissue dry with a crust of residues. If the tissue has been fixed in media containing chromic acid or osmic acid, it can be restored by washing for some days in either 10 per cent ethyl alcohol or 10 per cent formalin, or by washing the residue away with strong sulphuric acid, washing in water for some hours and then treating as usual.

2. Dry preservation of plant materials for transport has been suggested by Vaarama (1950). The objects are fixed, washed and stored in 70 per cent alcohol as usual. After 1 h in 70 per cent alcohol (at least three changes), they are dried at room temperature on blotting paper. According to him, they can be kept in this condition for a long period. For use later, the materials are soaked in 10 per cent ethyl alcohol for one to several days, treated in 50 per cent ethyl alcohol for 3 h and then the usual process is followed for embedding in paraffin and staining by iodine-crystal violet or Feulgen.

Dried material fixed in acetic–alcohol mixture (1 : 3) has not turned out to be suitable for squash preparations due to the fragility of chromosomes. Squash preparations can be attempted without applying any pressure, as otherwise the chromosomes undergo fragmentation.

REFERENCES

Adams, C. W. H. (1960).   *J. Histochem. Cytochem.* **8,** 262
Amano, M. (1962).   *J. Histochem. Cytochem.* **10,** 204
Allen, E. (1916).   *Anat. Rec.* **10,** 565
Amirkhanian, J. D. (1968).   *Stain Tech.* **43,** 167
Ammerman, F. (1950).   *Stain Tech.* **25,** 197
Anderson, H. D. and Hammick, D. L. (1950).   *J. chem. Soc.* **2,** 1089
Annéren, G., Berggren, A., Stahl, Y. and Kjessler, B. (1969).   In *Modern trends in human genetics.*
    London : Butterworths
Austin, A. P. (1959).   *Stain Tech.* **34,** 69
Badder, F. G. and Mikhail, H. (1949).   *J. chem. Soc.* **4,** 2927
Bahr, G. F. (1954).   *Exp. Cell Res.* **7,** 457
Bahr, G. F. (1955).   *Exp. Cell Res.* **9,** 277
Bahr, G. F. (1957).   *Acta radiol. Suppl.* 147
Baker, J. R. (1944).   *Quart. J. micr. Sci.* **85,** 1
Baker, J. R. (1950).   *Cytological technique.* London; Methuen
Baker, J. R. (1958).   *Principles of biological microtechnique* London; Methuen
Baker, J. R. (1966).   *Cytological technique,* 5th edn. London; Methuen
Battaglia, E. (1957).   *Caryologia* **9,** 368
Belar, K. (1929).   *Meth. Wiss. Biol.* **1,** 638
Bell, L. G. E. (1956).   *Physical techniques in biological research* **3,** New York; Academic Press
Benda, C. (1902).   *Anat. Hefte,* **12,** 743
Berg, W. (1927).   Articles 'Chromsaure', 'Chromsalze' and 'Osmiumsaure' in   *Enzyklopadie der*
    Mikroskopischen Technik, by Krause, R. Berlin; Urban and Schwarzenberg
Bloom, G. and Friberg, U. (1957).   *Acta morphol. neerl.-scand.* **1,** 12
Bowden, W. M. (1949).   *Stain Tech.* **36,** 171
Bowes, J. H. (1963).   *A fundamental study of the mechanism of deterioration of leather fibres.* Brit. Leather
    Manuf. Assoc. Rep.
Bowes, J. H. and Kenten, R. H. (1949).   *Biochem. J.* **44,** 142
Bullivant, S. (1965).   *Lab. Invest.* **14,** 1178
Burns, J. A. and Yang, S. J. (1961).   *Stain Tech.* **36,** 102
Burstone, M. S. (1969).   In *Physical techniques in biological research* **1,** New York; Academic Press
Calvery, H. O. (1938).   Section on 'The Isolation of the aminoacids from proteins' in *The Chemistry of*
    the aminoacids and protein, ed. by Schmidt, C. L. A., Springfield, Ill.; Thomas
Carnoy, J. B. (1886).   *Cellule* **3,** 1
Carnoy, J. B. and Lebrun (1887).   *Cellule* **13,** 68
Carothers, E. E. (1936).   *Biol. Bull.* **71,** 469
Carpenter, D. C. and Nebel, B. R. (1931).   *Science* **74,** 225
Casarett, G. W. (1953).   *Stain Tech.* **28,** 125
Catcheside, D. G. (1934).   *Ann. Bot., Lond.* **48,** 601
Catcheside, D. G. (1925).   *Genetica* **17,** 313
Champy, C. (1913).   *Arch. Zool. exp. gén.* referred to in Gatenby and Beams, 1950
Clements, R. L. (1962).   *Anal. Biochem.* **3,** 87
Cope, G. H. (1968).   *J. Roy. Micros. Soc.* **88,** 235
Criegee, R. (1936).   *Ann. Chem. Pharm. S.* **22,** 75
Cutter, V. M. (1946).   *Stain Tech.* **21,** 129
Dallam, R. D. (1957).   *J. Histochem. Cytochem.* **5,** 178
Darlington, C. D. and La Cour, L. F. (1960,1968).   *The Handling of chromosomes,* London; Allen and Unwin
Davidson, H. B. (1949).   *Stain Tech.* **25,** 145
De Nordwall, J. H. and Staveley, L. A. K. (1956).   *Trans. Faraday Soc.* **52,** 1061
Falck, B. and Owman, C. (1965).   *Acta Univ. Lund* Sect. II, No. 7
Fernandez-Moran, H. and Finean, J. B. (1957).   *J. biophys. biochem. Cytol.* **3,** 725
Finean, J. B. (1954).   *Exp. Cell Res.* **6,** 283

Fischer, A. (1899). *Fixierung, Farbung und Bau des Protoplasmas.* Jena; Fischer
Flemming, W. (1879). Referred to in Baker, 1950
Flemming, W. (1882). *Zellsubstanz, Kern and Zellteilung.* Leipzig; Vogel
Flemming, W. (1884). *Z. wiss. Mikr.* **1,** 349
Flesch, M. (1879). *Arch. mikr. Anat.* **16,** 300
French, D. and Edsall, J. T. (1945). *Adv. Prot. Chem.* **2,** 277
Freytag, A. H. (1964). *Stain Tech.* **39,** 167
Gatenby, J. B. and Beams, H. W. (1950). In *The Microtomists's Vade-mecum* by Lee, B. London; Churchill
Gersh, I. (1959). *The Cell* **1,** New York; Academic Press
Gibbons, I. R. and Bradfield, J. R. G. (1956). *Biochem. biophys. Acta* **22,** 506
Green, R. W. (1953). *Biochem. J.* **54,** 187
Hairston, M. A. (1956). *Cytologia* **21,** 179
Hake, T. (1965). *Lab. Invest.* **14,** 470
Hance, R. T. (1917). *Anat. Rec.* **12,** 371
Hayes, T. L., Lindgren, F. T. and Gofman, J. W. (1963). *J. Cell. Biol.* **19,** 251
Hermann, F. (1899). *Arch. mikr. Anat.* **34,** 58
Hill, H. D. and Myers, W. M. (1945). *Stain Tech.* **20,** 89
Holmes, W. (1944). Referred to in Baker, 1950.
Hughes, W. L. Jr. (1950). *Cold Spr. Harb. Symp. quant. Biol.* **14,** 79
Hyde, B. B. and Gardella, C. A. (1953). *Stain Tech.* **28,** 305
Jensen, W. A. and Kavaljian, L. G. (1957). *Stain Tech.* **32,** 33
Jones, R. M. (1920). Referred to in Baker (1950).
Julien, J. B. (1958). *Canad. J. Bot.* **36,** 607
Kahle, W. (1908). *Die Paedogenesis der Cecidomyiden.* Stuttgart; Schweizerbart
Karpechenko, G. D. (1924). *J. Genet.* **14,** 387
Karpechenko, G. D. (1927). *Bull. appl. Bot., St. Petersb.* **17,** 305
Kauzmann, W. (1959). *Adv. Prot. Chem.* **14,** 1
Khan, A. A., Riemersma, J. C. and Booij, H. L. (1961). *J. Histochem. Cytochem.* **9,** 560
Kihara, H. (1927). *Bot. Mag., Tokyo* **41,** 124
Kollmann, (1885). *Arch. Anat. Physiol., Lpz.* 296
La Cour, L. F. (1931). *J. R. micr. Soc.* **51,** 199
La Cour, L. F. (1944). *Proc. Roy. Soc. Edinb.* **62,** 73
Lang, A. (1878). *Anat. Anz.* **1,** 14
Langlet, O. F. J. (1948). Referred to in Gatenby and Beams, 1950
Lassek, A. M. and Lunetta, S. (1950). *Stain Tech.* **25,** 45
Law, A. G. (1943). *Stain Tech.* **18,** 117
Lesins, K. (1954). *Stain Tech.* **29,** 261
Levan, A. (1949). *Hereditas, Lund. Suppl.* Vol. 326
Levy, M. (1933). *J. biol. Chem.* **99,** 767
Lewitsky, G. A. (1931). *Bull. appl. Bot., St. Petersb.* **27,** 19 and 176
Lhotka, J. F. and Davenport, H. A. (1947). *Stain Tech.* **22,** 139
Lillie, R. D. (1954). *Histopathologic technic and practical histochemistry,* Philadelphia; Blakiston
Lison, L. (1960). *Histochimie et cytochimie animales.* Paris; Gauthiers-Villars
Lojda, Z. (1965). *Folia Morphol.* **13,** 65 and 84
Lotke, P. A. and Dolan, M. F. (1965). *Cryobiology* **1,** 289
Maaløe, A. and Birch-Andersen, A. (1956). *Bacterial anatomy,* p. 261. London; Cambridge University Press
McClintock, M. (1964). *Cryogenics,* New York; Reinhold
McClung, C. E. (1918). *Anat. Rec.* **45,** 265
McGoldrick, P. T., Bohn, G. W. and Whitaker, T. W. (1954). *Stain Tech.* **29,** 127
McManus, J. F. A. and Mowry, R. W. (1960). *Staining methods, histologic and histochemical.* New York; Harper
Makino, S. (1932–34). *J. Fac. Sci. Hokkaido Univ., Zool.* **2,** and **3,** 117
Makino, S., Udagawa, T. and Yamashina, Y. (1956). *Caryologia* **8,** 275
Malhotra, S. K. (1968). In *Cell structure and interpretation,* London; Edward Arnold
Mann, G. (1894). *Z. wiss. Mikr.,* **11,** 491
Marks, G. E. (1952). *Stain Tech.* **27,** 333
Marks, G. E. (1960). *Euphytica* **9,** 254
Marinozzi, V. (1963). *J. Roy. Micros. Soc.* **81,** 141

Metzger, R. L. and Leng, E. L. (1955).   *Stain Tech.* **30,** 41
Meves, F. (1908).   *Arch. mikr. Anat.* **72,** 816
Miller, M. W. and Colaiace, J. (1968).   *Stain Tech.* **43,** 303
Middlebrook, W. R. and Phillips, H. (1942).   *Biochem. J.* **36,** 294
Minouchi, (1928).   *Jap. J. Zool.* **8,** 219
Morrison, J. H., Leak, L. V. and Wilson, G. B. (1949).   *Trans. Amer. micr. Soc.* **76,** 358
Müller, A. (1859).   *Verh. phys.-med. Ges. Würzh.* **10,** 138 and 179
Nakamura, T. (1928).   *Mem. Coll. Sci. Kyoto* B, **55,** 1
Navashin, M. (1925).   Referred to in Gatenby and Beams, 1950
Newcomer, E. H. (1952).   *Stain Tech.* **27,** 205
Newcomer, E. H. (1953).   *Science* **118,** 161
Oguma, K. and Kihara, H. (1923).   *Arch. Biol.* **33,** 493
Okunuki, K. (1961).   *Adv. Enzymol.* **23,** 29
Opie, E. L. and Lavin, G. I. (1946).   *J. exp. Med.* **84,** 107
Painter, T. S. (1924).   *Anat. Rec.* **27,** 77
Paolillo, D. J., Jr. (1960).   *Stain Tech.* **35,** 152
Pearse, A. G. E. (1968).   *Histochemistry,* London; Churchill
Pease, D. C. (1966).   *J. Ultrastruct. Res.* **14,** 356
Perenyi. (1882).   *Zool. Anz.* **5,** 459
Perenyi. (1888).   *Zool. Anz.* **274,** 139
Pischinger, A. (1937).   *Z. Zellforsch.* **26,** 249
Porter, K. R. and Kallman, F. (1953).   *Exp. Cell Res.* **4,** 127
Prokofieva, A. (1934).   *Cytologia* **5,** 498
Prokofieva, A. (1935).   *Z. Zellforsch.* **22,** 254
Rabl, C. (1884).   *Morph. Jb.* **10,** 215
Randolph, L. F. (1935).   *Stain Tech.* **10,** 95
Rebhun, L. I. (1965).   *Fed. Proc.* **24,** S 217
Riemersma, J. C. (1963).   *J. Histochem. Cytochem.* **11,** 436
Riemersma, J. C. and Booij, H. L. (1962).   *Ibid.* **10,** 89
Rutherford, T., Hardy, W. S. and Isherwood, P. A. (1964).   *Stain Tech.* **39,** 185
Romeis, B.   Referred to in Gatenby and Beams, 1950
Sabatini, D. D., Bensch, K. and Barrnett, R. J. (1963).   *J. Cell Biol.* **17,** 19
Salem, L. (1962).   *Canad. J. Biochem. Physiol.* **40,** 1287
San Felice, F. (1918).   *Ann. Inst. Pasteur* **32,** 363
Schneider, W. C. (1945).   *J. biol. Chem.* **161,** 293
Schultze, M. and Rudneff, M. (1865).   *Arch. mikr. Anat.* **1,** 298
Sjöstrand, F. S. (1956).   *Int. Rev. Cytol.* **5,** 456
Sjöstrand, F. S. (1969).   In *Physical techniques in biological research* **3C,** New York; Academic Press
Smetana, K. (1967).   In *Methods in Cancer Research* **2,** New York; Academic Press
Smetana, K. and Busch, H. (1964).   *Cancer Res.* **24,** 537
Smith, E. S. (1935).   *J. Genet.* **49,** 119
Smith, S. G. (1943).   *Canad. Ent.* **75,** 21
Stephenson, J. L. (1956).   *J. biophys. biochem. Cytol.* **2,** 45
Strakov, I. P. (1951).   *Zhur. Prikled. Khim.* **24,** 142
Stockkenius, W. and Mohr, S. C. (1965).   *Lab. Invest.* **14,** 458
Taylor, W. R. (1924).   *Bot. Gaz.* **78,** 236
Telleyesniczky, K. (1905).   *Arch. mikr. Anat.* **66,** 367
Treffenberg, L. (1953).   *Arkiv. Zool.* **4,** 295
Turner, B. L. (1956).   *Amer. J. Bot.* **43,** 577
Vaarama, A. (1950).   *Stain Tech.* **25,** 47
Von Beneden, E. and Neyt. (1887).   *Bull. Acid. Sci. Belg.* **14,** 218
Walker, J. F. (1953).   *Formaldehyde.* New York; Reinhold
Webber, J. M. (1930).   *Univ. Calif. Publ. Bot.* **11,** 319
Wigglesworth, V. B. (1957).   *Proc. Roy. Soc. B.* **147,** 185
Wigglesworth, V. B. (1964).   *Quart. J. micr. sci.* **105,** 113
Wolman, M. (1955).   *Int. Rev. Cytol.* **4,** 79
Woodroffe, D. (1941).   *Fundamentals of leather science.* Harvey; Waddon
Young, J. A. (1935).   *Nature,* **135,** 823
Zacharias. (1888).   *Anat. Anz.* **3,** 24
Zenker, K. (1894).   *Münch. med. Wschr.* **41,** 532
Zirkle, C. (1928).   *Protoplasma* **4,** 201 and **5,** 511

4 Processing

After suitable fixation, the tissue is processed for further study. Different schedules are followed for block and smear preparations.

BLOCK PREPARATION AND MICROTOMY

In order to study materials which cannot be squashed or smeared, they have to be dehydrated where necessary and embedded in a suitable medium. By embedding, small, or delicate objects can be surrounded with some plastic substance which supports it on all sides and allows sections of the material to be cut without distortion. It is also useful for showing the arrangement of cells in a tissue and the sequence of the stages of meiosis in a testis or an anther, or mitotic division in the somatic cells. The embedding material not only fills each cell and interstice, but it also ensures that the position of the minutest detail of structure is retained.

The procedure entails the following operations: washing, dehydration, clearing, infiltration and embedding, microtome-section cutting, and removal of embedding material.

Washing

Except in special cases, the tissue is thoroughly washed to remove all traces of fixing chemical. Different periods of washing in running water, from 1 h to overnight, are employed, depending on the nature and the thickness of the tissue and the fixing fluid used.

Comparatively hard tissue, like flower-buds, root tips or bulk masses of animal tissue, are put into perforated porcelain thimbles, which are then corked and placed in a beaker under running water. If the tissue is very small or delicate, it is kept in the original tube. The fixing fluid is drained off and the tissue is washed in successive changes of warm (44 °C) water at half-hourly intervals until all traces of the fixative are washed off.

Alcoholic fixatives should be washed out with plain alcohol of approximately the same percentage as that of the original solution. Reagents containing picric acid should always be washed out with alcohol and never with water unless there is another constituent present in the fluid which fixes chromatin indissolubly.

In some cases, however, long washing in water is unnecessary. Randolph (1935) and later Upcott and La Cour (1936) omitted washing altogether in studying root tip chromosomes. They transferred the root tips directly from the fixative to the dehydrating agent, namely, 70 per cent alcohol.

Dehydration

Since the embedding material is usually immiscible with water and aqueous solutions, it is necessary to dehydrate the tissues before they can be embedded. Tissues should not be transferred directly from water or aqueous solution to the undiluted dehydrating agent because an unequal shrinkage and distortion of the tissue is caused. The tissue should be passed through a series of solutions, each containing a mixture of the dehydrating agent and water, with the concentration of the former increasing gradually. Finally the tissue passes on to the pure agent.

The most suitable and economical dehydrating agent is ethyl alcohol. It also has a hardening effect on the tissue and the schedules are usually so arranged as to utilise both dehydration and hardening effects. The tissue is generally passed through successive grades containing 30, 50, 70, 80, 90, 95 per cent and absolute alcohol, the period of treatment being variable, depending upon the nature and thickness of the specimen. For tissue of the size generally used for chromosome study, 1 h in each is quite long enough, while for plant tissue, overnight treatment in each of 70 per cent and absolute alcohol is found to be most effective. The tissue can be stored in 70 per cent alcohol, if necessary. The different alcohol grades can be prepared from rectified spirit (approx. 96 per cent ethyl alcohol) as it is cheaper than absolute alcohol; Table 3 gives the relative proportions of rectified spirit and distilled water required to prepare different concentrations of alcohol.

Variations of the alcohol-dehydration schedule can be made, depending on the nature of the tissue. The entire process can be speeded up for thin and

Table 3   PREPARATION OF ALCOHOL OF REQUIRED CONCENTRATION USING RECTIFIED SPIRIT (APPROX. 96 PER CENT)

Volume of rectified spirit taken	Volume of water used	Alcohol concentration obtained, %
31·2	68·8	30
52·1	47·9	50
62·5	37·5	60
72·9	27·1	70
83·3	16·7	80
93·5	6·5	90

delicate tissue, and some workers, instead of passing the tissue through different grades, prefer to increase the concentration of the medium containing the tissue gradually by adding drops of strong alcohol at fixed intervals.

The disadvantages of using ethyl alcohol as a dehydrating agent are: (a) the schedule is rather time-consuming, (b) if the fixed schedule is not suited to the tissue, excessive hardening and shrinkage of the tissue result, and (c) as alcohol does not mix with paraffin or celloidin, an intermediate 'clearing' agent is needed, which unduly lengthens the process.

Several alternative dehydrating agents have been used from time to time, some of which are miscible with the embedding material and so eliminate the use of the clearing agent. The most important of these chemicals are given below.

Tertiary butyl alcohol—This was used by Johansen (1940), who claimed that it was the least drastic of all dehydrating agents. He transferred the tissue, which had been dehydrated up to 30, 50 or 70 per cent alcohol, to a mixture of distilled water, ethyl alcohol and tertiary butyl alcohol, then passed it through a series of grades containing decreasing proportions of distilled water and alcohol and increasing proportions of butyl alcohol, till a mixture of butyl alcohol and ethyl alcohol in proportion 3:1 was obtained. Finally the tissue was given three changes in pure tertiary butyl alcohol (one overnight). According to Johansen, this treatment removes every trace of unbound water.

Dioxane (diethyl dioxide)—This mixes with water, alcohol and xylol, and dissolves balsam and paraffin wax, and can therefore be used as a substitute at any stage of the usual ethyl alcohol–xylol dehydrating and clearing schedule. Direct treatment is also possible. The material can be transferred directly from water to 60 per cent dioxane in distilled water, thence to 70 and 95 per cent and finally given at least two changes of pure dioxane: alternatively, the material can be transferred directly to pure dioxane and then given two changes. As no clearing is required, this can be followed immediately by embedding in paraffin wax. Dioxane was first used in microtomy by Graupner and Weisberger (1933). Later, La Cour (1937), Maheshwari (1939) and Johansen (1940) also used it. Baird (1936) used it effectively in animal tissues.

Though dioxane is very effective in short-term schedules, particularly for animal tissue, it should be used very cautiously as it has a markedly toxic and cumulative action and is an injurious substance which should be handled with extreme care. It is also non-specific in its action, and in addition is heavier than melted paraffin and therefore difficult to remove from tissue during infiltration.

n-Butyl alcohol—It can be used in both plant and animal materials (Zirkle, 1930; Margolena, 1932; Stiles, 1934; Randolph, 1935). The material can be transferred, after partial dehydration in a low concentration of ethyl alcohol, to a mixture of ethyl alcohol and butyl alcohol, with successive changes in mixtures containing increasing proportions of *n*-butyl alcohol, followed by changes in *n*-butyl alcohol alone, and kept overnight. This chemical, however, is not as satisfactory as ethyl alcohol. Pure chloroform causes brighter staining but more distortion that butyl alcohol (Rawlins and Takahashi, loc. cit.).

Iso- and normal propyl alcohols—These can also be used effectively as substitutes for ethyl alcohol. The schedule followed is the same as for ethyl alcohol (Bradbury, 1931). Hauser (1953) suggested the use of iso-propyl alcohol for dehydration, removal of paraffin and clearing before mounting in balsam. A schedule for infiltration includes treatment in 60 per cent, followed by three changes in 99 per cent isopropyl alcohol and two changes in molten paraffin (Doxtader, 1948). Reeve (1954) outlined two methods for dehydration:

 (i) through primary dehydration by glycerol.
 (ii) through isopropyl alcohol alone—60, 85 and 99 per cent. Tissues are placed over solid paraffin in a phial and heated to 56–58 °C.

Ethylene glycol—As 'Cellosolve' glycol mono-ethyl ether, ethylene glycol was used as a dehydrating agent by Frost (1935) and Thorp (1936). It is, however, expensive and inflammable, and rapidly absorbs water from the air.

Other chemicals—Other chemicals have also been used as combined dehydrating and clearing agents, like amyl alcohol (Hollande, 1918) and methylal–methylene dimethyl ether (Defrenoy, 1935, Banny and Clark, 1949), followed by paraffin oil.

Dehydration with alcohols is carried out in corked or glass-stoppered phials or in shallow dishes with ground glass tops. Care must be taken not to keep the liquids exposed to moisture in the air. In special cases, dry erythro-sine dye can be added in the last tertiary butyl alcohol solution in Johansen's (1940) method. The tissue is stained red superficially and can be demarcated during embedding in paraffin wax. For very small materials (Madge, 1936), eosin can be added to the 70 per cent alcohol or fuchsin in the final stages of the alcohol–chloroform clearing process to render the tissue conspicuous.

Glycerol—is found to cause less distortion than ethyl alcohol; 95 per cent alcohol removes some glycerol, sets the protoplasm and improves the staining (Rawlins and Takahashi, 1947).

Clearing

This step is required only when paraffin is the embedding material. For celloidin-embedding, no separate clearing is required, as clearing is done during dehydration itself.

Since paraffin does not mix with many dehydrating agents, an inter-mediate medium which is miscible with both the dehydrating agent and paraffin is generally necessary, this medium performing the function of ridding the tissue of the dehydrating agent. Most of these reagents render the tissues translucent, as their refractive indices are close to that of the proteins of the tissue and rays of light can pass through without refraction; therefore they are also called 'clearing' agents and the process 'clearing the tissue'. The term is, however, a misnomer, as it suggests that the tissue needs 'clearing', while it is not so. Apàthy (1912) used instead the term 'ante-medium' for those intermediaries, which is probably preferable.

The ideal antemedium should have rapid penetrating power and should mix equally well both with the dehydrating and embedding agents. An immediate transfer from the pure dehydrating agent to the pure ante-medium may cause shrinkage or distortion of the tissue and so mixtures of the two fluids should be used in varying proportions before transfer to pure antemedium. The most satisfactory and most widely used antemedium is chloroform (CHCl$_3$). The tissue is usually passed, after dehydration in ethyl alcohol, through a series of alcohol–chloroform grades, 3:1, 1:1 and 1:3, being kept for an hour, or more if necessary, in each, the period of treatment depending on the thickness of the tissue. Finally it is transferred to pure chloroform. Chloroform is an excellent de-alcoholising agent and does not render the tissue brittle within the scheduled period of treatment. The only drawback is that it is too volatile and all phials should be kept tightly corked.

Another effective clearing agent is benzene (C$_6$H$_6$). It can be used in a series of grades in combination with the dehydrating agent similar to chloroform.

Xylol is also used extensively, but has a tendency to shrink and harden the tissue (Romeis, 1928; Tarkhan, 1931). Amylacetate has been used for tissue containing yolk (Barron, 1934). Terpineol, toluol and methyl benzoate have also been used at different times (Wetzel, 1931).

Various organic oils, like Bergamot, cedarwood, clove and aniline oil have also been used as antemedia, and of these, cedarwood oil and oil of Bergamot have no hardening effect and can clear bulkier pieces of tissue better than most other reagents. The time of reaction in different grades has to be much longer with these oils than with more volatile liquids. With cedarwood oil, 2–3 h of treatment in each grade is quite enough. The only limitation of these two oils is that they are not good solvents for paraffin and are, therefore, not easily replaced from the tissue by paraffin.

For transferring the tissue through the different grades from dehydrating agent to antemedium, the tissue may be kept in a stoppered tube, and after each change can be kept close to the bottle containing the mixture which has been added, to identify the grade. The bottles can be arranged serially and labelled.

Infiltration and Embedding

As mentioned before, embedding methods are employed to surround small and delicate objects with some plastic substance that will support them without injury while sections are being cut (Firminger, 1950; Hale, 1952; Kuhn and Lutz, 1958). The fill up and infiltrate the entire tissue with the embedding substance, giving the tissues a firm consistency and ensuring the cutting of fine sections, retaining the minute details of the structure *in situ*. Two major methods for embedding are used in the study of chromosomes: the paraffin method and the celloidin or collodion method. A modified method of the above two, which is used, is the 'double' embedding method. If, however, the material is to be cut in a quite fresh condition, then the frozen sectioning technique can be employed (*see* page 82).

Paraffin method—This method is the one most extensively used for the study of chromosomes. It has several advantages over the others, such as: (*a*) it is quite easy and rapid; (*b*) materials embedded in paraffin can be kept for an indefinite period; (*c*) very thin serial sections can be obtained by the process, and (*d*) paraffin mixes easily with most antemedia. As the embedding mass is removed before staining, a wider selection of stains can be used. The chief drawback of paraffin embedding is that, in spite of all precautions, the lengthy procedure of dehydration, clearing and embedding in molten paraffin causes some amount of shrinkage and distortion, a drawback which also applies to the celloidin method.

The paraffin waxes generally used for embedding have melting points ranging from 46° to 60 °C, depending on the atmospheric temperature: very rarely are waxes with lower or higher melting point used. This material was first used for embedding by Klebs (1869) and later revived by Bowene (1882).

In general, the paraffin method involves three steps:

The *first step* of this process is the saturation with an antemedium, which is also a solvent of paraffin. This step was considered under the heading 'Clearing' (*see* page 76).

The *second step* is the gradual replacement of this solvent by paraffin, this process being called 'infiltration'.

Earlier workers (Apàthy, 1912) transferred the tissue directly from the antemedium to the melted paraffin, but this, however, hardens the tissue, so a gradual increase in the paraffin content, before adding molten paraffin, is necessary.

The most commonly used schedule is to dehydrate the tissue by ethyl alcohol and then clear it by passing it through alcohol–chloroform grades, until finally pure chloroform is reached. The tissue is kept in pure chloroform for 10–30 min, and then small chips of paraffin wax, of a melting point lower than the one desired for embedding, are added to the chloroform containing the tissue. The tube with the tissue and the chloroform with paraffin is kept on the hot plate (35 °C) for periods ranging from 2 h to overnight. The mixture forms a warm saturated solution of paraffin in chloroform and the period of infiltration depends on the nature and thickness of the tissue. Later, the tube is transferred to a warm bath (45 °C), where it may be kept overnight. It can also be stored in this temperature for an indefinite period, as the temperature is much below the melting point of the paraffin and too low to harm tissues fixed for studying their chemical nature.

The tissue is finally transferred to a hot bath maintained between 55° and 60 °C, at the same or higher temperature as the melting point of the embedding paraffin. As the paraffin–chloroform mixture melts, it is changed immediately with molten embedding paraffin. Two more successive changes are given with molten embedding paraffin at intervals of 15–30 min, the final change in molten paraffin being given only when no trace of the smell of chloroform is left.

This gradual change from a paraffin wax of low melting point to one of high melting point has been criticised by some workers but it has been found to be very effective in the authors' laboratory. An alternative method is gradually to warm, on a hot bath, the chloroform containing the tissue up to the melting point of the paraffin employed and, during warming, to add by degrees small pieces of paraffin to the chloroform. As soon as the bubbles of the tissue cease, the addition of paraffin may be stopped. This process, being a gradual one, minimises the danger of shrinkage, but cannot be recommended for tissues which are to be treated for chemical study.

Numerous alternative infiltration techniques exist to suit the different dehydrating and clearing schedules; and some are mentioned below.

In Johansen's (1940) technique of dehydration with tertiary butyl alcohol, described previously, the tissue is transferred from pure butyl alcohol to a mixture of paraffin oil and tertiary butyl alcohol (1:1) and is kept for 1 h. A container is filled three-quarters full with melted Parowax (of Standard Oil Company), the wax being allowed to just solidify, and the material is placed on the top of the Parowax, covered with the butyl alcohol–paraffin mixture and the whole placed in an oven at a temperature close to the melting point of Parowax. The material is gradually infiltrated and slowly sinks through the wax until it reaches the bottom. After 1 h the entire mixture is drained off and replaced by pure molten Parowax. The process is repeated twice and a final replacement is given with a rubber Parowax mixture or a paraffin wax of the required melting point. Care should be taken to see that: (*a*) the original Parowax is solidified but not allowed to get cold, as otherwise the glass vial might crack in the oven; (*b*) the temperature of the oven be such that the wax does not melt too quickly; (*c*) the oven be well ventilated so that the evaporating alcohol is blown away; and (*d*) before giving the final change in Parowax there should be no discernible odour of butyl alcohol.

The time of infiltration varies with the thickness of the tissue.

Either technique can be followed in dehydration and infiltration with other dehydrating and clearing agents, such as xylol, normal or secondary butyl alcohol or an essential oil. Another method of infiltration is to carve a small block of paraffin wax into the form of a long, narrow cone and place it in the container with the pointed end downwards, touching the bottom. It will sink slowly into the antemedium as it dissolves. An alternative method is to cut coarse wire gauze into square pieces, slightly larger in size than the diameter of the container used. The sides of the gauze are bent and the entire gauze is fitted into the bottle so that it forms a shelf, supported by the bent ends. Shavings of paraffin are placed on the shelf and, on dissolving the paraffin, covers the material under it.

The further steps of both these methods, regarding the replacement of the solvent by paraffin, are similar to those of the widely used chloroform–paraffin schedule, discussed previously (*see* page 78).

With dioxane as dehydrating agent, little chips of Parowax are added gradually to the dehydrated material, contained in pure dioxane. The mixture is kept in a warm bath till dioxane, which is not a good paraffin solvent, is saturated with Parowax. The later steps are similar to chloroform–paraffin infiltration.

Acetone can be used as a substitute for ethyl alcohol in dehydration, and clearing can be done through acetone–chloroform grades.

The *third step* is *embedding*, after the complete infiltration of the tissue with paraffin. This includes pouring the molten paraffin with the tissue into a suitable receptacle, arranging the tissue in a proper manner and rapidly cooling the paraffin with the material. (Melnyk, 1961.)

For infiltration and embedding in celloidin method, the following procedure is adopted:

The tissue is transferred to a mixture of absolute alcohol and ether (1:1) and kept overnight.

A hole of a suitable size is made in a piece of junket to hold a part of the tissue and both the tissue and junket are transferred to 2 per cent celloidin solution in a wide-mouthed jar and kept overnight.

In the next step, either of two methods can be employed; with the celloidin used cold or hot.

In the *first method*, the piece of junket is transferred to the wide-mouthed jar containing 4 per cent celloidin solution with grooved side upward, the tissue being fitted into the hole or groove. The junket with the tissue is later transferred through 6 and 8 per cent solutions to 10 per cent celloidin solution. The time of immersion in each concentration depends on the nature of the material and may vary from one to many days. The glass container is kept undisturbed away from sunlight till the celloidin solution forms a gel. The entire container is covered first with the glass cover and then with the bell jar.

In the *second method*, the bottle with the material in 2 per cent celloidin solution is corked very firmly and placed in a hot oven at a temperature between 45 and 55 °C. The changes in different concentrations are given at intervals of 24 h. The bottle is always cooled before opening.

After 10 per cent solution is reached, the gel is allowed to set and helped to solidify further by evaporation or by adding small chips of dry celloidin.

In section cutting, rocking microtome for celloidin sections and rotary microtome for paraffin sections are recommended (Morris, 1965; Wachtel *et al.* 1966; Collins, 1969). For fixation and mounting sections on slides, Mayer's adhesive is most commonly used (Mayer, 1883; Baker, 1933) but several other methods are also available (Cove and Schoenfle, 1946; Lewis, 1945; Giovacchini, 1958; Weaver, 1955).

A modification was devised for fixing the section on a slide with Haupt's medium (Lewis, 1945). For obtaining serial sections, the celloidin block is marked with an ink composed of suspension of lamp black in 2 per cent celloidin in ether–ethyl alcohol (1:1) after each section is cut (Melton, 1956).

Other embedding media—The sections are cut according to the procedure followed for paraffin section. A high relative atmospheric humidity is found to interfere with microtome cutting of tissues embedded in polythene glycol wax. With increase in atmospheric temperature, the tolerance of relative humidity decreases (Hale, 1952). For fixing Carbowax sections to slides, a jelly, of 15 g gelatin, 55 cm^3 distilled water, 50 cm^3 glycerol and 0·5 g phenol, is used (Giovacchini, 1958).

Frozen section technique—The frozen section technique is based on the principle of freezing the tissue directly to harden it, and cutting sections while the tissue is frozen. This possibility was first explored by Raspail (1825) and later by Stilling (1842). The method has several advantages, namely: (*a*) it takes up much less time than the paraffin and celloidin methods; (*b*) since the tissue is not dehydrated, the cells retain a life-like appearance with little shrinking; and (*c*) the tissues can be sectioned, if necessary, without any fixation at all. However, there are also some drawbacks which prevent this technique from being universally applicable, and chief amongst these are that serial sections cannot be cut and that for loose tissues, there is no satisfactory process of holding them together before freezing.

Special models of microtome are available for cutting frozen sections. In older models, ether was used for freezing and the tissue was usually soaked for some hours in gum, dextrine or sugar solutions to prevent ice crystal forming. In the modern technique, CO_2 jets are used on the microtome knife for cooling. The sections are attached to the slide by means of their own fluid. A vibratory microtome for cutting sections of living roots has been used by Persidsky (1953).

The material can be cut fresh or after fixation, and any one of the usual fixatives can be used, the tissue being washed thoroughly. The material may be kept overnight in thick gum-arabic, if necessary, or this step can be omitted. When cells adhere very closely together, as in vertebrate liver, the most successful sections can be cut.

The piece of tissue is trimmed into a size within 3×3 cm and is placed on a little water on the microtome table freezer to freeze the tissue to the table, the temperature depending on the nature of the tissue. Tissues with compact cells should be frozen at about -10 to $-15\,°C$ while others can be cut at -20 to $-30\,°C$. The knife is oriented with its edge close to the tissue at right angles, and the tap controlling the CO_2 cylinder is opened and closed several times until the tissue is congealed to the table and frozen right through. In order to regulate the freezing, only short jets of CO_2 should be allowed to escape at intervals on the material.

The tissue is left for a few minutes and then the knife is passed over it, till sections begin to cut. If the sections have fine cracks and are brittle or roll up it means that the tissue has been overfrozen and it should be allowed to stand and warm up before being cut again; but if, on the other hand, the sections are too soft and disintegrate or tear or fail to form, the tissue is underfrozen and must be exposed to further jets of CO_2 to freeze it properly.

As soon as the tissue is ready to be cut, sections should be taken very rapidly and allowed to accumulate in a mass on the knife, and they can then be transferred to a petri dish containing distilled water and their thickness tested. The loose sections can be lifted on clean slides and observed un-stained under a cover-glass.

A freeze-drying apparatus devised with liquid nitrogen, permitted section cutting within 5 h after the fresh tissue was obtained (Stowell, 1951). White and Allen (1951) designed a microtome for cutting frozen sections, in which the knife moves across the tissue, in contrast to the sledge type in which the tissue is moved while the knife is stationary. Reiner (1953) described fixation of fresh tissues by undiluted and un-buffered formalin, combined with heating to 56–60 °C, preliminary to freezing and cutting on the 'warm' knife freezing microtome. Woods and Pollister (1955) used cold ethyl alcohol as a dehydrating agent after freezing in partially frozen isopentane, cooled with liquid nitrogen for drying frozen plant tissues. In a modification for obtaining thin sections from unfixed tissues for histochemical staining, a microtome with an apparatus for simultaneous cooling of the knife with the freezing stage is used. The sections are processed according to a particular schedule, mounted and dried in warm air (Wachstein and Meisel, 1953). If a cellulose tape is pressed against a paraffin or frozen tissue block just before cutting each section, sections as thin as 1 μm can be obtained. The method has been recommended for very large, hard or brittle specimens (Palmgren, 1954). For cutting frozen sections on a paraffin microtome (Pauly, 1956), the specimen is fixed on the object holder in a drop of water by freezing it in dry ice in a box. Chips of dry ice are wedged between the metal disc and the object clamp of the microtome during cutting.

These sections can also be processed in two ways: they may be attached to slides by gelatin, or they may be stuck by their own coagulated juice to the slide.

Gelatin-coated slides are prepared by smearing them with a thin film of specially prepared gelatin (2 g of gelatin soaked in 100 cm^3 of distilled water and heated to 50–60 °C) and the slides are dried in a warm dust-proof chamber. The section is lifted out of the water by passing a gelatin-coated slide under it and then raising it so that the section is caught in the middle of the gelatined surface, and the water is run off from the sides of the slide. The section is then flattened out by pressing on it with a pad through a piece of wet cigarette paper, and the paper is peeled off. If not already fixed, the slide with the section is placed in a closed container with a plug of cotton wool soaked in formalin for 15–30 s. It is then kept overnight in 10 per cent formol saline before staining.

In a modification of this schedule, the frozen sections are soaked for 5 min or longer in a mixture of 1·5 per cent aqueous gelatin and 80 per cent ethyl alcohol (1:1), teased on to a slide and blotted with filter paper dampened in rectified spirit. The gelatin congeals, anchoring the section to the slide (Albrecht, 1954).

In the other method, the section is cut out on a dry clean cover-glass held in a pair of forceps just where the section is curling up during cutting. The cover-glass is gently removed with the section which thaws on it and is fixed immediately by exposing it to osmic or formalin vapour. It is then put in a fixing fluid.

Sections of ordinary mammalian tissues, after fixation, can safely be floated in distilled water, but sections of such tissues as mollusc ovotestis and mammalian testis tend to break up if floated and are best cut fresh and fixed dry directly on to slides as described in the second method. They can also be attached directly to gelatin-coated slides.

Removal of the Embedding Material

After cutting the sections and mounting them on slides, the next step is to remove the embedding material and gradually bring down the tissue to the medium in which the stain is dissolved, usually water. The steps followed are usually the reverse of the process leading to embedding through dehydration and infiltration and the schedules differ depending on the material used for embedding.

Sections embedded in paraffin—For removing paraffin wax from sections, the most effective chemical is xylol because it dissolves paraffin and also because it is cheap and not very volatile. After xylol, a mixture of xylol and absolute alcohol in equal proportions is usually used, followed by pure absolute alcohol and alcohol grades with decreasing percentage of alcohol to water.

All these chemicals are kept in covered wide-mouthed jars, with or without slide racks, the jars being labelled and arranged in a series. The slide is transferred carefully from one jar to the next so that the sections are not dislodged due to rough movement, the period of treatment in the chemicals depending upon the nature and the thickness of the tissue.

A typical grade of chemicals for removing paraffin is:

2 jars of xylol, I and II
1 jar of xylol and absolute alcohol mixture (1:1)
1 jar of absolute alcohol
1 jar of 95 per cent alcohol
1 jar of 90 per cent alcohol
1 jar of 80 per cent alcohol
1 jar of 70 per cent alcohol
1 jar of 50 per cent alcohol
1 jar of 30 per cent alcohol
Distilled water

The slides are usually kept in each of xylol I and II for 30 min, in absolute alcohol-xylol mixture for 30 min, in absolute alcohol for 15 min, and 5 min in all the remaining solutions. If necessary, the slides can be preserved for an indefinite period in 70 per cent alcohol. If bleaching is necessary for tissues fixed in heavy metallic fixatives containing osmium or platinum, the slides are transferred from 80 per cent alcohol to a jar containing hydrogen peroxide and 80 per cent alcohol mixed in equal proportions and kept overnight on a hot plate at 35 °C. The bleaching period can be prolonged if required. Afterwards the slides are passed through 70, 50 and 30 per cent alcohol and brought down to water.

For thin sections of animal tissue, the period taken in bringing the slides down to water can even be reduced to 10 min in xylol and 5 min in the subsequent grades.

Sections of woody tissues and those containing mucilage, fats, etc., usually get disengaged by the time the slide reaches water and float off, in spite of all precautions. The sections can be stained in a dye without staining celloidin, and in that case the sections and slides can be covered with a thin coating of celloidin. The slides are transferred from the xylol-alcohol stage to a mixture of absolute alcohol and ether (1:1), to which enough celloidin solution has been added to make it about 1 per cent. They are kept in the mixture for 5–10 min and then allowed to dry in air till the celloidin is almost dry, forming a whitish film. They are next immersed in 70 per cent alcohol for 5 min to harden the celloidin and are then, as usual, brought down to water.

For softening paraffin-embedded tissues, several procedures are available. Dilute hydrofluoric acid, used alone and with glycerol and ethyl alcohol, softens plant materials embedded in paraffin. Tannins and phlobaphene compounds can then be removed by treating the sections for 12–48 h in a mixture of aqueous chromic acid, potassium dichromate and glacial acetic acid (Foster and Gifford, 1947). Exposure to a mixture of glacial acetic acid and 60 per cent ethyl alcohol (1:9 or 2:8) for 2–5 days also gives very good results (Gifford, 1950). Alcorn and Ark (1953) advocate soaking paraffin-embedded plant specimens, after exposing one side of the tissue in a mixture of glycerol, 10 cm^3; Dreft 1 g and water, 90 cm^3 for 2–3 days at 37 °C.

Sections embedded in celloidin—The sections of materials originally embedded in celloidin are usually stained as such, without removal of the celloidin matrix, which does not interfere with staining. The slides with the sections or the individual sections themselves are brought down to water from 70 per cent alcohol and then stained. The matrix, if necessary, can be removed before final mounting in balsam. This method is described in Chapter 5.

SQUASH OR SMEAR PREPARATIONS

Within the last few years, the sectioning method of the tissue has been largely replaced by the smear or squash technique. This method has a great advantage over sectioning in that the entire process is rapid and in addition it is also much more suitable for critical observations. In properly prepared smears or squashes, one can carry out observation on separated single cells; moreover, the cell, being released of its compactness, undergoes much enlargement in volume, affording wider space for the chromosomes to become scattered. Owing to these advantages, it has more or less become the universal routine method in chromosome work. The only disadvantage of this method, when specially applied to somatic chromosomes, is that the individual cells, being released from one another, shift from their original site and the original topography is altered.

The terms 'smears' and 'squashes' are often loosely used, resulting in the worker often getting the impression that the two processes are identical, but, strictly speaking, they are not so.

In smears, the cells are directly spread over a slide prior to fixation, and in this process no treatment is necessary to secure cell separation. Pollen mother cells from anthers are the most convenient objects for smears.

In squashes, on the other hand, special treatments are needed for dissolution of the pectic salts of the middle lamella so that separated individual cells can be obtained from a compact mass of cells, this treatment being carried out after fixation or even after staining. After passing through the required steps, the softened bulk material or small tissue can be neatly squashed on a slide by generally applying pressure or tapping with a needle over the cover glass. The best way to study mitotic behaviour of chromosomes is by squashes of root tips.

The term 'smear' is commonly applied to cases where cells have been spread on the slides before fixation, while squashing, on the other hand, is used for the process performed after fixation or staining. Compared to this, the criterion of difference based on the treatment for cell separation as mentioned above seems to be more convenient and rational.

Smears

The general procedure for preparing smears of pollen mother cells is to squeeze out the fluid from the anther on to a dry slide (to ensure success, the slide should be moistened by breathing on it before smearing), spread it with the aid of a scalpel and immediately invert it in a smearing tray containing the fixative. The entire process should be rapidly executed and must not take more than 4–5 s. Quick handling is essential, as otherwise the fluid tends to dry up, resulting in chromosome clumping. The use of a scalpel aids in the addition of iron which acts as a mordant (*see* Chapter 5 for the stain).

In the case of very small anthers, they should be smeared complete, instead of the fluid which is difficult to squeeze out. In such cases it is always necessary, after fixation, to remove the anther debris with the aid of a needle, keeping only the pollen mother cells. For even smaller objects, such as flower buds of Amaranthaceae, Chenopodiaceae, etc., where it is difficult to take out the anthers from the young buds, the smear method should not be tried and recourse should be made to sectioning. This is one of the limitations of the smearing procedure.

There are certain methods in which pollen mother cells can be smeared or squashed in a fluid which serves the double purpose of fixing and staining. The best example of such a procedure is Belling's (1926) *Iron–Acetic-Carmine Schedule*, when the anther is directly smeared in a drop of a solution containing carmine dissolved in acetic acid. The acetic-carmine method can be applied both in smears and squashes and it has been applied in plant and animal tissues after modifications, the modifications principally involving intensification of colour, which are discussed in Chapter 5.

For certain materials it may, however, be necessary to fix the flower buds in Farmer's fluid (acetic–ethanol, 1:3) prior to smearing in carmine. This is performed to secure cytoplasmic clearing and for fixation. Before staining, the material is kept in 45 per cent acetic acid for 15–30 min to cause swelling, counteracting the effect of ethanol and to soften the tissue.

Squashes

Regarding squashes, the most important step is the softening of the tissue. The different schedules can be divided into two categories, namely. softening performed prior to staining, and softening, clearing and staining accomplished in the same fluid.

Within the first category are included the various types of chemical agents employed by different authors for this purpose, including the use of enzymes.

Chemical agents—The most important agent needed for softening the tissue is dilute hydrochloric acid. In Feulgen staining (discussed in Chapter 5), this step is essential to secure Schiff's reaction for aldehydes. In addition to liberating aldehydes of sugar, normal hydrochloric acid at 58 °C serves two more important purposes, namely dissolution of pectic salts of the middle lamella, thus helping in cell separation (*see* Sharma and Bhattacharjee, 1952), and clearing of the cytoplasm. These two properties of hydrochloric acid can be advantageously employed as well in other squash schedules after fixation. If dilute hydrochloric acid (such as 10 per cent) is used, for best results, the treatment should be carried out in a slightly warm temperature, such as 58–60 °C for 4–5 min until softening, the acid being washed off either in 45 per cent acetic acid solution or water before staining.

Softening and maceration of the tissue can also be achieved during fixation by the use of a mixture of equal parts of 95 per cent alcohol and concentrated hydrochloric acid as the fixative. No warming is needed and even after 5 min treatment the tissue becomes fixed and softened at the same time. If necessary, hardening the tissue for 10 min in Carnoy's fluid after this treatment can also be carried out (Warmke, 1935). For rather hard materials, alcohol and hydrochloric acid mixed in the proportion of 3:1 is more effective. This acidified alcohol treatment is specially effective for materials with thick walls, such as pollen grains, leaves, etc., where slowly penetrating fluids are ineffective.

Sharma (1951) observed that the purposes of cytoplasm clearing and softening of the tissue can even be achieved by trichloracetic acid. There are, however, two serious disadvantages in this: firstly, it reduces the stainability of the chromosomes, so a further mordanting in metallic acid may be necessary for intensification of colour (*see* Sharma, 1956); and secondly, acid hydrolysis may cause depolymerisation of DNA as well as a break in the nucleoprotein link if the treatment is prolonged, with a probability of causing shedding of DNA from the chromosome, amounting to absolute loss of stainability. Acid treatment therefore, should be performed under strictly controlled conditions.

Maceration by chromic acid also can be carried out after fixation, especially with osmium-containing fluids, but this procedure is time consuming and may take at least 24 h to soften the tissue. Since a strong concentration of chromic acid may injure chromosome parts, prolonged treatment with a dilute solution (1 per cent) (Singleton, 1953; Day, Boone and Keitt, 1956; Elliott, 1956) is generally preferred. For a critical study of chromosome morphology, this procedure is generally not recommended.

In addition to acids, other reagents may also bring about softening of the tissue. Hydrogen peroxide is used with a trace of sodium or lithium carbonate added, and the middle lamella is initially attacked. Satisfactory dissolution of the middle lamella has been obtained by using a mixture of saturated aqueous solution of ammonium oxalate and hydrogen peroxide mixed in equal proportions for a few minutes. Ford (referred to in Darlington and La Cour, 1960) has recommended it for meristematic tissue fixed in osmic acid. However, for comparatively stiff materials, this reagent is not suitable for softening. Alkali treatment, prior to squashing, was used by Tandler (1959) in plants. The tissue, fixed in acetic-ethanol (1:3), is brought to water, hardened in 10 per cent formalin for 4–6 h, treated in a 2 per cent NaOH for 12 h at 25–37° C, washed, treated in 10 per cent acetic acid and stained by carmine or Feulgen squash method.

Enzyme treatments—The most reliable method of softening and clearing without causing injury to cellular parts is 'enzyme treatment'. Fabergé (1945) used *cytase*, together with other enzymes from the stomach extract of snails, *Helix pomatia*, for the purpose, and this method has been found to be very useful in a wide variety of plants, including fungi (McIntosh, 1954). The use of pectinase for dissolution of pectic salts of the middle lamella has been applied by McKay and Clarke (1946), Setterfield *et al.* (1954) as well as Chayen and Miles (1953), the latter authors using a 5 per cent solution in 1 per cent aqueous peptone for the purpose. This procedure is rather time-consuming and at least 2–5 h of treatment is necessary. Harris and Blackman (1954) treated Feulgen-stained roots in 2 per cent pectinase solution (pH 6·6) for 12 h, followed by a commercial pectin product 'Certo', and then suspended the cells by suction and expulsion through a pipette. The present authors have, however, noted that a 2 per cent aqueous solution of pectinase, if applied for half an hour at 37 °C, results in considerable softening of plant materials (Sharma and Chatterjee, 1971). The limitations of the enzyme treatment (as pointed out previously) lie in the difficulty of securing a pure form of enzyme preparations. If it can be obtained, this is the most reliable method of cell separation. Treatment of anthers in 1 per cent clarase solution or an extract prepared by grinding the contents of flask cultures of certain fungi for 10 min to several hours and squashing in acetic-carmine destroys some of the elasticity of the cytoplasm, so that on pressing, the chromosomes spread out (Emsweller and Stuart, 1944).

Of all the methods so far devised for cell separation and softening, treatment with dilute HCl, in spite of its limitations, is most commonly employed because temperature and period of treatment can be varied as necessary, and principally because of the low cost, easy availability and rapidity of the schedule.

In addition to the above-mentioned methods of softening, there are a number of schedules in which the softening is carried out together with staining. In this procedure the tissue is heated, after fixation, over a flame for a few seconds in a mixture of one of the acetic dyes and hydrochloric acid. The commonly used acetic solutions of dyes are acetic-orcein (La Cour, 1941), acetic-lacmoid (Darlington and La Cour, 1960) and acetic-carmine (*see* Chapter 5). Normal hydrochloric acid is mixed with a 2 or 1 per cent solution of the acetic-dye, usually in the proportion of nine parts of dye to one part of acid. The proportion of acid may be increased if the tissue proves difficult in squashing, the material then being directly squashed under a cover-slip in the mounting medium which is either 45 per cent acetic acid solution or 1 per cent solution of the dye dissolved in 45 per cent acetic acid solution (*see* Chapter 6). This method, though suitable for somatic tissues of plants, may be too drastic for extremely soft animal tissues which may undergo dissolution. It has, however, been observed that as acetic-carmine contains acetic acid as one of its components, mere heating (Markarian, 1957) in acetic-carmine may serve the purpose even for comparatively soft plant tissues.

Summarising the different aspects of smearing and squashing, the situation is as follows. Smearing is a comparatively easier schedule than squashing and should be performed prior to fixing only on cells lying in a fluid medium, such as pollen mother cells. Squashing, on the other hand, involves different steps, such as pre-treatment, fixation and softening. Pre-treatment can be performed with any of the agents mentioned in Chapter 2. Softening can be carried out either by acid, alkali or enzyme treatment after fixation but prior to staining in most cases. In a number of schedules, staining and softening are carried out in the same fluid and following staining, the materials are either first teased with the needle in the mounting medium prior to mounting or mounted directly in the medium under the cover-slip. Final squashing is performed by applying pressure over the cover-glass on a blotting paper before sealing for observation.

The methods of pre-treatment as mentioned above have mostly been devised for plants, though these have been tried in animals as well. Animal materials for chromosome studies provide less difficulty in squashing because of the soft nature of the tissue, and the absence of a cellulose wall in meiotic material.

In order to study meiotic material of animal tissues, such as insects, the testis is generally dissected out in normal saline solution, as dissection in the body fluid itself is difficult to perform. The fats should be removed and the material can be directly squashed in acetic-carmine solution or fixed in Carnoy's fluid before squashing. For permanent smear preparations, spreading on the slide and inverting it in the fixing fluid should be done as for plants; other staining methods too can be applied as for plants. For avian and reptilian testes, the tissues are generally fixed directly in the fixative without going through normal saline solution. In a method devised for amphibia, 0.5 cm^3 of a mixture of 60 mg of desiccated mammalian pituitary and 3 cm^3 of isotonic amphibian Ringer's solution is injected intraperitoneally into the animal. After 48 h, 0.5 cm^3 of 0.0015 per cent colchicine is injected. The gonads are removed after 2 h and smeared in acetic-orcein (Walsh and Archambault, 1954).

The study of divisional stages from eggs is difficult to carry out because of the very low frequency of divisional cells, and they lie mostly in an arrested condition. For the study of early oögonial divisions, immature eggs are generally taken and the procedure is the same as for testes. For oöcyte division, mature eggs before laying are necessary and the shell should be punctured for smearing. To secure a large frequency of metaphase stage, feeding the animal with colchicine and honey 24 h prior to dissection has been effective. With the exception of colchicine, other pre-treatment agents have very rarely been tried on animals and ample scope for investigation still lies in this direction. Owing to the poisonous nature of colchicine, a search for other and more suitable anti-mitotic agents for animals is necessary.

Different methods have been prescribed for the suppression of the yolky material which interferes with staining (Roberts, 1967). Cather (1959) suggested staining in diluted Gomori's chrome alum-haematoxylin following hydrolysis at room temperature in HCl, in marine gastropods. In a method adopted for American Easter oyster, fixation in Carnoy's fluid (3:1) was followed by extraction for 2 h in a microsoxhlet apparatus with chloroform–methyl alcohol (1:1) and squashing in orcein, A thimble with a glass end, pore size 40 µm, is used as a receptacle for the eggs in the soxhlet apparatus (Longwell and Stiles, 1968). Menzel and Menzel (1965) suggested the use of phase contrast studies in addition to staining for egg cell chromosomes in clams.

The study of both mitosis and meiosis from mammalian tissues, including those of man, deserves independent treatment, due to the vast amount of progress in this field within the last decade. These methods have been discussed in detail in a separate chapter.

The mitotic division of animal material, other than mammalian and human, can be studied from larval tails, ganglia, spermatogonial cells, etc., but before squashing in acetic-carmine solution it is always necessary to fix the tissue, preferably in a fluid having a strong penetrating capacity such as acetic–methanol mixture. The tissue must also be teased out on the slide as a prerequisite for squashing.

Chromosomes of mosquito, ants, Drosophila, larval ganglion cells, etc. have also been studied through modifications of the squash technique (Lewis and Riles, 1960; French *et al.*, 1962; Imai, 1966).

On the other hand, for mitotic divisions of mammalian tissues, such as bone marrow, endometrium, cervical tissue, cornea, etc., certain specialised methods are generally adopted. Marrows are generally extracted with a hypodermic syringe by puncturing the sternum (Ford and Hamerton, 1956) and aspirated continuously in sodium citrate solution to cause swelling and to prepare a fine suspension before fixing. In the case of solid tissue, homogenisation in sodium citrate solution is needed, followed by incubation at a slightly warm temperature for a few minutes (Smith, 1943; Manna, 1956) to break and soften the tissue. Nuclei in a mass both from bone marrows and solid tissues are then obtained by centrifuging. For soft endometrial tissue, or even marrows, mere treatment in coumarin for a few minutes, hypotonic salt solution or even water may result in swelling of the cells and scattering of chromosomes.

Details of the schedules adopted for mammalian chromosomes, including human, are discussed in a separate chapter.

Whole Mount Technique tor Chromosome Study

It is difficult to embed and section small organisms for the study of their chromosomes, so a method was originally devised by Schmuck and Metz (1931) and later modified by Whiting (1950), in which small organisms such as small nematodes, insect eggs and embryos, and insect ovaries, can be mounted in entirety and fixed and stained within a short period. The material is fixed in shell vials; if it settles immediately, fluids are poured over it and removed by pipettes while floating materials are kept wrapped in lens paper sacks till ready for mounting. Invertebrate material is fixed in Kahle's fixative (distilled water, 30; rectified spirit, 15; formalin, 6; glacial acetic acid, 1) for 30 min to 24 h. Insect eggs are usually punctured. The material is rinsed in 2 changes of distilled water for 1 h each and then hydrolysed in N HCl at room temperature for 10 min and then at 60 °C for 10 min, followed by staining for 2 h in leuco-basic fuchsin solution. After a dip in SO_2 water it is washed in two changes of distilled water for 20 min, then run through two changes of triethylphosphate and one of triethylphosphate and xylol mixture (1:1), keeping 15 min in each. It is then transferred to xylol and mounted in balsam.

Squash techniques have been applied to whole embryos of animals by certain workers. For fish embryo, Simon (1964) suggested fixation immediately after hatching in acetic–ethanol (1:3) in the cold for 24 h. The blastodiscs are removed, treated in acetone for 5 min, rinsed in two changes of acetic–ethanol (1:3), stained for 10–15 min in propionic acid: 2 per cent acetic–orcein mixture (1:9) and squashed. In Atlantic salmon, embryos were fixed in 45 per cent acetic acid and stained in 4–6 per cent acetic-carmine at 60 °C for 15–20 min (Boothroyd, 1959). The squash method was not very efficient for studies on implanted embryos of laboratory mammals (Bomsel-Helmreich and Thibault, 1962; Hungerford, 1958).

The air drying method, described in Chapter 15, as utilised by Tarkowski (1966) for pre-implantation stages of mouse eggs, was not successful for implanted embryos. The technique was used by Butcher and Fugo (1967) on 11·5 day-old rat embryos. A fairly large initial amount of material was required, due to loss during trypsinisation, homogenisation or centrifugation. A further modification, developed by Wroblewska and Dyban (1969), described in the chapter on mammalian chromosomes, utilises dissociation after fixation, followed by air-drying.

For studying the chromosomes of adult trematodes, the entire animal is fixed in 40 per cent acetic acid or acetic–orcein for 24 h, transferred to either Gilson's fluid for sections, or to acetic–ethanol (1:3) for squashes, and hydrolysed in 10 per cent perchloric acid at 25 °C for 12 h or N HCl at 60 °C for 10 min prior to staining in Feulgen (Bergan, 1955). Avian microchromosomes can be studied following the squash schedule as well (Krishan, 1962).

There are special methods of making permanent slides from squash preparations; these are discussed in Chapter 6.

Modifications have been developed for both smear and squash techniques to obtain better preparation, better preservation and easier schedules. Preservation of plant tissues by storage at or below $-10\,°C$ after fixation in acetic–ethanol (1:3) and chloroform–ethanol–acetic acid (4:3:1) mixture gives very good staining with carmine squash even up to 6 months (Davies, 1952). For mounting histological materials, squashing the stained root tip in a drop of stain between two pieces of plastic, and lamination in a electric press eliminates dehydration (La Croix and Press, 1960). Glass cover slips can be replaced in squashes by cellophane by placing the specimen on an albumin-coated slide, and squashing under a wet square of cellophane. The slide is exposed to formalin vapour for 45 min before stripping off cellophane and staining (Murin, 1960). In another method, the coverslip is directly glued on a slide by a rubber solution or 10 per cent solution of polyvinyl acetate in acetone–ethanol (1:1) mixture. After smearing, fixation and staining on the cover slip, it is detached from the slide by a suitable solvent (Östergren, 1963). Smears can be made on 35 mm photographic film, instead of a glass slide, and fixed and stained as usual by handling on a photographic film developing reel. The slides are sprayed with a plastic cement and examined, using a special holder to keep the film base flat (Sommer and Pickett, 1961).

REFERENCES

Albrecht, M. M. (1954). *Stain Tech.* **29,** 89
Alcorn, S. M. and Ark, P. A. (1953). *Stain Tech.* **28,** 55
Apàthy, A. V. (1912). *Z. wiss. Mikr.* **29,** 449
Baird, T. T. (1936). *Stain Tech.* **11,** 13
Baker, J. R. (1933). *Cytological technique.* 1st edn. London; Methuen
Baker, J. R. (1943). *Stain Tech.* **18,** 113
Banny, T. M. and Clark, G. (1949). *Stain Tech.* **24,** 223
Barron, D. H. (1934). *Anat. Rec.* **59,** 1
Belling, J. (1926). *Biol. Bull.* **50,** 160
Bergan, P. (1955). *Stain Tech.* **30,** 305
Bomsel-Helmreich, O. and Thibault, C. (1962). *Ann. Biol. Anim. Bioch. Biophys.* **2,** 265
Boothroyd, E. R. (1959). *Canad. J. Genet. Cytol.* **1,** 161
Bowene, (1882). In *The Microtomist's Vade-mecum,* ed. Lee. London; Churchill
Bradbury, Q. C. (1931). *Science* **74,** 225
Butcher, R. L. and Fugo, N. W. (1967). *Fertil. Steril.* **18,** 297
Cather, J. N. (1959). *Stain Tech.* **34,** 146
Chayen, J. and Miles, U. J. (1953). *Stain Tech.* **29,** 33
Collins, E. M. (1969). *Stain Tech.* **44,** 33
Cove, H. M. and Schoenfle, A. (1946). *Amer. J. Clin. Path. Sect.* **10,** 31
Darlington, C. D. and La Cour, L. F. (1960). *The Handling of chromosomes.* London; Allen and Unwin
Davies, E. (1952). *Nature* **169,** 714
Day, P. R., Boone, D. M. and Keith, G. W. (1956). *Am. J. Bot.* **43,** 835
Defrenoy, J. (1935). *Science* **82,** 335
Doxtader, E. (1948). *Stain Tech.* **23,** 1
Duval, M. (1879). *C. R. Soc. Biol. Paris* **1,** 35
Elliott, C. G. (1956). *Symp. Soc. Gen. Microbiol.* 279
Emsweller, S. E. and Stuart, N. W. (1944). *Stain Tech.* **19,** 109

Fabergé, A. C. (1945).   *Stain Tech.* **20,** 1
Firminger, H. I. (1950).   *Stain Tech.* **25,** 121
Ford, C. E. and Hamerton, J. L. (1956).   *Stain Tech.* **31,** 247
Foster, A. S. and Gifford, E. M. Jr. (1947).   *Stain Tech.* **29,** 129
French, W. L., Baker, R. H. and Kitzmiller, J. B. (1962).   *Mosquito news* **22,** 377
Frost, H. F. (1935).   *Watson's micr. Rec.* **34,** 19
Gifford, E. M. Jr. (1950).   *Stain Tech.* **25,** 161
Giovacchini, R. P. (1958).   *Stain Tech.* **33,** 247
Graupner, V. H. and Weisberger, A. (1933).   *Zool. Anz.* **102,** 39
Hale, A. J. (1952).   *Stain Tech.* **27,** 189
Hance, R. T. (1933).   *Science* **77,** 353
Harris, B. J. and Blackman, G. E. (1954).   *Nature* **173,** 642
Hauser, H. (1953).   *Microscope* **9,** 207
Hollande, A. C. (1918).   In *The Microtomist's Vade-mecum,* ed. Lee, B. London; Churchill
Hungerford, D. (1958).   *J. Morph.* **97,** 497
Imai, H. T. (1966).   *Acta Hymenopterologica* **2,** 119
Johansen, D. A. (1940).   *Plant microtechnique.* New York; McGraw-Hill
Klebs. (1869).   *Arch. mikr-Anat.* **5,** 164
Krishan, A. (1962).   *Stain Tech.* **37,** 335
Kuhn, G. D. and Lutz, E. L. (1958).   *Stain Tech.* **33,** 1
La Cour, L. F. (1937).   *Bot. Rev.* **5,** 241
La Cour, L. F. (1941).   *Stain Tech.* **16,** 169
La Croix, J. D. and Press, S. K. F. (1960).   *Stain Tech.* **35,** 331
Lewis, E. B. and Riles, L. S. (1960).   *Drosophila Inform Serv.* **34,** 118
Lewis, L. W. (1945).   *Stain Tech.* **20,** 138
Longwell, A. C. and Stiles, S. S. (1968).   *Stain Tech.* **43,** 63
McIntosh, D. L. (1954).   *Stain Tech.* **29,** 29
McKay, H. H. and Clarke, A. E. (1946).   *Stain Tech.* **21,** 111
Madge, M. (1936).   *Ann. Bot. Lond.* **50,** 677
Maheshwari, P. (1939).   *Cytologia* **10,** 257
Manna, G. K. (1956).   *Stain Tech.* **31,** 45
Margolena, L. A. (1932).   *Stain Tech.* **7,** 9
Markarian, D. (1957).   *Stain Tech.* **32,** 147
Mayer, P. (1883).   *Mitt. Zool. Sta. Neapel* **4,** 521
Melnyk, J. (1961).   *Stain Tech.* **36,** 202
Melton, H. D. (1956).   *Stain Tech.* **31,** 96
Menzel, R. W. and Menzel, M. Y. (1965).   *Biol. Bull.* **129,** 181
Morris, J. E. (1965).   *Stain Tech.* **40,** 215
Murin, A. (1960).   *Stain Tech.* **35,** 351
Östergren, G. (1963).   *Hereditas* **50,** 414
Palmgren, A. (1954).   *Nature* **174,** 46
Pauly, J. E. (1956).   *Stain Tech.* **31,** 35
Persidsky, M. D. (1953).   *J. Lab. Clin. Med.* **42,** 468
Randolph, L. F. (1935).   *Stain Tech.* **10,** 395
Raspail, S. V. (1825).   *Ann. Sci. Nat.* **6,** 224
Rawlins, T. E. and Takahashi, W. N. (1947).   *Stain Tech.* **22,** 99
Reeve, R. H. (1954).   *Stain Tech.* **29,** 81
Reiner, L. (1953).   *Lab. Invest.* **2,** 336
Roberts, F. L. (1967).   *Prog. Fish-Culturist* **29,** 75
Romeis, B. (1928).   *Tashenbuch der Mikroskopischen Technik.* München; Oldenbourg
Schmuck, M. L. and Metz, C. W. (1931).   *Science* **74,** 600
Setterfield, G., Schreiber, R. and Woodward, J. (1954).   *Stain Tech.* **29,** 113
Sharma, A. K. (1951).   *Nature* **167,** 441
Sharma, A. K. (1956).   *Bot. Rev.* **22,** 665
Sharma, A. K. and Bhattacharjee, D. (1952).   *Stain Tech.* **22,** 20
Sharma, A. K. and Chatterjee, T. (1971).   *Biologia* **26,** 309
Simon, R. C. (1964).   *Stain Tech.* **39,** 45
Singleton, J. R. (1953).   *Amer. J. Bot.* **40,** 124
Smith, S. G. (1943).   *Canad. Ent.* **75,** 21

Sommer, J. R. and Pickett, J. P. (1961).   *Arch. Path.* **71,** 669
Stiles, C. (1934).   In *The Microtomist's Vade-mecum,* ed. Lee, B. London; Churchill
Stilling, J. (1842).   In *The Microtomist's Vade-mecum,* ed. Lee. B. London; Churchill
Stowell, R. E. (1951).   *Stain Tech.* **26,** 105
Tandler, C. J. (1959).   *Stain Tech.* **34,** 234
Tarkhan, A. A. (1931).   *J. R. micr. Soc.* **51,** 387
Tarkowski, A. K. (1966).   *Cytogenetics* **5,** 394
Thorp, R. H. (1936).   *Watson's micr. Rec.* **38,** 22
Upcott, M. B. and La Cour, L. F. (1936).   *J. Genet.* **33,** 352
Wachstein, M. and Meisel, E. (1953).   *Stain Tech.* **28,** 135
Waddington, C. H. and Kriebel, T. (1935).   *Nature* **136,** 685
Walsh, M. P. and Archambault, W. V. (1954).   *Stain Tech.* **29,** 69
Warmke, H. R. (1935).   *Stain Tech.* **10,** 101
Weaver, H. L. (1955).   *Stain Tech.* **30,** 63
Wachtel, A. E., Grettner, N. E. and Ornstein, L. (1966).   In *Physical techniques in biological research* **3,**
    New York; Academic Press
Wetzel, G. (1931).   *Z. wiss. Micr.* **48,** 360
White, R. T. and Allen, R. A. (1951).   *Stain Tech.* **26,** 137
Whiting, A. R. (1950).   *Stain Tech.* **25,** 21
Woods, P. S. and Pollister, A. W. (1955).   *Stain Tech.* **30,** 123
Wroblewska, J. and Dyban, A. P. (1969).   *Stain Tech.* **44,** 147

5 Staining

The structure and behaviour of chromosomes can be studied only after they are made visible under the microscope. In order to keep up normal activity at the time of observation it is best to mount the dividing cells in the body fluid of the organism and to observe chromosome movements under a phase contrast microscope. The evaporation of the fluid can be prevented by paraffin oil, which has the added advantage of being oxygen-solvent and non-toxic to the tissue (Ris, 1943). In addition to body fluid, which is not always easy to secure, physiological solutions have often been applied for the same purpose with success (Belar, 1929). Duryee (1937) used calcium ions in the medium which improved the visibility of chromosomes. A phase microscope is needed because the refractive indices of the resting nuclei and dividing chromosomes differ so little that it is difficult to distinguish them under an ordinary lens (for details *see* 'Phase contrast and interference microscopy', p. 277).

No doubt the movement of chromosomes, as well as the structures related to them, such as the spindle, can best be studied through phase contrast lenses, but phase microscopy has not yet attained that stage of refinement required in cytogenetic and cytochemical analysis which would allow complete visibility of all the chromosome segments in detail. Even today, therefore, stained preparations, in spite of their limitations, have a decided advantage over unstained ones, especially where chromosome studies are concerned.

Staining, as performed on chromosomes, can be classified as 'vital' and 'non-vital'.

In the case of vital staining, non-toxic dyes are applied to the living tissue so that the latter can be studied without being killed. If the stain is applied to isolated cells, such as blood, bone marrow, etc., the staining is called supravital. Of the different vital stains so far applied to the study of chromosomes, methylene blue is the only stain which has been found to be effective in demonstrating cell division in tissue culture. It is a basic dye of the thiazine group, $C_{16}H_{18}N_3SCl$, and is soluble in water.

$$\left[\text{(CH}_3)_2\text{N} \cdots \text{N(CH}_3)_2 \right]^+ \text{Cl}^-$$

Ludford (1935) observed the advantages of methylene blue as a vital stain in that, in tissues, dividing cells do not stain as intensely as the metabolic cells. Darlington and La Cour (1960), however, stated that even with this staining there is every possibility of the chromosomes undergoing a certain amount of injury. Ludford (1936) observed that spindle formation can be prevented by auramine. Mollendorff (1936) noted that neutral red interfered with mitosis. Brilliant cresyl blue can also cause chromosome breakage (D'Amato, 1951). All this evidence indicates that although vital staining is essential for the study of mitochondria, golgi bodies, etc., it cannot be applied effectively in the study of chromosome structure.

In non-vital staining, the coloration of the chromosomes in the killed tissue is caused by certain chemical agents which are insoluble in the chromosome substance. The principal kinds of dyes that are used to stain chromosomes are synthetic organic dyes, derived from coal tar. The colour of a dye is due to certain chemical configurations, known as *chromophores*; similarly, the tissue must retain the colour which is due to certain chemical configurations in the dye molecule itself, known as the *auxochromes*.

The best example of a chromophoric group is the quinonoid ring. The coloration due to quinonoid arrangement is exemplified by the conversion of colourless hydroquinone to yellow quinone. Auxochromes, responsible for the adherence of the dye to the tissue, are mostly —NH$_2$ and —OH groups which convert the non-dyeing coloured substance into a form which undergoes electrolytic dissociation in water and is capable of forming salts with acids or bases (Baker, 1950).

Hydroquinone (colourless)   Quinone (yellow)

The mere presence of auxochromes cannot confer colour: for example, aniline, having the chemical structure:

is colourless, but if aniline is mixed with *p*-toluidine in the presence of mercuric chloride, basic fuchsin can be obtained. It possesses the quinonoid ring, responsible for its own colour, and also has auxochromes (NH$_2$) which allow it to be retained in the tissue.

Basic fuchsin
(*p*-rosaniline chloride)

The dyes are generally termed basic or acidic on the basis of their chemical nature and behaviour. In an *acidic* or *anionic* dye, the balance of the charge on the dye ion is negative. In a *basic* or *cationic* dye, however, the dye ion charge is positive. Most of the acid dyes are prepared as metallic salts and are generally neutral or slightly alkaline in reaction, but they react with, and stain substances with, a basic reaction. A basic dye, on the other hand, is manufactured as a salt of mineral or aliphatic organic acids, and it stains substances which are acidic. Several of the dyes used in the study of the chromosomes are amphoteric, such as orcein; they behave both as acids and bases. The majority of the chromosome dyes are triphenyl methane or aniline derivatives, though other dyes have also been found to stain chromosomes (cf. Kasten, 1967).

In order to stain the chromosomes at specific loci, the general procedure is to over-stain it, followed by the removal of the excess stain—a process called 'differentiation'. Staining can also be performed progressively by gradually increasing the intensity of the colour. The process of differentiation allows the stain to adhere to specific sites of the chromosomes, as often basic dyes stain the cytoplasm as well. Similarly, acidic dyes may stain chromatin by proper differentiation, where quite likely the dye reacts with the protein moiety of the chromosome.

Most of the acid dyes are salts of potassium or sodium, whereas basic dyes are mostly available as chlorides or sulphates. For chromosome staining, basic dyes are applied, chromatin being strongly acidic. Acid dyes colour the cytoplasm which is predominantly basic. The terms basophilic and acidophilic are based on the affinity for basic or acidic dyes, chromosomes being basophilic and cytoplasm acidophilic.

Staining has been claimed to be principally a process of adsorption (*see* Baker, 1950). Adsorption has been defined as 'a process by which a substance accumulates at a boundary surface of two contiguous phases in a concentration higher than that in which it exists in the interior of these two phases' (Michaelis, 1926). It is better to consider, as Baker has pointed out, that staining may simultaneously be a chemical reaction and physical adsorption.

Though chromosome staining is the product of a chemical reaction, yet the intensity of the stain fades with age. The fading may be attributed to the effect of ultra-violet light through continued exposure to daylight, to progressive acidity of the mounting medium or to retention of contamination of elements during the process of staining.

The adherence of a dye to the tissue may also be accelerated through the process of mordanting. Though the term 'mordant' is loosely used, it should only be applied to the salts of di- and tri-valent metals (Lison and Fautrez, 1939). These metallic hydroxides form compounds with the dye which attach the dye to the tissue and are called the 'lake' for the particular dye. The *mordant* is the term applied to the salt used.

In cases where the 'lake' formed is insoluble in water, the general practice is to dip the tissue in the mordant first, followed by staining. In carmine staining, iron salts are often used as mordants and are added to the stain itself. Baker has dealt with the way in which the mordant forms a double compound, one with the dye and the other with the recipient tissue. Alizarin, a dye having more or less a structure similar to carmine, combines with aluminium hydroxide mordant in the following manner:

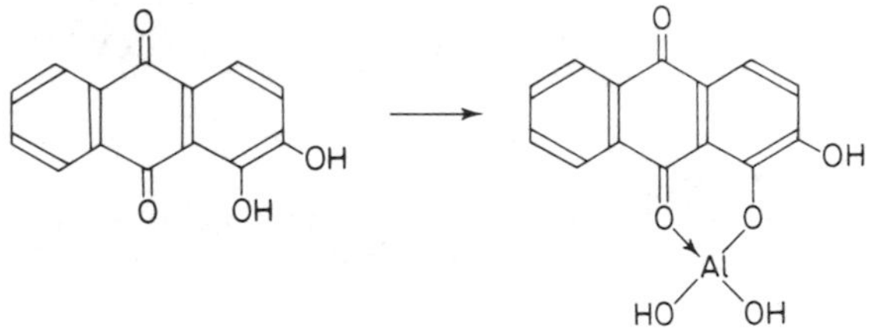

$Al(OH)_2$ combines here on the one hand with the dye alizarin and on the other hand with the tissue. The process of combination, as represented between the dye and the mordant, is known as 'chelation', due to the resemblance to the two claws of a crab. —OH groups of $Al(OH)_2$ are free to react with the tissue.

In cytological studies, whenever the dye or the mordant is used separately, the purpose of the mordant is either to modify the isoelectric point of the tissue or to form a chemical link between the stain and the chromosome. In principle, it changes the surface conditions of the fixed chromosomes.

Several types of mordants are used in cytology, though according to Baker, as mentioned above, the term should be rather limited. Chromium trioxide, iodine in alcohol, iron alum (ferric ammonium sulphate), ammonium molybdate and picric acid are the more commonly used mordants. In cases where mordants are used prior to staining, they evidently modify the chromosome surface in such a manner that the dye adheres strongly to the chromosomes. Chromium trioxide mordanting before crystal violet staining and iron alum mordanting before haematoxylin staining, are applied for this purpose. Mordanting in iodine–alcohol, picric acid and even iron alum solution is used after staining. Post-mordanting not only helps to retain the stain for a prolonged period, but also clears the cytoplasm. The latter effect is caused by removing the undesirable stain at non-specific sites from the cell. This effect is due to the acidity of the mordant, such as iodine in ethyl alcohol, which, being higher than the cytoplasm, removes the stain from its surface. Chromatin, on the other hand, having a stronger acidity, retains the stain. Iron alum, as mentioned before, is applied both before and after staining. During pre-mordanting or mordanting

prior to staining, the excess alum, which does not react with the chromosome, remains in the cytoplasm. This extra alum, during staining, forms a
dye–alum compound not bound to the tissue. Post-mordanting in iron
alum removes this component. Chromium trioxide is also used after staining as a post-mordant and helps to retain the stain for a prolonged period;
however, the cytoplasm takes up a yellow colour, which makes post-
mordanting in chromium trioxide undesirable. The purpose of post-
mordanting is more or less to bleach out the undesirable elements. An
oxidising post-mordant oxidises the dye, present at certain sites, to a colourless substance.

Certain authors (Darlington and La Cour, 1960) consider even acid
hydrolysis needed for Feulgen staining as mordanting, but in our opinion
the term should be used more critically. Hydrolysis in normal HCl helps
to liberate aldehyde groups (*see* 'Feulgen staining', page 101) without the
acid itself combining with the dye or the tissue. The application of the term
as such is undesirable.

The term 'mordant' should preferably be restricted to those agents
which are applied before staining and which form a complex with the dye
or the tissue. Agents, when applied after staining, act more as differentiating
chemicals than as mordants, therefore this epithet is not suitable for iron
alum or iodine in ethyl alcohol, applied after staining. Unless the term
'mordant' is used strictly in the sense already denoted, it may merge with
other chemicals like clove oil, etc., which are used solely for differentiation.

OUTLINE OF DIFFERENT TYPES OF STAINING PROCEDURES AND PRINCIPLES

Fuchsin

Of all the different staining methods employed for the study of chromosomes, the Feulgen reaction is considered to be the most effective with
regard to chromosome staining, the chemistry having been well worked
out by different authors (Feulgen and Rossenbeck, 1924; Hillary, 1939;
Gulick, 1941; Li and Stacey, 1949; Overend and Stacey, 1949; Overend,
1950; Lessler, 1951, 1953; Kurnick, 1955; Swift, 1955) and considerable
knowledge, even of the factors controlling the process, has been gained.
Meischer first isolated the nuclear material 'nuclein' from pus cells in 1869,
but the cytological and chemical demonstration of its acidic component,
nucleic acid, was not possible for a long time. In 1924, Feulgen and Rossenbeck devised a method based on the Schiff's reaction for aldehydes which
stains the nucleic acid of the chromosomes specifically and, as such, has
been effectively employed for the visualisation of chromosomes.

Feulgen solution or, more precisely, fuchsin sulphurous acid is prepared
from the dye, basic fuchsin, and its preparation involves several steps. Its
composition and properties are now outlined.

Basic fuchsin—This dye belongs to the triphenyl methane series and is
magenta red in colour. It can be derived, as mentioned before, by combining
a few cubic centimetres of aniline with *p*-toluidine in the presence of

mercuric chloride. The commercially obtained 'basic fuchsin' is a mixture of three compounds, namely p-rosaniline chloride (Michrome No. 722), basic magenta (rosaniline chloride, Michrome No. 623), and new magenta (new fuchsin, Michrome No. 624) (Conn, 1953; Gurr, 1960).

The molecular weights of these compounds are 328·815 for p-rosaniline chloride ($C_{19}H_{18}N_3Cl$), 227·841 for basic magenta ($C_{20}H_{20}N_3Cl$) and 365·893 for new magenta ($C_{22}H_{24}N_3Cl$). It is evident that all these three compounds are characterised by quinonoid arrangements within the molecule; in fact, the chief constituent of basic fuchsin is p-rosaniline or triaminotriphenyl methane chloride; the quinonoid structure makes the dye unstable and it undergoes oxidation easily. All these different compounds can be separately obtained as chlorides or acetates, though in chromosome

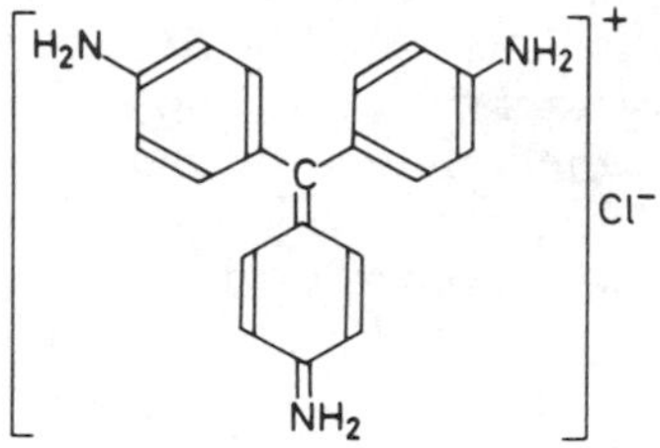

p—Rosaniline chloride

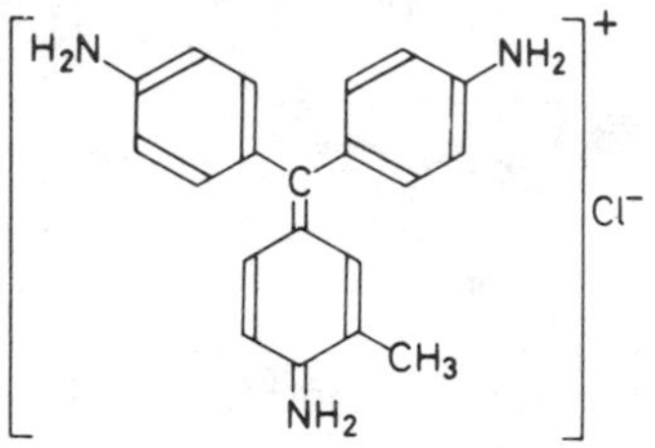

Rosaniline chloride

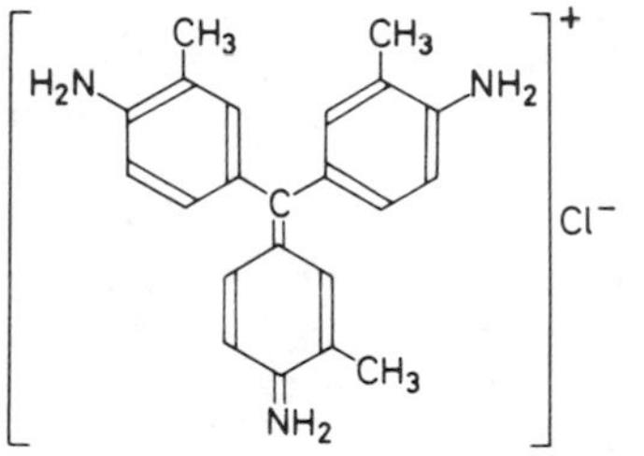

New Magenta

studies the demand for basic fuchsin is much more than any of these compounds. For critical work, however, p-rosaniline chloride is occasionally used instead of basic fuchsin. Increasing methylation adds to the bluish shade in the colouring of the dye, and therefore p-rosaniline chloride with no methyl group has less colour than basic magenta or new magenta. Complete replacement of the hydrogen of amino groups by methyl groups results in violet coloration of the dye. Fuchsin base is the colourless carbinol base of the dye.

Basic fuchsin is easily soluble in water and alcohol, and for the preparation of Feulgen reagent, 0·5 per cent solution is prepared in boiling distilled water.

Preparation of Feulgen solution—The principle of preparing the Feulgen or Schiff's reagent is to treat the basic fuchsin solution with sulphurous acid, the product obtained being colourless fuchsin sulphurous acid. This reagent is the Schiff's reagent, utilised by Feulgen and Rossenbeck (1924) for the demonstration of the DNA component of chromosomes. The method of preparation has been modified by different workers (Coleman, 1938; Mallory, 1938; Tobie, 1942; Rafalko, 1946; Ely and Ross, 1949; Lillie, 1951; Atkinson, 1952; Longley, 1952; Casselman, 1959, etc.). The procedure involves, in short, the preparation of a basic fuchsin solution in warm water, followed by cooling at a particular temperature and the subsequent addition of hydrochloric acid and potassium metabisulphite, needed for the liberation of SO_2, prior to storage in a sealed container in a cool, dark place. The last two steps can be replaced by passing SO_2 water (Tobie, 1942; Rafalko, 1946), which gives not only a colourless solution but a very sensitive reagent as well. The addition of activated charcoal, as suggested by Coleman (1938), Longley (1952) etc., removes the yellowish impurities and, as such, a transparent colourless reagent can be obtained. However, as activated charcoal often removes a small amount of acid as well, it should be added in strictly limited quantity. The colour of Schiff's reagent varies, depending upon the type of dye used, the hydrogen ion concentration and SO_2 content; but without using activated charcoal, the solution should be straw-coloured. The reagent is unstable, as it loses SO_2 on continued exposure to air and becomes coloured again. This coloured product is, however, different from the product obtained on reaction with aldehydes (Shriner and Fuson, 1948). The colour is, therefore, non-specific. Though Atkinson (1952) noted that the particular reducing agent used in preparing Schiff's reagent has a marked effect on the potency of the solution, Casselman's (1959) re-investigation did not reveal any such effect whatsoever. For this reason it should be kept in a sealed container wrapped in black paper away from light. In order to avoid any contamination, used Schiff's reagent should not be mixed with the stock solution.

The preparation of the reagent is one of the most critical steps in the execution of Feulgen reaction. It must be carried out under strictly controlled conditions and in addition to the factors mentioned above, the temperature in the different steps of the reaction, as well as during the period of bleaching, also requires vigilance.

The principle underlying the preparation of Schiff's reagent from *p*-rosaniline is that, during this preparation, the basic fuchsin solution, or more precisely, *p*-rosaniline chloride solution, undergoes conversion to leucosulphinic acid, which is colourless. This conversion is caused by the addition of sulphurous acid across the quinonoid nucleus of the dye. Sulphurous acid is obtained through the action of HCl on potassium

p-Rosaniline chloride

N,N Sulphinic acid derivative of p-rosaniline

mctabisulphite. The excess of SO_2 undergoes reaction with leucosulphinic acid to produce bis-N-aminosulphinic acid, popularly known as Schiff's reagent.

Procedure for Feulgen reaction—In order to carry out the test, the procedure in outline involves hydrolysis of the fixed tissue in normal HCl at 56–60 °C, for a period varying from 4 to 20 min before immersing the material in Schiff's reagent. The colour develops within a short time and the chromosomes take up a magenta colour and can be observed after mounting in 45 per cent acetic acid. If the test is performed strictly according to the recommended procedure, the chromosomes appear to be specifically coloured against a clear cytoplasmic background. Prolonged keeping in Schiff's reagent is undesirable, as further hydrolysis may take place due to the reagent being an acid (Serra, 1948); moreover, a rinse in sulphite solution or SO_2 water is often helpful in order to remove excess of colour, if any, in the cytoplasm.

The chemical basis of the reaction, as outlined by Feulgen (1926), Levene and Bass (1931), Gulick (1941), etc., includes two principal steps:

1. By hydrolysis with normal HCl, the purine-containing fraction of deoxyribonucleic acid (DNA) is separated from the sugar, unmasking the aldehyde groups of the latter.

2. The reactive aldehyde groups then enter into combination with fuchsin sulphurous acid to yield the typical magenta colour. Feulgen reaction is, therefore, based essentially on the Schiff's reaction for aldehydes. After removal of the base, carbon atom 1 of the furanose sugar is so arranged as to form a potential aldehyde, capable of reacting with fuchsin sulphurous acid. The ribose sugar, with an —OH in place of —H at carbon 2, is not hydrolysed by N.HCl and so does not react with fuchsin sulphurous acid. In the pyrimidine–sugar linkage, on dissociation, the aldehyde groups are not free to react, unlike the open and reactive aldehydes obtained after breakdown of the purine–sugar linkage.

2–deoxyribofuranose
–3–phosphate

D–ribose

Kissane and Robins (1958), through fluorescent assay, suggested a fluorescent-quinaldine reaction in Feulgen hydrolysis which, however, has not been confirmed (Stoward, 1963). The reaction of the aldehyde with bis-N-aminosulphinic acid is not yet fully disclosed. Wieland and Scheuing (1921) suggested that two molecules of aldehydes react to yield a reddish purple complex. Following the addition of these two molecules, the addition complex undergoes a molecular rearrangement to form a complex having quinonoid arrangement. The reaction, as outlined by Wieland and Scheuing, is shown on p. 102.

If this interpretation of the reaction of two aldehyde molecules with one of leucofuchsin is accepted, then it implies that two nucleotides enter into combination with one leucofuchsin molecule.

The coloured aldehyde-addition complex, formed as a result of Feulgen reaction, is entirely different from the original basic fuchsin, and therefore the term 'restoration of colour' should never be used. Stowell and Albers (1943) demonstrated that the absorption spectra of the original basic fuchsin and the aldehyde addition complex are different.

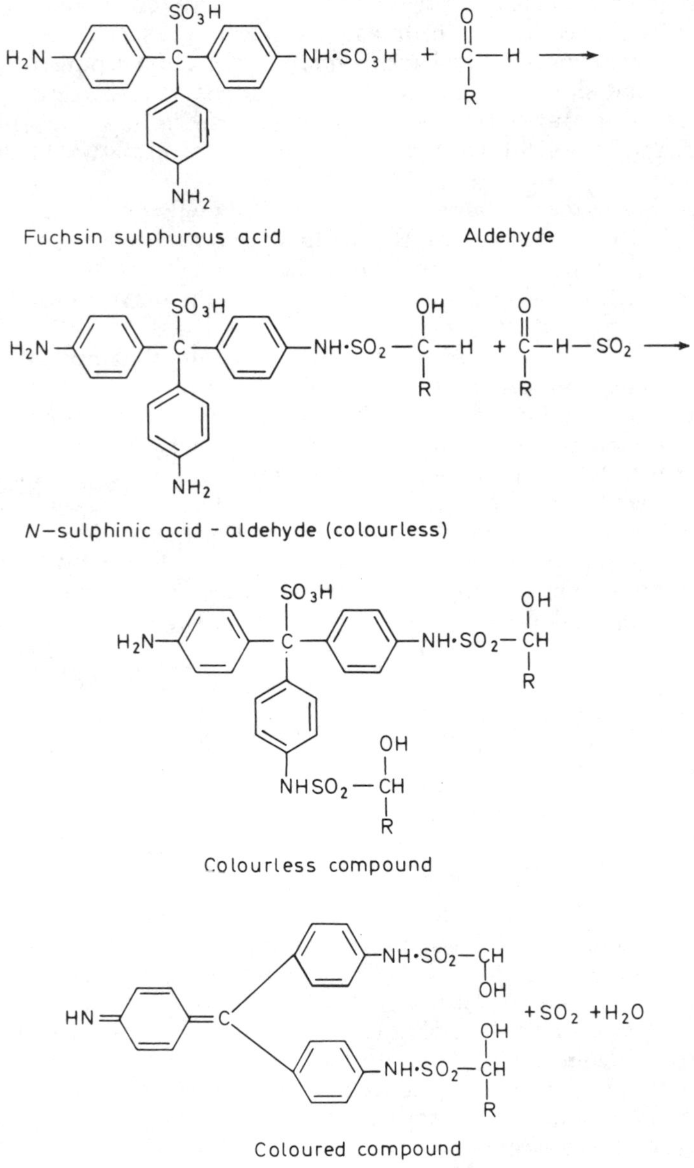

Fuchsin sulphurous acid   Aldehyde

N—sulphinic acid - aldehyde (colourless)

Colourless compound

Coloured compound

Lessler (1953) suggested, on the basis of the original concept of alternate arrangement of purines and pyrimidines in the nucleic acid molecule, that the aldehydes reacting with the Schiff's reagent must be present in alternate nucleotides, but the demonstration of the two-strand nature of DNA (Watson and Crick, 1953) reveals that each nucleotide pair contains one purine and one pyrimidine group. Evidently, two purines are available in adjacent pairs, though located on complementary strands and therefore the two aldehyde groups, as visualised by Wieland and Scheuing, are available from two adjacent strands to react with one molecule of leucofuchsin. This interpretation is commonly accepted (Kasten, 1960).

Rumpf (1935) and Hörmann *et al.* (1958) suggested an entirely different structure for Schiff's reagent, which, according to them, fits in well with the reaction for aldehydes. According to them, instead of sulphinic acid, alkyl sulphonic acid is formed.

Barka and Ornstein (1960) and Hiraoka (1960) criticised Wieland and Scheuing's hypothesis. Sloane–Stanley and Bowler (1962) suggested that in Schiff's reagent, for every molecule of fuchsin, there are three aldehydes, not two. Lacoste and Martell (1955) claimed that even two —H atoms of one amino group can be replaced by two methylene sulphonic acids. The view that alkylsulphonic acid ($NH.CHR.SO_3H$) is the product formed is also supported by Hardonk and van Duijn (1964). They observed that staining intensity was identical both with pararosaniline and Schiff's reagent, after rinsing with sulphite and then water, thus refuting the Wieland–Scheuing hypothesis. Stoward (1966), however, was of the opinion that these observations would even fit azomethine or *N*-sulphinic acid as the end product. He regarded the electron–deficient central methane carbon atom as the essential chromophore.

With regard to the unmasking of aldehyde groups responsible for the Feulgen reaction, it was claimed by certain authors (Li and Stacey, 1949; Overend and Stacey, 1949) that all the purine-containing fractions

need not necessarily be separated from the sugar to yield the magenta colour. They showed that if sperm deoxyribonucleic acid is depolymerised by dialysis of a solution of its sodium salt against tap water, a white crystalline non-fibrous powder can be obtained which gives intense Feulgen reaction while retaining its original purine contents. Apparently, therefore, the breakdown of all purine–sugar linkage is not an essential step in executing the Feulgen reaction, and only a few of the aldehyde groups are necessary for the colour reaction, and the maximum intensity of colour can be obtained with the optimum period of hydrolysis. The suggestion is therefore that the breakdown of nucleic acid components takes place in two distinct steps. At the initial step, the sugar linkages engaged in polymeric bonding are attacked, and this is followed by a second step involving rupture of the glycosidic linkage between sugar and bases. The deoxyribose components revealed thereby still remain attached through phosphate linkages at C_3 and C_5 positions of the main nucleic acid chain; thus they are held strongly in the furanose form, which immediately undergoes transformation into the aldehyde form. This revealed that the aldehyde group undergoes Feulgen reaction as suggested by Wieland and Scheuing to form nucleic acid–aldehyde groups: p-rosaniline–SO_2 dye.

If the two steps above are taken into consideration, Feulgen test under optimum conditions should be interpreted as the reaction of fuchsin sulphurous acid with aldehydes of the deoxyribose sugar liberated after the breakdown of polymeric and glycofuranosidic linkages by acid hydrolysis to yield the magenta coloured addition complex.

In addition to normal hydrochloric acid, several other acids have also been tried out by different workers to bring about hydrolysis. Citric acid was used by Widström (1928); perchloric acid by Di Stefano (1948, 1952); trichloracetic acid by Sharma (1951); phosphoric acid by Hashim (1953), and chromic, nitric and sulphuric acids and bromine by Barka (1956). Hot acids, which may sometimes cause removal of the tissue from the slide, have been substituted by Itikawa and Ogura (1954) with prolonged keeping in concentrated acid at room temperature, but hydrochloric acid has been seen to give the best possible results compared with others tried so far.

Attempts have been made to substitute Schiff's reagent with other aldehyde reagents (Pearse, 1951; Pearse, 1960, 1968). Casselman (1959) claimed that none of them gives any significant advantage over Schiff's reagent.

In order to demonstrate the Feulgen reaction, it is always necessary to keep controlled unhydrolysed sections, as pointed out by Bauer (1932, 1933). Free aldehydes may be present in the cytoplasm as lipids (Gerard, 1935; Gomori, 1942; Lessler, 1953, and also *see Int. Rev. Cytol*, 1961). Similarly, higher aliphatic aldehydes, released from acetal phosphatides, may be obtained in the cytoplasm (Cain, 1949; Hayes, 1949). Polysaccharides, after oxidation by chromic acid fixation, may release aldehydes to react with the Feulgen solution (Bauer, 1932, 1933). All these aldehydes are available without hydrolysis and can undergo colour reaction. Chayen and Norris (1953) suggested the possibility of false localisation of DNA by acid hydrolysis and the presence of cytoplasmic particles which may be digested by desoxyribonuclease.

Moreover, Lison (1932) claimed that Schiff's reaction may occur with aliphatic ketones, unsaturated compounds, amino-oxides and catalytic oxidisers in addition to aldehydes. Hydrolysis removes all these components together with aldehydes, thus allowing only specific nuclear reaction. In plants, lignin, suberin, etc., also undergo Feulgen reaction but adequate control for their tests should be maintained in parallel sets in cases where such materials are present. Cytoplasmic aldehydes are removed, not only through acid hydrolysis but also through the complicated process of dehydration in alcohol, embedding in paraffin, etc., involved in block preparation (Danielli, 1949).

Semmens (1940) criticised the aldehyde reaction in the Feulgen test and suggested that colour may be due to purine components of the nucleic acid. This was substantiated by the fact that if piperidine and pyridine are added to the Feulgen reagent, development of colour is observed. Barber and Price (1940) claimed that this development of colour, which is attributable to their basic properties, cannot be compared with the typical Feulgen reaction.

Factors controlling reaction and its intensity—The critical preparation of materials through Feulgen staining depends on several factors, especially hydrolysis, method of fixation and type of tissue used. In addition, the concentration of DNA is no doubt one of the principal factors.

Several authors, namely Bauer (1932), Hillary (1939), Di Stefano (1948) noted that the time required for hydrolysis depends on the type of fixation applied. With acetic–ethanol fixation, a very short period of hydrolysis is necessary, whereas with metallic fixatives, like chromic acid, a considerably longer period of hydrolysis may be required. Osmium- and platinum-fixed materials, if not properly bleached, do not yield a good colour reaction. As pointed out by Darlington and La Cour (1960) and also confirmed from this laboratory, fixation with formalin, which is often recommended, leads to cytoplasmic staining if formalin is not properly washed out. Sibatani and Fukuda (1953) observed the minimum loss of DNA in fixed tissues after formalin or ethanol–formalin fixation. Helly's fixative has been recommended by Murgatroyd (1968) for mouse pancreas. Swift (1950) noted that formaldehyde fixation yields more intense colour than acetic–alcohol fixed materials. Lower concentrations of formalin, on the other hand, result in brighter colour than that obtained with higher concentrations (*see* Sibatani and Fukuda, 1953). Post-mortem changes, prior to fixation, may affect the intensity of the colour (Lhotka and Davenport, 1951). Acetic–alcohol fixation yields the maximum intensity in the Feulgen reaction, as observed by the present authors in plant materials, though the formation of acetaldehyde due to fixation after-effect, which is removed with mild acid hydrolysis, is another limitation of this process. However, it has been observed that alcoholic fixative causes violet coloration whereas the colour that appears after fixation in a mixture containing chromium is red (Darlington and La Cour, 1960). Penetration of the fixative is also another important factor, as, in a block of tissue, Swift (1955) noted stronger reaction in the periphery as compared to that in the centre. No variation in intensity was observed in frozen–dried tissue preparations in different parts.

The development of the actual magenta colour may be related to the type of aldehyde undergoing reaction. Pearse (1960) suggested that, though colour itself indicates a positive reaction, yet departure from the normal colour may have interesting significance. Variation in colour in different types of aromatic and aliphatic aldehydes has been demonstrated by Gomori (1950) and Casselman (1959). Hydrogen ion concentration also plays an important role in this process (Dutt, 1968, 1971). Moreover, it has been shown that SO_2 content is an important factor in colour formation. On aldehydes combining with low SO_2–Schiff, the colour is reddish whereas with high SO_2–Schiff, it is bluish (Elftman, 1959; Barka and Ornstein, 1960).

The temperature is another factor to be taken into consideration, and results are more quickly obtained if the hydrolysis is carried out for a short period at a higher temperature; also the duration affects the specificity of the colour to a significant extent. With hydrolysis for a period shorter than the recommended one, the cytoplasm shows a diffuse colour, indicating that other aldehyde components of the cytoplasm have not been removed, due to inadequate acid treatment. On the other hand, if hydrolysis is carried out for a longer period than recommended, a similar diffuse colour in the cytoplasm is observed, cytoplasmic coloration in this case being attributed to the free depolymerised nucleic acid molecules lying in the cytoplasm on being detached from the nucleoprotein component of the chromosome. That longer hydrolysis in hydrochloric acid frees the chromosome of its nucleic acid component has been observed by Taylor, Woods and Hughes (1957) and others, and thus an optimum reaction can only be obtained under strictly controlled hydrolysis. Jordanov (1963) suggested hydrolysis with 5N HCl at room temperature. The importance of temperature in hydrolysis has also been stressed by other authors (Aldridge and Watson, 1963). The time of hydrolysis has been found to play a role in dye binding of normal and tumour cells, where Böhm and Sandritter (1966) visualised the presence of two types of DNA against one type in mouse liver as claimed by Kasten (1965). It may be mentioned that Lima-de-Faria (1962) suggested the presence of non-Feulgen-positive DNA in the cytoplasm of several species of amoeba, *Lilium* and *Gryllus*.

Different species as well as different tissues often require different periods of hydrolysis for the optimum reaction and development of the maximum intensity of colour. Nuclei of *Spirogyra* require a much longer period as compared to other plant cells (Hillary, 1939); sea urchin eggs require a shorter period than that necessary for most animal cells, and root tips of *Chrysanthemum* need an unusually long period for a proper development of their colour (Dowrick, 1952). It is easier to stain a testicular tissue than a fatty one, possibly because the former permits easy penetration. Thymus tissue, in general, has been found to be more resistant to hydrolysis than other tissues (De Lamater, Mescon and Barger, 1950). Brachet and Quertier (1963) used hydrochloric acid in ethanol in a modified Feulgen reaction to localise oöcyte cytoplasmic DNA (cf. Cowden, 1965).

The differential response to hydrolysis in different species and tissues clearly suggests that nuclear and cytoplasmic constitutions play an important role in the manifestation of the Feulgen reaction, and that organs and species differ in the constituents of the cytoplasm. Inhibiting agents, which may not allow the colour to develop, may be present in the cytoplasm, and in some instances the associated protein may interfere with the Feulgen reaction (Shinke, Ishida and Ueda, 1957). The association of other nuclear components with DNA may also protect the latter against hydrolysis. In view of these considerations, it is always necessary to adjust the specific requirements of the Feulgen reaction in every species and organ.

The intensity of the colour may also depend on the amount of DNA present. Haploid, diploid and higher polyploid cells differ in the intensity of the colour due to the amount of DNA present (Vendrely and Vendrely, 1956). In fact, because of the specific nature of the stain and proportionate increase of the staining intensity with the amount of DNA, quantitative estimation through microspectrophotometry of Feulgen-stained tissues is commonly carried out, and in this connection it is important that the thickness of the sections should be considered.

Validity of the Feulgen test—Although the Feulgen nuclear reaction has been widely accepted as a specific test for localising DNA *in situ*, some authors have persistently raised objections regarding the specificity of the test. These objections have been reviewed in detail by different workers (Stowell, 1946; Gomori, 1952; Sharma, 1952; Lessler, 1953; Kurnick, 1955; Kasten, 1956, 1960; Pearse, 1960, 1968). Only the outlines of the objections, pertinent to the test, will be mentioned here.

It was first pointed out by Carr (1945) that the colour of Schiff's reagent is regenerated by adsorption on the chromosome surface. He further suggested that nuclear coloration depends on the destruction of cytoplasmic constituents by acid hydrolysis. An excess of sulphur dioxide does not block the reaction, which would happen if the reaction depended on the presence of aldehydes.

This objection has been refuted by different workers. Dodson (1946) showed that chromosomes are not mere adsorbants of the Schiff's reagent, and further claimed that hydrolysis causes only a negligible loss of cytoplasm. That the Feulgen reaction is a typical aldehyde reaction has been established by Lessler (1951), who observed that after acid hydrolysis, if aldehyde-coupling reagents such as sodium bisulphite, trimethyl amino-acetohydrazide, semi-carbazide, phenylhydrazine or hydroxylamine are added, the Feulgen reaction is completely checked due to blocking of aldehydes.

A serious criticism of the Feulgen test is made by Stedman and Stedman (1943, 1948). They do not object to the aldehyde reaction in the Feulgen test, but hold that the latter is incapable of localising DNA *in situ* and according to their interpretation, DNA is the principal component of the nuclear sap. After acid hydrolysis and the application of fuchsin sulphurous acid, the colour develops outside the chromosome. The dye formed is a diffusible and water-soluble basic substance which is adsorbed on the chromosome surface by an acidic protein of the chromosome, which they term 'chromosomin'. On this basis, the Feulgen reaction is not fundamentally different from other staining procedures and differs only in the fact that in other cases a coloured solution is used as the staining medium whereas in the Feulgen test, the dye solution is prepared inside the cell from a decolorised compound. In support of their claim, they cited Choudhuri's (1943) observation that chromosome staining can be secured by the coloured compound obtained by interaction of Feulgen reagent and hydrolysed DNA. Stedman and Stedman (1950) also claimed that hydrolysis in the Feulgen test causes a profound change of the original product in extracted nuclei, and that considerable loss of nucleic acid, phosphorus and histones from the cell following acid hydrolysis also occurs. The diffusion of fragments containing a considerable amount of phosphorus into the hydrolysis fluid, giving a strong Mollisch reaction, was also noted, and they are of the opinion that similar behaviour occurs within the cell following acid hydrolysis, and diffusion products are formed both in the extracted nuclei as well as at an intracellular level. On the basis of these results, the unsuitability of the Feulgen test for *in situ* localisation and for quantitative estimation of DNA has been asserted.

The objections by Stedman and Stedman were refuted by several workers. Danielli (1947) pointed out that staining of squashes by the Feulgen technique is entirely different from the staining obtained by the action of the pre-formed Schiff's base, as in Choudhuri's material. This statement is based on the fact that the cytoplasm becomes stained in the latter but remains uncoloured in the former. Baker and Sanders (1946) suggested that if the products can diffuse through the tissues away from the place where they have first been formed, they can also diffuse into the fluid in the staining jar and thus be lost. Stedman and Stedman, however, consider that the diffusion of substances within the nuclear fluid is much less than in water. Overend and Stacey (1949) pursued the problem in detail by synthesising a number of 2-deoxysugars and studying their properties, and the results obtained by them have led them to conclude that aldehyde forms an actual complex at the chromosome level with decolorised *p*-rosaniline.

Lessler (1953) brought out several fallacies in Stedman's arguments. He stated that the reaction between free hydrolysed DNA and Feulgen reagent may produce a soluble coloured product, but it is washed off during acid hydrolysis. He also points out that nucleic acid, at intranuclear levels, may behave very differently from the isolated nucleic acid of Stedman as the nuclear nucleic acid exists as nucleoprotein complex. Caspersson (1944) also observed, through spectroscopic studies, that nucleic acid is present in the metaphase chromosome and is not a component of the nuclear sap. Kasten (1956, 1960) demonstrated that the Feulgen reaction is a specific test for DNA and is not a simple staining reaction as suggested by Stedman and Stedman.

Further evidence of the specificity of the test was secured by Brachet (1947). If the nuclei are treated with thymonucleodepolymerase, which removes DNA, the Feulgen reaction becomes negative for the nucleus. Similar observations were reported by Catcheside and Holmes (1947). They observed that the deoxyribonuclease treatment removed Feulgen-positive bands from the salivary gland chromosomes of *Drosophila*.

All the above evidence no doubt points out the validity of the reaction as a specific test for precise localisation of DNA *in situ*, but as different factors control the development of the colour it is always necessary to keep a strict control over temperature and duration of hydrolysis and the type of fixation. With a check through the control, the specificity of the test is unquestionable. The possibility, suggested by Pearse (1968), that there are several structural arrangements depending on the availability of SO_2 as well as the dye and aldehyde molecules, cannot be ruled out.

Carmine

One of the most widely used dyes for chromosome staining is carmine. It is prepared from the ground-up dried bodies of the cochineal insect. The term cochineal is applied to dried females of *Coccus cacti*, a tropical American Homoptera living on the plant *Opuntia coccinellifera*. In view of the fact that the product yielded by this insect may often vary because of the different species used, the carmine of commerce is generally variable in quality.

This dye is a crimson-coloured product prepared by adding compounds of aluminium or calcium to cochineal extract. It is not truly a definite chemical compound but a mixture of substances, the composition of which often varies on the basis of the method of manufacture. The active principle of carmine, to which its staining property is due, is carminic acid. If this is applied in a pure form, it does not give any better staining than carmine; but it may be applied in a critical study, as its exact chemical composition is known.

Carminic acid (Michrome No. 214) can be obtained by extracting cochineal with boiling water followed by treatment with lead acetate and decomposition of lead carminate with sulphuric acid (Gatenby and Beams, 1950). This dye belongs to the anthraquinone group and has the formula $C_{22}H_{20}O_{13}$, the molecular weight being 492·38. Its chemical structure is

The chromophoric property is attributed to its quinonoid linkage and auxochromes are also present. It is soluble in water in all proportions (Gurr, 1960), and is a dibasic acid and claimed to be nearly insoluble at its isoelectric point, pH 4–4·5 (Baker, 1950). If it is dissolved on the acid side of its isoelectric point it acquires a positive charge, behaves like a basic dye and stains chromatin, but if dissolved in alkaline solution it can behave as an acid dye. Carminic acid is not used as such for nuclear studies, except in the form of carmalum as recommended by Mayer (cited in Baker, 1950) for animal tissues. Carmalum mixture is composed of carminic acid, potassium aluminium sulphate and water, with sodium salicylate as the preservative. Evidently the purpose of potash alum here is to form a lake (*see* page 97). This mixture, however, is now obsolete.

In chromosome studies, carmine is used in solution with 45 per cent acetic acid, and the stain thus prepared is known as acetic-carmine (Belling, 1921). This solution serves the double purpose of fixation and staining, as acetic acid is a good fixative for chromatin and is a rapidly penetrating fluid. In the original schedule of Belling, which is widely followed, 1 per cent solution of the dye is prepared in hot 45 per cent acetic acid, whereas certain authors (Schneider as quoted in Gurr, 1960) prefer even 5 per cent solution. Belling suggested the addition of ferric hydroxide in acetic-carmine during its preparation and the purpose was evidently to allow the formation of a lake needed for the intensification of colour. According to Gatenby and Beams (1950), the best way of adding iron is in the form of a solution of ferric hydrate in 50 per cent acetic acid. The addition of a few drops of ferric chloride or ferric acetate solution also serves to intensify the colour, but iron must not be added in too heavy an amount as an excess of iron causes blackening of the entire cytoplasm.

The common procedure of using acetic-carmine as a stain is to squash the tissue in a drop of the dye solution. Warmke (1935) recommended the use of warm carmine on even smears. In the case of bulk compact tissues, such as root tips, leaf tips, etc., materials can be treated in hot acetic-carmine and hydrochloric acid mixture which serves the double purpose of softening and staining. Materials which provide difficulty in staining due possibly to inadequate fixation can be fixed in Carnoy's fluid prior to acetic-carmine staining; also, if needed, the use of a 2 per cent iron alum solution for a few minutes prior to staining may serve the purpose of mordanting and thus help in the intensification of colour. While squashing, the best way of adding iron is to tease the tissue in a drop of carmine with the help of a scalpel, or penetration can be aided by slight warming. Being present in the form of an acetic solution, carmine is not a suitable stain for sectioned materials. In certain cases acetic acid is substituted by propionic acid—the stain thus prepared is called propionic-carmine—and by this means the excessive swelling effect of acetic acid is generally eliminated; but prior to staining, fixation in acetic–alcohol is substituted by propionic–alcohol. This method has been found to be suitable for grass chromosomes (Swaminathan and colleagues, 1954). Hydrochloric acid–ethanol mixture can be also used as a solvent for carmine for staining tissues in bulk (Snow, 1963).

Occasionally, both plant and animal tissues which present difficulties in Feulgen staining are mounted in acetic-carmine (Schreiber, 1954) after Schiff's reaction. In such cases, hydrolysis in normal HCl as well as treatment with fuchsin sulphurous acid clears the cytoplasm, allowing specific coloration of the chromosomes.

The application of carmine as a chromosome stain is widespread. Starting from the lower groups of plants like algae (Godward, 1948), it can be applied to all other advanced groups, including all animal and human tissues. Even in the study of special chromosomes like the salivary gland chromosomes of *Drosophila*, its effect is remarkable.

In order to obtain permanent preparations from carmine squashes, McClintock's schedule or its modifications are commonly used in addition to the dry ice or vapour method (*see* Chapter 7), but in all these cases one of the serious limitations lies in the gradual fading of colour following prolonged keeping. No doubt, if the acidity of the mounting medium and other conditions are controlled, the fading may be checked to a certain extent, but the presence of acetic acid itself in the dye makes it liable to be acidic in time. In spite of this limitation, the convenience in chromosome staining and rapidity of the schedule make it the most widely used stain in the study of chromosomes.

Orcein

Orcein was first employed as a chromosome stain by La Cour in 1941. The dye has a molecular weight of 500·488, the formula being $C_{28}H_{24}N_2O_7$ (Michrome No. 375), but its exact chemical structure is unknown. It is a deep purple-coloured dye, obtained from the action of hydrogen peroxide and ammonia on the colourless parent substance *orcinol*.

$$CH_3$$

HO   OH

Orcinol is 3,5-dihydroxytoluene, having a molecular weight 160·166 and the formula $C_7H_8O_2$. It is available both in natural and synthetic forms. In nature, it is obtained from the two species of lichens, *Rocella tinctoria* and *Lecanora parella*. As a chromosome stain, Conn (1953) indicated that synthetic orcein is not so effective as natural orcein. Engle and Dempsey (1954), from studying the physical and chemical properties of four fractions of orcein, separated by chromatography, concluded that both orcein and its fractions are valuable chromosomal stains.

Orcein is soluble in water as well as in alcohol. Under certain conditions, it can behave as an amphoteric dye (Gurr, 1960). Fullmer and Lillie (1956), working exhaustively on orcein staining, demonstrated its basic properties. In the study of chromosomes it is used in the form of acetic-orcein, that is, 1 per cent solution in 45 per cent acetic acid. It can be used in the same way as acetic-carmine and has the added advantage that no iron mordanting is necessary. In our experience the intensity of the stain, especially for meiotic materials, is not as good as carmine, though it is effective where carmine staining fails. It has been found to be a very effective stain for salivary gland chromosomes (Darlington and La Cour, 1960) as well as the chromosomes of mosses (Vaarama, 1949). For the study of root tip and leaf tip chromosomes, the use of a stronger hot solution of acetic-orcein and normal HCl, mixed in a specific proportion, is necessary for softening the tissue before mounting in a dilute solution of acetic-orcein (Tjio and Levan, 1950; Sharma and Sharma, 1957). In various species of fungi, especially those belonging to Ascomycetes, hydrolysis in normal HCl for a few minutes at 60 °C, after fixation and prior to staining, has been found to be very effective (Singleton, 1953; Elliott, 1956). Just like carmine, acetic-orcein can be substituted by propionic-orcein (Cotton, 1959), which has been found to be useful in studying the chromosomes of *Heteroptera*.

Although not widely applicable on meiotic materials, it is found to be especially useful in the study of somatic chromosomes of plants. It has, however, to be applied with extreme caution, since overheating in orcein–HCl mixture has been found to induce chromosome breakage (Sharma and Roy, 1955). Therefore, at least for the study of chemical effects on chromosomes, orcein staining should perferably not be applied as its effect may often mask the effect of chemicals. Sen (1965), while working out the chemical basis of orcein staining and chromosome breakage, has shown that staining involves condensation of the phenolic dye at the point of polypeptide linkage of chromosome protein. Following prolonged heating under acidic conditions, tertiary amine involved in polypeptide linkage results in the production of ammonium chloride and breakage of the link, causing chromosome breakage.

Chlorazol Black

A solution of chlorazol black E in alcohol has been applied by Nebel (1940) as an auxiliary stain for chromosomes, along with acetic-carmine. This dye was applied after fixation prior to acetic-carmine staining and proved effective for species of Rosaceae, such as apple, pear, peach, plum, etc., where ordinary acetic-carmine stain was ineffective. It has even been applied as a stain by itself and has been found to be effective for the study of root tip chromosomes of plants (Nebel, 1940; Conn, 1943).

Chlorazol black E, however, is an acid dye of the trisazo group (Cannon, 1937) and has a molecular weight of 781·738 (Michrome No. 92). Its formula is $C_{34}H_{25}N_9O_7S_2Na_2$. It has the following structure:

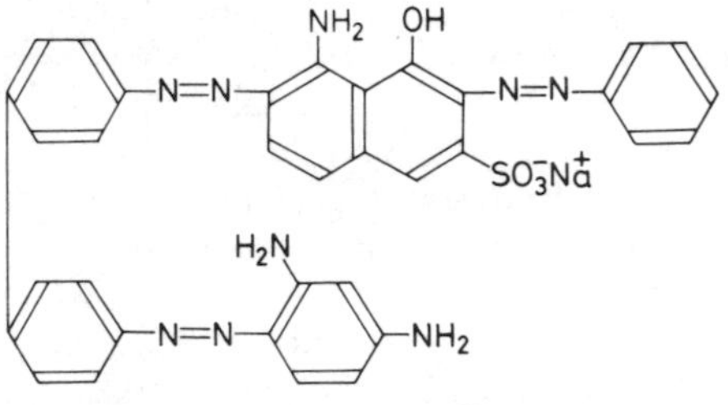

Chlorazol black E

It is highly soluble in water and sparingly in alcohol. Being an acidic dye, the basis of its stainability with chromosomes is not clear, but it is probable that it stains the protein component. Chlorazol black is possibly effective in materials where protein components of the chromosome are high, which accounts for the limited application of the dye. More evidence is necessary to substantiate this suggestion.

Crystal Violet

The discovery of crystal violet as a stain for chromosomes is attributed to Newton (1926). In fact, in his studies Newton used gentian violet (Michrome No. 417) which is a mixture of crystal violet and tetra- and penta-methyl *p*-rosaniline chlorides. Crystal violet itself is hexamethyl *p*-rosaniline chloride. Gentian violet is a basic dye belonging to the triphenyl methane series. The structures of the three compounds constituting gentian violet are shown on p. 113.

The molecular formulae and weights of the three compounds are $C_{23}H_{26}N_3Cl$: 379·919; $C_{24}H_{28}N_3Cl$: 393·945, and $C_{25}H_{30}N_3Cl$: 407·971 respectively. Gentian violet is soluble in both water and alcohol.

Crystal violet (Michrome No. 103), which is supposed to be one of the most adequate stains for chromosomes, is a bluish violet dye. The term itself is derived from the fact that it forms large crystals with nine molecules of water. The dye is closely allied to basic fuchsin from which it can be derived by the replacement of the six hydrogen atoms of three amino groups by six methyl groups. It is soluble in both water and alcohol. In chromosome studies, aqueous 1 per cent solution is used.

In Newton's crystal–iodine technique, after the application of the stain to the sections or smears, the excess dye is first washed off in water. Then the slides are processed through iodine and potassium iodide in alcohol mordant, followed by dehydration in alcohol; differentiation in clove oil and clearing in xylol before final mounting in balsam.

Tetramethyl *p*–rosaniline chloride   Pentamethyl *p*–rosaniline chloride

Hexamethyl *p*–rosaniline chloride (crystal violet)

The use of iodine as a mordant after crystal violet staining, is based on the Gram effect on bacteria. During dehydration, crystal violet can easily be washed off in alcohol, but de-staining can be prevented if iodine is added to bacteria. The same principle holds good for chromosome staining; in fact, after the addition of iodine, the stain becomes bluish-black from violet. Baker (1950, 1958) stated that the liability of crystal violet to be removed by alcohol is actually lessened by iodine treatment, which results in the chromosomes retaining the dye against a clear cytoplasmic background. In order to obtain a proper coloration of the chromosomes in difficult materials, iodine mordanting, which is normally carried out for a few seconds, should be further reduced, but in no case should this step be omitted, as acidic components of the cytoplasm also take up the colour which is removed by iodine in alcohol (for details, *see* page 152). Although Baker has recommended the omission of the iodine step in certain materials, in the experience of the present authors, the shortened schedule is ineffective.

Incidentally, it may be mentioned that the closely related dyes or rosanilines, which can be obtained by combining one molecule of aniline, one of *o*-toluidine and one of *p*-rosaniline, do not show this Gram effect

(Baker, 1950, 1958). Schmidt (1944) suggested that iodine causes the p-rosaniline dyes to undergo a molecular aggregation, which is possibly interfered with by the methyl group of o-toluidine.

Differentiation in clove oil is an essential step in the crystal violet technique. Due to the rapid passage through alcohol, dehydration remains incomplete and it is finally completed in a slowly differentiating fluid, clove oil.

Clove oil is an essential oil, yielded by the flower buds of *Eugenia caryophyllata*. It consists principally of eugenol, a guaiacol derivative with the formula

In addition, it also contains other organic compounds, such as methyl alcohol, furfural, etc. Differentiation in clove oil completes the dehydration, clears the cytoplasmic background and imparts a crisp colour to the chromosomes.

All the alcohol in clove oil should be removed by continually stirring the slide in the fluid. The retention of alcohol may cause cloudiness and lessening of the stain on mounting in balsam.

Final clearing in xylol, $C_6H_4(CH_3)_2$, or dimethyl benzene is an essential step, as clove oil must be completely removed before final mounting. The retention of this oil ultimately leads to the fading of colour.

For materials that are difficult to stain, slides can be mordanted in 1 per cent chromic acid and washed prior to staining. In order to secure complete cytoplasmic clearing in materials having a heavy cytoplasmic content, the slides can be further mordanted in chromic acid in between the different alcohol grades, after mordanting in iodine. The underlying principles are discussed in the chapter on mordanting.

Crystal violet is widely used as a chromosome stain for plants, animals and lower organisms. It is most effective on pollen mother cell smears or for mitotic and meiotic studies from sectioned tissues. Unfortunately it cannot be applied effectively on tissues to be squashed after staining. This disadvantage may be attributed to two main reasons: (1) Being an aqueous stain, its rate of penetration is very slow and, as such, the different cell layers will have different intensities of colour. Makino and others used crystal violet dissolved in acetic acid for smears of mammalian chromosomes. (2) As it is a non-specific basic dye, acidic components of cytoplasm also take up the colour. Washing of the cytoplasmic colour through iodine mordanting from bulk tissue is rather difficult.

Barring these limitations, crystal violet can be safely recommended for the staining of chromosomes.

Certain other basic dyes of the triphenyl methane series are able to replace basic fuchsin, to some extent, like dahlia violet, magenta roth, methyl violet, brilliant green, malachite green and light green as observed in our laboratory. The methods of preparation and staining are similar to that followed for leucobasic fuchsin. A direct correlation is observed between the colour of the different dyes and the active groups present in them. The amino group, in general, produces mauve and magenta colour, while ethyl and methyl groups principally impart violet to greenish coloration. In basic fuchsin, amino groups impart mauve coloration in the nuclei whereas their replacement by methyl groups in crystal violet results in completely violet coloration. Magenta roth is a suitable substitute for basic fuchsin and results in a magenta colour. With crystal violet and methyl violet, the chromosomes colour violet in Feulgen reaction. Brilliant green shows better effect than the other green dyes and the ethyl group is more easily reacted upon than the methyl, as shown in the following table:

Dye	Ml No.	No. of batches	Active group	Colour of nuclei
Basic fuchsin	421	3	$-NH_2$	Mauve
Dahlia violet	105	2	$-C_2H_5$	Mauvish violet
Magenta roth	624	1	$-NH_2$	Magenta
Methyl violet 6B	180	2	$-CH_3$	Violet
Crystal violet	103	2	$-CH_3$	Deep violet
Brilliant green	406	1	$-C_2H_5$	Blue-green
Malachite green	315	1	$-CH_3$	Green
Light green		1	$-CH_3$	Yellow-green

(Golechha, 1968; Golechha and Sharma, in press)

Haematoxylin

This natural colouring substance is obtained from the heartwood of *Haematoxylin campechianum*, a native of Mexico. As the heartwood is supplied in the form of small logs, the term 'log wood' is often applied. The crude product is generally manufactured by boiling small chips of log wood in water followed by filtration and evaporation to dryness. Extraction of this crude product with ether yields *haematoxylin*.

The dyeing property of haematoxylin is attributed to its oxidation product, haematein (Michrome No. 360), as haematoxylin is itself not a dye. Haematoxylin has the molecular formula, $C_{16}H_{14}O_6$, the molecular weight being 302·272. The molecular formula of haematein is $C_{16}H_{12}O_6$, the molecular weight being 300·256. The structural formulae of the two are:

Haematoxylin   Haematein

The structural formula of haematoxylin was first worked out by Perkin and Everest (1918), and Baker (1950) rearranged the structural formulae of the two as outlined above so that the presence of quinonoid arrangement in haematein and its absence in haematoxylin is clear. The process of oxidation, which is otherwise known as 'ripening', may take several weeks, spontaneously, but this process may be hastened by the use of oxidising agents such as sodium iodate, hydrogen peroxide, chloral hydrate, potassium permanganate, etc. Slow atmospheric oxidation is, however, preferred to the use of oxidising agents, as too much oxidation may make haematein quite unfit for the purpose of staining.

In view of the necessity of oxidation in the preparation of haematein from haematoxylin, the aqueous solution of haematoxylin is prepared and allowed to ripen for several weeks (Heidenhain, 1896). For rapid oxidation, a small quantity of sodium iodate is often added if the solution has to be used immediately, but without sodium iodate, at least a few weeks are necessary for the colour to ripen. Shortt (1923) suggested the addition of carbolic acid to haematoxylin solution, which apparently acts as a preservative.

Without the use of a mordant, haematoxylin solution is entirely ineffective in staining chromosomes. Commonly used mordants are potassium aluminium sulphate, iron alum and ammonium alum. They form lakes which become positively charged and behave as basic dyes. For chromosome studies, potassium aluminium sulphate (Mayer, 1903) and iron alum (Benda, 1896) are widely used, the latter being more effective. The potash alum lake of haematoxylin is used for progressive staining, whereas iron alum is utilised in regressive staining (*see* page 97). Progressive staining implies gradual addition of the stain till the maximum colour is obtained, whereas regressive staining involves overstaining the material and subsequently washing off the excess stain.

When aluminium alum is used as the mordant, the dye and the mordant solution can be kept mixed together forming the lake; but when iron alum is used, it cannot be kept mixed with the dye because of the possibility of heavy iron precipitation.

Haematein, after ferric mordanting, has a strong tendency to accumulate around densely stained materials. For this reason, it has most often been used in chromosome studies. On the basis of Heidenhain's schedule, the sections or smears in water are first mordanted in a strong solution of iron alum (4 per cent), followed by washing in water and staining in haematoxylin. Differentiation is carried out in a dilute solution of iron alum or picric acid to wash off the excess stain from the cytoplasm (*see* page 153). In properly controlled differentiated preparations, chromosomes appear intensely black. After washing once more in water, the tissue is dehydrated through alcohol, cleared in xylol and mounted in balsam.

As haematoxylin acts as a non-specific basic dye, cytoplasmic components and spindle initially become coloured. As the differentiation is gradual in iron alum or picric acid solution, it can be adjusted to retain the spindle stain, if necessary. However, so far as plant chromosomes are concerned, haematoxylin staining is not very effective, due to the heavy cytoplasmic content. Animal materials (such as testes, smears of grasshopper and other insects, very thin sections, etc.) can be stained with haematoxylin, and being devoid of any strong acid or clove oil in the staining schedule, the chromosomal stain, once obtained, does not fade in permanent slides. It has been superseded in recent years by crystal violet, acetic-carmine and other chromosomal stains due to the limitations of non-specific coloration, complicated procedure of preparation and the time-consuming process of differentiation observed in the haematoxylin schedule.

Lacmoid

Lacmoid (otherwise known as resorcin blue) is a blue acidic dye of the oxazine series. According to Conn (1953), the exact molecular structure of lacmoid is not fully worked out, but Gurr (1960) has given the following structure:

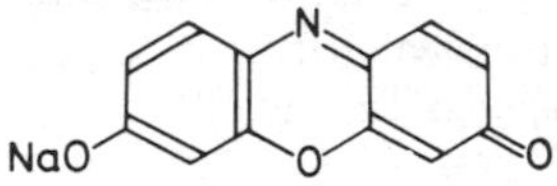

Oxazine

Its empirical formula is $C_{12}H_6NO_3Na$, the molecular weight being 235·173. It can be obtained by heating resorcinol with sodium nitrate until the smell of ammonia is no longer present. Similar to carmine, it can be used as an acid–base indicator and, when dissolved in acetic acid, it behaves as a basic dye. Unlike carmine, it is fairly soluble both in water and alcohol.

Darlington and La Cour (1942, 1960) used it in place of carmine, and acetic-lacmoid solution has been found to be very effective for the chromosomes of root tips, embryo sacs and pollen grains. Koller utilised this method for the study of the chromosomes of tumours. For comparatively compact tissues of plants, like root tips, similar to orcein, heating in acetic–lacmoid–HCl mixture is needed prior to squashing for dissolution of the pectic salts of the middle lamella. Cedarwood oil and euparal are recommended as mounting media.

However, acetic-lacmoid as a stain cannot be universally applied like acetic-carmine. It has a comparatively limited application and may be tried on those materials where other stains have failed.

Brazilin

Belling (1924) and Capinpin (1930) used brazilin for staining chromosomes. It is a dye which is extracted from Brazil wood. Similar to haematoxylin, it

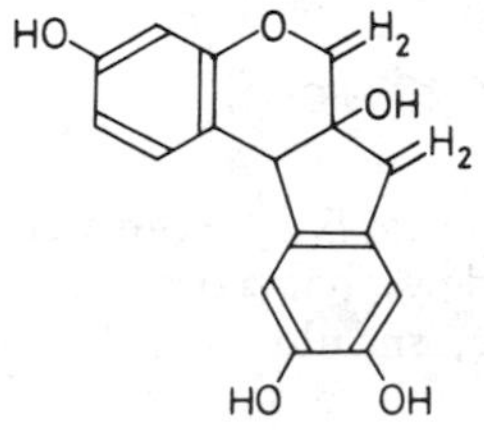

Brazilin 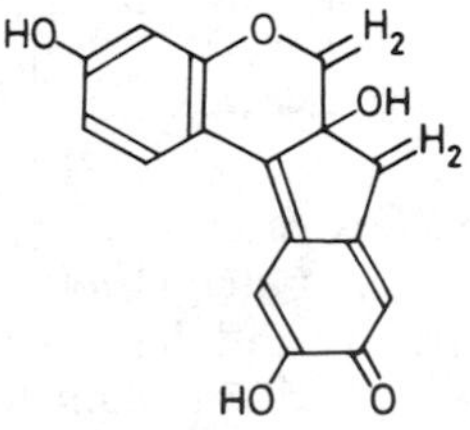   Brazilein

is not a dye in itself and its tinctorial property is due to brazilein—its oxidation product. In its preparation, like haematoxylin, it needs ripening.

The empirical formula of brazilin is $C_{16}H_{14}O_5$ whereas that of brazilein is $C_{16}H_{12}O_5$, the respective molecular weights being 286·272 and 284·256. Brazilin is highly soluble both in water and in alcohol. Under certain conditions, brazilin can act as an amphoteric dye. Belling (1928, as mentioned by Conn, 1953), used brazilin instead of carmine for staining plant chromosomes. Like haematoxylin, it has the defect of staining the cytoplasm as well, and even in the chromosome the stain is much weaker.

Azure A (Michrome No. 718)

Azure A (a basic dye) is a member of the thiazine series and is violet blue in colour. It can be prepared by oxidation of methylene blue with potassium dichromate. It has the empirical formula $C_{14}H_{14}N_3SCl$, the molecular weight being 291·799. The structure is:

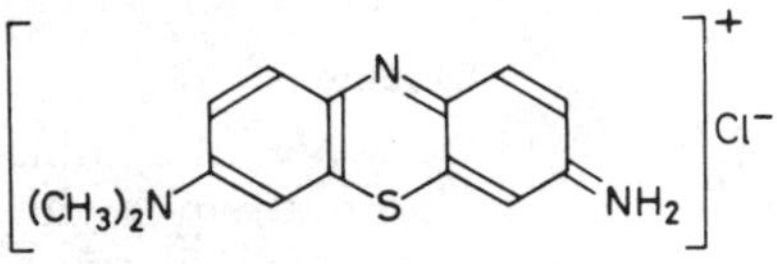

Himes and Moriber (1956) prepared leuco-azure A just like Schiff's reagent by substituting azure A in place of basic fuchsin. The aldehyde reaction in DNA of chromatin was obtained. De Lamater (1951) demonstrated nuclear staining by azure A in the presence of SO_2, after hydrolysing the tissue in normal HCl at 60 °C. Evidently, in such cases also, typical aldehyde reaction is demonstrated. Though it has been employed extensively in stomach, intestine, thyroid tissues and root tips of plants, the aldehyde reaction for the staining of chromosomes is much brighter with basic fuchsin than with azure A.

Azure B (Michrome No. 357)

Azure B is also a member of the thiazine series and is prepared by the oxidation of methylene blue with potassium dichromate, though the method employed is slightly different from that of azure A. It is blue violet in colour and has the chemical structure:

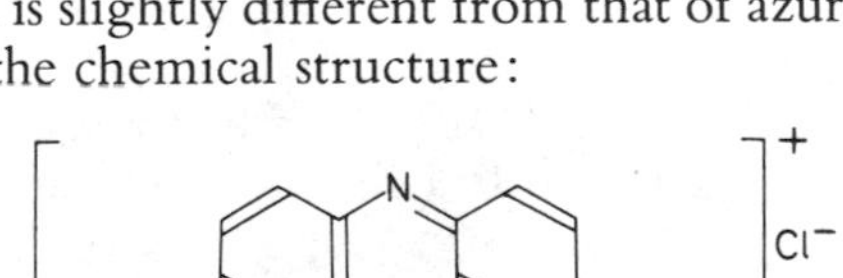

Although in a few cases staining with azure B is recommended more than that with azure A, Gurr (1960) considers that azure A staining, which is more convenient and less expensive, should be followed. However, Saez (1952) employed this stain for securing differential coloration of the sex chromosomes at the prophase stage of meiosis, and according to him, by hot water treatment, heterochromatic sex chromosomes can be made to stain *m*-chromatically instead of *o*-chromatically. Acid pH is, however, necessary. Flax and Himes (1952) obtained metachromatic staining of testes tissue. Their method involves treatment of the Carnoy-fixed tissue in tepid water for a few minutes, followed by staining for 3 h in Azure B solution (0·2–0·1 mg/cm^3 in potassium citrate buffer, pH 4·0) and dehydrating in tertiary butyl alcohol before mounting.

Orange G–Aniline Blue

Orange G is an acid dye belonging to the monoazo series, with the structure:

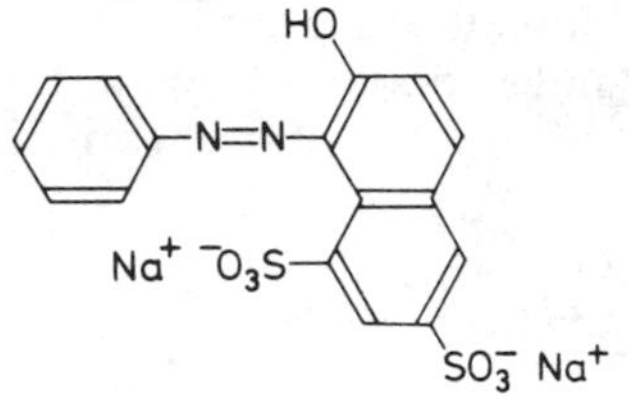

It is golden orange in colour, the molecular formula being $C_{16}H_{10}N_2O_7S_2Na_2$, and the molecular weight is 452·382. It is highly soluble in water and fairly soluble in alcohol. Mallory was the first to use this technique in combination with aniline blue. The latter is a blue acid dye belonging to the triphenyl methane series, its molecular weight being 737·736, and its formula $C_{32}H_{25}N_3O_9S_3Na_2$. It is used dissolved in water and has the structure:

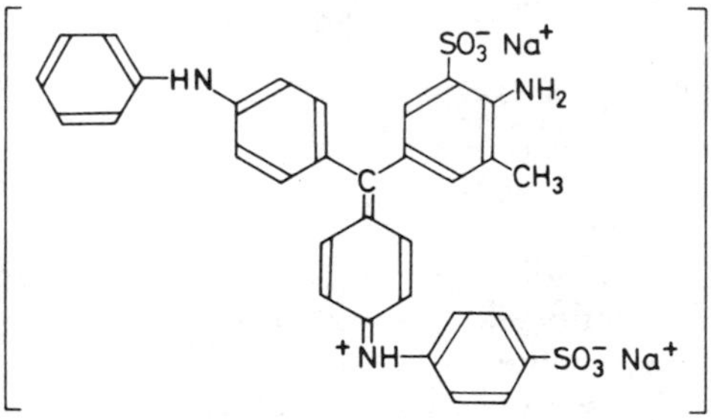

La Cour (1958 in Darlington and La Cour, 1960, 1968) utilised the combination of these two dyes for staining heterochromatin at the resting stage and telophase, and chromosomes in other stages of mitosis. The basic principle of this staining is however not clear, though the dyes, being acidic in nature, may be considered as staining the protein component of the chromosomes. The protein stainability may be attributed to either of these two dyes.

Toluidine Blue (Michrome No. 641)

Toluidine blue, a basic dye of the thiazine series, is bluish violet in colour. Its molecular weight is 305·825, the formula being $C_{15}H_{16}N_3SCl$. The structure is:

Robinson and Bacsich (1958) suggested the preparation of a lake of toluidine blue with mercuric chloride or potassium iodide. The dried dye which is already mordanted, evidently yields a very intense colour, staining both types of nucleic acids, and it can be used selectively for DNA if normal hydrochloric acid hydrolysis is performed for a very short period prior to staining. Pelc (1956) utilised toluidine blue for staining through film in autoradiographic procedure, it being applied in an aqueous solution which is soluble both in water and alcohol. In view of its restricted application, it is not recommended as a general stain for chromatin.

Giemsa (Michrome No. 144)

Giemsa is not a single dye but a mixture of several dyes, namely methylene blue and its oxidation products, the azures as well as eosin Y. The quality of the stain varies with regard to the proportion of the dyes used.

Eosin y

Eosin Y itself is an acidic dye, rose pink in colour, and belongs to the xanthene series; for methylene blue and azures, reference should be made to the earlier part of this chapter. The combination stain, Giemsa, is generally prepared by dissolving the powdered mixture in glycerin and methyl alcohol, and in staining, chromatin is stained red and cytoplasm blue.

Giemsa has been used extensively for staining different types of bacteria (Robinow, 1941), and later in the study of nuclei and chromosomes of yeast by Lindegren, Williams and McClary (1956), McClary and colleagues (1957), Ganesan and Swaminathan (1958) and mammals as well. Darlington and La Cour (1960) have cited Belar as recommending Giemsa staining (Gelei, 1921) for animal or bacterial materials. In this modification, mordanting in ammonium molybdate is necessary before staining.

Though Giemsa staining is quite useful, especially for the bacterial and fungal nuclei, its only disadvantage is that it can stain both types of nucleic acid. So, unless RNA is completely removed from the cytoplasm and other constituent parts, it is difficult to obtain specific staining of chromosomes. Though normal hydrochloric or perchloric acid treatment is generally performed to remove RNA, a proper control is necessary to ensure that RNA is completely removed.

SOME COMMON STAINS AND THEIR PREPARATION

Feulgen Reagent (Fuchsin Sulphurous Acid)
  Materials required

Basic fuchsin	0·5 g
N HCl	10 cm^3
Potassium metabisulphite	0·5 g
Activated charcoal	0·5 g
Distilled water	100 cm^3

Preparation—Dissolve 0·5 g of basic fuchsin gradually in 100 cm^3 of boiling distilled water. Cool at 58 °C. Filter, cool the filtrate down to 26 °C. Add it to 10 cm^3 of N HCl and 0·5 g of potassium metabisulphite. Close the mouth of the container with a stopper, seal with paraffin, wrap the container in black paper and store in a cool dark chamber. After 24 h, take out the container. If the solution is transparent and straw-coloured, it is ready for use. If otherwise coloured, add to it 0·5 g of charcoal powder, shake thoroughly and keep overnight in cold temperature (4 °C). Filter and use.

Alternatively, after dissolving the dye, bubble a stream of SO$_2$ through the solution. Filter and store. (Itikawa and Ogura, 1954.)

Precaution—Always keep the container sealed after use and store in cool temperature away from light. According to Lhotka and Davenport (1949), Feulgen reagent in sealed containers, kept at 0–5 °C, retains its efficiency for 6 months.

Modifications—1. Newcomer (1959) doubled the amount of potassium metabisulphite in a Schiff's reagent for staining tissues prior to embedding.

2. *Acetic-basic fuchsin* is prepared by dissolving 1 g of basic fuchsin in 50 cm^3 of 40 per cent acetic acid at 50 °C, cooling to 25–30 °C and filtering. The staining process involves hydrolysis at 60 °C in N HCl for 15–60 s, followed by staining for 1–3 h and squashing in 30 per cent acetic acid (Tanaka, 1961).

3. *Carbol fuchsin for human chromosomes*

0·3% soln. of basic fuchsin in 5% phenol	45 cm^3
Glacial acetic acid	6 cm^3
37% aq. formaldehyde soln.	6 cm^3

It is found to be effective for human tissue cultures. The period of staining is 2–5 min (Carr and Walker, 1961).

4. Tissues stained following Feulgen reaction can be squashed in 1 per cent acetic-carmine solution instead of 45 per cent acetic acid. It has been employed for studying chromosomes of eggs (Schreiber, 1954) and is effective for plant chromosomes as well.

5. Fuchsin solution can be prepared in solvents other than water. 1 per cent basic fuchsin in 30 per cent alcohol (2 min) or 0·2–0·4 per cent fuchsin in

5–10 per cent lactic acid (5–20 min) have been employed on animal chromosomes (Makino and Nishimura, 1952).

6. In a simplified method, Schiff's reagent can be prepared by adding fuchsin (1 g) and sodium metabisulphite (1·9 g) to 100 cm^3 of 0·15N HCl, shaking mechanically for 2 h, decolorising with fresh activated charcoal and filtering (Lillie, 1951).

7. Normal hydrochloric acid is substituted by trichloracetic acid (Sharma, 1951; Bloch and Godman, 1955); 5N hydrochloric acid and 5N nitric acid (Itikawa and Ogura, 1954); perchloric acid (Cassel, 1950; Di Stefano, 1952) as well as by bromine in carbon tetrachloride (Barka, 1956) and phosphoric acid (Hashim, 1953).

An alternative method (Berger and DeLamater, 1948) involves dissolving 1 g of basic fuchsin in 400 cm^3 of boiling water. After cooling and filtration, 1 cm^3 of thionyl chloride is added. The solution is kept in the dark in a sealed chamber overnight. Shaking with activated charcoal (2 g) for 1 min, followed by filtering, yields a clear solution. The method is based on the reaction $SOCl_2 + H.OH = SO_2 + 2HCl$.

Purification of parafuchsin (pararosaniline C.I. 42500) (Gabler, 1965)— Dissolve 4 g of parafuchsin in 800 cm^3 ethanol at room temperature. Add 8 g of activated charcoal, boil and filter immediately. Repeat this procedure four times, each time taking 8 g of charcoal. Evaporate in vacuum with slight warming. After purification, dissolve in ethanol–chloroform mixture (1:1) and pass through an aluminium oxide column. The region to be eluted is indicated by a broad dark-red zone, which on drying in vacuum, after elution, yields a dye that can be decolorised by the usual procedure.

8. Both plant and animal materials, after fixation, hydrolysis, and staining in leuco-basic fuchsin solution, can be stored in 45 per cent acetic acid for an indefinite period at a temperature of 14 °C. On squashing, the results are as good as fresh material. If the stain is faint, the material can be squashed in 1 per cent acetic-carmine or acetic-orcein solutions (Ford and Hamerton, 1956). Storage of somatic material in distilled water between Feulgen staining and squashing is, however, preferable to storage in 70 per cent alcohol between fixation and staining (Flagg, 1961).

9. Some modifications of the use of basic fuchsin for fungi are (De Lamater, 1948):

(a) Staining in 0·25 per cent aqueous basic fuchsin for 5–15 min.
(b) Mordanting the hydrolysed cells in 2 per cent formalin for 2–4 min before staining.
(c) Combining mordant and stain in a solution of 0·5 per cent basic fuchsin and 2 per cent formalin in 0·04 N HCl.

Acetic-carmine, Acetic-orcein, Acetic-lacmoid

Materials required
(a) For 2% solution:

Carmine, orcein or lacmoid	2 g
Glacial acetic acid	45 cm^3
Distilled water	55 cm^3

(b) For 1% solution:

The same, except for 1 g of the dye

Preparation—Add distilled water to glacial acetic acid to form 45 per cent acetic acid solution. Heat the solution in a conical flask to boiling. Add the dye slowly to the boiling solution, stirring with a glass rod. Boil gently till the dye dissolves. Cool down to room temperature. Filter and store in a bottle with a glass stopper.

Precautions—Keep the mouth of the flask covered with a piece of filter paper while the solution is being heated. Store acetic-orcein as 2·2 g dissolved in 100 cm^3 glacial acetic acid. Dilute as needed.

Procedure—Acetic-carmine: use 1 per cent solution directly for staining and squashing; acetic-orcein or acetic-lacmoid: use 1 per cent solution directly for staining. Alternatively, heat the tissue for a few seconds in a mixture of 2 per cent solution and normal hydrochloric acid (9:1) and then squash in 1 per cent solution.

Modifications—1. In the most common modification of acetic-carmine, variations of Belling's (1921) iron–acetic-carmine are used. Iron is added to 1 per cent acetic-carmine solution as ferric acetate, ferric chloride or ferric hydroxide in different proportions.

2. A mixture of acetic-carmine (10·5 parts), 45 per cent acetic acid (4·5 parts), N HCl (2 parts) and 1 per cent fast green FCF solution in rectified spirit (1 part) forms a good stain for studying nuclear structures in protozoa (Dippell, 1955).

3. For polyploids in *Bromus*, 1 per cent orcein in 45 per cent acetic acid is used for heating without any HCl (Markarian, 1957).

4. For perithecia of Ascomycetes, an effective stain is made up by adding 3 g of carmine to 50 cm^3 of acetic acid and 90 cm^3 of absolute alcohol. Boil gently for 6 h, make up to original volume with absolute alcohol. Filter. Expose the stain to air and light for one week and again make up to original volume with alcohol (Cutter, 1946).

5. For studying Avian chromosomes from stored material, Newcomer (1952) devised a boiled mixture containing:

Vinyl acetate	10 cm^3
Tertiary butyl alcohol	10 cm^3
Propionic acid	50 cm^3
Distilled water	75 cm^3
Carmine	0·25 g
Orcein	0·25 g

Stored materials are stained and squashed in this mixture.

6. For *Zaprionus* glands, pre-staining acid hydrolysis gives good results (Hartmann-Goldstein, 1961). The stain used is:

1% orcein in conc. lactic acid	50 cm^3
30% acetic acid	50 cm^3

The period of treatment is for 1 min in saturated acetic-carmine solution, followed by 5 min in the above stain and squashing in the same mixture. This stain has also been used for mouse embryos (Wroblewska and Dyban,

1964). A solution of 1 g orcein in a mixture of 28 cm^3 of 85 per cent lactic acid and 22 cm^3 glacial acetic acid has been recommended for chromosomes of ants, Formicinae and Myrmicinae (Imai, 1966).

7. An alternative stain for the same material is 2·5 per cent orcein in 60 per cent acetic acid, the period needed being 15–20 min.

8. Treatment in 4–5 per cent aqueous pectinase solution for 45 min to 1 h before maceration in 1 per cent acetic-carmine solution gives good effects in pollen tube and p.m.c. smears (Pandey and Henry, 1959).

9. In orchids, 2 per cent acetic-lacmoid solution in glacial acetic acid has been used, diluted with an equal quantity of water. The roots are macerated in 10–20 per cent HCl for 5–15 min before staining (Wimber, 1956). Carmine solution, prepared by gently boiling in a mixture of acetic acid, 85 per cent lactic acid, and water (25:20:40) with a trace of rusty iron, for 1 h has been used effectively for orchid chromosomes. The root-tips, after fixation in acetic-lactic-ethanol (1:1:4) for 4–16 h, are heated in conc. HC1 with 50 per cent ethanol (1:4), followed by treatment in 50 per cent ethanol for 5–10 min, then heating three times (at intervals of 5 min) in the carmine stain mixed with an equal amount of 50 per cent ethanol, and keeping for 1–2 h before squashing (Freytag, 1963).

10. For *Cucurbita*, a stain was used containing 45 per cent acetic acid, 1 per cent acetic-carmine and brown storage solution (95 per cent alcohol with a flake of rusty iron), 1:1:1 (McGoldrick, Bohn and Whitaker, 1954).

11. The preparation and application of propionic-carmine and propionic--orcein are similar to those of acetic-carmine and acetic-orcein except that propionic acid is used instead of acetic acid.

12. Carmalum (Mayer)

Carminic acid	0·5 g
5% aq. potash alum soln.	100 cm^3

13. Lithium carmine

Saturated aq. lithium carbonate soln.	100 cm^3
Carmine	5 g

Alternatively:

Lithium carmine powder	2 g
Distilled water	50 cm^3

Heat to boiling, cool and filter.

14. Lactic-propionic-orcein, prepared by dissolving 2 g of natural orcein in 100 cm^3 of lactic and propionic acids mixture (1:1) and diluted to 45 per cent with water, is very effective for p.m.c.s. For mitotic preparations, maceration in 1 N HCl at 60 °C for 5 min between fixation and staining is necessary (Dyer, 1963).

15. For *Solanum* microsporocytes, Matsubayashi (1963) advocated keeping the anther in 1 per cent acetic-carmine solution in 45 per cent acetic acid for several days till the contents were fully stained.

16. A 1 per cent acetic-carmine: 2 per cent acetic-orcein: N HCl mixture (9:9:1) is effective for meiotic chromosomes of *Ipomea* (Raghuvanshi and Joshi, 1963).

17. In a modification devised by Snow (1963), 4 g of carmine is gently boiled in 15 cm^3 of distilled water to which 1 cm^3 of conc. HCl has been added. After cooling 95 cm^3 of 85 per cent ethanol is added and the solution is filtered. This stain can be used for tissues in bulk.

Crystal Violet (Newton's Schedule)

Materials required

Crystal violet	1 g
Distilled water	100 cm^3

Preparation—Dissolve the dye in water with constant stirring and boiling. Filter. Allow it to mature for a week before use.

Modifications—1. If difficulty is experienced in staining with crystal violet schedule, dissolve 5 g of commercial crystal violet in 10 cm^3 of absolute ethyl alcohol. Keep the solution in an open watch-glass and allow the alcohol to evaporate. On complete drying, weigh the stain and prepare an aqueous solution as usual.

2. Infiltrate fixed root tips with chloroform, stain them for 15–30 min in leuco-basic fuchsin, wash in water for 1–4 h and double stain with crystal violet (Bowden, 1949).

Giemsa

The staining solution is prepared by triturating 3·8 g of the powder with 250 cm^3 of pure glycerin and 250 cm^3 of absolute methyl alcohol.

10 drops of Giemsa stain in 5 cm^3 phosphate buffer stains mycelia of *Helminthosporium sativum* after 2 h treatment (Hrushovetz, 1956).

To obtain a Giemsa stain of uniform composition, oxidise methylene blue, neutralise and precipitate with eosin (Lillie, 1943).

Bismarck Brown

Can be used as a counterstain for Feulgen reaction for mast cells in mitosis. Fix mesentery of rat in 10 per cent formalin for 1 h. Stain first in Feulgen solution and then in 0·5 per cent Bismarck Brown dissolved in 50 per cent ethanol containing 0·2 per cent acetic acid (Meggers and Allen, 1962).

Phenyl Diamine and Quinone Diimine (Meyer, 1948)

(a) *p*-Phenyl diamine	0·2 g
20% aq. acetic acid soln.	20 cm^3
30% aq. hydrogen peroxide soln.	0·1 cm^3

Dissolve the dye in hot acetic acid, add H_2O_2 and cool the mixture. Treatment for 1 h stains *Drosophila* salivary glands effectively.

(b) Quinone diimine	10 mg
70% aq. acetic acid soln.	1 cm

This solution also has a similar effect.

Trypan Blue

Trypan blue	0·2 g
Cresol	10 cm³
Absolute alcohol	60 cm³
Distilled water	30 cm³

This mixture has been recommended for plant smears (Hoffmeister, 1953).

Nigrosine

It is soluble in alcohol and is a complex basic dye of the azine series. It can be used as 1 per cent solution in 70 per cent alcohol for studying salivary gland chromosomes of *Drosophila*. (Pares, 1953.)

Acetic–dahlia (Ehrlich)

Dahlia is a reddish violet basic dye of the triphenyl methane series. A 2 per cent solution in 45 per cent acetic acid is useful in staining animal tissues. It has been used for squashing testes of moth larva and is used frequently in staining ascitic suspension.

Aniline Blue
(a) *Aqueous aniline blue*

Aniline blue	1 g
Distilled water	100 cm³

(b) *Mallory's aniline blue—Orange G*

Aniline blue	0·5 g
Orange G	2 g
Phosphomolybdic acid	1 g
Distilled water	100 cm³

Gallocyanin (Einarson)

Gallocyanin	0·3 g
Chrome alum	10 g
Distilled water	200 cm³

Methylene Blue (Loeffler)
(a) *Alkaline (Loeffler)*

Methylene blue	3 g
Absolute ethyl alcohol	30 cm³
1% aq. potassium hydroxide soln.	1 cm³
Distilled water	99 cm³

(b) *Acidic (Lillie)*

Methylene blue	1 g
0·5% acetic acid in 20% alcohol	100 cm³

It can be used as counterstain after carbol fuchsin for bacteria.

Methyl Violet (Jensen)

Methyl violet 6 B	0·5 g
Distilled water	100 cm³

Neutral Red (Jensen)

1% aq. neutral red soln.	10 cm³
1% aq. acetic acid soln.	0·2 cm³
Distilled water	100 cm³

Celestine Blue B with Iron (Gray and colleagues, 1956)

Conc. sulphuric acid	0·5 cm³
Celestine blue B	1 g
2·5% aq. ferric alum soln.	100 cm³
Glycerol	14 cm³

Add the acid to the stain and dissolve the mass in the mixture of ferric alum solution and glycerol.

Sudan Black B

This is used in various combinations with some of the lower fatty acids and related acids (Cohen, 1949). For chromosomes of mouse, mix together equal parts of 1 per cent Sudan black B solutions in propionic acid, in 85 per cent lactic acid, in 20 per cent formic acid and water, and filter. Mix drops of cell suspensions of neoplasma; stain or mince the solid tissue and mix with the stain for 10–60 min. Dehydrate in alcohol vapour and mount in euparal or diaphane (Bunker, 1961).

Chlorazol Black E

Chlorazol black E	1 g
Distilled water	100 cm³

It has been used effectively on root tips and the chromosomes take up a deep black stain.

Brilliant Cresyl Blue

Brilliant cresyl blue	2 g
Glacial acetic acid	45 cm³
Distilled water	55 cm³

Dissolve the stain in the mixture of acid and water. It has been used for root tip chromosomes (Stewart and Schertiger, 1949).

Cresyl violet

1 per cent solution in 50 per cent acetic acid is found to be effective for both mosquito chromosome spreads and root tip smears (Amirkhanian, 1964, 1968).

Haematoxylin

(a) *Heidenhain's stain*
    Materials required

Haematoxylin	0·5 g
96% ethanol	10 cm^3
Distilled water	90 cm^3

Preparation—Dissolve the stain in the mixture of 96 per cent alcohol and distilled water. Filter and ripen for 6–8 weeks.

(b) *Delafield's stain*
    Materials required

Ammonium alum	10 g
Haematoxylin	1 g
Absolute ethyl alcohol	6 cm^3
Distilled water	100 cm^3
Glycerol	25 cm^3
Methyl alcohol	25 cm^3

Preparation—Dissolve ammonium alum in distilled water to prepare a saturated solution. Dissolve haematoxylin in absolute alcohol. Add the latter solution slowly to the former. Expose to air and light for one week. Filter. Add 25 cm^3 of glycerol and 25 cm^3 of methyl alcohol. Allow to stand, exposed to air, until the colour darkens. Filter. Store in a tightly closed container. Allow the solution to ripen for a month before use.

(c) *Ehrlich's acid haematoxylin*
    Materials required

Absolute ethyl alcohol	100 cm^3
Glycerol	100 cm^3
Glacial acetic acid	10 cm^3
Haematoxylin	2 g
Distilled water	100 cm^3
Alum	in excess

Preparation—Dissolve haematoxylin in alcohol, add acetic acid, then glycerol and water. Allow the mixture to ripen in light, with occasional admission of air, till it acquires a dark red colour. For quick ripening, use either potassium permanganate or chloramine T and barium peroxide.

(d) *Cole's iodine haematoxylin (1943)*
  Materials required

Haematoxylin	0·5 g
1% iodine soln. in 95% ethyl alcohol	50 cm^3
Warm distilled water	250 cm^3
Saturated aq. ammonium alum soln.	700 cm^3

Preparation—Dissolve haematoxylin in warm water. Add iodine and alum solutions. Bring to boil and allow to cool.

Modifications—1. For quickly ripening haematoxylin stain, bubble pure oxygen through the freshly prepared stain (Hance and Green, 1959).
  2. *Groat's modification of haematoxylin (1949)*

Ferric ammonium sulphate	1 g
Sulphuric acid	0·80 cm^3
95% ethyl alcohol	50 cm^3
Haematoxylin	0·5 cm^3

Add these chemicals successively to 50 cm^3 of distilled water. Filter and use. The best period for staining is after 5 h and before 4–8 weeks of standing. Staining time is 3–30 min. De-staining can be done in a mixture of water (50 cm^3), rectified spirit (50 cm^3) and sulphuric acid (0·18 cm^3).
  3. Haematoxylin as nuclear stain was double stained by neutral red, followed by differentiation in aniline–xylol in paraffin sections (Duffett, 1949).
  4. *Haemalum (Harris)*

Haematoxylin	0·5 g
Mercuric chloride	0·25 g
Potash alum	5 g
Distilled water	100 cm^3

It has been used on tails of Urodele larvae (Finkhauser, 1945).
  5. *Haemalum (Mayer)*

Haematein	0·1 g
Absolute alcohol	5 cm^3
Potash alum	5 g
Distilled water	100 cm^3

  6. *Chrome haematoxylin (Gomori)*

5% aq. haematoxylin soln.	100 cm^3
Chrome alum	1·5 g
Potassium dichromate	0·1 g
Conc. sulphuric acid	0·1 cm^3

After staining for 15 min at room temperature, differentiation in 1 per cent HCl in ethanol for 1 min and counterstaining with phloxine is suitable for sections. For tissues, staining at 60 °C for 40 min and differentiation in 45 per cent acetic acid for 30 min before squashing is recommended (Melander and Wingstrand, 1953).

7. Harris's modification of haematoxylin

10% haematoxylin in absolute alcohol	5 cm³
Mercuric oxide	0·25 g
10% aq. potash alum soln.	100 cm³
Glacial acetic acid	4 cm³

Mix the haematoxylin and alum solutions and heat to boiling. Add mercuric oxide and when the solution turns deep purple, turn off the heat. Cool and add acetic acid.

8. Regaud's haematoxylin

10% haematoxylin in absolute alcohol	10 cm³
Glycerol	10 cm³
Distilled water	80 cm³

9. Weigert's haematoxylin

(a) 10% ripened haematoxylin in absolute alcohol	10 cm³
Absolute alcohol	90 cm³
(b) 30% aq. ferric chloride soln.	4 cm³
Hydrochloric acid	1 cm³
Distilled water	95 cm³

Wittmann's acetic-iron-haematoxylin (1962)

Chrome alum	0·1 g
Iron alum	0·1 g
Iodic acid	0·1 g

Add to 6 cm³ of a mixture of HCl and ethanol (1:1). Treat plant material fixed in acetic–ethanol (1:3) in the mixture for 10 min, then in Carnoy's fluid (6:3:1) for 10–20 min, squash in a drop of stain containing 4 per cent haematoxylin and 1 per cent iron alum in 45 per cent acetic acid and heat gently. Lowry (1963) utilised this stain for studying basidia of agarics (*see* chapter on schedules).

A further adaptation for materials not requiring hydrolysis, like leucocytes, ascites cells, etc. contains (Wittmann, 1965):

Stock solution:

> Haematoxylin 4 g; Iron alum 1 g;
> 45 per cent acetic acid 100 cm³.
> Ripen for 1–7 days.

Stain (working solution):

> Stock solution, 5 cm³: Chloral hydrate 2 g.
> (For schedule, *see* chapter on human chromosomes)

This technique was modified by Serra and Vincente (1960) but does not have a wide application.

10. *Modification by Henderson and Lu (1968)*
Stock: (a) 2% haematoxylin
       (b) 0·5% iron alum, both in 50 per cent propionic acid. Mix, keep
for 1 day and use as carmine or orcein after fixation, or as a
fixative-cum-stain. With unripened A, equal proportions of
A and B are needed, but with increased ripening of A, the
amount of B needed decreases.

Some Miscellaneous Double Stains

Gallocyanin and other stains—Tissues warmed in gallocyanin solution
for 2–4 min can be counterstained in Biebrich scarlet, phloxine or eosin Y
(Cole, 1947).
Safranin O and aniline blue—Root tips are stained 15 min in 1 per cent
aqueous safranin O and rinsed in distilled water. They are then stained in
1·0 per cent aniline blue W.S. in 95 per cent alcohol for 2 min (Darrow, 1944).
Carbol fuchsin and methylene blue—Seminal fluid can be stained in a
mixture of carbol fuchsin and rectified spirit (1:1), followed by a rinse in
water and staining for 2 min in 1·3 aqueous methylene blue soln. (Isenberg,
1949.)
Orange G and aniline blue—has been used for both mitotic and meiotic
chromosomes. Sections fixed in chromic–formalin (1:1) are rinsed in
potassium citrate buffer, stained in a mixture containing 2 g orange G and
0·5 g aniline blue dissolved in 100 cm^3 of potassium citrate buffer for up to
3 min, washed in the buffer, dehydrated and mounted (La Cour and Chayen,
1958).
Ruthenium red and Orange G after fuchsin staining—Stem-tips, after 30
min hydrolysis, are stained in fuchsin solution for 24 h, rinsed, stained in
aqueous ruthenium red solution for 30 min, dehydrated, stained for
$1\frac{1}{2}$ min in orange G in absolute ethanol and clove oil, run through clove oil
and xylol and mounted in balsam. Chromosomes take up deep purple stain
and resting nuclei less intense stain (Flint and Matzke, 1948).
Feulgen and a Schiff-type reagent—To check differential contraction by
different pre-treatments, a common sample treated in two chemicals is
stained separately, one in Feulgen and the other in Schiff type reagent,
Toluidine blue, Azure A or Chrysoidin yellow. After smearing under one
cover glass, the contrast in colour can be used as an aid in analysis (Savage,
1967).

Mixed aniline blue–eosin B

5% aq. aniline blue soln.	2 parts
5% aq. eosin B soln.	1 part
1% aq. phenol	1 part

Liquefied semen can be stained in the above mixture at 40–60 °C for
5–7 min (Casarett, 1953).

Croceine scarlet and Celestine blue

Croceine scarlet	0·38 g
Celestine blue B	114 cm^3

The dispersion of the former dye in the latter gives results similar to haematoxylin–eosin (Gray and colleagues, 1958).

Crystal violet and eosin

(a) Crystal violet 1 per cent aqueous solution.
(b) 3 per cent eosin–HCl in 70 per cent ethanol, prepared by adding 10 per cent HCl to 5 per cent aqueous solution of eosin Y, filtering, washing and drying the precipitate before dissolving in ethanol. It is used for selective staining of mitosis in follicle bulbs of sheep skin. Hydrolyse paraffin sections, after bringing down to water for 10 min in N HCl at 45 °C, rinse in water, stain for 1 min in crystal violet solution, rinse, treat with Lugol's iodine (1 g iodine, 2 g pot. iodide, 100 cm^3 water) for 1 min, rinse. Counterstain with eosin–HCl in 70 per cent ethanol for 5 s, rinse, keep in 70 per cent ethanol for 3 min, dehydrate, clear and mount (Clarke and Maddocks, 1963).

Staining Tissue with Small Quantities of Stain (Engle and Souders, 1952)

In order to stain tissues for long periods on slides with small quantities of dye, paraffin is painted on the free edge of a small spotless glass beaker of about 5 cm^3 capacity. The stain is collected in the beaker. The slide to be stained is heated slightly in an oven and placed on the open end of the beaker so that the tissue to be stained is in the beaker. The beaker and slide are pressed together till the paraffin hardens. The slide and beaker are then inverted so that the tissue is immersed in the dye for the required period. Several slides can be stained with the same small quantity of dye.

REFERENCES

Aldridge, W. G. and Watson, M. L. (1963). *J. Histochem. Cytochem.* **11**, 773
Amirkhanian, J. D. (1964). *Nature* **201**, 319
Amirkhanian, J. D. (1968). *Stain Tech.* **43**, 167
Atkinson, W. B. (1952). *Stain Tech.* **27**, 153
Baker, J. R. (1950). *Cytological technique.* London; Methuen
Baker, J. R. (1958). *Principles of biological microtechnique.* London; Methuen
Baker, J. R. and Sanders, F. K. (1946). *Nature, Lond.* **158**, 129
Barber, H. N. and Price, J. R. (1940). *Nature, Lond.* **146**, 355
Barger, J. D. and De Lamater, E. D. (1948). *Science* **108**, 121
Barka, T. (1956). *J. Histochem. Cytochem.* **4**, 208
Barka, T. and Ornstein, L. (1960). *J. Histochem. Cytochem.* **8**, 208
Bauer, H. (1932). *Z. Zellforsch.* **15**, 225
Bauer, H. (1933). *Z. Micro-anat. Forsch.* **33**, 143
Belar, K. (1929). *Meth. Wiss. Biol.* **1**, 638
Belling, J. (1921). *Amer. Nat.* **55**, 573
Belling, J. (1924). Referred to in Conn, 1953
Benda, C. (1896). *Verh. physiol. Ges. Berl.* 562
Bloch, D. P. and Godman, G. C. (1955). *J. Biophys. Biochem. Cytol.* **1**, 17

Böhm, N. and Sandritter, W. (1966).   *J. Cell Biol.* **28,** 1
Bowden, W. M. (1949).   *Stain Tech.* **24,** 171
Brachet, J. (1947).   *Symp. Soc. exp. Biol.* **1,** 207
Brachet, J. and Quertier, J. (1963).   *Exp. Cell Res.* **32,** 410
Bunker, M. C. (1961).   *Canad. J. Genet.* **3,** 355
Cain, A. J. (1949).   *Quart. J. micr. Sci.* **90,** 75
Cannon, H. G. (1937).   *Nature* **139,** 549
Capinpin, J. M. (1930).   *Science* **72,** 370
Carr, D. H. and Walker, J. E. (1961).   *Stain Tech.* **36,** 233
Carr, J. G. (1945).   *Nature* **156,** 143
Casarett, G. W. (1953).   *Stain Tech.* **28,** 125
Caspersson, T. (1944).   *Nature* **153,** 499
Cassel, W. A. (1950).   *J. Bact.* **59,** 185
Casselman, W. G. B. (1959).   *Histochemical technique.* London; Methuen
Catcheside, D. and Holmes, B. (1947).   *Symp. Soc. exp. Biol.* **1,** *Nucleic Acid,* 225
Chayen, J. and Norris, K. P. (1953).   *Nature* **171,** 472
Choudhuri, H. C. (1943).   *Nature* **152,** 475
Clarke, W. H. and Maddocks, I. G. (1963).   *Stain Tech.* **38,** 252
Cohen, I. (1949).   *Stain Tech.* **24,** 177
Cole, W. V. (1943).   *Stain Tech.* **18,** 135
Cole, W. V. (1947).   *Stain Tech.* **22,** 103
Coleman, L. C. (1938).   *Stain Tech.* **13,** 123
Conn, H. J. (1953).   *Biological stains.* Geneva, N.Y.; Biotech. Publications
Conn, J. E. (1943).   *Stain Tech.* **18,** 183
Cotton, J. (1959).   *Nature, Lond.* **183,** 128
Cowden, R. R. (1965).   *Histochemie* **5,** 441
Cutter, V. M. (1946).   *Stain Tech.* **21,** 129
D'Amato, F. (1951).   *Caryologia* **3,** 299
Danielli, J. F. (1947).   *Symp. Soc. exp. Biol.* **1,** 101
Danielli, J. F. (1949).   *Quart. J. micr. Sci.* **90,** 67
Darlington, C. D. and La Cour, L. F. (1942).   *The Handling of chromosomes,* 1st ed. London; Allen & Unwin
Darlington, C. D. and La Cour, L. F. (1960 and 1968).   *The Handling of chromosomes,* 3rd and 5th ed. London; Allen & Unwin
Darrow, M. A. (1944).   *Stain Tech.* **19,** 65
De Lamater, E. D. (1948).   *Mycologia* **4,** 423
De Lamater, E. D. (1951).   *Stain Tech.* **26,** 199
De Lamater, E. D., Mescon, H. and Barger, J. D. (1950).   *J. invest. Derm.* **14,** 133
Di Stefano, H. S. (1948).   *Chromosoma* **3,** 282
Di Stefano, H. S. (1952).   *Stain Tech.* **27,** 171
Dippell, R. B. (1955).   *Stain Tech.* **30,** 69
Dodson, E. O. (1946).   *Stain Tech.* **21,** 103
Dowrick, G. J. (1952).   *Heredity* **6,** 365
Duffett, R. E. (1949).   *Stain Tech.* **24,** 73
Duryee, W. R. (1937).   *Arch. Exp. Zellforsch.* **19,** 171
Dutt, M. K. (1968).   *Experienta* **24,** 615
Dutt, M. K. (1971).   *Nucleus* **14,** 4
Dyer, A. F. (1963).   *Stain Tech.* **38,** 85
Elftman, H. (1959).   *J. Histochem. Cytochem.* **7,** 93
Elliott, C. G. (1956).   *Symp. Soc. gen. Microbiol.* 279
Ely, J. O. and Ross, M. H. (1949).   *Anat. Rec.* **104,** 103
Engle, R. L. Jr. and Dempsey, E. W. (1954).   *J. Histochem. Cytochem.* **2,** 9
Engle, R. L. Jr. and Souders, M. J. (1952).   *Stain Tech.* **27,** 339
Feulgen, R. (1926).   *Handb. biol. Arb. Meth.* **213,** 1055
Feulgen, R. and Rossenbeck, H. (1924).   *Hoppe-Seyler's Z. physiol. Chem.* **135,** 203
Finkhauser, G. (1945).   *Quart. Rev. Biol.* **20,** 20
Flagg, R. O. (1961).   *Stain Tech.* **36,** 95
Flax, M. H. and Himes, M. H. (1952).   *Physiol. Zool.* **25,** 297
Flint, T. J. and Matzke, E. B. (1948).   *Science* **108,** 191

Ford, C. E. and Hamerton, J. L. (1956).   *Stain Tech.* **31,** 297
Freytag, A. W. (1963).   *Stain Tech.* **38,** 290
Fullmer, H. M. and Lillie, R. D. (1956).   *J. Histochem. Cytochem.* **4,** 64
Gabler, W. (1965).   *Acta Histochem.* **21,** 387
Ganesan, A. T. and Swaminathan, M. H. (1958).   *Stain Tech.* **33,** 115
Gatenby, J. B. and Beams, H. W. (1950).   In *The Microtomist's Vade-mecum* by Lee, B. London;
    Churchill
Gelei, J. (1921).   *Arch. Zellforsch.* **16,** 88
Gerard, P. (1935).   *Bull. Histol. Tech. micr.* **12,** 274
Godward, M. B. E. (1948).   *Nature* **161,** 203
Golechha, P. (1968).   Proc. Int. Seminar on chromosomes, *The Nucleus,* Suppl., Calcutta
Gomori, G. (1942).   *Proc. Soc. exp. Biol., N.Y.* **51,** 133
Gomori, G. (1950).   *Ann. N.Y. Acad. Sci.* **50,** 968
Gomori, G. (1952).   *J. nat. Cancer Inst.* **13,** 222
Gray, P., Bereezky, E., Maser, M. D. and Nevsimal, C. (1958).   *Stain Tech.* **33,** 215
Gray, P., Pickle, E. M., Maser, M. D. and Haywater, L. J. (1956).   *Stain Tech.* **31,** 141
Groat, R. A. (1949).   *Stain Tech.* **24,** 157
Gulick, A. (1941).   *Bot. Rev.* **7,** 433
Gurr, E. (1960).   *Encyclopaedia of microscopic stains.* London; Leonard Hill
Hance, R. J. and Green, F. J. (1959).   *Stain Tech.* **34,** 237
Hardonk, M. J. and van Duijn, P. (1964).   *J. Histochem. Cytochem.* **12,** 533, 748
Hartmann-Goldstein, I. (1961).   *Stain Tech.* **36,** 309
Hashim, S. A. (1952).   *Stain Tech.* **28,** 27
Hayes, R. E. (1949).   *Stain Tech.* **24,** 1923
Heidenhain, M. (1896).   *Z. wiss. Mikr.* **13,** 186
Henderson, S. A. and Lu, B. C. (1968).   *Stain Tech.* **43,** 233
Hillary, B. B. (1939).   *Bot. Gaz.* **101,** 276
Himes, M. and Moriber, L. (1956).   *Stain Tech.* **31,** 67
Hiraoka, T. (1960).   *J. biophys. biochem. Cytol.* **8,** 286
Hoffmeister, E. R. (1953).   *Stain Tech.* **28,** 309
Hörmann, H., Grassman, W. and Fries, G. (1958).   *Liebig's Ann. Chim.* **616,** 125
Hrushovetz, B. (1956).   *Canad. J. Bot.* **34,** 321
Imai, H. T. (1966).   *Acta Hymenopterologica* **2,** 119
Isenberg, H. D. (1949).   *Amer. J. Clin. Path.* **18,** 94
Itikawa, O. and Ogura, Y. (1954).   *Stain Tech.* **29,** 9
Jordanov, J. (1963).   *Acta histochem.* **15,** 135
Kasten, F. H. (1956).   *J. Histochem. Cytochem.* **4,** 310
Kasten, F. H. (1960).   *Int. Rev. Cytol.* **10,** 1
Kasten, F. H. (1965).   *J. Histochem. Cytochem.* **13,** 13
Kasten, F. H. (1967).   *Int. Rev. Cytol.* **21,** 142
Kissane, J. M. and Robins, E. (1958).   *J. biol. Chem.* **233,** 184
Kurnick, N. B. (1955).   *Int. Rev. Cytol.* **4,** 221
Lacoste, R. G. and Martell, A. E. (1955).   *J. Amer. Chem. Soc.* **77,** 5512
La Cour, L. F. (1941).   *Stain Tech.* **16,** 169
La Cour, L. F. and Chayen, S. (1958).   *Exp. Cell Res.* **14,** 462
Lessler, M. A. (1951).   *Arch. Biochem. Biophys.* **32,** 42
Lessler, M. A. (1953).   *Int. Rev. Cytol.* **2,** 231
Levene, P. A. and Bass, L. (1931).   *Nucleic acids.* New York; Chemical Catalogue Co.
Lhotka, J. F. and Davenport, H. A. (1949).   *Stain Tech.* **24,** 237
Lhotka, J. F. and Davenport, H. A. (1951).   *Stain Tech.* **26,** 35
Li, C. F. and Stacey, M. (1949).   *Nature* **163,** 538
Lillie, R. D. (1943).   *Publ. Hlth. Rept. Wash.* **58,** 449
Lillie, R. D. (1951).   *Stain Tech.* **26,** 123 and 163
Lima-de-Faria, A. (1962).   *Prog. Biophys.* **12,** 281
Lindegren, C. C., Williams, M. A. and McClary, D. O. (1956).   *Leeunwenhoek, Ned.* **22,** 1
Lison, L. (1932).   *Bull. Histol. Tech. micr.* **9,** 177
Lison, L. and Fautrez, J. (1939).   *Protoplasma* **33,** 116
Longley, J. B. (1952).   *Stain Tech.* **27,** 161
Lowry, R. J. (1963).   *Stain Tech.* **38,** 199

Ludford, R. J. (1935–6). *Arch. exp. Zellforsch.* **18,** 411
McClary, D. O., Williams, M. A., Lindegren, C. C. and Ogwe, M. (1957). *J. Bact.* **73,** 360
McGoldrick, P. T., Bohn, G. W. and Whitaker, T. W. (1954). *Stain Tech.* **29,** 127
Makino, S. and Nishimira, I. (1952). *Stain Tech.* **27,** 1
Mallory, F. B. (1938). *Pathological technique.* Philadelphia; Saunders
Markarian, D. (1957). *Stain Tech.* **32,** 147
Matsubayashi, M. (1963). *Stain Tech.* **38,** 264
Mayer, P. (1903). *Z. wiss. miks.* **20,** 409
Meggers, D. E. and Allen, A. M. (1962). *Stain Tech.* **37,** 221
Meischer, F. (1869). Referred to in Baker, 1950
Melander, Y. and Wingstrand, K. G. (1953). *Stain Tech.* **28,** 217
Meyer, P. L. (1948). *Proc. Soc. exp. Biol. N.Y.* **68,** 664
Michaelis, J. F. (1926). Referred to in Baker, 1950
Mollendorff, W. (1936). Referred to in Gurr, 1960
Murgatroyd, L. B. (1968). *J. Roy. Micr. Soc.* **88,** 133
Nebel, B. R. (1940). *Stain Tech.* **15,** 69
Newcomer, E. H. (1952). *Stain Tech.* **27,** 205
Newcomer, E. H. (1959). *Stain Tech.* **34,** 349
Newton, W. F. C. (1926). *J. Linn. Soc. (Bot.)* **47,** 339
Overend, W. G. (1950). *J. chem. Soc.* **27**: 69
Overend, W. G. and Stacey, M. (1949). *Nature* **163,** 538
Pandey, K. K. and Henry, R. D. (1959). *Stain Tech.* **34,** 19
Pares, R. (1953). *Nature* **174,** 1151
Pearse, A. G. E. (1951). *Quart. J. Micr. Sci.* **32,** 393
Pearse, A. G. E. (1960). *Histochemistry, theoretical and applied.* London; Churchill
Pearse, A. G. E. (1968). *Histochemistry* 3rd ed. London; Churchill
Pelc, S. R. (1956). *J. appl. Rad. Isotopes* **1,** 172
Perkin, A. G. and Everest, A. E. (1918). *The Natural organic coloured matters.* London; Longman
Rafalko, J. S. (1946). *Stain Tech.* **21,** 91
Raghuvanshi, S. S. and Joshi, S. (1963). *Stain Tech.* **38,** 311
Ris, H. (1943). *Biol. Bull.* **85,** 164
Robinow, C. (1941). *Proc. Roy. Soc. B.* **130,** 299
Robinson, R. L. and Bacsich, P. (1958). *Stain Tech.* **33,** 71
Rumpf, P. (1935). *Ann. Chim.* **3,** 327
Saez, F. A. (1952). *Anat. Rec.* **113,** 571
Savage, J. R. K. (1967). *Stain Tech.* **42,** 19
Schmidt, G. M. J. (1944). Referred to in Baker, 1950
Schreiber, J. (1954). *Stain Tech.* **29,** 285
Semmens, C. S. (1940). *Nature* **146,** 130
Sen, S. (1965). *Nucleus* **8,** 79
Serra, J. A. (1948). *Bol. Soc. Broteriana* **17,** 203
Serra, J. A. and Vincente, M. J. (1960). *Rev. Port. Zool. Biol. Geral.* **2,** 219
Sharma, A. K. (1951). *Nature* **167,** 441
Sharma, A. K. (1952). *Portug. acta biol.* **3,** 239
Sharma, A. K. and Roy, M. (1955). *Chromosoma* **7,** 275
Sharma, A. K. and Sharma, A. (1957). *Stain Tech.* **37,** 167
Shinke, N., Ishida, M. R. and Ueda, K. (1957). *Proc. Int. Genet. Symp. Cytologia Suppl.* p. 156
Shortt. (1923). *Indian J. med. Res.* **23**
Shriner, R. L. and Fuson, R. C. (1948). *The Systematic identification of organic compounds.* New York; Wiley
Sibatani, A. and Fukuda, M. (1953). *Biochem. biophys. Acta* **10,** 93
Singleton, J. R. (1953). *Amer. J. Bot.* **40,** 124
Sloane-Stanley, G. H. and Bowler, L. M. (1962). *Biochem. J.* **85,** 34
Snow, R. (1963). *Stain Tech.* **38,** 9
Stedman, E. and Stedman, E. (1943). *Nature* **152,** 267
Stedman, E. and Stedman, E. (1948). *Biochem. J.* **43,** 23
Stedman, E. and Stedman, E. (1950). *Biochem. J.* **47,** 508
Stewart, W. N. and Schertiger, A. M. (1949). *Stain Tech.* **24,** 39
Stowell, R. E. (1946). *Stain Tech.* **31,** 137

Stoward, P. J. (1963).   *D. Phil. thesis,* Oxford
Stoward, P. J. (1966).   *J. Histochem. Cytochem.* **14,** 681
Stowell, R. E. and Albers, V. M. (1943).   *Stain Tech.* **18,** 57
Swaminathan, M. S., Magoon, N. L. and Mehra, K. L. (1954).   *Indian J. Genet.* **14,** 87
Swift, H. (1950).   *Physiol. Zool.* **23,** 169
Swift, H. (1955).   *The Nucleic Acids. Chemistry & Biology,* Chap. 17 by Chargaff, E. and Davidson, J. N.
    New York; Academic Press
Tanaka, R. (1961).   *Stain Tech.* **36,** 325
Taylor, J. H., Woods, P. S. and Hughes, W. L. (1957).   *Proc. nat. Acad. Sci., Wash.* **43,** 122
Tjio, J. H. and Levan, A. (1950).   *An Estac. exp. Aula Dei* **2,** 21
Tobie, W. C. (1942).   *Industr. Engng. Chem. (Anal.)* **14,** 405
Vaarama, A. (1949).   *Portug. acta biol. A.* 47
Vendrely, R. and Vendrely, C. (1956).   *Int. Rev. Cytol.* **4,** 269
Warmke, H. E. (1935).   *Stain Tech.* **10,** 101
Watson, J. D. and Crick, F. H. C. (1953).   *Nature* **171,** 737
Widström, G. (1928).   *Biochem. Z.* **199,** 298
Wieland, H. and Scheuing, G. (1921).   *Ber. dtsch. chem. Ges.* **54,** 2527
Wimber, D. E. (1956).   *Stain Tech.* **31,** 124
Wittmann, W. (1962).   *Stain Tech.* **37,** 27
Wittmann, W. (1965).   *Stain Tech.* **40,** 161
Wroblewska, J. and Dyban, A. P. (1969).   *Stain Tech.* **44,** 147

6 Mounting

After staining, the tissue, whether in the form of section or smear, is mounted in a suitable medium for observation under the microscope. The nature of the medium depends on the type of stain used and the type of preparation required, and the process of mounting also depends on the medium used. If the tissue is observed under the microscope without mounting, it will usually dry up and be rendered opaque. The chief aims of mounting are to render the tissue transparent, to increase its visibility under the microscope, to hold it with the protecting cover-glass firmly in place and to preserve it for a period of time or indefinitely.

MOUNTING MEDIA

All preservative media may be used for mounting, though the only media that afford absolutely sure preservation of soft tissues are the resinous ones.

The prerequisites of a good mounting medium are:

1. It should have a refractive index which is the same or slightly higher than that of the glass slide and almost the same as the tissue. If the tissue is impregnated with the medium, a light ray passing through the almost homogeneous mass of glass slide, embedding medium with tissue and cover-glass, will not be lost through refraction. This will aid in the optical visibility.

2. It should harden quickly in contact with air so as to fix the cover-glass firmly to the slide.

3. It should check de-staining and prevent acidity in the preparation.

4. It should be stable and not decompose with storage or changes of temperature.

A number of substances are used as mounting media depending on the type of preparation desired. Table 4 lists the commonly used ones with their refractive indices.

In addition to these chemicals, temporary squashes can be mounted

Table 5   REFRACTIVE INDICES OF MOUNTING MEDIA*

Resin		Solvents		Water-soluble media	
Balsam (dry)	1·535	Absolute ethyl alcohol	1·367	Abopon	1·4372
Balsam (in xylol)	1·5322	Aniline oil	1·580	Apáthy-Lillie	1·4189
Balsam (60:40 in xylol)	1·5300	Castor oil	1·490	Distilled water	1·336
		Cedarwood oil (thickened)	1·520	Farrant's glycerol gum arabic	1·4417
Clarite X (do)	1·5352	Cinnamon oil	1·567		
Dammar resin (do)	1.5317	Clove oil	1·533	Fructose syrup	1·4362
Diaphane (colourless)	1·4777	Creosote	1·538	Glycerol jelly	1·4353
Euparal	1·483	Glycerol	1·4674	Glycerin (50% aq. soln.)	1·397
Harleco resin (in xylol)	1·5202	Liq. paraffin	1·471	Sea water	1·343
Lucite (in xylol)	1·4962	Methyl alcohol	1·323	Solution of white of egg	1·350
Mahady micro-mount (solid resin)	1·4879	Methyl benzoate	1·517		
		Oil of aniseed	1·557		
		Oil of bergamot	1·464		
Permount (solid resin)	1·5376	Oil of turpentine	1·473		
Polystyrene (in xylol 1:1)	1·5378	Olive oil	1·473		
		Polyvinyl alcohol	1·386		
Polystyrene (solid resin)	1·6279	Terpinol	1·484		
		Toluene	1·4956		
		Xylol	1·4982		

* Refractive index of Crown Glass: 1·518

either in stain or its solvent. For permanent mounts, some of these chemicals are more widely used than others. The better-known ones are now described.

Balsam

Balsam is probably the soundest mounting medium optically. When dissolved in xylol, its refractive index (1·53) is very close to that of glass

(1·518). When quite dry, the refractive index is approximately 1·535. It also has almost the same optical dispersion as glass.

Preparation— It is an oleoresin collected from blisters formed in the bark of *Abies balsamea* found chiefly in North America. An oleoresin is an essential oil in which resins, themselves oxidation products of essential oils, are dissolved. The balsam used in cytological work, commonly known as Canada balsam, is a very thick, light yellow, transparent liquid, 24 per cent of which is essential oil. The remaining 76 per cent is composed chiefly of a resin soluble in both xylol and ethanol, and partly of a resin which is soluble in xylol but insoluble in ethanol. Hence Canada balsam is fully soluble in xylol but partially soluble in ethanol. On drying, the essential oils are removed.

Advantages—Canada balsam fulfils most of the conditions necessary for a perfect mounting medium. It has almost the same refractive index and optical dispersion as glass and is transparent and almost colourless. It is thick and affixes the cover-glass firmly to the slide and in the presence of air forms a very hard medium. It is stable and does not decompose or granulate in heat or after long storage.

Drawbacks—The chief drawback of Canada balsam is that it darkens and becomes acidic with time by slowly oxidising xylol, in which it is dissolved, to toluol and phthalic acids. The acidity causes certain basic stains to fade. Various neutral resins have been devised, such as neutral balsam in sealed tubes and 'Clarite X', to eliminate this possibility. Canada balsam and the allied modified media are usually used dissolved in xylol, though occasionally toluol, benzol, dioxane or trichlorethylene can also be used. The tissue, after staining, is usually dehydrated in ethyl alcohol and passed through xylol grades before mounting in balsam, but xylol can be replaced by any of the solvents mentioned above, both as ante-medium and as solvent for the mounting medium. Hillary (1938, 1939, 1940) recommended dioxane and dioxane balsam for dehydration and mounting.

Precautions—The essential precautions in using Canada balsam are: it should never be heated for melting, and a piece of clean marble can be kept inside the bottle to prevent acidity, replacing it occasionally with a fresh piece.

Dammar Balsam or Gum Dammar

Dammar balsam or gum dammar is sometimes used for microscopical preparations, and is prepared by melting the solid gum over a hot flame in a container and then pouring it into the solvent, which is usually benzol. It is then carefully filtered and used. It is composed of unsaturated resin acids and a little ester (m.p. 100 °C).

Euparal

Euparal is used extensively. It is a mixture of camsal, sandarac, eucalyptol and paraldehyde and has a higher refractive index than Canada balsam.

Camsal is a liquid formed by the mutual solution of the two solids, sodium salicylate and camphor, in the proportion of 3:2. Gum sandarac is an unsaturated acid resin, composed typically of sandaracolic acid and callitrolic acid.

Advantages—The advantages of euparal are: it does not dry too rapidly and, as it is soluble in butyl and ethyl alcohols, in addition to xylol, toluol, etc., the tissue can be mounted directly from ethyl alcohol in euparal, thus shortening the absolute ethyl alcohol, alcohol–xylol schedule. Its use helps in keeping the material on the slide without involving the risk of being washed off.

Drawbacks—The drawbacks of euparal as a mounting medium are that it has a slight solvent action on celloidin, but this property can be taken advantage of in flattening out curled or too stiff sections; it has a tendency to cloud readily, but the cloudiness can be dispelled by slight warming.

Diaphane is a similar semi-synthetic mixture, allied to euparal.

Lactophenol

Lactophenol is a mixture of equal parts of phenol crystals, lactic acid, glycerin and distilled water, but the proportion of glycerin can be increased as necessary. The application of this medium is rather limited, however. Tissues can be mounted in it after staining or from any of the alcohol grades. The preparations have to be sealed as lactophenol is unable to attach cover-glasses firmly to the slides.

Cedarwood Oil

The only advantage of this medium is that its refractive index is very close to that of glass. It also preserves stains well. After mounting, however, cedarwood oil hardens only along the edges of the cover-glass, the inside remaining liquid. The tissue has to be dehydrated and then passed through xylol before mounting in this oil, as xylol is its most commonly used solvent.

Glycerin Jelly

Glycerin jelly is probably the most widely used aqueous mounting medium. The medium is a mixture of gelatin, used as adhesive, glycerin, used for increasing the refractive index, cresol or phenol, used as a preservative, and distilled water.

Preparation—The mixture can be prepared, according to Baker, as follows:
5 g of gelatin are soaked in 25 cm^3 of distilled water for an hour and then warmed in a hot bath until the gelatin is completely dissolved; 35 cm^3 of glycerin, 40 cm^3 of distilled water and 0·25 cm^3 of cresol are mixed together and kept in the hot bath. The two fluids are mixed

thoroughly and filtered through a muslin strainer in a large incubator and the medium is poured into its container whilst still warm. When needed for mounting the mixture is warmed so that it melts, and on cooling, or through evaporation, it sets into a gel and attaches the cover-glass firmly to the slide.

Alternative medium by Evens (1961):

Stock: (a) Gelatin 4 g, glycerin 60 cm^3, water to make 100 cm^3.

   (b) Formalin 10 cm^3, glycerin 60 cm^3, water to make 100 cm^3. Mix in equal proportions before use. Add precipitated chalk for alkalinisation.

Another medium for use after 1 per cent acetic carmine contains:

   water 50 cm^3, gum arabic 30 g,
   chloral hydrate 200 g, glycerin 16 cm^3
   (Beeks, 1955).

Uses—This medium is used for most temporary squash techniques, and the tissue can be transferred to it directly from water after staining.

The medium should be divided into several small containers as it deteriorates on continuous remelting.

Various Other Synthetic Media

Various other synthetic media have also been devised from time to time, some of which are naturally occurring resins and others synthetic substitutes.

Styrene—$C_6H_5CH = CH_3$, can be obtained as a plant product or can be synthesised. It is polymerised on heating.

Distrene 80—This is a derivative of styrene and its exact formula is not disclosed by its manufacturers. It is readily soluble in xylol and is a good mounting medium for tissues which have been dehydrated and brought through ethanol–xylol grades. It has, however, a tendency to shrink on drying, which can be prevented by adding a plasticiser, tricresylphosphate. This mixture is known as Kirkpatrick and Lendrum's D.P.X. and is prepared by mixing 7·5 cm^3 of tricresylphosphate with 40 cm^3 of xylol and adding to it 10 g of distrene.

D.P.X. mixture can be used in the same way as Canada balsam and its advantages are that the pH does not change on storage and it preserves the colour of basic dyes, and that the medium, being completely colourless, aids visibility.

Alternatively, dibutylphthalate can be used as a substitute for tricresylphosphate in D.P.X.

Clarite X—This is another synthetic medium and has the following advantages over Canada balsam: it is colourless, neutral, inert, homogeneous and dries quickly. It cannot, however, be used for acetic-carmine preparations and becomes milky with ethanol. An 80 per cent solution in xylol is used for mounting plant specimens and a 60 per cent solution in toluol for animal tissues. The tissues have to be thoroughly dehydrated and passed through ethanol–xylol grades before mounting in clarite.

Several other resins are also available, like Abopon, but they are not so universally effective.

Oils

Several oils, like clove oil, castor oil, bergamot oil, aniseed oil, turpentine oil, etc., have been used as mounting media. Optically, they are quite effective, but their chief defect is that they thicken very slowly and do not attach the cover-glass firmly.

Polyvinyl Alcohol

Polyvinyl alcohol is also frequently used as a mounting medium. It is prepared by adding 15 g of polyvinyl alcohol (Elvanol H-24 from Du Pont) to 100 cm^3 of cold distilled water, which is then stirred and heated till the resultant substance thickens, and is then filtered and used as stock solution; 56 cm^3 of the stock solution is mixed with 22 cm^3 of lactic acid and 22 cm^3 of phenol to give the required mounting medium. Its refractive index is 1·386, which increases with the evaporation of alcohol. The sections can be mounted in it directly from 70 per cent ethanol. However, it is not very good for eosin-stained material.

PROCESSES OF MOUNTING

Depending on the medium used and on the type of preparation required, the different schedules, used after staining, can be classified into the following groups:
1. Temporary mounts.
2. Permanent mounts, further divisible into:
    (a) Permanent mounts of sections.
    (b) Permanent mounts of squashes and smears directly processed for permanency.
    (c) Permanent mounts prepared indirectly from temporary squashes and smears.
3. Miscellaneous.

For observations under phase contrast, Crossman (1949) suggested temporary mounting in aqueous glycerin or butyl carbitol to find out the proper index liquid. Later, a solid mounting medium having a close refractive index is substituted.

The different processes are discussed below separately:

Temporary Mounts

Temporary mounts are made mostly of squashes and smears. The medium for mounting is either the stain itself or its solvent. In the former case, the process of staining and mounting are done simultaneously, the tissue then being lifted on to the slide containing a drop of the stain and squashed as described in Chapter 5. In the latter, the tissue is stained and squashed on a slide in a solvent of the stain. In both cases, the cover-glass is ringed with paraffin wax and then observed.

The usual media are the stains themselves, for example:

1 per cent acetic-orcein solution—Tissues heated in 2 per cent acetic-orcein–N HCl mixtures are squashed in 1 per cent acetic-orcein solution and the preparations are sealed for observation, for example, root tips, tumour tissue, endosperm, leaf tip, etc. Alternatively, the tissues are directly placed in a drop of 1 per cent acetic-orcein solution, squashed as usual and sealed—for example, anther cells, testes, etc.

1 per cent acetic-carmine solution—Anthers, testes, etc., can be directly squashed in 1 per cent acetic-carmine solution and sealed. Root tips and similar tissue, after heating in 2 per cent acetic-orcein–N HCl mixture or 2 per cent propionic-carmine–N NCl mixture can be squashed in 1 per cent acetic-carmine solution (Zirkle, 1939).

1 per cent propionic-carmine solution—This can be used instead of 1 per cent acetic-carmine in similar types of tissue, particularly where background cleansing becomes necessary.

1 per cent propionic-orcein solution—This is used in cases similar to 1 per cent acetic-orcein for squashing and mounting tissues which have been heated in 2 per cent propionic-orcein–N HCl mixture.

Solvents are, for example:

Distilled water—For immediate observation, tissues can be mounted in distilled water after staining and the cover-glass ringed with paraffin. However, the tissue cannot be kept intact for any long period.

Sea water—Sea water is also suitable as a temporary mounting medium after staining in any of the aqueous stains.

45 per cent acetic acid—This is used for squashing stained tissues. For example, tissues stained in Feulgen solution or in 2 per cent acetic-orcein–N HCl mixture can be squashed and mounted in 45 per cent acetic acid. Alternatively, smears of tissues on slides, stained in Feulgen solution, can also be mounted in 45 per cent acetic acid.

Subsequent to ringing the cover-glass with paraffin, the slides can be kept for nearly a week, after which the material tends to dry up. Dentists' sticky wax can also be used for ringing coverslips (Conger, 1960).

Permanent Mounts

Permanent mounts indicate mounting the tissue, after suitable processing, so that the preparations can be kept for a long period, often for several years, without appreciable distortion of the structure or intensity of stain.

Permanent mounts of sections from paraffin blocks—The different steps in the process usually depend on the medium in which the stain is dissolved. In general, the entire process is based on first dehydrating the tissue, then impregnating it with the solvent of the mounting medium and finally mounting with the medium chosen.

The most commonly used dehydrating agent for paraffin sections is ethyl alcohol, though acetone and various other alcohols are also used. As the tissue has already been embedded in paraffin and therefore has attained a permanent shape, dehydration does not have to be done in gradual stages. If the stain is dissolved in water, the sections can be transferred directly to absolute alcohol after staining and mordanting. If, however, a counter-stain in a lower grade of alcohol is applied, the tissue has to be passed through the required grade before transference to absolute alcohol. Usually two or three jars containing absolute alcohol are kept and the slides are kept for 4–5 s in each.

The slides can be transferred directly from absolute alcohol to the mounting medium, e.g. to euparal. They may otherwise be transferred to a mixture of the solvent of the mounting medium and absolute alcohol in equal proportions, then to the pure solvent, as in the case of preparations stained in Feulgen and mounted in Canada balsam. Alternatively, they may be transferred to a differentiating medium for removing excess of the stain before transference to the solvent of the mounting medium, for example, tissues stained in crystal violet and mounted in Canada balsam.

The choice of differentiating or clearing medium usually depends upon the stain used. The most suitable differentiating agent for crystal violet or gentian violet is clove oil. It removes superfluous stain from the cytoplasm, thus rendering the stain in the chromosomes brighter, and also completing the dehydration. Usually the slides are kept for 2–5 min in the clearing medium. As soon as the surplus stain is washed off, they are transferred to the pure solvent for the mounting medium. Often two jars of clove oil are kept as, after some time, clove oil gets slightly coloured. The clearing agent may be a solvent for the stain or a mordant, as iron alum solution for haematoxylin stain.

The next step is to transfer the tissue to a solvent of the mounting medium. This step can be omitted for mounting media which are soluble in absolute ethanol, like euparal and lactophenol. For Canada balsam, which is partially insoluble in ethanol, an intermediate step of transfer to xylol is needed. The sections may be kept from half an hour to an indefinite period in xylol. Usually two or three successive jars of xylol are used, as the xylol in the jar immediately after the clove oil grades tends to become coloured. In such cases, the slides are kept for the longest period in the xylol immediately preceding the mounting medium, to remove all traces of clove oil from sections along with clearing.

The last step is to mount the sections in the mounting medium. Rectangular cover-glasses are generally used for mounting sections. The instruments required are a needle and a pair of forceps. The slide is lifted out and the xylol drained off. The under-surface of the slide is wiped on a piece of cloth or blotting paper and it is placed face upwards on a sheet of blotting paper. A drop of the mounting medium is placed near the end of the slide at the left side of the sections, and a clean cover-glass is lifted up by means of the forceps and is placed with its left edge touching the slide—the right edge of the cover-glass is kept tilted upwards with the forceps or with a needle. The cover-glass is allowed to settle down slowly on the drop of medium by gently lowering the needle in the right hand, and as the cover-glass descends (supported at the two sides by the left forefinger and thumb of the operator) it slowly pushes the medium ahead and helps it to spread uniformly throughout the area of the cover-glass. After the entire cover-glass has settled down, the forceps or needle supporting it is removed and the cover-glass is pressed slightly from the top to remove any air bubbles that might have lingered within the medium. The slides are then kept, face upwards, on a hot plate between 35 ° and 45 °C for a day or two and finally, when partly dry, they are kept in slide cabinets meant for the purpose.

The necessary precautions which have to be adopted during mounting are:

The cover-glasses must be absolutely clean and dry. They should be kept in rectified spirit, thoroughly wiped with a clean cloth before use and kept on a clean piece of paper in a tilted condition. If necessary they are passed quickly through an ethanol flame for dehydrating. They are handled with a pair of fine forceps or, if fingers are used, the cover-glasses should always be held across the edges to avoid any smudges. The mounting medium must be placed on the slide as quickly as possible after lifting it out of xylol. Other-wise the sections may dry up. The cover-glass is placed very gently on the medium to prevent air bubbles entering the preparation.

The size of the drop of mounting medium depends on the thickness of the medium and the size of the cover-glass. If after mounting it is found that the medium does not cover the entire area within the cover-glass, a small drop of medium can be inserted under one edge of the cover-glass, holding the slide in a slanting position. If, however, the medium is in excess, the top of the slide and cover-glass are wiped carefully with a soft cloth moistened in xylol to remove the excess medium.

When the cover-glass is first applied, it should be held in the exact position desired. After settling down, it should never be removed sideways in any direction, as otherwise the sections may be damaged.

For sections embedded in celloidin, a slightly different procedure is followed. Since celloidin has a suitable refractive index, it can be retained in the final mount, because it holds the cells together. For this purpose, dehydration with ethyl alcohol is omitted, as celloidin is soluble in ethyl alcohol. The slide, after staining, is passed through 90 per cent alcohol to alcohol–chloroform mixture (1:1) and kept in it for 2 min. It is next transferred to a mixture of absolute alcohol, chloroform and xylol (1:1:1) and kept for 2 min. Finally the slide is given two changes in pure xylol and mounted in Canada balsam or in one of the synthetic resins.

Permanent mounts of squashes and smears directly processed for permanency —These more or less follow the same procedure as that used for permanent mounts of paraffin sections. With Canada balsam and synthetic resins the usual steps of dehydration in absolute ethyl alcohol are followed, the period being reduced to 2–3 dips in each jar. Differentiation in clove oil, if necessary, is done for 5 min. The period of treatment in xylol is from 1 h to overnight.

With euparal or other mounting media, clove oil and xylol stages are not necessary.

Permanent mounts prepared indirectly from temporary squashes and smears
—Several techniques have been devised to render the temporary squashes
and smears, as described previously, permanent. They can be divided into
two general categories: methods involving the removal of the cover-glass,
and methods without removing the cover-glass.

The methods involving the removal of the cover-glass are quite numerous. They generally face two difficulties, namely: the problem of attaching
the tissue to the slide after the cover-glass has been removed, and the
problem of retaining the stain. Some of the more commonly used techniques are discussed here.

The acetic–alcohol schedule was first devised by McClintock (1929), and
modified forms of this method are used even now. This technique is based
on the principle of detaching the cover-glass in a solvent, dehydrating and
clearing before mounting in the desired medium.

The solvent used for detaching the cover-glass is either an aqueous
solution of the acid in which the medium used in mounting the tissue has
been dissolved, or a mixture of the acid and the dehydrating agent.

For tissues stained and mounted in stains dissolved in 45 per cent acetic
acid, like acetic-carmine, acetic-orcein or acetic-lacmoid, 10 per cent
acetic acid or a mixture of glacial acetic acid and absolute ethyl alcohol
(1:1) is used. On the other hand, for propionic-carmine or propionic-
orcein, 10 per cent propionic acid or a mixture of propionic acid and
absolute ethyl alcohol (1:1) is used. Similarly for tissues mounted in 45 per
cent acetic acid, like those stained by the Feulgen reaction, 45 per cent
acetic acid or acetic ethanol (1:1) is a good solvent.

The paraffin seal around the cover-glass of the temporary preparation is
removed carefully by wiping the top of the cover-glass with a piece of
clean cloth soaked in xylol. The slide is now inverted in a covered tray
containing the solvent. The cover-glass detaches and falls to the bottom of
the tray after half an hour. A part of the material adheres to the cover-glass
and comes off with it. The slide and cover-glass are transferred to a mixture
of the acid and absolute ethyl alcohol (1:1) and treated for 15 min. Later
they are passed through acetic-ethanol mixture (1:3) and two grades of
absolute alcohol, keeping them 5 min in each. If needed, slides with tissues
stained in 1 per cent acetic-carmine solution are mounted in euparal, while
those with tissues stained in acetic-orcein or acetic-lacmoid are mounted in
cedarwood oil or euparal.

To mount the preparation in Canada balsam, however, the slide with
tissue and cover-glass are transferred from absolute ethyl alcohol to alcohol–
xylol mixture (1:1), then to alcohol–xylol (1:3) and finally to pure xylol,
keeping them in each for about 5 min. They can be kept in pure xylol for
30 min and then mounted in Canada balsam. The slide and the cover-glass
are mounted separately.

Different variations of this basic schedule are available, depending mainly
on the periods of treatment and the different stages of dehydration.

The drawbacks of this system are that the process is rather time consuming, particularly if the ethyl alcohol–xylol grades are incorporated, and a
lot of the material is lost while passing the slide through the grades. In spite

of careful handling, the material is liable to be washed away while transferring the slides. An airtight container lined with absorbent paper which is saturated with ethanol for 24 h can be employed for dehydration (Bridges, 1937).

The tertiary butyl alcohol schedule—This technique involves the use of tertiary butyl alcohol as a solvent as well as a dehydrating agent.

After the seal has been removed from the cover-glass, the slide with the material is inverted in a covered tray containing tertiary butyl alcohol and kept as such overnight. The cover-glass, which has fallen off, and the slide are now mounted separately in euparal.

This technique is applicable to tissues stained with almost all stains. It has the advantage over the acetic–ethyl alcohol schedule of being much less complicated and, hence, the loss of material is also much reduced. However, it has been found in the authors' laboratory that the stain of the tissue made permanent through the tertiary butyl alcohol technique shows a tendency to diffuse.

Alternatively, the slide may be passed through glacial acetic acid–butyl alcohol grades (1:1, 1:3) till the cover-slip is detached, before treating in butyl alcohol and mounting in euparal (Celarier, 1956).

Dry ice technique—This method was devised by Conger and Fairchild in 1953. It is at present the most convenient technique for making temporary preparations permanent and is based on the principle of freezing a temporary preparation on dry ice to remove the cover-glass and subsequently dehydrating it before mounting.

Slides are frozen at the stage at which dehydration would ordinarily begin or at the time at which slide and cover-glass would be separated. Acetic- or propionic-stained slides are frozen while they are still in the stain, while Feulgen-stained tissue is frozen after squashing in 45 per cent acetic acid.

The slide is laid on a flat block of dry ice with the cover-glass facing upwards. It is then pressed against the ice with weights or with a hard scalpel or pencil so that the material is completely frozen. The usual time for freezing is 30 s, but a longer period is not harmful. The cover-glass is prised off by inserting a blade between it and the slide.

The frozen slide and cover-glass are lifted off the block of dry ice and placed immediately, before thawing, in 95 per cent or absolute ethyl alcohol. After treatment for 5 min, they are transferred to another jar of absolute ethyl alcohol, where they can be kept from 10 min to 5 h. They are then mounted in euparal. Temporary preparations, in which the stain has partly diffused, can be cleared by adding 5–20 per cent acetic acid to the first ethyl alcohol.

The advantages of this technique are its speed, the ease with which the cover-glass is separated, prevention of the loss of material and the retention of the original stain in the permanent slides.

The precautions to be observed during this process are: the frozen slides must be transferred to 95 per cent ethyl alcohol immediately, before thawing; the slides should be removed from the last alcohol without draining off the excess liquid; a large drop of euparal should be used in mounting, and the excess mounting medium allowed to dry and not pressed out.

The only drawback so far observed is that the cells occasionally collapse on being transferred to the mounting medium from absolute ethyl alcohol.

Liquid air can be used as an alternative for dry ice for freezing materials mounted in slides and covered with a cover-glass. Liquid CO_2 can also be used. The slides are clamped to a 22 mm square hole on a freezing microtome specimen holder. The cover-slip is removed, the specimen dehydrated and mounted (Bower, 1956). A surgical microtome freezing head can be modified to allow rapid freezing (Johnson and Janick, 1962). In order to remove cover slips from old preparations, the slide is frozen for 10 min, the cover slip removed and the slide remounted with new balsam (Simms, 1957). Instead of CO_2 or liquid air, a Freon-aerosol mixture can be sprayed onto the slide, with a nozzle spray tube, for a few seconds, to cover the underside (Elston and Sheehan, 1967).

For permanent mounts, after staining mammalian tissue for 5 min in acetic-orcein at 60 °C, the tissue is rinsed in chilled ethanol–acetone mixture (1:1), held for 2 h or more at 20 °C by dry ice. It is differentiated for 30 s at room temperature in 95 per cent ethanol with 1 per cent HCl, counter-stained in 0·01 per cent fast green FCF in ethanol for 10 s, passed through xylol and mounted (Peary, 1955).

In all techniques where removal of the cover-glass is involved, it can be prepared by smearing thinly with Mayer's albumin and drying over the flame for a few seconds. It may otherwise be coated with silicon (Dri-film 9987 from General Electric). These treatments facilitate its removal.

Alternative schedules

After staining with haematoxylin, coal tar dyes or other stains, the coloured pieces are immersed in glycerin and squashed. The glycerin is removed with water by capillarity and later by ethanol. The slide is mounted in an alcohol-soluble resin (Serra, 1947).

A commercial preparation, Clearcol, can be added in the final stage of any acetic-carmine squash schedule to make the stain permanent, thus omitting dehydration and clearing (Zuck, 1947).

The technique for making a preparation permanent without removing the cover-glass is based on dehydrating and mounting the tissue by allowing a few drops of the fluid to run beneath the cover-glass, without dislodging it.

The paraffin seal is first wiped off with a piece of cloth soaked in xylol. A drop of the stain in which the tissue has been mounted is placed at the end of the cover-glass and allowed to penetrate under it. The slide is placed on an end in an alcohol vapour jar and left for 4–6 h. A few drops of absolute ethyl alcohol are placed at the edge of the cover-glass and allowed to flow under it; then it is pressed out. Euparal is now applied all around the edge of the cover-glass and the slide is placed with the cover-glass in an ethyl alcohol vapour container in which the atmosphere is *not* alcohol-saturated. (For this 10–15 drops of absolute alcohol may be used in a closed pair of Petri dishes of $10 \times 10 \times 1·5$ cm dimensions.) The slide is removed after 24 h and the mounting medium allowed to harden at room temperature.

After acetic-iron haematoxylin squashes, the cover slip can be ringed with Karo corn syrup or the squash mounted in the syrup, after removing the cover slip, by CO_2 freezing (Wittmann, 1963).

Miscellaneous

Some media have been invented for staining, fixing and making permanent mounts in a single step for different objects. Of these, two methods devised by Traub are effective.

Traub's T-101 method—This medium can be used with the Feulgen reaction or when acetic preparations are squashed in 45 per cent acetic acid. It is prepared by mixing 1 g of pure dry arabinic acid with 1·2 g of 83 per cent sorbitol syrup ('Artax') and 9·8 cm^3 of 45 per cent acetic acid, and allowing to stand in the dark at 25 °C until the arabinic acid dissolves. The period may be for several months. The clear supernatant liquid is decanted and used.

After staining, the tissue is squashed in one drop of the T-101 medium on a slide. The excess medium is blotted off and the acetic acid in the medium within the cover-glass is allowed to evaporate in a low temperature, causing the medium to solidify.

Traub's T-103 method—This can be used for natural orcein and carmine stained preparations but is not recommended for synthetic orcein. The medium contains 1 g of arabinic acid, 1·2 g of 83 per cent sorbitol syrup and 98 cm^3 of Belling's 1 per cent ferric acetic-carmine solution. It is prepared in the same manner at T-101.

The tissues, after staining, are placed in a drop of the medium on a slide and squashed. The slide is gently heated and the cover-glass pressed down tightly. The acetic acid soon evaporates at room temperature, causing the medium to solidify. If the stain of the tissue tends to fade, the proportion of the dye in the medium may be increased.

Venetian turpentine mounting medium—Venetian turpentine—a natural resin mixed with a small amount of water and carmine—was used as a combined staining and mounting medium (Zirkle, 1940). The carmine was frequently precipitated however and the resin was insufficient to prevent drying. A modification by Wilson (1945) omitted the carmine and iron mordant was used. A recent medium contains venetian turpentine 50 cm^3; 90 per cent liquid phenol 45 cm^3; 98 per cent propionic acid 35 cm^3; glacial acetic acid 10 cm^3, and water 15 cm^3 (Haunold, 1968). Root tips stained in Feulgen are macerated in a minimum amount of propionic-carmine. A drop of the mounting medium is mixed with carmine, the cover slip applied and firmly squashed.

REFERENCES

Beeks, R. M. (1955). *El Aliso* **3,** 131
Bower, C. C. (1956). *Stain Tech.* **31,** 87
Bridges, C. B. (1937). *Stain Tech.* **12,** 51

150   *Mounting*

Celarier, R. P. (1956).   *Stain Tech.* **31,** 155
Conger, A. D. (1960).   *Stain Tech.* **35,** 225
Conger, A. D. and Fairchild, L. M. (1953).   *Stain Tech.* **28,** 289
Crossman, G. C. (1949).   *Stain Tech.* **24,** 241
Elston, R. N. and Sheehan, J. F. (1967).   *Stain Tech.* **42,** 317
Evens, E. D. (1961).   *J. Quekett micr., Cl.* **5,** 444
Haunold, A. (1968).   *Stain Tech.* **43,** 153
Hillary, B. B. (1938).   *Stain Tech.* **13,** 161
Hillary, B. B. (1939).   *Stain Tech.* **14,** 97
Hillary, B. B. (1940). *Bot. Gaz.* **102,** 225
Johnson, K. W. and Janick, J. (1962).   *Turtox News* **40,** 282
McClintock, B. (1942). *Stain Tech.* **4,** 53
Peary, J. Y. (1955).   *Stain Tech.* **30,** 249
Serra, J. A. (1947).   *Stain Tech.* **22,** 157
Simms, H. R. (1957).   *Turtox News* **35,** 118
Wilson, G. (1945).   *Stain Tech.* **20,** 133
Wittmann, W. (1963).   *Stain Tech.* **38,** 217
Zirkle, C. (1939).   *Science* **85,** 528
Zirkle, C. (1940).   *Stain Tech.* **15,** 139
Zuck, R. K. (1947).   *Stain Tech.* **22,** 109

7 Representative Schedules of Treatment for Higher Plants and Animals (except Mammals)

Some sample schedules for the study of chromosomes in different tissues are considered in this chapter.

SCHEDULES FOR THE STUDY OF MITOTIC CHROMOSOMES

IN PLANT MATERIALS

IN ROOT TIPS

Schedule A. From Paraffin Sections Dehydrated through Alcohol Chloroform Grades and Stained in Crystal Violet

Material used—Fresh young roots of *Allium cepa*. The following steps are involved:

1. *Fixation*—Cut fresh root tips of *Allium cepa*, about 1 cm long, wash away dirt particles with water and a brush and transfer the root tips to a tube containing a mixture of 1 per cent chromic acid and 10 per cent formalin (CF 1:1). Keep for 12–24 h.

2. *Washing*—Wash the roots in a porcelain thimble in running water for 3 h.

3. *Dehydration*—Transfer the roots to a glass phial containing 30 per cent ethyl alcohol and keep for 1 h; then to 50 per cent alcohol, keeping for 1 h; to 70 per cent alcohol, treating overnight; through 80, 90 and 95 per cent alcohol, keeping the roots for 1 h in each. Finally keep overnight in absolute alcohol.

4. *Clearing*—Keep in alcohol–chloroform mixtures (3:1, 1:1 and 1:3) successively for 1 h in each.

Transfer to pure chloroform and keep for 10 min; then change the used chloroform with pure chloroform, and add small shavings of paraffin to the chloroform containing the material.

5. *Infiltration*—Keep overnight on a hot plate at 35 °C, then remove the stopper, add a little more wax and keep the phial with contents in an oven at 45 °C for 2 days and transfer to another oven maintained at 60 °C. Change the wax containing the material with fresh molten wax at intervals of half an hour for 2 h.

6. *Embedding*—Pour the molten paraffin with the roots into a paper tray and add some more molten wax, then orient the roots in groups of three with their tips pointing to the same side and kept at the same level. After the wax has cooled slightly, plunge the block into cold water.

7. *Section-cutting*—Trim the block and cut transverse sections of the root tips 14 μm thick on the microtome. Cut the ribbons into suitable segments and mount serially in water on a slide previously coated with Mayer's adhesive. For coating a slide, put a tiny drop of Mayer's adhesive, about the size of a mustard grain, on the slide and smear it three-fourths over the surface of the slide with the little finger, then place the slide on a hot plate and help the ribbons to stretch with a pair of needles. Drain off the water and keep the slide overnight on the hot plate to dry.

8. *Bringing to water*—Place the slide with sections in pure xylol grades I and II, keeping in each for half an hour. Transfer the slide to a jar of ethyl alcohol–xylol (1:1) and keep for 15 min, then pass through absolute ethyl alcohol, 95, 90, 80, 70, 50 and 30 per cent alcohol, keeping in the first three for 10 min each, and 5 min in each of the rest, and then transfer to water.

9. *Pre-mordanting*—Keep the slide in 1 per cent aqueous chromic acid solution overnight. Wash in running water for 3 h.

10. *Staining*—Stain in 0·5 per cent aqueous crystal violet solution for 20 min. Rinse in water.

11. *Mordanting*—Keep in 1 per cent iodine and 1 per cent KI mixture in 80 per cent alcohol for 30–45 s, then dip in absolute alcohol for 2 s.

12. *Dehydration*—Pass through 3 successive grades of absolute alcohol, dipping in each for 2 s.

13. *Differentiation*—Transfer to clove oil I, then differentiate under the microscope after keeping in clove oil II for 2 min.

14. *Clearing*—Transfer to xylol (UP) grade I and keep for half an hour; pass through pure xylol II and III, keeping 1 h in the former and 30 min in the latter.

15. *Mounting*—From xylol III, mount in Canada balsam under a coverglass. Allow the slide to dry overnight on the hot plate.

Alternative fixative—Any one mixture of the list of fixatives given in the chapter on fixatives can be used, depending on the material.

Alternative clearing agent—n-Butyl alcohol can be used instead of chloroform in the clearing process.

Alternative pre-mordanting—For materials which stain easily, premordanting in 1 per cent chromic acid solution can be omitted; for materials difficult to stain, Navashin's fluid A can be used instead of 1 per cent chromic acid solution in pre-mordanting.

Alternative mordanting—Gram's aqueous solution of potassium iodide and iodine can be used instead of the solution KI and I_2 in 80 per cent alcohol.

For materials in which the background retains stain even after mordanting in KI and I_2 solution, keep the slide in 1 per cent chromic acid solution after step 11 and before step 12 for 15 s.

Alternative differentiation—For high accuracy in staining, after dehydration by dipping in grades of absolute alcohol, transfer the slides to very thin terpineol, keep for 1–2 min, then rinse in xylol and transfer to clove oil. Differentiate as usual.

Schedule B. From Paraffin Preparations Dehydrated by Rapid Dioxane Methods and Stained in Crystal Violet

This method is suitable only for root tips (La Cour, 1937) and differs from the usual procedure only with regard to dehydration and embedding.

1 and 2. *Fixation and washing*—As in the previous schedule.

3. *Dehydration*—Transfer the roots through aqueous solutions of dioxane, 25 and 75 per cent, keeping 2 h in each. Keep in pure dioxane overnight.

4. *Infiltration*—Transfer the phial to an oven at 60 °C, adding paraffin of low melting point at intervals of 30 min for 4 h, then add pure molten wax and keep the roots in it for 2 h before embedding.

The remaining steps are the same as the previous schedule.

Schedule C. From Paraffin Sections Stained in Haematoxylin

Materials used—Root tips of *Pisum sativum*. The initial stages, from fixation to block preparation and bringing the sections down to water, are the same as followed in Schedule A, steps 1–8.

Of the numerous schedules followed for staining, only two are given below:
In an older method:

9. *Mordanting*—Mordant the slide in alum (3 per cent ferri-ammonium sulphate) solution for at least 3 h.

10. Rinse thoroughly in water.

11. Stain for 24 h in 0·5 per cent aqueous haematoxylin solution.

12. Rinse in water.

13. Differentiate and de-stain in alum for 5 min or more.

14. Rinse for 15 min in running water.

15. Dehydrate by passing through ethyl alcohol series 50, 70, 90, 95 per cent and absolute, keeping for 5 min in each.

16. Pass through alcohol–xylol (1:1) and pure xylol I, keeping in the former for 15 min and in the latter for 1 h.

17. Mount in Canada balsam.
A later and more rapid technique is:

9. Mordant in 4 per cent alum solution for 10–20 min.

10. Rinse in running water for 10 min.

11. Stain in 0·5 per cent haematoxylin (ripened) solution for 5–15 min.

12. Wash in water and de-stain for 5–20 min in saturated aqueous picric acid.

13. Keep for 1 min in a jar containing water with 1 or 2 drops of 0·88 per cent ammonia. Rinse in running water for 30 min.

14. Dehydrate through ethyl alcohol grades 20, 60, 80 per cent and absolute. The remaining steps are similar to the previous method.

Basic Fuchsin Method (Arzac, 1950)

Reagents

Hydrochloric acid	1 N
Sulphuric acid	1 N solution in 96% ethyl alcohol

Reagent a

Basic fuchsin in 10% ethyl alcohol	0·05 cm³
Absolute ethyl alcohol	5 cm³
Phenol crystals	3 g
Distilled water	95 cm³

Dissolve the phenol in distilled water and add fuchsin solution and alcohol.

Reagent b

10% aq. potassium meta- bisulphite solution	2 vols
1 N aq. sulphuric acid solution	1–2 vols

Procedure

1. Bring the sections of materials (originally fixed in alcoholic or formalin-mixed fixatives) down to distilled water.

2. Hydrolyse in 1 N HCl for 5 min at room temperature.

3. Keep in 1 N HCl at 60 °C for 15 min.

4. Immerse in 1 N HCl at room temperature.

5. Wash in water. Stain for 2–3 min in reagent *a*.

6. Transfer immediately, without washing, to reagent *b* and keep for 5 min.

7. Immerse in a second lot of reagent *b* for another 5 min. Rinse in water.

8. Immerse in 1 N H_2SO_4 in 96 per cent alcohol, for 3–5 min. Wash thoroughly in water.

9. Dehydrate through alcohol and xylol grades and mount in Canada balsam.

10. DNA takes up magenta colour.

ALDEHYDE ALKALINE SILVER REACTION

Korson (1964) and de Martino *et al.* (1965) demonstrated aldehyde of DNA with alkaline silver reaction involving hydrolysis in molar citric acid or N HCl at 60 °C, followed by treatment with hexamine silver solution.

Schedule D. From Pre-treated Squash Preparations

(a) *Stained in Feulgen*
    Materials used—Root tips of *Hemerocallis fulva*.
    1. *Pre-treatment*—Cut fresh root tips, remove the dirt and treat them in 0·02 aqueous oxyquinoline solution at 10–12 °C for 3 h.
    2. *Fixation*—Transfer to acetic ethanol (1:1) mixture and keep for 1 h.
    3. *Washing*—Rinse in distilled water.
    4. *Hydrolysis*—Hydrolyse the root tips in n HCl at 60 °C for 12 min.
    5. *Washing*—Rinse in water.
    6. *Staining*—Transfer root tips to leuco-basic fuchsin solution and keep in it for 30 min to 1 h till the tips are magenta coloured.
    7. *Squashing*—Transfer each tip to a drop of 45 per cent acetic acid on a slide, cut out the tip region only and discard the other tissue. Place a cover-glass over the tip and squash it, applying uniform pressure with the thumb on a piece of blotting paper placed on the whole.
    8. *Observation*—The preparation can be ringed with paraffin wax and observed.
    9. *Making permanent*—Invert the slide in a closed tray containing glacial acetic acid–ethanol (3:1) mixture. After the cover-glass falls off, pass both slide with material and cover-glass through acetic–alcohol (1:1, 1:3) mixtures, pure alcohol, alcohol–xylol (1:1) mixture and xylol I and II, keeping 5 min in each. Mount in Canada balsam.
    For making permanent, the other schedules for mounting (*see* Chapter 6) can also be followed.
    Modifications—Pre-treatment chemicals include *p*-dichlorobenzene, acenaphthene, coumarin, aesculine, isopsoralene etc., for different materials.
    Root tips of sugar cane are difficult materials for squash schedules (Li *et al.*, 1954; Stevenson, 1965). Pre-treat in sat. α-bromonaphthalene solution in 0·05 per cent saponin for 3 h in the cold, fix in acetic-ethanol (1:3) for 48–72 h, hydrolyse in n HCl at 60 °C for 7–8 min, treat with 3 per cent pectinase solution in pH 3·6 acetate buffer for 60–90 min and stain according to the Feulgen technique (Sisodia, 1968).
    An earlier method by Bhaduri and Ghosh (1954) for cereals suggests: soak root tips for 1–2 h at 18–20 °C in saturated aqueous α-bromonaphthalene, 1–2 h in water, and 0·5–1 h at 10–14 °C in a mixture of 1 per cent chromic acid 5 cm^3, 2 per cent osmic acid 1 cm^3, and aqueous 0·002 M OQ 1 cm^3. Treat successively in water for 1–2 min, 1 per cent sulphuric acid solution for 10–15 min, water for 1–2 min, 1 per cent chromic acid solution for $\frac{1}{2}$–1 h, and squash in acetic-carmine.
    Treatment in aqueous α-bromonaphthalene followed by successive washing in water and 22 per cent acetic acid, fixation in n HCl: 22 per cent acetic acid mixture (1:12), prior to usual hydrolysis and Feulgen staining schedule can be effectively used for members of Triticinae (Upadhya, 1963).

(b) *Stained in acetic-orcein solution*
    Material used—Root tips of *Aloe vera*.
    1. *Pre-treatment*—Cut fresh root tips about 1 cm long and, after washing

away dirt in water with a brush, keep in a corked glass phial containing a saturated aqueous solution of *p*-dichlorobenzene (PDB) for $2\frac{1}{2}$–3 h at 12–14 °C.

2. *Fixation*—Transfer to glacial acetic acid–ethyl alcohol mixture (1:2) and keep for 30 min to 2 h, followed by treatment in 45 per cent acetic acid for 15 min.

3. *Staining*—Place the root tips in a glass phial containing 2 per cent acetic-orcein solution and N HCl mixture in the proportion of 9:1, and heat gently over a flame for 5–10 s, taking care that the liquid does not boil.

4. *Squashing*—Lift a root tip from the mixture and put it in a drop of 1 per cent acetic-orcein solution on a slide and cut off and remove the older part of the tip. Place a cover-glass on the tip and squash by applying uniform pressure on the cover-glass with the thumb through a piece of blotting paper.

5. *Observation*—Ring the cover-glass with paraffin and observe under the microscope for temporary preparation.

6. *Making permanent*—Invert the slide with cover-glass after squashing in a covered tray containing tertiary butyl alcohol and keep till cover-glass is detached. Mount slide and cover-glass separately in euparal.

Alternative pre-treatment chemical—OQ, coumarin, aesculine, α-bromo-naphthalene, etc.

Alternative stain and fixative—2 per cent propionic-orcein, 1 per cent propionic acid and propionic–ethanol can be used instead of 2 per cent acetic-orcein, 1 per cent acetic-orcein and acetic–alcohol solutions in the respective steps. Treatment in 45 per cent acetic acid in step 2 is optional.

To intensify stain—A drop of aqueous ferric chloride solution can be added to the acetic-orcein–N HCl mixture or the tissue can be kept in the staining mixture for a period extending up to 12 h and then mounted in 45 per cent acetic acid.

(c) *Stained in acetic-lacmoid solution*

1 and 2. *Pre-treatment and fixation*—As for the acetic-orcein schedule.

3. *Staining*—Stain the tissue by transferring the root tips to a glass phial containing 10 cm of standard acetic-lacmoid solution and 1 cm^3 N HCl, and heat for 5–10 s over a flame, taking care not to boil the fluid, then leave for 10 min.

4. *Mounting*—Transfer a root tip to a drop of standard acetic-lacmoid solution on a slide and squash as usual as in previous schedules.

Steps 5 and 6 are similar to the acetic-orcein schedule.

IN LEAF TIPS

Fixation for paraffin blocks is rather ineffective. Usually squashes made after pre-treatment yield the best results. A sample schedule is given below, Sharma and Mookerjea (1955):

Material—Leaf tip of *Cestrum nocturnum*

1. *Pre-treatment*—Dissect out very young leaf tips of *Cestrum nocturnum*, wash in water and immerse in a corked glass phial containing

saturated aqueous solution of aesculine and keep at 12–14 °C for 15 min to 24 h.

2. *Fixation*—Fix the materials in acetic–ethanol (1:1) mixture for at least 3 h, the period being extended, if necessary, up to 24 h.

3. *Staining*—Remove the tips and place them in a mixture of 2 per cent acetic-orcein solution and N HCl (9:1), heat over a flame for 3–4 s, then leave the tips in the mixture at 30 °C for 30 min.

4. *Squashing*—Squash the tips on a dry slide in a drop of 1 per cent acetic-orcein solution with a cover-glass, applying uniform pressure with the help of blotting paper.

5. *Making permanent*—Similar to root tip smears.

Alternative pre-treatment chemicals used for root tips can be used here.

The period of fixation in acetic–ethanol should be increased, if necessary, to remove the chlorophyll completely. Acetic–alcohol mixture (1:2) or (1:3), chilled 80 per cent ethanol or acidulated ethanol (conc. HCl—95 per cent ethanol 1:3) can be used instead of acetic–alcohol (1:1) mixture.

Modifications

1. Pre-treatment in saturated solution of α-bromonaphthalene in 0·05 per cent saponin for 3 h in the cold gives good results with grass leaf chromosomes. Expose young leaf, sever top of the shoot and cut slits into the remaining tube for penetration (Latour, 1960; Bennett, 1964).

2. For *Saccharum* leaf, pre-treatment with 0·2 per cent colchicine at room temperature for 2 h is effective (Price, 1956, 1962).

3. For tea leaf, pre-treatment in aqueous saturated *p*-dichlorobenzene (2–3 h) at 4–10 °C, fixation for 6–12 h in a mixture of propionic acid, chloroform and ethanol (1:3:6), staining with 2 per cent propionic-orcein at 80 °C and squashing in 1 per cent propionic-orcein is recommended (Bezbaruah, 1968).

4. For grass leaf, soak longitudinal sections of leaf shoots for 2–4 h in aqueous 0·002 M OQ at 25 °C, blot and fix in ethanol, chloroform and acetic acid mixture (3:4:1). Macerate at 45 °C for 30 min in a pectinase solution before staining and squashing (Powell, 1968).

IN POLLEN GRAINS

For studying the mitotic division in pollen grains, the following steps are followed:

Material—Flowers of *Nothoscordum fragrans*.

1. *Dissection*—Dissect out an anther from a flower bud of suitable size, put in a drop of 1 per cent acetic-carmine solution on a dry slide and smear the anther with a clean scalpel, cover with a cover-glass and observe under the microscope. In flower buds of a suitable size, mitotic division is observed in the pollen grains.

2. *Smearing*—Dissect out the remaining anthers from the flower bud in

which pollen grain division was observed. Place the anthers on a clean dry slide, cut off the edges of each anther with a clean scalpel, squeeze out the inner fluid and reject the empty anther lobes, then smear the fluid with a clean scalpel.

3. *Fixation*—Immediately invert the slides in a covered tray containing Navashin's A and B fixatives, mixed in the proportion 1:1 and keep overnight.

4. *Staining*—Wash the slides in running water for 3 h, then stain in 0·5 per cent aqueous crystal violet solution for 30 min and rinse in water.

5. The subsequent steps, namely *mordanting, dehydration, differentiation, clearing* and *mounting* are similar to the corresponding steps 11, 12, 13, 14 and 15 of the technique followed in staining root tip sections cut from paraffin blocks (*see* Schedule A, page 151).

Modifications

1. For wheat pollen, treat anthers at 18–20 °C in 0·2 per cent aqueous colchicine and 0·002 M aq. oxyquinoline solutions at 10–14 °C for 1 h, fix in Carnoy's fluid for 6 h, wash, hydrolyse in HCl, stain in leucobasic-fuchsin and smear in 1 per cent acetic-carmine (Bhaduri and Majumdar, 1955). For *Saccharum* and related genera, pre-treat in 0·5 per cent aqueous colchicine for 1 h, wash, treat in 0·002 M aqueous OQ for 1 h, wash, fix in a mixture of methanol 60 cm^3; chloroform 30 cm^3; water 20 cm^3; picric acid 1 g and mercuric chloride 1 g, for 24 h. The remaining schedule is similar to that adopted for wheat (Jagathesan and Sreenivasan, 1966).

2. For studying chromosomes from herbarium sheets of *Impatiens*, soak anthers overnight in a saturated solution of iron acetate in 45 per cent acetic acid, rinse and smear in dilute acetic-carmine. Heat several times to boiling and seal (Khoshoo, 1956).

Precautions

The operation of smearing the dissected anthers should be carried out swiftly so that the fluid does not dry before inversion in the fixative. Smearing should be carried out away from any direct air current which may dry the fluid within the anthers.

Alternative methods

1. The slides can be made permanent immediately after acetic-carmine smearing by alcohol-vapour or tertiary butyl alcohol techniques.

2. If the stain taken by the chromosomes is not satisfactory, premordanting in 1 per cent chromic acid solution, as in the case of root tip sections, described before, is followed.

IN ENDOSPERM

For the study of the mitotic division from endosperm tissue, two different methods can be followed, in addition to others described elsewhere (*see* Chapter 8).

Feulgen Squash Method (Rutishauser and Hunziker, 1950)

1. *Fixation*—Dissect out very young developing seeds and fix them in acetic–alcohol mixture (1:2) for 1–2 h and keep overnight in 95 per cent alcohol.

2. *Washing*—Run the seeds through 70, 50 and 30 per cent alcohol, keeping them for 10 min in each, then wash in running water in a porcelain thimble for 10 min.

3. *Hydrolysis*—Hydrolyse in N HCl at 60 °C for 8–12 min.

4. *Staining*—Rinse in water and stain in leucobasic fuchsin solution for 2 h, then wash for 10 min in two changes of tap water.

5. *Mounting*—Dissect out the endosperm on a clean dry slide in a drop of 45 per cent acetic acid solution, under a dissecting microscope using tungsten needles pointed in molten $NaNO_2$. Using Mayer's adhesive, film a cover-glass and dry it by passing over a flame, then squash the dissected endosperm under the cover-glass, exerting strong but uniform pressure under a piece of blotting paper.

6. *Making permanent*—The slides can be made permanent following any one of the schedules described under the chapter on mounting.

Modifications—For intensifying the stain, 1 per cent acetic-orcein solution can be used instead of 45 per cent acetic acid solution as the mounting medium.

Acetic-orcein Squash Method

This method was developed in the author's laboratory (Sharma and Varma, 1959).

Material used—Endosperm of *Cestrum nocturnum*.

1. *Pre-treatment*—Dissect out the very young developing seeds with the help of a dissecting needle under a dissecting microscope and place them immediately in a saturated solution of aesculine and keep at 10–12 °C for 3 h.

2. *Fixation*—Fix in acetic–ethyl alcohol mixture (1:1) at room temperature for 2 h.

3. *Staining*—Heat in a mixture of 2 per cent acetic-orcein solution and N HCl (9:1) over a flame for 9 or 10 s, removing the tube at intervals so that the fluid does not boil. Keep for 30 min.

4. *Squashing*—Remove each seed, put it on a clean slide in a drop of 1 per cent acetic-orcein solution, cut it into 2 or 3 pieces with a scalpel and keep the pieces at a little distance from each other. Squash the whole under a long cover-glass, exerting strong and uniform pressure and ring with paraffin.

5. *Making permanent*—If necessary, the slides can be made permanent following any one of the methods described previously.

Notes

The seeds selected should be as young as possible, preferably taken within a week of fertilisation.

If the seeds are very small the entire seed should be squashed, but if, however, the seeds are comparatively large, the endosperm should be dissected out before mounting, under a dissecting microscope, and then squashed. The endosperm can be excised, placed directly in acetic-carmine, macerated and brought to boiling, the nuclei and mitotic figures are separated by centrifugation (Persidsky and Duncan, 1957).

IN POLLEN TUBE

For the study of mitosis from the division of the generative nucleus in germinating pollen grains, the methods are divided into two parts: (*a*) germination of the pollen tube, and (*b*) treatment for observation of the chromosomes. Of the numerous techniques followed, three representative ones are described here.

Hanging Drop Culture Method

Material—Seeds of *Papaver* sp.

1. *Pollen grain culture*—Fit a ring on a slide and smear both rims of the ring with vaseline so that it is attached to the slide at one end. Place a drop of 3 per cent sugar solution on a clean cover-glass, then dust pollen grains from an opened flower into the sugar solution. Invert the cover-glass with the drop of sugar solution on it and attach it to the other vaselined rim of the ring, so that the drop with pollen grains hangs in the closed chamber enclosed by the ring. Growth of the pollen tubes can be noted by observing the ringed slide under the microscope. After about 3h, remove a few pollen tubes for observation at intervals of 1 h till the optimum time is reached.

2. *Treatment for chromosome study*—Lift out the pollen tubes with a brush, put them in a drop of 1 per cent acetic-orcein solution on a clean slide, warm slightly and squash as usual under a cover-glass applying uniform pressure.

3. *Permanent preparation*—The slides can be made permanent, if the dividing time is found to be correct, by the alcohol vapour technique.

Modifications

For accumulating metaphase, 0·05 per cent colchicine can be added to the sugar solution.

For controlling humidity of the hanging drop chamber, add a drop of water on the slide or place a drop of sugar solution beside the hanging drop.

Coated Slide Technique (Conger, 1953) and modifications

Material—Tradescantia virginiana
1. *Preparation of medium*—Weigh out 12 g of lactone, 1·5 g of agar and 0·01 g of colchicine. Heat the lactone and agar with 100 cm³ of distilled water in a double boiler until the agar is dissolved, then add colchicine when the medium has cooled to 80–60 °C. Keep the medium at about 60–70 °C.
2. *Preparation of slides*—Coat slides cleaned in alcohol with a thin layer of egg white. Dip the slide in a beaker containing the medium at 60 °C until it warms up, then withdraw the slide, drain off the medium, and wipe off the back with a piece of clean cloth.
3. *Preparing the pollen*—When the medium has set, but is not dry, pick up the pollen with a brush, and dust a thin film on the medium. Immediately place the slide sown with pollen in a moist growing box, the box being a horizontal glass staining dish lined with damp blotting paper on the two sides and top, and keep the temperature between 20° and 25 °C. Observe at intervals till the optimum period for maximum division is found.
4. *Fixation and staining*—Add a drop of 1 per cent acetic-carmine solution to the pollen tube, squash under a cover-glass and observe.
5. *Making permanent*—The alcohol vapour technique can be used.
Modification—1. The slides can also be stained by the Feulgen schedule after 12 min hydrolysis in N HCl at 60 °C. They must be handled carefully, as otherwise the medium may be washed off.
2. For species with binucleate pollen, germinate in a medium containing H_3BO_3, 0·01 g; $Ca(NO_3)_2 . 4H_2O$, 0·03 g; $MgSO_4 . 7H_2O$, 0·02 g; KNO_3, 0·01 g; sucrose, 10 g; water, 100 cm³, on a slide resting on moist filter paper in a closed Petri dish for 24 h. Place crystals of acenaphthene on filter paper. After 24 h, squash pollen tubes in propionic-orcein solution (Dyer, 1966).
3. For convenient handling of germinated pollen during Feulgen and autoradiographic procedure, grow pollen on an autoclaved membrane filter (Millipore AA WP 025 00) in contact with a sterilised medium which contains agar 0·5–1 per cent; sucrose 0·1–0·5 per cent, and boric acid 0·01 per cent, for 2 h to overnight at 2–4 °C on a filter paper with a mixture of OsO_4 1 g; CrO_3, 1·66 g and water, 233 cm³. For *Persica* pollen, add 10 per cent acetic acid. Wash in water, bleach in a mixture of 3 per cent hydrogen peroxide and saturated aqueous ammonium oxalate solution on filter paper. Hydrolyse with 5N HCl for 18 min at room temperature, stain in Feulgen, wash in three changes of 2 per cent $K_2S_2O_5$ at pH 2·3 (KH_2PO_4, 1·4 g; conc. HCl, 0·35 cm³; dist. water, 100 cm³) by placing membrane on filter paper wet with the respective fluids. Transfer pollen to glacial acetic acid, squash and process (Jona, 1967).
4. Sow *Tradescantia* pollen on lactose-agar medium at 38–39 °C for 16 h. Fix slides in acetic-ethanol (1:3) for 1–3 h, hydrolyse in N HCl at 60 °C, treat with water at 65 °C. Delaminate upper layer of medium in cold water. Flatten and fix remaining single layer of pollen tubes to the slide by pressing under a cover glass by quick freeze technique, stain in Feulgen and mount as usual (Ma, 1967).

5. For palm chromosomes, sow pollen in a medium containing H_3BO_3, 100 p.p.m.; colchicine, 0·04 per cent, lactose, 12 per cent, gelatin, 5 per cent and egg albumen, 1 drop in 10 cm^3 (Read, 1964).

Collodion Membrane Technique (Savage, 1957)

1. *Formation of collodion membrane*—Mix 1 part of collodion (necol collodion solution from BDH, England) with 3 parts acetone. Prepare 3 per cent aqueous sugar solution in a Petri dish and warm and put a drop of collodion–acetone mixture on the sugar solution.

This drop rapidly spreads out in a thin film over the surface and the film hardens into a thin membrane as acetone evaporates. Cover the open dishes lightly by filter paper and leave in a warm dry place for 3 h till acetone evaporates completely.

2. *Germinating the pollen*—Dust the pollen with a brush or directly from the anther on to the smooth areas of the floating membrane towards the centre, replace the lid on the Petri dish and transfer it to an incubator.

3. *Fixation and staining*—After the pollen tubes have germinated, use a pair of scissors to cut out a piece of the membrane, about 1 cm in diameter, with the pollen tubes, while it is still floating on the sugar solution; lift it with a needle on to a clean dry slide, add a drop of 1 per cent acetic-orcein solution and squash under an albuminised cover glass, applying uniform pressure. Ring with paraffin for observation.

4. *Permanent mounts*—Invert the slide, after wiping off the paraffin with xylol, in a covered tray containing acetic–ethanol mixture (1:3) and after the cover glass is detached with the membrane run it through 2 changes of absolute alcohol, alcohol–xylol (1:1) mixture and pure xylol, keeping for 10 min in each. Mount the cover glass, pollen side down, in a drop of Canada balsam on a clean slide.

Modification—Feulgen staining is used in an alternative method.

1. The earlier steps are similar, up to the germination of pollen.

2. Float the cut membrane with pollen tubes on a slide and blot off excess solution. Add a drop of acetic–alcohol mixture (1:3) to the pollen and cover with an albuminised cover glass. Press gently and invert in a tray containing acetic–alcohol mixture.

3. After $1\frac{1}{2}$ h, remove the cover glass with pollen tubes to 95 per cent alcohol and treat overnight.

4. Bring down to water through 80, 70, 50 and 30 per cent alcohol, keep 2 min in each, rinse and hydrolyse for 12 min at 60 °C in N HCl.

5. Rinse in water and stain in leuco-basic fuchsin solution for 30 min and squash in 45 per cent acetic acid on a clean slide.

6. For making permanent, the usual techniques described before can be followed.

Metaphase Arrest Technique

This method completely excludes nutrient medium in the germination of the pollen tube.

Material used—Tradescantia virginiana

1. *Pollen tube germination*—Line both bottom and cover of a pair of Petri dishes with well-moistened filter paper and place a clean slide in the Petri dish and dust on it pollen grains from a newly opened flower. Spread 50–100 mg of fine acenaphthene crystals on the filter paper close to the slide and cover and keep at 20–22 °C for 24 h.

2. *Staining*—Add a drop of 1 per cent acetic-carmine solution to the pollen tubes on the slide and squash under a cover glass.

IN ANIMAL MATERIAL

Squash Preparation

The usual sources of material are: larval tails, ganglia, and spermatogonial cells.

Material—Urodele larva

1. Cut off the tail of a growing larva of suitable size immediately behind the anus.

2. Fix in acetic–ethanol mixture (1:1) for 2–24 h in a glass phial.

3. Remove the tissue from the phial and place it on a clean slide in a drop of 1 per cent acetic-carmine solution to which a trace of iron has been added. The tissue may be teased by means of two needles before squashing.

4. Squash by placing a cover glass on the preparation and pressing gently so as not to crush the cells. Ring with paraffin and observe.

5. If necessary, the slide can be made permanent by any one of the techniques outlined previously.

Modifications

Acetic-lacmoid solution can be used instead of acetic-carmine.

For Feulgen staining, after fixation in step 2, hydrolyse the material in N HCl at 60 °C for 12 min. Modify the treatment if necessary, rinse, stain in leuco-basic fuchsin solution for 30 min and squash in 45 per cent acetic acid solution.

Carnoy's fluid or Newcomer's fluid can be used as a fixative instead of acetic–alcohol mixture.

This method is applicable to all the other tissues.

Paraffin Block Preparation

This method is principally recommended for the study of divisional figures.

Material—Urodele larva

1. *Fixation*—Cut out the growing tail of a larva near the anus and fix it in San Felice's fluid (1 per cent chromic acid: 16 cm^3, 40 per cent formaldehyde: 8 cm^3, and glacial acetic acid: 1 cm^3) for 12 h.

2. *Washing*—Transfer the tissue to a porcelain thimble and wash in running water for 3 h.

3. *Dehydration*—Transfer the tissue to a phial containing 30 per cent ethyl alcohol and keep for 1 h. Pass the tissue to 50 per cent alcohol, keeping it in for 1 h; then to 70 per cent alcohol and keep overnight; to 80, 90 and 95 per cent alcohol, keeping 1 h in each; and finally store overnight in absolute alcohol.

4. *Clearing*—Transfer the tissue through ethyl alcohol–chloroform grades (3:1, 1:1 and 1:3), keeping in each for 1 h, then to pure chloroform for 10 min. Give another change in pure chloroform, and add shavings of paraffin wax.

5. *Infiltration*—Keep the phial containing the tissue in the mixture of chloroform and molten wax on a hot plate at 35 °C for 24 h, transfer to an oven kept at 45 °C, remove the cork and keep for 24 h, then transfer to another oven maintained at 60 °C. Change the wax with fresh molten paraffin wax, at half-hourly intervals, for 2 h until the smell of chloroform has completely disappeared.

6. *Embedding*—Pour the molten paraffin with the material into a paper tray, add some more molten wax, orient the material, placing each piece at a distance of 1 cm from its neighbours, and after the wax has cooled slightly, plunge the block into cold water.

7. *Section cutting*—Trim the block, attach it to the holder and cut sections 12 μm thick on a Spencer's rotary microtome. Cut the ribbons into segments of equal size. Put a minute drop of Mayer's adhesive on a clean slide and rub it into a thin film along three-quarters of the area of the slide. Put a few drops of distilled water on the slide and float the ribbons serially on the water. Place the slide on a hot plate and stretch the ribbons with needles. Drain off the water, arrange the sections in neat rows and allow the slide to dry overnight on the hot plate.

8. *Bringing down to water*—Run the slide with sections through pure xylol I and II, keeping 30 min in each, pass through xylol–ethyl alcohol (1:1) mixture and absolute alcohol grades, the period of immersion being 15 min in each, and gradually transfer the slide through 95, 90, 80, 70, 50 and 30 per cent alcohol grades, keeping 2–3 min in each, finally rinsing thoroughly in water.

9. *Mordanting*—Keep the slide with material in 4 per cent iron alum solution for 10–20 min.

10. *Washing*—Rinse in running water for 10–15 min.

11. *Staining*—Stain in 0·5 per cent haematoxylin solution (ripened for 1–2 months) for 5–15 min.

12. *De-staining*—Rinse in water and de-stain for 5–20 min in saturated aqueous solution of picric acid.

13. *Blueing*—Blue the stain by placing the slide in 100 cm^3 of distilled water containing 1 or 2 drops of 0·88 per cent ammonia for 1 min.

14. *Dehydration*—Pass through 30, 60, 80 per cent and absolute alcohol grades, keeping the slide for 2 min in each.

15. *Differentiation*—Transfer to clove oil and keep for 2–3 min, then lift out the slide, wipe off the extra oil from the back of the slide and observe

under a microscope to judge the stain.

16. *Clearing*—If stain is satisfactory, transfer the slide to pure xylol and keep in pure xylol I for 30 min, xylol II for 1 h and xylol III for 30 min to overnight.

17. *Mounting*—Finally mount in Canada balsam or clarite X.

Modifications

This technique can also be applied to ganglia and other somatic tissue, except bone marrow cells.

For difficult materials, the tissue can be pre-fixed in acetic–alcohol (1:3) mixture for 1 h before fixing in San Felice's fluid as usual.

Some alternative fixatives which can be used are Flemming's strong fluid, Champy's fluid and Minouchi's fluid (*see* list of fixatives).

For haematoxylin staining, the more prolonged schedule as followed for plants in Schedule C (page 153) is also applicable in the case of animals.

The Newton's crystal violet staining schedule already described can also be applied to animal materials.

Smear Preparations

This technique was devised for the bone marrow cells of mammals after pre-treatment (Ford and Hamerton, 1956). However, it can also be applied to thymus, spleen, cornea and other tissues.

1. *Injection*—Inject 0·5 cm^3 of 0·025 per cent (w/v) colchicine solution intraperitoneally in the animal and leave for 1 h or more. Too much toxicity should be avoided. Prolonged keeping may result in polyploidy.

2. *Kill*—Remove femurs and cut off epiphyses. Wash out the marrow into a small phial with warm 0·12 per cent aqueous solution of sodium citrate, using a hypodermic syringe with a fine needle.

3. *Gently aspirate* the marrow in and out of the syringe till it breaks up into a fine suspension, and keep the phial, with the suspension, in a water bath at 37 °C for 10 min.

4. *Filter* by centrifuging the suspension through nylon bolting cloth in a bacterial infiltration tube to obtain a fine clean suspension without any debris.

5. *Fix* in chilled acetic–alcohol (1:3) mixture for 30 min to 2 h.

6. *Transfer* to 30 per cent alcohol and keep for 15 min, then put in water and rinse for 15 min.

7. *Hydrolyse* in 1 N HCl at 60 °C for 4 min, then transfer the tissue to chilled water and keep for 1–2 min.

8. *Stain* in leuco-basic fuchsin solution for 1 h.

9. Place the material on a clean slide in 45 per cent acetic acid and squash under a cover-glass. Ring with paraffin for observation.

10. The slide can be made permanent following the dry ice or alcohol vapour schedules and mounted in euparal.

Modifications

Materials which fail to stain in leuco-basic fuchsin solution can be restained by squashing them in 1 per cent acetic-orcein solution instead of in 45 per cent acetic acid solution.

The animal can be killed by cervical dislocation or by chloroform.

The period of hydrolysis is varied according to the material.

Colcemid from C.I.B.A. Laboratories Ltd. can be used instead of colchicine and it is less toxic.

Tissues other than bone marrow should be chopped into very small pieces in the hypotonic citrate solution and then the rest of the schedule from step 5 should be followed.

In an alternative technique (Sparano, 1961), transfer the suspension to a mixture of 2 per cent acetic-orcein and N HCl (9:1), heat for a few seconds and mount in 45 per cent acetic acid. The steps from 5–9 of the sample schedule can be omitted.

Permanent Smear Preparation

Smith (1943), for insects, modified by Manna (unpublished) in mammalian chromosomes (also applied to other materials).

1. Take any tissue with division, e.g. cornea, bone marrow, liver testes, etc.

2. If the cells are in suspension as in bone marrow, follow the procedure recommended by Ford and Hamerton (1956).

If the tissue is solid, macerate it by means of a homogeniser in sodium citrate solution (for swelling) and incubate the suspension in 37 °C and leave for 15 min.

3. Centrifuge gently at about 1000 rev/min for 5 min.

4. Decant off the supernatant fluid, add acetic–methyl alcohol mixture (1:3) to the sediment and flush it with a pipette, then fix for 15 min or more (it can be kept overnight in cold). If the fixed cells settle down at the bottom, decant; if they remain in the supernatant, gentle centrifuging may also be applied.

5. Add a drop of 45 per cent acetic acid solution and make it a milk emulsion.

6. Take a very small drop, place it on a slide and squash by a cover glass, with gentle pressure by means of a thumb (uniform layer).

7. Dry the squashed slide over a small flame (preferably spirit lamp) for at least 45 min or less till the edge of the cover glass appears to be dry. Avoid overheating.

8. Place the slide in 50 per cent alcohol and allow the cover-glass to detach itself. Otherwise it may be removed by the edge of a sharp scalpel. The slide with detached cover glass can be stored in 70 per cent alcohol and can be stained with any standard cytological stain.

SCHEDULES FOR THE STUDY OF MEIOTIC CHROMOSOMES

IN PLANT MATERIALS

Meiotic chromosomes are studied usually from pollen mother cells and occasionally from embryosac mother cells.

STUDY FROM POLLEN MOTHER CELLS

Temporary Squash Technique

Material—Flower buds of *Solanum torvum*
1. Take flower buds serially from an inflorescence, starting from the smallest and working up to the largest, until the correct bud having divisional stages is found.
2. Dissect out a single anther from a bud with a needle. Place it on a clean slide.
3. With a clean scalpel, smear the entire anther on the slide and add a drop of 1 per cent iron–acetic-carmine solution to it immediately. Remove the debris.
4. Heat slightly over a flame. Cover with a cover glass and ring with paraffin.

Modifications

Instead of fresh anthers, anthers fixed in acetic–alcohol (1:1) mixture or in Carnoy's fluid, and later stored in 70 per cent alcohol can also be observed following this method. If stored in 70 per cent alcohol, keeping 1 h in each of acetic–ethanol and 45 per cent acetic acid solutions is necessary before smearing in 1 per cent acetic-carmine solution. For materials taking bright stain, treatment in acetic–alcohol can be omitted.

Instead of iron–acetic-carmine solution, 1 per cent acetic-carmine solution can be used and a trace of iron added by rubbing a rusty needle in the drop of stain on the slide.

In this schedule, 1 per cent acetic-carmine solution can be replaced by acetic-orcein, acetic-lacmoid, nigrosine, etc., solutions.

The slides can be kept as such in a refrigerator for a few weeks and then made permanent, following either of the schedules described earlier.

Very small sized buds in tight inflorescences can be fixed in acetic–alcohol mixture or in Carnoy's fluid for 1 h, hydrolysed at 60 °C for 5–10 min in N HCl, rinsed in water and stained for 1–3 h in leuco-basic fuchsin solution. They should be rinsed in two or three changes of 45 per cent acetic acid solution. Single anthers are to be dissected out, squashed in 45 per cent acetic acid solution and made permanent as usual.

Maceration in a mixture of 15 per cent chromic acid, 10 per cent nitric acid, 5 per cent HCl (2:1:1) for 5–7 min and hardening in ethanol-propionic acid (1:1), between fixation and staining has been used in *Gossypium* microspores (Bernardo, 1965). Restoration of deteriorated temporary acetic-carmine preparations involves replacing the acetic-carmine under the cover glass first with 2 N HCl and then with 1 per cent acetic-carmine (Persidsky, 1954).

Permanent Smear Technique

Material—Flower buds of *Datura fastuosa*

1. *Determination of size*—Take flower buds of different sizes. Dissect out a single anther from each bud and observe by squashing in 1 per cent iron–acetic-carmine solution as given in the last temporary technique until, in the bud of a particular size, meiotic divisional figures are observed.

2. *Smearing and fixation*—Dissect out the remaining anthers of the bud showing division and place each on a clean slide, cut off one end with a clean scalpel and squeeze out the contents by pressing with the left thumb. Discard the empty anther lobe. Quickly draw the fluid into a thin smear on the slide with a clean scalpel and immerse immediately in a tray containing Navashin's fluids A and B, freshly mixed in equal proportion. Keep in the fixative for 3–12 h.

3. *Washing*—Wash the slide in running water for 1 h.

4. *Staining*—Stain in 0·5 per cent aqueous crystal violet solution for 20 min or more and rinse in water.

5. *Mordanting*—Mordant in 1 per cent solution of I_2 and KI in 80 per cent ethyl alcohol for 45 s.

6. *Dehydration*—Dehydrate by passing the slide through absolute alcohol grades, I, II and III, keeping about 2–3 s in each.

7. *Differentiation*—Transfer the slide to clove oil I, keep for 2–3 min, take out the slide and observe the staining under a microscope. If found satisfactory, transfer the slide to clove oil II and keep for 2–3 min.

8. *Clearing*—Pass the slide through xylol grades I, II and III, for 30 min, 1 h and 30 min respectively.

9. *Mount*—Mount in Canada balsam or clarite X.

Modifications

Materials difficult to stain should be pre-mordanted overnight in 1 per cent chromic acid solution after fixation and washing, i.e. after step 3. They are then to be washed in running water for 3 h before staining.

In an alternative method, after smearing on a clean slide, immerse the material in acetic–alcohol mixture and keep for 1 h. Treat the slide with material in 45 per cent acetic acid for 15 min and hydrolyse in N HCl at 60 °C for 15 min and, finally, rinse it in water, stain in leuco-basic fuchsin solution for 1–2 h and mount with 45 per cent acetic acid.

Permanent Paraffin Section Technique

Material—Flower buds of *Allium cepa*

1. *Determination* of size—Test single anthers of individual flower buds until a bud of suitable size with divisional figures is obtained. Collect several buds of approximately the same size.

2. *Fixation*—Dip each bud, holding it with a pair of fine forceps, in Carnoy's fluid and keep for 2–3 s. Drop it in a container of water and rinse thoroughly, then transfer the bud to a phial containing a mixture of Navashin's A and B fluids (1:1) and keep for 24 h.

3–7. *Washing, dehydration, clearing, infiltration and embedding*—Are to be carried out in the same way as described under the corresponding steps in the paraffin section technique for root tips (Schedule A, page 151). While embedding, orient the flower buds in groups, depending on their size, about 1 cm away from each other.

8. *Section-cutting*—The process is similar to that followed for root-tips already described. Cut longitudinal sections 12 μm thick. Mount the ribbons as given before.

9–16. *Bringing down to water, pre-mordanting in 1 per cent chromic acid solution, staining in crystal violet, mordanting, dehydration, differentiation, clearing and mounting*—Are similar to the methods followed in the case of root tips.

Modifications

Haematoxylin staining can be done after step 9, bringing the slides down to water, following the schedule given for root rips (Schedules A–C, pages 151–153).

Pre-mordanting in 1 per cent chromic acid can be omitted for materials that are easy to stain.

In plants with very small flowers, like members of Araceae, the entire inflorescence is cut into equal segments and fixed as such. Transverse sections of the inflorescence are cut serially for study.

In plants with very large flowers, the anthers are dissected out and fixed.

For scattering chromosomes, 0·002M oxyquinoline solution can be mixed with the fixative (1:1) (Sharma and Ghosh, 1951).

Instead of Navashin's A and B fluids, other regular fixatives can also be used (*see* Chapter 3).

STUDY FROM EMBRYOSAC MOTHER CELLS

Squash Technique for Young Embryosac

1. Dissect out ovules from the ovary and fix in Carnoy's fluid for 1 day.

2. Keep in 95 per cent ethyl alcohol for 1–2 days. Run through 90, 80, 70, 50 and 30 per cent alcohol, keeping in each for 5–10 min. Rinse in water.

Two separate staining schedules can be followed:

(a) *Staining in Feulgen solution:*

3. Hydrolyse for 8–10 min at 60 °C in N HCl.

4. Rinse in water and stain in leuco-basic fuchsin solution for 2 h.

5. Intensify the stain by keeping in water for 15 min.

6. Transfer to a drop of 45 per cent acetic acid on a clean slide and squash under a cover-glass, applying uniform pressure.

7. Dehydrate by inverting in tertiary butyl alcohol and mount in euparal.

(b) *Staining in acetic-orcein solution:*

3. Transfer the ovules from water to a mixture of 2 per cent acetic-orcein solution and N HCl (9:1). Heat gently for 5–10 s without boiling the fluid.

4. Keep for 20 min in the mixture.

5. Transfer to a drop of 1 per cent acetic-orcein solution on a clean slide and squash under a cover-glass, exerting uniform pressure.

6. Make the slide permanent by any of the schedules described in the chapter on mounting.

Modifications

1. In plants with very small ovules, the entire ovary can be cut into tiny pieces and treated.

2. Acetic-lacmoid solution can be used instead of acetic-orcein solution in staining.

3. Acetic–alcohol mixture (1:1) can be used instead of Carnoy's fluid.

Squash Technique for Mature Embryosac

1. Fix the ovary in Carnoy's fluid for 2 days.

2. Transfer to a mixture containing 10 drops of a saturated solution of iron acetate in 45 per cent acetic acid and 10 cm^3 of 4 per cent iron alum solution. Keep the phial containing the material in the mixture in a water bath at 75 °C for 3 min.

3. Give two changes with distilled water heated to 75 °C, keeping for 2 min in each on the water bath.

4. Transfer to cold water and keep for 2–3 min.

5. Hydrolyse in 50 per cent HCl for 10 min.

6. Rinse in several changes of distilled water for 20 min.

7. Transfer the ovary to a drop of 1 per cent iron–acetic-carmine solution on a clean slide. Dissect out the ovules with needles into the stain and remove the rest of the ovary.

8. Tap the ovules with a flat-bladed scalpel until the cells are separated.

9. Apply a cover glass and heat the slide gently.

10. Squash and seal. The slide can be made permanent according to the usual schedules.

Paraffin Block Preparation for Embryosac Mother Cells

1. *Fixation*—Fix the dissected ovules overnight in La Cour's 2BX fixative (2 per cent chromic acid, 2 per cent potassium dichromate, 2 per cent osmic acid, 10 per cent acetic acid, 1 per cent saponin, distilled water in ratio 10:10:12:6:1:5).

2. Wash overnight in running water.

The later steps, namely, dehydration, clearing, infiltration, embedding and section cutting are similar to the corresponding steps in paraffin preparations of root tips. Sections are cut 15–20 µm thick.

3. Run the slides with sections through xylol I and II grades, keeping in each for 30 min. Pass through alcohol–xylol and absolute ethyl alcohol grades (15 min in each). Keep in 95 per cent alcohol for 10 min.

4. *Bleaching*—Keep in a mixture of 80 per cent alcohol and hydrogen peroxide (3:1) for 24 h. Observe, and if the background is not clear, keep for another 24 h.

5. Transfer to 70 per cent alcohol, then 50 and 30 per cent alcohol, keeping 5 min in each. Rinse thoroughly in water.

The next steps in the schedule, pre-mordanting in 1 per cent chromic acid, staining in crystal violet, mordanting, dehydration, differentiation, clearing and mounting are done as in the case of paraffin preparations of root tips described in earlier schedules.

RESTAINING SCHEDULES FOR BOTH MITOTIC AND MEIOTIC PREPARATIONS

Re-staining Procedure

Permanent slides kept for a long period usually lose the brightness of their stain, as almost all stains fade in certain environmental conditions. The principal factors responsible for this fading are: (*a*) the progressive acidity of the mounting medium, chiefly Canada balsam—this defect can be remedied by using neutral mounting media; (*b*) exposure to ultra-violet light by leaving the slides lying about carelessly or by exposure to arc lamp projectors; and (*c*) failure to remove all extraneous chemicals. If dehydration and later clearing in xylol are insufficient and traces of any solvent of the stain are carried over into the mounting medium, the stain fades quite rapidly.

Preparations, in which the stain has faded, can be re-stained if necessary. For re-staining, the original stain itself or some other stain which suits the fixative in which the tissue was originally fixed is used. The process consists of the following steps:

1. *Removal of the mounting medium*—The slides are placed in a solvent of the mounting medium, usually xylol, till the cover glass is detached. They are usually given a change in xylol to remove the mounting medium completely.

2. *Bringing down to water*—The slides are then brought down to water

through alcohol–xylol mixture (1:1), absolute alcohol, 95, 90, 80, 70, 60, 50 and 30 per cent alcohol grades, keeping 10 min in each.

3. *Staining*—The original schedule is followed. Both pre-mordanting and post-mordanting in 1 per cent chromic acid are done for slides being stained by crystal violet schedule, e.g. slides originally fixed in Flemming's fluid and stained in crystal violet can be restained following Feulgen schedule.

4. *Making permanent*—For any staining, the usual procedure is followed.

The re-staining procedure is also applied to slides which have not taken satisfactory stain in the original schedule and have been rejected during differentiation. For example, during the crystal violet schedule, if on differentiation in clove oil the tissue is seen to have taken insufficient stain, the slide is transferred directly to down grade xylol I and allowed to remain overnight. Afterwards it is brought down to water, pre-mordanted and stained as usual.

The only exception to the re-staining process is Feulgen stain. Slides with insufficient stain cannot be re-hydrolysed and re-stained in leuco-basic Fuchsin solution. However, several other alternatives can be suggested:

1. For tissues originally fixed in acetic fixatives, acetic stains, like acetic-orcein or acetic-lacmoid solution, can be used.

2. Tissues fixed in aqueous fixatives can be stained following the crystal violet schedule.

3. After fixation in alcoholic fixatives, haematoxylin staining is employed.

4. For slides which have taken insufficient Feulgen stain, the chromosome stain can be brightened by immersing the slide in 1 per cent acetic-carmine or acetic-orcein solution for 2–5 min.

IN ANIMAL MATERIAL

STUDIES OF MEIOTIC STAGES FROM TESTES

Temporary Squash Schedules

(a) *In grasshopper testes*

Dissect out the testes of a male grasshopper in 0·75 per cent normal saline solution by pulling with two fine forceps at the two ends, separate the head region from the rest of the body, trailing out the intact salivary glands and the anterior portion of the alimentary canal. Remove the glands by breaking off the duct attaching them with a pair of forceps.

For smear—Take a few lobules on a clean cover-glass and remove the excess saline by touching the edge of a filter paper. Cut the tip of the lobules by means of a cataract knife, and the fluid, which will come out immediately, should be spread quickly in uniform layers over the slide. Invert the slide immediately on a tray containing acetic-ethanol solution or any other fixative. Staining can be carried out in the usual way.

Acetic-carmine squash—Put testes lobules in a drop of acetic-carmine solution for 5 min and squash. The bulk testes should be fixed in acetic-alcohol mixture (1:3) and then put in acetic-carmine solution for 5–15 min.

Squash, according to Smith (1943), is performed on the tissue prior to staining.

Squash after staining: Feulgen staining—Bring bulk tissue fixed in acetic–alcohol mixture down through alcohol grades to water. Wash in water and put in cold HCl (N). Hydrolyse in N HCl for 12 min (or adjust accordingly) at 60 °C. Rinse in cold HCl and put in Feulgen solution (15 min–1 h). Take a few lobules of the stained tissue in a small drop of 45 per cent acetic acid on a slide and squash gently by applying pressure on a cover glass. Seal and observe. For temporary observation, materials can be sectioned 25 μm thick.

(b) *In amphibian testes*

1. *Dissection*—Dissect out the testes from the newt, *Trituridus* sp. Cut into very small pieces.

2. *Fixation*—Fix in acetic–ethanol mixture (1:3) or, in Carnoy's fluid in a glass phial for 2–3 days.

3. *Staining and squashing*—Lift a piece of the tissue on a drop of 1 per cent iron–acetic-carmine solution on a clean slide. Squash under a cover glass, warm and seal with paraffin wax.

4. The preparation can be made permanent following one of the usual schedules.

(c) *In Culex testes*

1. *Dissection*—Place the pupa on a clean slide in a drop of Ringer solution A (0·65 g of NaCl, 0·025 g of KCl, 0·03 g of $CaCl_2$, 0·02 g of Na_2CO_3 in 100 cm^3 of distilled water) and observe under a dissecting microscope. Pull the head and tail of the larva with two needles, breaking it in two. Lift out small translucent testes with a needle.

2. *Fixation*—Place the testes in a phial containing acetic–alcohol (1:3) mixture or Carnoy's fluid and fix for 2 min.

3. *Staining*—Transfer to a drop of 1 per cent iron–acetic-carmine solution on a clean slide, warm slightly, cover with a cover glass and squash through blotting paper.

4. The slides can be made permanent following one of the usual schedules.

Notes—1. This technique can be followed for most small insects.

2. Acetic-carmine can be replaced by 1 per cent acetic-orcein solution.

3. For Feulgen staining, transfer the testes to N HCl after fixation and hydrolyse at 60 °C for 4 min. Rinse in water and transfer to leuco-basic fuchsin solution. After 20 min, squash under a cover glass on a clean slide in 45 per cent acetic acid. The entire operation can be carried out on a slide.

(d) *In mammalian testes*

1. *Dissection*—Remove the testes entire. Cut into very thin sections.

2. *Fixation*—Fix directly in acetic–alcohol mixture (1:3) or 80 per cent chilled alcohol at 10–12 °C for 1 h.

3. *Staining*—Lift up a section, put it in 1 per cent iron–acetic-carmine solution, tease out the tubules; warm slightly, and squash under a cover glass and ring with paraffin wax. This method, though recommended, is however not very satisfactory.

For details, *see* chapter on mammalian chromosomes.

(e) *In avian and reptilian testes*

The techniques for avian and reptilian testes are similar to those for mammals and have been discussed in detail in the chapter on mammalian chromosomes.

Note—Invertebrate testes are usually dissected out in Ringer's solution A, but this step is not necessary for vertebrates.

Paraffin Section Schedules for Animal Testes in General

1. *Dissection*—Dissect out the testes. If they are large, cut them carefully into small pieces.

2. *Fixation*—Fix the tissues overnight in any of the fixatives given in Chapter 3.

The remaining steps, namely washing, dehydration, infiltration, embedding, section cutting, bringing down to water, staining, mordanting, differentiation and mounting are similar to those followed for the somatic tissue described previously. Sections should be cut between 10 and 16 μm thick.

For haematoxylin staining, Schedule C, page 153, is used.

STUDIES OF MEIOTIC STAGES FROM EGGS

Oogonial Division

1. For early oogonial divisions, when the ovary is relatively immature, the techniques used for study are similar to those for testes.

2. For studying older ovaries, both squash and paraffin section techniques are followed.

Oocyte Division

Take mature eggs before laying, and follow the usual schedule.

(a) *Squash technique*

This can be applied to those eggs which have a thin shell and little yolk.

1. Puncture the shell of the egg. Smear and fix the contents in acetic–ethanol mixture (1:1) for a few hours, hydrolyse in N HCl and stain following the Feulgen schedule.

2. Puncture the wall to bring out the contents and allow them to dry on a slide. Stain the dried eggs with aqueous Bismarck brown.

3. Certain insects can be fed with a mixture of honey and 1 per cent colchicine solution for 24 h. Dissect out the ovary, keep them in 0·5 per cent colchicine solution for 5–10 min and squash in 1 per cent acetic–carmine solution.

The meiotic studies from eggs are difficult to perform; the divisional stages are rarely found and lie in an arrested condition.

(b) *Permanent paraffin block preparations*

1. *Dissection and fixation*—Dissect out the ovary and fix in Smith's modification of Kahle's fluid (100 cm^3 absolute alcohol, 7 cm^3 glacial acetic acid and 40 cm^3 chloroform) for 2 h.

2. *Dehydration*—Pass the tissue through progressive ethyl alcohol–n-butyl alcohol and phenol grades (1 h in water-ethyl alcohol–n-butyl alcohol–phenol mixture 3:5:2:0, 24 h in 1:50:35:4, 1 h in 0:5:40:55 and 1 h in 0:1:3:0). Transfer to 4 per cent phenol in n-butyl alcohol. Change once after 10 min.

3. *Infiltration*—To the material in 4 per cent phenol in n-butyl alcohol, add an equal quantity of paraffin wax and keep in an oven. Change to pure molten wax after 16 h.

4. *Embedding*—Give two more changes in molten wax and embed as described in earlier schedule.

5. *Section cutting*—Trim the block. Cut off one end to expose the material and soak it in water for 24 h, then cut sections described before and bring the slide down to water.

6. Hydrolyse in N HCl for 6 min and stain in leuco-basic fuchsin solution for 1 h. Mount in 45 per cent ethyl alcohol.

7. The slides can be made permanent following any of the techniques previously described.

Note—If necessary, the eggs should be punctured with a needle to permit the fluid to enter.

(c) *Alternative schedule for marine invertebrate eggs (Cather, 1958)*

1. Place the eggs in a drop of water on an albuminised slide.

2. Add a few drops of fixative (absolute ethyl alcohol 1·5, tertiary butyl alcohol 1, acetic acid 1) to the eggs, allow it to flow over the eggs and drain off.

3. Similarly add 70 and 50 per cent alcohol and distilled water, treating in each for 10 min and then drain off.

4. Hydrolyse at room temperature with three changes of HCl—1 N HCl for 5 min, 5 N HCl for 15 min and again 1 N HCl for 3–5 min.

5. Stain in a jar containing Gomori's chromalum haematoxylin solution at 60 °C for 30 min.

6. Mordant in 1 N HCl for 3–5 min.

7. Keep in water for 10 min.

8. Rinse in SO$_2$ water for 5 min. Treat with chilled 45 per cent acetic acid for 5 min.

9. Wash in water for 5 min.

10. Treat in 1 per cent aqueous papain solution for 10 min. This causes the tubules to shrink.

11. Wash in water and transfer to 60 per cent acetic acid. The cells swell to greater than the original size.

12. Place the cover glass on the material and squash under blotting paper, applying uniform pressure.

13. The slides can be made permanent, following one of the usual schedules.

14. Change through 30, 50, 70, 95 per cent and absolute alcohol, keeping 10 min in each. Treat in pure xylol for 30 min and mount in Canada balsam.

REFERENCES

Arzac, J. (1950).   *Stain Tech.* **25**, 187
Bennett, E. (1964).   *Euphytica* **13**, 44
Bernardo, F. A. (1965).   *Stain Tech.* **40**, 205
Bezbaruah, H. P. (1968).   *Stain Tech.* **43**, 279
Bhaduri, P. N. and Ghosh, P. N. (1954).   *Stain Tech.* **29**, 269
Bhaduri, P. N. and Majumdar, B. R. (1955).   *Stain. Tech.* **30**, 93
Cather, J. N. (1958).   *Stain Tech.* **33**, 146
Conger, A. D. (1953).   *Stain Tech.* **28**, 289
De, D. (1958).   *Stain Tech.* **33**, 57
Dyer, A. F. (1966).   *Stain Tech.* **41**, 277
Ford, C. E. and Hamerton, J. L. (1956).   *Stain Tech.* **31**, 247
Jagathesan, D. and Sreenivasan, T. V. (1966).   *Stain Tech.* **41**, 43
Jona, R. (1967).   *Stain Tech.* **42**, 113
Khoshoo, T. N. (1956).   *Stain Tech.* **31**, 31
La Cour, L. F. (1937).   *Bot. Rev.* **3**, 241
Latour, G. de (1960).   *New Zealand J. Sci.* **3**, 293
Li, H. W., Ma, T. H. and Shang, K. C. (1954).   *Taiwan Sug.* **1**, 13
Ma, T. (1967).   *Stain Tech.* **42**, 285
Manna, G. K. (1957).   *Proc. zool. Soc. Beng.* Mukerji Mem. Vol. p. 95
Persidsky, M. D. (1954).   *Stain Tech.* **29**, 278
Persidsky, M. D. and Duncan, R. E. (1957).   *Stain Tech.* **32**, 117
Powell, J. B. (1968).   *Stain Tech.* **43**, 135
Price, S. (1956).   *Proc. IX Cong. Ins. Soc. Sug. Tech.* 780
Price, S. (1962).   *Proc. XI Cong. Int. Soc. Sug. Tech.* 583
Read, R. W. (1964).   *Stain Tech.* **39**, 99
Rutishauser, A. and Hunziker, H. R. (1950).   *Arch. Klaus-Stif. vererbForsch.* **25**, 477
Savage, J. R. K. (1957).   *Stain Tech.* **32**, 283
Sharma, A. K. and Ghosh, C. (1951).   *Sci. Cult.* **16**, 528
Sharma, A. K. and Varma, B. (1960).   *Cytologia* **24**, 498
Sharma, A. K. and Mookerjea, A. (1955).   *Stain Tech.* **30**, 1
Sisodia, N. S. (1968).   *Stain Tech.* **43**, 139
Smith, S. G. (1943).   *Canad. Ent.* **75**, 21
Sparano, B. M. (1961).   *Stain Tech.* **36**, 41
Stevenson, G. C. (1965).   *Genetics and breeding of sugar cane.* London; Longman's
Swanson, C. P. (1940).   *Stain Tech.* **15**, 49
Tjio, J. H. and Whang, J. (1962).   *Stain Tech.* **37**, 17
Upadhya, M. D. (1963).   *Stain Tech.* **38**, 293

8 Methods for Special Materials

The special techniques for the study of the differential nature of chromosome segments can be put into groups according to the nature of the chromosomes—1, spiral structure; 2, centromere; 3, secondary constriction; 4, heterochromatin; 5, salivary gland chromosomes; 6, lamp brush chromosomes; 7, pachytene chromosomes; 8, prochromosomes; 9, pollen grains; 10, embryosac mother cells; 11, endosperm; 12, study of nucleolus. The methods adopted for studying chromosomes of thallophytes have also been included in this chapter.

SPIRAL STRUCTURE

The spiral nature of chromosomes was first observed by Baranetzky in 1880 in metaphase plates of *Tradescantia*. Sakamura in 1927 demonstrated internal spirals in the same plant during meiosis by fixation in boiling water. Gradually various schedules were developed for the study of the spirals, and, in general, they fall into three groups: those based on uncoiling the chromosome threads through some shock; those based on dissolving the outer envelope of nucleic acid, exposing the inner thread; and those based on causing the chromosomes to swell, by differential hydration of the segments.

These methods have been applied in both plants and animals for the study of mitosis and meiosis.

FIRST GROUP

Several agents are used for uncoiling the chromosome thread, including (a) acid fumes, (b) ammonia vapour, (c) cold temperature and (d) boiling water. They can be applied to both mitotic and meiotic chromosomes.

Acid fumes—For mitotic cells, expose the tissue directly in the living state to the fumes of concentrated nitric acid, hydrochloric acid or acetic acid for 1–2 min (for example, bulbs of *Allium cepa* with roots intact are held with the roots near the mouth of a jar containing the acid). Fix the tissue (cut root tips in case of *Allium cepa*) in acetic–ethanol (1:1) for $\frac{1}{2}$ h and wash successively in 70 per cent alcohol and water. Hydrolyse in N HCl for 10 min at 58–60 °C. Wash in water, stain in leuco-basic fuchsin solution for 30 min (Feulgen schedule), squash in 45 per cent acetic acid and seal with paraffin.

For meiotic cells, smear the tissue, expose the slides on the smeared side to the acid fumes for 1 min, fix overnight in La Cour's 2BE or Navashin's A and B fluids, wash in running water, stain following the crystal violet schedule (*see* Chapter 7), and mount in Canada balsam.

Ammonia vapour—Expose the tissue, in its medium, to fumes of ammonia for 5–15 s. (For example, pollen mother cells are kept in 3 per cent sucrose solution during exposure to ammonia vapour.) Fix the tissue in the requisite fixative, stain and mount. Kuwada and Nakamura (1934) smeared flower buds after exposure in 1 per cent acetic-carmine solutions; pollen mother cells can also be exposed after smearing and fixed and stained according to the schedule followed for exposure to acid fumes. Both roots and anthers can also be fixed in acetic–alcohol after exposure, and hydrolysed and stained according to Feulgen technique as given with the acid fumes (La Cour, 1935).

Cold temperature—Root tips, on being frozen for 3–5 days, show nucleic acid starvation. If they are then fixed and stained following the Feulgen schedule, the major spirals can be seen (Callan, 1942).

Micro-incineration—Barigozzi (1937) and Über (1940) used micro-incineration for demonstrating the spiral nature of chromosomes.

SECOND GROUP

The comparatively more effective schedules based on dissolving the outer nucleic acid envelope are:

Ammonia–Alcohol Schedule

Treat the tissue with ammonia in 30 per cent alcohol (6 drops in 50 cm^3) for 5–20 s. Wash in water and fix overnight in Flemming's fluid. Wash in running water for 3 h, stain in crystal violet (*see* Chapter 5), and mount in Canada balsam. This method is effective for studying *meiosis* in both plants and animals (Sax and Humphrey, 1934; Creighton, 1938), but for mitosis an alternative method is to fix the tissue after exposure in acetic–alcohol and to stain following the Feulgen schedule, as given before.

Weak Alkali

Weak alkaline solvents of nucleic acid are used, like NaOH, NaHCO$_3$ or NaOH in 1/100 g mol solutions. Place the tissue in the alkali for 15 s to

5 min, depending on the material, and fix in Flemming's fluid and stain in crystal violet or fix in acetic–alcohol and stain by the Feulgen schedule as given before (Oura, 1936: Kuwada, Shinke and Oura, 1938; Hillary, 1940). The pollen mother cells can be smeared before treatment on a dry slide and the smears subjected to treatment.

Precipitation of DNA

This method is based on the principle of precipitating the nucleic acid as a metallic salt and subsequently partially digesting the protein.

Treat the tissue (slides with smears in case of meiotic cells) in 1/100 g mol solution of NaCN for 30 s. Fix overnight in Flemming's fluid and then in water for 3 h, and keep the tissue in 0·1 per cent lanthanum acetate for 12 h to precipitate the nucleic acid as lanthanum salt. Digest in 1 per cent trypsin solution containing a trace of lanthanum acetate for 24 h at 37 °C, and hydrolyse and stain, following the usual Feulgen schedule (Hillary, 1940).

An alternative method for root tips is to treat them in ammonia followed by thorium nitrate solution (Nebel, 1934). The tissue can also be treated in 0·00005 g mol solution of NaCN, followed by acetic–alcohol fixation and Feulgen schedule, as given previously (Coleman, 1940).

THIRD GROUP

Treatment of the tissue, particularly of root tips, in any of the pre-treatment chemicals for a specified period at a requisite cold temperature results in changes in the cytoplasm. These changes cause a differential hydration and subsequent swelling of the chromosome arms, and often the major spirals can clearly be observed in the swollen chromosomes.

For this purpose, treat the root tips in a pre-treatment chemical like aesculine, *p*-dichlorobenzene, or 8-oxyquinoline for a certain period (1 h for aesculine, 4 h for *p*DB, 3 h for OQ) at cold temperature (10–15 °C). Fix in acetic–alcohol, heat in 2 per cent acetic-orcein–N HCl (9:1) mixture and squash in 1 per cent acetic-orcein solution.

CENTROMERE

The centromere has been the object of close scrutiny for many years (Östergren, 1947; Lima-de-Faria, 1949). Carothers (1936) considered this as composed for temporarily modified chromomer. Matsuura (1941) regarded the centromeric segment as a specialised portion of genonema. Darlington (1939) visualised a fluid consistency having a number of parallel micelles oriented in it. Schrader (1939) and Sharp (1943) suggested spherular constitution of the centromere. According to the accepted view of Tjio and Levan (1950), the structure of the centromere is composed of:

(a) An outermost zone connecting the chromomeres of the arm with the centromere—two thin fibrillae are observed in this zone.
(b) The chromomeric zone consisting of two pairs of centromeric chromomeres—these two pairs of chromomeres are disposed in longitudinal sequence and each pair is attached to the two fibrillae of the outer zone.
(c) The interior zone comprising the space between the two pairs of centromeric chromomeres. Two thin fibrillae can be observed in it.

These centromeric bodies show the same divisional cycle as the chromatids, and in order to study the structure of the centromere, the gap should be increased. For this purpose, the material is first treated at cold temperature in a pre-treatment chemical, fixed and stained. The most satisfactory pre-treatment chemical has been found to be 8-oxyquinoline (OQ) solution.

The principle underlying the process is that the pre-treatment chemical when cold, changes the viscosity of the cytoplasm, causing the spindle to break. The chromosome segments undergo differential hydration and contract while the more or less rigid plasma keeps the chromosome arms in their original position, causing a contraction towards the middle of each chromosome arm. As a result, the centromeric gap is exaggerated and the structure of the centromeric apparatus is brought out clearly.

A general schedule for centromere study is given below (Tjio and Levan, 1950):

1. Cut fresh root tips (for example, those of *Agapanthus umbellatus*) and treat in aqueous solution of 8-oxyquinoline (0·002M) at 12–18 °C for 3 h.

2. Heat the root tips in a mixture of 2 per cent acetic-orcein solution and N HCl (9:1) for 3–4 s.

3. Squash the tips in 1 per cent acetic-orcein solution, seal and observe.

Lima-de-Faria (1949) found that fixation in acetic–alcohol mixture (1:1) tends to reduce the stainability of the centromere.

For the study of *centromere in flower buds*, Lima-de-Faria (1948) dissected out the anthers, placed them in a drop of 1 per cent acetic-carmine solution, added a trace of iron by macerating them with a needle, heated slightly, squashed and sealed. In some cases, he fixed the buds in acetic–alcohol (1:4) for 4 h, transferred to 95 per cent alcohol (overnight), stored for some days in 70 per cent alcohol and followed the schedule outlined above.

Lima-de-Faria has also outlined an alternative schedule for staining meiotic centromere in Feulgen solution. The technique, used on flower buds of *Agapanthus umbellatus*, is as follows:

1. Fix the buds in acetic–alcohol mixture (1:4) for 7–9 h. Transfer to 95 per cent alcohol (overnight) and store in 70 per cent alcohol for 2–3 days.

2. Cut the anthers and squeeze out the pollen mother cells into a drop of 45 per cent acetic acid on a slide.

3. Squash by placing a cover glass over the pollen mother cells.

4. Gently heat the slide once over a flame.

5. Invert the slide in 1 per cent aqueous acetic acid.

6. After the cover glass falls off, keep both slide and cover glass in a tray containing Belling's modification of Navashin's fixative for 3 days.

7. Wash in distilled water for 1 h.

8. Immerse in N HCl at 60 °C for 6 min.

9. Wash in distilled water (1–2 min) and immerse in fuchsin sulphurous acid for $3\frac{1}{2}$ h.

10. Pass through fresh SO_2 water, three changes of 10 min each.

11. Rinse in distilled water, pass through 20, 50, 70, 95 per cent and absolute alcohol, keeping in each for 3 min. Mount separately in euparal and seal.

The chromosomes are staind with the characteristic violet red colour. The centromeric chromomeres, being intercalated among longer fibrillae than the chromomeres of the arms, can be clearly distinguished.

All these techniques have been carried out on plant chromosomes. They can, if necessary, be applied to animal materials with modifications. Thus there is ample scope for improvement.

SECONDARY CONSTRICTION

Secondary constrictions are those constrictions observed in a chromosome which lack a centromere, are heterochromatic in nature and do not generally exhibit allocycly. They are the loci where nucleoli are organised, as suggested by Heitz (1931), McClintock (1934) and others. Their universal nucleolar nature has, however, been disputed by Darlington (1926) and Darlington and La Cour (1942), *see* Lettré and Siebs (1961).

The secondary constriction is generally represented as a distinct gap during the divisional cycle. When the constriction is present at the end and the distal chromatic part appears as a knob, it is known as a *satellite*. Often a thin chromatic thread, called a satellite thread, is seen to bridge the gap. Sometimes more than one secondary constriction is observed on the same chromosome, being called *supernumerary constrictions*.

For the study of secondary constrictions, the gap has to be exaggerated. This can chiefly be brought about by two means: cold treatment, and pre-treatment by chemicals.

Cold Treatment

The phenomenon underlying this technique is the nucleic acid starvation observed at cold temperatures. Keep the plant overnight in a cold chamber at 3–7 °C. Cut out the root rips, wash in water and fix in acetic–ethanol (1:1) for 2 h, heat in a mixture of 2 per cent acetic-orcein and N HCl for a few seconds, squash in 1 per cent acetic-orcein under a cover glass, and seal. The secondary constrictions are observed as unstained gaps in the chromosome.

Pre-treatment

Pre-treatment in the various cold pre-fixing fluids tends to cause viscosity change in the cytoplasm and differential hydration and dehydration of the chromosome segments. The secondary constriction regions are de-spiralised and expand, and the contraction of the swollen arms on either side tends to exaggerate this region, which becomes very pronounced.

Treat cut root tips with a saturated solution of 8-oxyquinoline for 4 h at 10–15 °C. Fix in acetic–alcohol mixture (1:1) for 1 h and follow the acetic-orcein squash technique as given in the previous treatment. The secondary constrictions appear as pronounced unstained gaps.

The Feulgen schedule can be employed in lieu of acetic-orcein after fixation in acetic–alcohol.

HETEROCHROMATIN

Heitz (1931) originally defined heterochromatin as that chromatic substance which retains maximum metaphase condensation throughout the divisional cycle. Pontecorvo (1944) regarded heterochromatin as that chromosome segment which stains during mitosis differently from the rest of the chromosome, and suggested that duplication of gene tandem along the chromosome would give rise to regions different from the others in their nucleic acid content. Darlington and La Cour (1960) have defined it as that part of a chromosome which fails to maintain the maximum nucleic acid cycle at mitosis, and considered allocycly as the identifying characteristic of heterochromatin. For details of heterochromatin, please see Brown (1966) and Wolf and Wolf (1969).

Summing the various ideas, heterochromatic segments may be defined as those segments which differ in any respect from euchromatic ones. They may show allocycly (e.g., prochromosomes), may be positively heteropyknotic throughout the division cycle (sex chromosomes) or may be negatively heteropyknotic in all the stages. They are of a heterogeneous nature and include the centromeric heterochromatin (called 'chromocenters'), intercalary heterochromatin, telomeric heterochromatin and also the heterochromatin in the secondary constriction regions. Entire chromosomes may be heterochromatic, like sex chromosomes in *Drosophila*, supernumerary chromosomes and B chromosomes in maize. Heterochromatic regions exhibit, in general: (a) extreme susceptibility to external conditions, like cold; (b) heteropyknosity and allocycly, in most cases; (c) property of heterochromatising adjacent euchromatic segments and (d) genetic inertness (!)—a debated issue.

In this chapter only the methods for demonstrating the heterochromatic segments of a chromosome, other than the chromocenters and secondary constriction regions, are described, as the methods for studying centromere and secondary constriction have already been discussed under their respective headings. The differential staining of the euchromatic and heterochromatic regions is based on the assumption that possibly there is an accumulation of DNA in the heterochromatin which is liberated from the euchromatic segments. Under normal conditions, during metaphase, heterochromatin adjacent to centromere shows Feulgen-negative behaviour due to allocycly, but owing to treatment with mercuric salt, DNA protein linkage in euchromatin becomes broken, and the DNA thus set free adheres to heterochromatic segment, the latter becoming Feulgen positive. With further treatment and prolonged digestion, DNA from heterochromatin also comes out, and chromosomes become entirely negative. Two techniques, digestion with trichloracetic acid, and digestion with mercuric nitrate, have given best results. For other methods, please see Chapters 13 and 15, using fluorescence technique.

Treatment with Trichloracetic Acid

This method is based on the initial digestion of the tissue in a particular concentration of TCA for a definite period, resulting in differential removal of DNA from the chromosomes, followed by staining (Sharma, 1951).

Fix the cut tissue in acetic–alcohol mixture (1:1) or 80 per cent alcohol in cold (10 °C) for 2 h to overnight. Treat the tissue in 0·25 M TCA at 60 °C for 40 min. Wash in distilled water. Keep the tissue in leucofuchsin solution for 30 min, and squash under a cover glass in 45 per cent acetic acid. The heterochromatin, particularly that of the centromeric and telomeric regions, stains sharply in leucofuchsin, the rest of the chromosome remaining unstained. The secondary constrictions do not show any basophilia.

Treatment with Mercuric Nitrate (Levan, 1946)

The tissue is initially treated in an inorganic salt for a period in cold temperature followed by fixation and staining, when the chromosome arms show differential staining in the heterochromatic regions (Levan, 1948, 1949, 1952). The most effective salt solution is mercuric nitrate.

Wash the tissue (for example, cut fresh root tips of *Allium cepa*) and pre-fix in 0·005 mol aqueous mercuric nitrate solution for 4 h at 10–12 °C. Transfer the roots to a mixture of equal parts of Navashin A and B solutions, mixed just before use, and keep overnight. Wash, embed in paraffin, cut longitudinal sections and stain and mount following the usual crystal violet staining schedule.

During the divisional cycle, the chromosome is de-stained and becomes hyaline except for the heterochromatic regions on the two sides of the centromere and the telomeres. The maximum stain is retained in the chromocenters.

SALIVARY GLAND CHROMOSOMES

Balbiani in 1881 first observed giant chromosomes in the salivary gland cells of some Dipteran species. These chromosomes, also found in some other gland tissue, remain in a state of permanent prophase. Later, Kostoff (1930), Heitz and Bauer (1933) and Painter (1933, 1934) studied their structure and nature in detail.

These giant chromosomes are the largest available for chromosome study, being approximately 100 times the length of somatic metaphase chromosomes. The homologous chromosomes show a type of close synapsis, and also reveal a distinctive pattern of bands, consisting of alternating chromatic and achromatic areas. These bands differ in the details of structure so distinctly that on this basis each chromosome can be accurately mapped throughout its euchromatic length.

Structurally each giant chromosome shows a 'polytene' or 'multiple' nature (Metz, 1941; Painter, 1941). The chromosome remains in a permanent prophase condition, in which the chromomeres and chromonemata continue to duplicate without separation, the result being a multi-stranded structure, visible under a hand lens when stained. The aggregation of the homologous chromosomes produces the transverse chromatic bands, and the numerous chromonemata, which may be as many as 1024, according to Painter (1941) and Swift and Rasch (1955), are associated in parallel lines. Others, like Kodani (1942) and Ris and Crouse (1945) disagree with the polytene concept, regarding the chromonemata to be swollen by lateral enlargement of the interband regions, and the bands to be coiled structures.

The salivary gland nuclei are held to be polyploid due to the giant size of the chromosomes, their multiple nature and their total DNA content, which is many times higher than that of ordinary nuclei.

The salivary gland chromosomes present a constant pattern of linear differentiation which aids in their study with relation to genetics and systematics. Caspersson (1940) utilised them in microchemical studies of cell structure, while Pavan and Ficq (1957) used the chromosomes of *Rhynchosciara angelae* for autoradiographic experiments with H^3-labelled thymidine for studying genic metabolism.

For the study of these chromosomes, the salivary glands of different Dipteran larvae are the most suitable material, though they also occur in the cells of malphigian tubules, fat bodies, ovarian nurse cells and gut epithelia. For general experiments, the insects most commonly used are *Chironomus* (Poulson and Metz, 1938), *Bibio* (Heitz and Bauer, 1933), *Drosophila* (Demerec, 1950), *Sciara* (Metz, 1935); and *Rhyncosciara* (Pavon and Da Cunha, 1969).

Old fat larvae, bred on rich yeast food at 16–20 °C, show the largest chromosomes. The diet varies with the type of larva used. The number of the cells in a salivary gland ranges from 28–32 in *Sciara*, 28–44 in *Chironomus* to 100–120 in *Drosophila*, all in different stages of development.

For the study of the chromosomes, the following schedule may be followed:

1. Place the full-grown larva on a slide in a drop of Ringer's solution or 0·73 per cent isotonic salt solution. Cut off the head with a scalpel in the right hand and press the body with a needle in the left hand at the same time. Remove the pressure. The salivary glands float out. Transfer them to a clean slide with the dissecting fluid using a pipette. The material is now ready for staining. For storage, add a few drops of paraffin oil to the material and place a cover glass over the whole. It will remain fresh for 24 h (Bauer, 1935).

2. Staining may be done either in acetic-carmine or acetic-orcein solution. For *Drosophila*, the most effective stain is 2 per cent acetic-orcein in 70 per cent acetic acid; while 1 per cent acetic-orcein in 45 per cent acetic acid, plus 1 cm³ chloroform may be used for *Sciara*; and 2 per cent orcein in 50 per cent acetic acid for *Chironomus*.

Transfer the tissue to a drop of the stain on a clean slide with a pipette. Leave for 5–10 min, depending on the material.

3. Prepare the cover glass by smearing thinly with Mayer's albumin and dry over a flame for 1–3 s.

4. Add the cover glass on the tissue, drain off the excess stain with filter paper, blot the preparation with filter paper, applying uniform pressure over the cover glass to flatten out the chromosomes and remove more stain. If the glands do not rupture, press with a blunt needle over the cover glass. Seal and observe.

5. For permanent preparations, keep the slide with squashed material and cover glass in a closed trough, lined with filter paper soaked in 95 per cent alcohol. After 24 h remove the cover glass and mount in euparal. Alternatively, the slides may be inverted in *n*-butyl alcohol until the cover glass is detached, and mounted separately.

Modifications of the schedule given above were suggested by many workers. Zirkle (1937) suggested a mixture of acetic acid (50 cm³), water (50 cm³), glycerin (1 cm³), powdered gelatin (10 g), dextrose (4 g), $FeCl_3$ (0·05 g) and carmine to saturation, boiled and filtered, to be used instead of the usual acetic-orcein stain. Heilborn (1937), in an alternative method, transferred the glands from Ringer's solution to 50 per cent acetic acid, fixed for 4–8 min and then transferred to a *dry slide*, covered with a cover glass, moistened with glycerin and squashed, suspended in 95 per cent alcohol until the cover glass detached, washed in water, stained overnight in 1 per cent acetic-carmine, washed again and then dehydrated following alcohol-xylol grades and mounted as usual.

LAMP BRUSH CHROMOSOMES

In a large number of vertebrates, within the developing oöcytes during diplotene, the chromosomes undergo a phase called 'lamp brush', characterised by a great increase in length and the formation of numerous side loops radiating from the chromosomes. These loops grow in number and size up to diplotene but decrease and disappear before metaphase. They have also been found in the spermatocytes of some invertebrates (Ris, 1945). The most satisfactory materials, are, however, the shark *Pristiurus*, some birds and amphibians (Duryee, 1941, 1950; Gall, 1952, 1954, 1956; Alfert, 1954; Callan, 1963, 1966; Izawa *et al.*, 1963; MacGregor, 1965; Miller, 1965). The longest lamp brush chromosomes are found in the newt, *Triturus viridescens*, ranging from 350 to 800 μm in length.

Duryee regarded a lamp brush chromosome as being composed of a cylindrical body in which chromatic granules are embedded at specific loci, the granules generally, but not always, paired within a bivalent. One to nine loops may arise from one chromomere, their length ranging from

9·5 μm in the frog to 200 μm in the newt. Duryee considered the loops to be chromatin material synthesised by the chromomeres for use by the oöcyte. Ris (1945) regarded them as an integral portion of the chromonemata. Gall (1956) supported this view, as his observations through the electron microscope show that the loops are a part of the chromonemata and that they become invisible during metaphase because they lose their covering of nucleic acid. Rogers (1965) analysed the components of the nuclear sap of salamander oöcytes through polyacrylamide gel electrophoresis technique.

Lamp brush chromosomes are very elastic and may be pulled to many times their length without injury; however the lateral coils, though elastic, are more fragile. The main axes and loops are DNA-positive and resistant to several chemical treatments. Obviously, the long lateral loops are formed at definite loci due to metabolic activity of genes at these regions (Callan and Lloyd, 1956, 1960).

The lamp brush chromosomes are of special interest, due to their great length and extreme fragility. The principal purpose of using lamp brush chromosomes is that the synthesis of ribonucleoprotein in the metabolically active DNA loop can be studied. For observation under a living constitution, the use of a phase contrast microscope is recommended.

A general schedule for study is given below (Gall, 1966):

1. Dissect out the ovary or several oöcytes, making a small incision at the ventral side of the animal (for example, the newt) anaesthetised with light ether or 0·1 per cent MS 222 solution for 15 min, and place it in a dry watch-glass containing the coelomic with adequate moisture, seal and keep at 4 °C for up to 2 days. Transfer a piece of oöcyte or ovary to a mixture of 5 parts 0·1M KC1 and 1 part of 0·1M NaCl buffered to pH 6·8–7·2 with 0·01M phosphate. This ratio may vary in different species (Callan and Lloyd, 1960; Naora *et al.*, 1962). With a pair of fine micro-forceps, break the oöcyte and take out the translucent nucleus. Puncturing the cell, followed by squeezing with the needle also results in ejection of the nucleus. In both cases, the nucleus starts swelling on isolation. Transfer the nucleus immediately to the observation chamber filled with the same 5:1 solution.

3. For preparation of the observation chamber, bore a 6·4 mm hole through the centre of a 76·2 × 25·4 mm slide and place a cover glass across the hole and seal with paraffin wax. Fill it with S11 solution before transferring the nucleus.

4. In the observation chamber, remove the nuclear membrane carefully, under a dissecting microscope with the help of a pair of very fine forceps and a tungsten needle pointed by dipping in molten sodium nitrate. The chromosomes are then liberated and confined in the hole in the slide on the cover glass forming its bottom. It can be kept in the cold for observation for a day in the unfixed condition. For fixation, it is preferable to keep the slide 3–5 min in a formaldehyde chamber or to add a drop of conc. formalin. After fixation, the preparations can be kept for several weeks. Cover the slide on the other side with another cover glass and seal.

5. Invert the optical train of the phase contrast microscope and observe the preparation under it. The chromosomes will be observed as coiled loops in a state of continuous movement.

Care must be taken to handle the preparation very gently. Microphotographs are taken in flashlight.

For making permanent preparations

Use a slide instead of a cover slip at the bottom of the observation chamber so that the chromosome can be isolated on the slide or alternatively, the cover slip may be used.

1. After isolation and clearing of the nucleus in 5:1 medium (calcium free), pass the nucleus quickly through a watch glass containing 0.1M 511 and 0.5×10^{-4} M $CaCl_2$ and transfer to the observation chamber again.

2. Remove the nuclear membrane, isolate the chromosome as above and transfer to a moist chamber for an hour, which later on becomes transformed to a formaldehyde chamber till the chromosome gets attached to the slide (in about 15 min). The pH may be lowered to facilitate attachment.

3. Transfer the set-up to a jar containing 2–4 per cent of 5–10 per cent formalin, buffered to pH 4–5, and keep for 1 h.

4. With a scalpel, separate the slide from the observation chamber. Dehydrate in ethanol, removing paraffin or vaseline with which the slide was attached and follow the usual procedure for staining in Feulgen solution for DNA and Azure B at pH 4 for RNA.

Notes

1. The medium should be diluted, if necessary, especially for the study of the chromomeres and the operation should be performed very quickly, to prevent the sap from becoming completely liquefied.

2. To check sap dispersal, osmium fixation is recommended, for which the chromosomes should be isolated in saline solution, dispersed in water and a moist chamber, fixed in osmium vapour and finally attached with 1 per cent aqueous acetic acid or vapour. Ethanol vapour may also be tried.

For autoradiography

In securing autoradiographs, lamp brush chromosomes are convenient materials, since the incorporation of precursors is very rapid. Specific radioactive precursors, such as, tritium-labelled uridine (100–300 mCi–specific activity 0.5–5.0 Ci mmol^{-1}) may be injected into the body cavity. If the explanting is carried out at 48–72 h, profuse labelling indicating synthesis of RNA or DNA loops can be obtained. But for short term and rapid incorporation, explanted ovaries can be treated in concentrated radioactive precursors taken in a watch glass. Both stripping film and emulsion coating can be performed as outlined in the chapter on autoradiography.

For electron microscope studies

Lamp brush chromosomes are not convenient materials due to the thickness of the chromosomes and the difficulty in interpreting the isolated material

(Gall, 1966). However, chromosomes attached to coverslips can be dehydrated and then polymerised, after inverting over a gelatin capsule filled with plastic monomer and the coverslip separated out, using the freezing technique. In whole mounts, mounting directly on steel or platinum grids coated with carbon can be performed (Lafontaine and Ris, 1958), as given in the chapter on electron microscopy. Gall (1966) has also devised a method for agar spreading of lamp brush chromosomes.

PACHYTENE CHROMOSOMES

Due to their configuration during the pachytene stage in pollen mother cells, the chromosomes of a large number of plants present very good material for the study of their individual structure and the nature of pairing. Different techniques have been devised for the study of the various structures and a few of the schedules followed are now considered.

For Chromosome Study

Lima-de-Faria, in 1948, employed a modification of the iron–acetic-carmine technique to study the B-chromosomes of rye at pachytene. Later, in 1952, he used this method for chromomere analysis of rye. In brief, the method includes the following steps:

1. Fix a complete spike of rye with awns cut close in acetic–alcohol mixture (1:4) for 3–4 h. Keep overnight in 95 per cent alcohol and pass on to 70 per cent alcohol. Dissect out the 3 anthers of each flower.

2. Transfer one anther to a drop of iron–acetic-carmine solution (2 drops of iron acetate in 10 cm^3 of acetic-carmine solution). Under a wide-field binocular microscope, cut the anther into two pieces, press each half gently to squeeze out the pollen mother cells, and remove all pieces of anther wall and tapetum with a needle, leaving only the pollen mother cells.

3. Observe under the microscope if the nuclei are in the pachytene stage. If so, add another drop of stain, place a cover glass on the material and heat gently over a flame three or four times.

4. Apply uniform pressure vertically with a U-shaped needle-point, blotting off excess stain from the side; check under a microscope. If stronger pressure is wanted, add more stain, heat again and press. Careless application of pressure may break up the chromosomes.

5. Heat again and invert the slide whilst still warm in 10 per cent aqueous acetic acid and after the cover glass falls off, pass both slide and cover glass through acetic–alcohol and alcohol–xylol grades and mount in Canada balsam.

6. The chromosomes are drawn with camera lucida. They appear as comparatively deeply stained dots arranged pair by pair throughout the length of the chromosomes. Their relative positions and structure can be observed.

For the Study of General Structure

For a general analysis of the chromosomes of barley, Sarvella, Holmgren and Nilan (1958) tried out a modification of Lima-de-Faria's technique. Fix the anthers in 3 parts 95 per cent alcohol: 1 part acetic acid. Tease out the pollen mother cells and place in iron–acetic-carmine solution. Place the slides over a hot water bath, and before final squashing, tap gently on the cover glass to separate the pollen mother cells. The slides are made permanent by Conger's dry ice technique and mounted in balsam.

For the Study of Heterochromatin

Brown (1949) devised a modified technique involving the use of iron alum mordanting before staining in 1 per cent acetic-carmine solution as usual. He applied this method in *Lycopersicum esculentum*. Thomas and Revell (1946) employed another technique for studying the heterochromatic bodies in *Cicer arietinum*.

Several Modified Techniques

Several modified techniques have been employed by different workers at various periods for the study of pachytene chromosomes from different aspects in many angiosperms. Some of them are:

1. Darlington's (1933) technique for studying the centromere structure in *Agapanthus*.

2. Brown's (1954) technique for the study of diffuse centromere in *Luzula*.

3. McClintock's (1934) technique for studying the nucleolar organiser region in *Zea mays*.

4. Pairing in diploid plants of *Zea mays* observed by McClintock (1931, 1933).

5. Techniques devised by Darlington (1929), Upcott (1939) and Gottschalk (1955) for the study of pairing in tetraploids of different plants.

6. Lately, Ford (1969) has studied in detail the chromomere patterns of human chromosomes.

PROCHROMOSOMES

The term 'prochromosome', though a misnomer, gives an indication of its relationship with the chromosomes. It is applied to the heterochromatic blocks of the interphase stage and the appearance of these blocks is due to the positive stainability of the heterochromatin lying at both sides of the centromere. The region appearing as a stained block during interphase remains de-stained during metaphase, representing true allocycly. That each block is the fused product of the two segments adjacent to each centromere is proved by their bilobed appearance maintained in certain cases. The number of prochromosomes may be equal to, or if fused, less than, the number of chromosomes in the complement. The polyploid nature of chromosome threads in the differentiated nuclei was first observed by Huskins through the prochromosome counts.

For the study of prochromosomes, the pre-treatment and Feulgen schedule adopted for centromere study, as described earlier, is equally effective. An ideal material is the root tip of *Scilla sibirica*. The prochromosomes appear as magenta-coloured bodies in the nuclei at resting stage.

POLLEN GRAINS

The pollen grains of Angiosperms are utilised for three purposes: (*a*) for studying mitotic division in pollen grains; (*b*) for studying the fertility of the grains, and (*c*) for studying the mitotic division in the generative cell inside the growing pollen tube.

Study of Mitotic Division in Pollen Grains

For the study of the first mitotic division in the pollen grain the methods followed are similar to those adopted for the study of meiosis in pollen mother cells (*see* chapter on representative schedules).

Study of Fertility of Pollen Grains

In order to study the apparent fertility of pollen grains two sets of methods are available: (*a*) based on staining the contents of the grains, and (*b*) based on the germination of the pollen tube.

Staining the contents of the grains—The first set of experiments involves staining the grains with different dyes and counting the percentage of empty and coloured grains. A number of staining media are available.

A very convenient method is to stain the pollen grains in Müntzing's mixture of glycerol and 1 per cent acetic-carmine solution for some hours. The filled grains take up the stain while the empty ones do not. The acetic acid evaporates, but glycerol remains. By ringing with paraffin, the preparation can be stored for a long time.

Another method is staining with Owzarzak's methyl green phloxine. The medium preserves the slides for several months. The walls of all the grains take colour but only the filled grains take up both colours.

These methods show the frequency of filled pollen grains, but they only give an indication of their fertility percentage, since only the empty grains are certainly sterile, while the filled grains are not necessarily able to germinate.

Germination of the pollen tube—The second set of methods, based on the actual germination of the pollen tubes, is a sure indication of the actual fertility of the pollen grains. They are described, in connection with pollen tube mitosis, later.

Study of Mitotic Division in Generative Cells inside Pollen Tubes

For the study of pollen tube mitosis the practice is to study the mitotic division of the generative nucleus into the two sperm nuclei. The division usually takes place between 2 and 48 h after germination of the pollen tube.

The methods generally include two main stages:
1. germination of the pollen tube, and
2. study of the chromosome structure.

The chief factors necessary for the artificial germination of pollen tubes are temperature, humidity and culture media. According to Bishop (1949), however, pollen tubes can be germinated without media. The chief principle underlying the preparation of artificial media is to provide a condition as closely approaching the normal secretion of the stigma as possible.

The different kinds of culture media available are:
1. Cane sugar: 3–30 per cent solution in water.
2. Agar or gelatin: 2 per cent solution in water.
3. Extract of style and placenta or stigma alone in water.
4. Stigmatic secretion in normal conditions.

The optimum temperature for germination is 20 °C and the optimum humidity is that approaching saturation. However, the optimum concentration of the medium for a particular plant depends on the species or the individual studied and partly on the temperature. The concentration of the medium is inversely proportional to the temperature of germination.

There are several methods for pollen tube germination. They vary chiefly with regard to the details adopted for bringing the pollen grains in contact with the medium. Some of the general methods are now given.

The hanging drop method—This method is based on growing the pollen tube in a hanging drop of medium within a moist chamber. It is most suitable for growing the tubes in an atmosphere of controlled humidity. The floor and slides of the chamber are formed by placing a ring coated with vaseline on a slide. A drop of the medium is placed on a clean cover glass and pollen dusted on it. The cover glass is inverted on the ring so that the drop with pollen hangs inside the chamber formed by the ring. The humidity is controlled by placing a drop of water on the slide or a drop of agar inside the chamber. In an alternative method, the ring is not coated with vaseline and the entire preparation is kept inside a desiccator containing an aqueous solution of glycerol to control humidity. After the pollen tubes have grown to a suitable length, they can be stained in staining-*cum*-fixing fluids or fixed in a metallic mixture and stained in crystal violet.

The floating cellophane method—This was devised by La Cour and Fabergé (1943) and is based on growing pollen grains on a square of cellophane paper (about 2 cm × 2 cm) which is floated in sugar solution in a Petri dish.

The pollen is dusted onto the upper surface of cellophane paper which is free of solution. The Petri dish is covered and kept at 20 °C.

The germinated tubes can be stained in a staining-*cum*-fixing mixture and observed; otherwise they can be fixed in acetic–ethanol mixture for 2–24 h, hydrolysed in N HCl for 6 min and stained in fuchsin sulphurous acid solution.

Precautions which should be observed are: the cellophane must not be thicker than 0·04 mm and should be of the non-waterproof type, and for accumulating metaphase, 0·05 per cent colchicine solution can be added to the sugar solution or acenaphthene crystals scattered on the Petri dish.

This method, while generally satisfactory, has two limitations: cellophane stains readily with acetic stain fixatives and so details in temporary mounts are obscured within a short time; and a considerable proportion of pollen is washed off during staining and mounting.

The coated slide method—This technique was first worked out by Conger and Fairchild (1952) and Conger (1953). The slide is coated with the medium containing sugar, agar and colchicine dissolved in distilled water. The pollen is dusted on the medium and grown in humid conditions. The slides, with the pollen tubes, are stained directly.

The advantages of this technique are: the material is handled easily, being attached to the slide itself, and the loss of materials during the process of transfer and staining is minimised.

The chief drawback, however, lies in the fact that the presence of the medium in the final preparations slightly distorts the visibility. Variations of this technique have been evolved by Dyer (1966), Jona (1967) and Ma (1967), details being given in the chapter on representative schedules.

The collodion membrane technique—This technique (Savage, 1957) is a modification of the cellophane technique, in which the processing has been omitted and the disadvantages rectified. The technique is based on the formation of a membrane when a drop of collodion in acetone is allowed to spread over warmed sugar solution.

A drop of collodion–acetone solution, when put on warmed sugar solution in a Petri dish, spreads out into a thin membrane on the surface, which hardens on evaporation of the acetone. The most satisfactory solution contains approximately 1 part collodion in 3 parts acetone. After complete evaporation of acetone, pollen is dusted on the smooth areas of the membrane and allowed to germinate at 20 °C. When necessary for observation, a piece of the membrane with pollen tubes is cut out and floated on a slide and observed by adding a drop of staining-*cum*-fixing mixture and covering with a cover glass. Since collodion does not stain in acetic-stain, the temporary preparation can be kept for a long time or made permanent by the acetic–alcohol schedule.

An alternative method for Feulgen staining of these pollen tubes has been devised by the same author (*see* schedules).

The pollen tube germination without culture medium, Swanson (1940) and De (1958)—This is based on the idea that a germinating medium is not necessary for the growth of the pollen tube, provided suitable temperature and humidity are given. The pollen tubes are grown in a closed moist chamber in the presence of crystals of acenaphthene, fumes from which disturb the spindle mechanism and provide contracted chromosomes in metaphase for cytological analysis (*see* Chapter 7).

An advantage is that no culture medium for the growth of the pollen tube is required; therefore it yields very satisfactory preparations for the study of the male gametophyte with the electron microscope.

Another method for germination in the absence of medium is given by Thomas (referred to in Darlington and La Cour, 1960). It consists of dusting the dry pollen over a cover glass and inverting over the hanging drop chamber containing a small piece of wet filter paper.

In all these techniques for accumulating metaphase plates, sometimes 0·05 per cent colchicine solution is added to the medium or acenaphthene crystals can be scattered on the Petri dish (Eigsti, 1942; Read, 1964). These methods are used in studying the haploid chromosome set from the pollen tube mitosis. Several modifications of the squash schedule are available for different materials (Morrison, 1953; Östergren and Heenan, 1962), including the use of pectinase for chromosome separation in wheat (Bhowal, 1963). They can also be employed to study the effect of x-rays or other rays on pollen and also the effect of oxygen in chromosome division (Bishop, 1949; Conger and Fairchild, 1952).

STUDY OF DIVISION IN EMBRYOSAC MOTHER CELLS

Meiotic division can also be studied from the embryosac mother cells; it usually takes place after the meiotic division in anthers, sometimes as long as 3 weeks afterwards.

The preliminary step in the process is to expose the ovules to the treatment fluids, having first dissected them out, this being easy for large ovules. For the very large ovules it is necessary, to expose them, to remove the ovary wall. In a number of monocotyledonous plants, the ovules can be dissected out on entire strings and are easily handled, but small ovules are more difficult to handle, as they tend to get lost if dissected out. Ovule squash technique, the small ovules being detached from the placenta just before squashing, involves maceration in N HCl and staining with Feulgen reaction (Hillary, 1940) or by the acetic-orcein, acetic-carmine or acetic-lacmoid schedule (Darlington and La Cour, 1942; Bradley, 1948, Haque, 1954). The former technique is more useful for larger ovules while the latter can be used for both small and large ones, the unwanted tissue being dissected out before final squashing.

For squash preparations, the most effective fixative is a modification of Carnoy's fluid having 4 parts chloroform, 3 parts absolute ethyl alcohol and 1 part glacial acetic acid. The high proportion of chloroform is necessary to keep the pliability of cell structures (Bradley, 1948). The period of fixation may vary from 2 days to 3 weeks. Whole mounts can be prepared by fixing in bulk for 24 h followed by hydrolysis and staining in Feulgen solution. The ovules are then dissected out (Paolillo, 1960).

For the study of small ovules and difficult materials paraffin blocks are prepared; the fixative used should be strong and should penetrate rapidly the covering tissues—La Cour's 2BX is effective (Darlington and La Cour, 1960).

The sections may be stained following the haematoxylin or crystal violet schedules.

The temporary squash technique takes up less time and shows better chromosome morphology than the paraffin method; however, the squash preparations do not preserve the original orientation of the different tissues (*see* Chapter 7).

STUDY OF ENDOSPERM CHROMOSOMES

For the study of endosperm chromosomes, different methods of pretreatment and staining are available which are more or less similar to those followed for other mitotic chromosomes (*see* Chapter 7). Lately, however, interesting developments have been made in the study of endosperm chromosomes in the living state, clarification of their birefringent property, the effect of chemicals on them, as well as their cinemicrographic analysis (Bajer, 1955; Bajer and Molé Bajer, 1956; Östergren and Bajer, 1958; Inoué and Bajer, 1961). Two such representative methods are now outlined.
Permanent Preparation of Endosperm Cells Flattened in the Living State (Östergren and Bajer, 1958)
Endosperm cells of *Haemanthus katherinae* serve as good material.

1. Select a suitable ovule and take out the contents of the embryo sac. Press the contents on to a cover glass smeared with a thin layer of a mixture of 0·5 per cent agar, 0·5 per cent gelatin and 3·5 per cent glucose and remove any excess of endosperm fluid. Surface tension aids in flattening the cells. Arrange the preparation in a moist chamber so as to avoid drying before fixation.

2. Fix cover glass with the material in chrome–acetic–formalin fixative diluted in equal parts with distilled water for at least 12 h.

3. Rinse in water and treat in a mixture of decinormal potassium cyanide and 2 per cent magnesium sulphate mixed in equal parts for 1 h. Wash off the cyanide by several changes in water.

4. Hydrolyse the cover glass with material for 4 h at room temperature in a mixture of rectified spirit and concentrated hydrochloric acid mixed in the proportion of 3 : 1.

5. Dip in water and stain in Feulgen solution for 12 h.

6. Wash the preparation in sulphur dioxide water, dehydrate as usual, pass through xylol and mount in balsam.

Study of Birefringence in Endosperm Mitosis (Inoué and Bajer, 1961)

1. With the stem material, follow step 1 as above without using gelatin, and use cover glass originally ringed with Vaseline–paraffin.

2. Place the cover-glass with the material on another glucose–agar coated cover glass and seal with ringed Vaseline–paraffin.

3. When the desired flattening of the cells is secured, tilt the preparation to allow excess liquid to drain on to the lower cover glass. The excess liquid can also be drained off by inserting a filter paper in between the two cover-glasses.

4. By proper adjustment, flatten the endosperm cells to obtain the chromosomes in one plane. Break the liquid contact between the top and bottom cover glasses and stabilise the preparation.

5. Study birefringence in a special polarising microscope with non-rectified coated stain free 25×0.65 N.A. Leitz oil immersion objective in conjunction with 10×0.25 N.A. American optical coated objective as condenser. Pure green light (546 nm) from a high-pressure mercury arc lamp can be isolated with a multilayer high transmission interference filter.

6. Birefringence in kinetochore region of the chromosome fibril can be observed.

7. It may be mentioned that Wada (1966) has studied in detail the spindle fibres and their arrangement.

STUDY OF NUCLEOLUS

In order to study chromosome–nucleous relationship, some of the different schedules developed are outlined below:

1. Staining of nucleolonema (Estable and Sotelo, 1952): (*a*) Silver impregnation procedures and iron–pyrogallic stain for fixed material, and (*b*) phase contrast microscopy, dark-field illumination and oblique transillumination for fresh material have been used for studying the filamentous structures within the nucleolus.

2. Acetic-carmine staining: For staining nucleoli, treat tissues with rectified spirit, 2; formalin, 1; 5 per cent glacial acetic acid, 1; hydrolyse with HCl at 60 °C for fixed periods and squash in 1 per cent acetic-carmine solution (Rattenbury, 1952).

3. Feulgen-light green schedule (Semmens and Bhaduri, 1941): Bring down sections fixed in a fixative without acetic acid to water. Treat for 2–3 h in 75 per cent ethyl alcohol. Wash in distilled water. Hydrolyse for 10 min in N HCl at 60 °C. Stain for 2 h in leuco-basic fuchsin solution. Wash in two changes of SO_2 water, keep for 10 min in each. Rinse successively in distilled water, 50 and 70 per cent alcohols. Mordant for 1 h in 80 per cent alcohol saturated with Na_2CO_3. Dip in 80 and 95 per cent alcohol. Stain for 20–25 min in filtered saturated alcoholic solution of light green with 2–3 drops of aniline oil. Drain. Rinse in a saturated solution of Na_2CO_3 in 80 per cent alcohol, 10 cm^3 and 90 cm^3 80 per cent alcohol. Differentiate in 95 per cent alcohol. Dehydrate through absolute alcohol, alcohol–xylol and xylol grades and mount in balsam. For smears and squashes, the initial treatment in 75 per cent alcohol is omitted. In squashes, squash cells after Feulgen staining in 45 per cent acetic acid. Separate cover glass in 40 per cent alcohol and proceed as usual.

4. Toluidine blue—molydate method (Löve, 1962; Löve and Walsh, 1963; Löve *et al.*, 1964), based on the principle of gradual blocking of

—NH$_2$ group of nucleoprotein and unmasking of phosphate groups of nucleic acids binding cationic dyes.

Toluidine Blue—molybdate method for ribonucleoprotein

According to Löve (1962) and Löve and Walsh (1963) and using tissue cultures or smears after wet fixation, with treatment in TCA and formal sublimate.

(a) The dyes giving satisfactory results are: Coleman Bell CU-3, National Aniline NU-2 and NU-17, Harleco NU-14, Matheson-Coleman Bell CU-9 and Biological Stain Commission NU-19. The dyes are dissolved in McIlvaine's buffer at pH 3·0 at 20 °C.

(b) Wash tissue culture in 0·85 per cent saline for 10 s before fixation.

(c) Treat two slides in 5 per cent aqueous trichloracetic acid for 10 min. Wash in distilled water.

(d) Fix one slide for 5 min and another for 10 min in formal sublimate, containing 40 per cent formaldehyde, 1 part, and 6 per cent aqueous mercuric chloride, 9 parts.

(e) Rinse in tap water and treat, with Lugol's iodine for 5 min.

(f) Immerse the slides in 5 per cent sodium thiosulphate for 5 min and rinse in water.

(g) Stain for 30 min in toluidine blue.

(h) Treat with 4 per cent aqueous ammonium molybdate solution for 15 min.

(i) Wash in tap water.

(j) Dehydrate in tertiary butyl alcohol, clear in xylol and mount in a synthetic resin.

(k) The *pars amorpha* of the nucleolus takes up bluish-green stain while the nucleolini are bright purple.

5. Chromic acid treatment has been used in mammalian materials (*see* Studzinski *et al.*, 1967). A combined schedule was developed by Morrison *et al.* (1959).

STUDY OF CHROMOSOMES FROM THALLOPHYTES

I. ALGAE

Chromosome study in algae has made tremendous progress within the last twenty years. Research has been initiated in nearly all groups of algae, some of it yielding crucial data on the evolution of chromosome structure. A critical analysis of the chromosome methodology of the group has been presented by Godward (1966).

A feature in common with the members of Tracheophyta is the wide occurrence of polyploidy and aneuploidy amongst the algae; certain groups, like the Chlorophyceae, showing more aneuploids than polyploids. The successful survival of these cytological variants may possibly be attributed to their aquatic environment, which provides a plentiful supply of nutrition, thus eliminating the severity of competition and struggle for survival between the normal and abnormal forms. The organisation of the body structure of an alga, is, in addition, very well suited for the maximum utilisation of the nutrition supplied by the environment.

Chromosome structure in algae is of particular interest since the group presents a variable pattern. In most of the forms, including a majority of Chlorophyceae, chromosome morphology is comparable to that of higher plants, whereas in others, like Conjugales, Euglenophyceae, and especially Dinophyceae, the structure demonstrates a number of unusual characteristics. In Conjugales (King, 1960; Godward, 1961; Brandham, 1964; Arnold and Newnham, 1965), the chromosomes are devoid of localised centromeres, comparable to some extent with the structure reported in *Luzula* of angiosperms. Euglenophyceae shows certain features of special interest, like the absence of a typical equatorial plate and centromere (Leedale 1958, 1962; cf. Saito, 1961) and the persistence of RNA-containing endosome throughout mitosis, as well as a quite different type of chromosome aggregation and movement from higher plants. Lastly, in Dinoflagellates (Grasse *et al.*, 1965; Grell and Schwalbach, 1965; Dodge, 1966), the cytochemical and ultrastructural data have shown the chromosomes to be sausage-shaped structures (Ris, 1962) composed of continuous fine fibrils of DNA only, without any basic protein—a structure finding a parallel with the genophore of procaryotes, that is, bacteria and blue green algae, but differing fundamentally from the latter in having a nucleus. Several authors have already suggested that this group represents a step linking the procaryotes with eucaryotes and the term Mesocaryota has been attributed to it (Godward, 1966).

Intensive research on the study of the flagellate chromosomes may ultimately lead to a complete understanding of the steps starting from the gene-bearing structure, or genophore, of procaryotes to the well-organised chromosomes of eucaryotes. Cytological studies on Phaeophyceae (Inoh and Hiroe, 1956; Subrahmanyan, 1957; Naylor, 1958; Kumagae *et al.*, 1960; Yabu, 1965) and Rhodophyceae (Iyengar and Balakrishnan, 1950; Drew, 1956; Svedelius, 1956; Austin, 1960; Yabu and Tokida, 1963; Magne, 1964) also show immense possibilities. The existence of several alternating generations in them is of special significance in relation to differentiation. Detailed cytochemical, structural and ultra-structural analyses of the chromosomes from different alternating phases may yield facts of fundamental significance regarding the genic control of differentiation.

Advance in the knowledge of the fine structure of chromosomes and their chemical make up in different biological groups has been phenomenal in recent years. This progress has led to the realisation that increasing complexity in the nature and seat of the gene-bearing structure has been an associated feature of evolution. This complexity of both chemical make up and behaviour has been a natural prerequisite for providing guidelines in the life cycle of organisms with extremely complex physical organisation and sequential and phasic growth and development. Amongst the algal groups, leaving aside the blue-green forms, one extreme is represented by the flagellates and the other by a miniature complex form observed in the Rhodophyceae. The scope and possibilities of their chromosome analysis cannot be over-estimated.

The methods of chromosome study in algae principally include two separate stages: culture of the algae to obtain suitable divisional figures, and processing to observe the chromosomes.

Culture is usually necessary, in the case of most forms of algae, for acquiring a sufficient number of metaphase plates in polar view for chromosome analysis. Algae growing in the wild condition have been found to show both somatic and germinal divisional figures, as seen in marine *Cladophora*, and *Ulva* and most forms of red algae. In the simplest form, the alga is grown in its natural medium, but progressively complex media with different proportions have been devised for obtaining better growth and synchronisation of mitosis to collect the highest number of mitotic stages. The other major factor is light period and in most cases, artificial light and dark photoperiods are provided.

For successful culture of any alga, and green algae in particular, the bulk material after collection is to be teased out under dissecting/compound microscopes to separate pure filaments from mixed ones. Media ranging from water of the habitat to different types of synthetic media have been devised. However, only limited success has been gained though new media are still being obtained (*see* Hutner *et al.*, 1950; Pringsheim, 1951; Provasoli, 1958; Droop, 1959; Nizam, 1960; Patel, 1961; Myers, 1962; Abbas, 1963; Provasoli and Hutner, 1964). Soil, peat or leaf extract added to a solution of mineral salts is sometimes effective. Trace elements and vitamins, and occasionally chelating agents are used, as for species of *Chlorella*. Amino-acids and DNA precursors, added to the medium, serve to increase the growth rate. By adjusting the photoperiod, synchronisation of mitosis is possible, e.g. through an artificial light photoperiod of 18 h and a dark period of 6 h in a culture chamber, mitosis takes place during the dark period. As many as 14 per cent cells in mitosis can be obtained by changes in the length of photoperiod restricting the mitosis to the dark period (Leedale, 1958, 1966; James and Cook, 1960; Newnham, 1962; Tamiya, 1963; Brandham, 1964). In Chlorophyceae and certain other algae, such synchrony in division can be obtained even in culture without any special treatment. Godward (1966) suggested the possibility of synchronisation of division through temperature control in continuous light with bubbled air enriched with 4 per cent carbon dioxide and by continuous culture in medium which is continuously renewed. She has also stressed the possibility of the evolution of variant races in cultures maintained for a long time by mitotic changes or mutations. It has been noted, however, that where extensive works have been carried out, algae, and Chlorophyceae in particular, are highly resistant to the action of colchicine and x-rays as compared to higher plants (Sarma, 1958, 1969; Godward, 1962); with colchicine, a concentration below 0·5 per cent is rarely effective. Similarly in *Spirogyra*, even 20 kR dosage of x-rays allows the plants to remain alive. Further research is necessary to work out the genetic basis of this resistance.

Smaller members of marine Dinophyceae can be cultured in sea water, to which nitrate, phosphate and soil extracts have been added (50 cm^3 soil extract; 0·2 g sodium nitrate and 0·03 g sodium phosphate, both in 10 cm^3 of water added per litre of sea water). Both soil extract and sea water have been replaced by artificial components by some workers (Provasoli, McLaughlin and Droop, 1957). The cultures are kept under fluorescent lighting between 1076–2152 lx at a temperature of 15–25 °C for 12–18 h. Cultures on fresh water forms are not so successful (*see* Dodge, 1966).

Culture of gametophytes and sporophytes of the Laminariales is influenced by nutrient, temperature and light. Sea water enriched by various nutrients has been used effectively (Schreiber, 1930; Harries, 1932). Naylor (1956) modified the nutrient recommended by Schreiber by including soil extract which is named Erdschreiber solution.

The most suitable temperature is between 10–16 °C. Harris (1932) advocated an indirect light intensity of 538–3228 lx as the most suitable, but never direct sunlight. Two commonly used nutrients are:

Nutrient A 0·01 M KH$_2$PO$_4$, 0·5 cm^3
0·01 M KNO$_3$, 0·5 cm^3, added to 25 cm^3 of sea water every 14 days.
Nutrient B 10^{-5} M KH$_2$PO$_4$ ($\frac{1}{20}$th of nutrient A stock);
5×10^{-5} M KNO$_3$ ($\frac{1}{4}$th of nutrient A stock)
0·5 cm^3 of each added to 25 cm^3 of sea water every 14 days.

ASP$_2$ solution is a more complex nutrient. These media, under different conditions of light and temperature, have been successfully used in culturing different members of Laminariales, for vegetative and gametophytic growth (for details, *see* Evans, 1966).

For a number of varieties belonging to the Phaeophyceae, Roberts (1966) used the same culture technique. Large pieces of the fruiting region are collected, washed in jets of boiled sea water and immersed in boiled sea water. After the zoospores are released, the suspension is poured in a flat dish containing sterilised slides or coverslips. After 30 min, the coverslips are removed with forceps and placed in a culture chamber. Roberts (loc. cit.) suggests the use of slides instead of cover slips, which may be placed back to back in cellophane covered glass pots. Slides are easier to handle during transport and subsequent processes of fixation, staining and cooling. Rapid growth is ensured by long hours of daylight or by permanent illumination with three 80 W fluorescent tubes at about 15 °C. As growth medium, the Erdschreiber medium is a very suitable one (for details, *see* Roberts 1966). In the Rhodophyta, culture methods have only a limited application; the habitat medium, with frequent changes, is sufficient in most cases. Complex media developed by Provasoli *et al.*, (1957 and *see* Fries, 1963) are often not necessary for cultures required for cytological study alone. Small quantities of soil extract or other nutrients, however, show a beneficial effect. A major factor in the culture of the members of Rhodophyta is contamination by epiphytes, like diatoms. Several methods are adopted for their elimination (*see* Dixon, 1966).

Chromosome studies

As in the higher plants, schedules for studying the chromosomes of algae involve mainly fixation, staining and mounting. Pre-treatment is given in specific cases, as are also different maceration techniques. There is considerable variation in response to fixatives and stains, not only between the different groups, but also between different genera and species, and in some cases between different parts of the same plant. Decolorisation of the pigment is an important factor in the choice of a fixative.

FIXATION

Among the different fixatives used are:

1. *Formalin–ethanol*—used in different proportions. In Rhodophyceae, it is suitable for haematoxylin or brazilin stains but useless for feulgen or acetic–carmine. For green algae, however, Godward (1966) finds fixatives containing formalin or mercuric chloride to be unsatisfactory.

2. *Formalin–acetic–ethanol*—also used in different formulae, the one by Westbrook (1935) being glacial acetic acid, 2·5 cm^3; 40 per cent formaldehyde, 6·5 cm^3, and 50 per cent ethanol, 100 cm^3. It is recommended for immediate use in Rhodophyceae since storage may cause shrinkage of the nucleus and later disintegration of the material (Magne, 1964).

3. *Amongst the metallic fixatives*—chromic acid–acetic acid mixtures have been used in a large range of variations, a very common one being the modification of Karpechenko's fluid given by Papenfuss (1946). It includes: Solution A containing chromic acid, 1 g; glacial acetic acid, 5 cm^3; sea water, 65 cm^3 and Solution B with 40 per cent formaldehyde, 40 cm^3 in 35 cm^3 sea water. The two solutions are mixed immediately before use. Sea water is replaced by distilled water for fresh-water forms and the use of formaldehyde is optional. Naylor (1957) used chrom-acetic-formalin as fixative for large parenchymatous Phaeophyceae, followed by softening in sodium carbonate and occasional bleaching with H_2O_2. Osmic acid vapour, nitric acid vapour, 2 per cent osmic acid solution, Belling's Navashin solution enriched with osmic acid, iodine water, bromine water and chromic acid solution have all been successfully used in different members of Chlorophyceae (Leedale, 1958; Sarma, 1958; King, 1960; Patel, 1961; Sinha, 1963; Brandham, 1964; Thakur, 1964; Dodge, 1966; Godward, 1966). Fixation in osmic acid must be a brief one and the osmium must be washed out before it forms a black deposit.

4. *Acetic acid: ethanol and acetic-acid: methanol mixtures*—with different concentrations and proportions of the constituents are widely used. Mixtures of glacial acetic acid and 95 per cent ethanol (1:1, 1:2, 1:3) are effective for Rhodophyceae, the two former giving better results. However, they can cause shrinkage in large-celled delicate forms and disintegration of calcified materials. Smaller thalli need only a few minutes treatment, but 1–6 h immersion is recommended for cartilaginous forms. The materials should be processed immediately after fixation and storage should be avoided wherever possible (for details, *see* Austin 1959; Dixon, 1966).

Within Phaeophyceae, filamentous forms can be fixed in acetic acid–ethanol mixture (1:3) for up to 24 h, preferably changing the fixative after 1 h. Evans (1963a) used lithium chloride for large parenchymatous plants after fixation as a softening agent. Decolorisation is usually satisfactory after 24 h.

Acetic–ethanol (1:3) is equally satisfactory for members of the Dinophyceae which are large and easy to handle. Acetic acid–methanol, in different proportions, gives very good results. Fixation is carried out in centrifuge tubes and the cells collected by centrifugation. Addition of a few drops of saturated ferric acetate (in acetic acid) to the fixative 1 h before staining is beneficial for certain species.

Various modifications have been used in different types of green algae. Mixtures of glacial acetic acid and 95 per cent ethanol (1:1, 1:2, 1:3) and also ethanol or methanol alone, gave good results in different forms (*see* Godward, 1966).

PROCESSING

1. Whole mounts are possible for uniseriate filamentous thalli in the Rhodophyceae as also for various members of the Dinophyceae and Chlorophyceae.

2. Serial sections of paraffin-embedded material are utilised only in cases of certain Rhodophyceae following the usual block preparation schedule (*see* page 154).

3. Squash preparations are most commonly used in studying algal chromosomes. The fixed material can be squashed directly or it may require a softening process, depending upon the hardness of the material. The softening agent and the period and conditions of treatment depend mainly upon the material but also on the fixative and stain used. Dilute solutions of acid or alkali in water or ethanol are used for softening, the more common ones being HCl or NaOH in concentrations ranging from 1 to 50 per cent (Papenfuss, 1937; Drew, 1945; Norris, 1957; Cole, 1963). Magne (1964) advocated the use of dilute solutions of sodium carbonate.

Since in the Rhodophyceae, the carpogonium or young carposporophyte is encased by a large amount of unwanted tissue, sections of the fertile axis (50–100 μm) are cut on a freezing microtome with dilute gelatin as the supporting medium and these sections are then squashed.

Since the brown algae are rather tough and resilient, they require special softening pre-treatments. The pre-treatment chemical used may be sodium carbonate (Naylor, 1959), lithium chloride (Evans, 1963a) or a mixture of ammonium oxalate and hydrogen peroxide (Lewis, 1956).

STAINING

The stains commonly applied to higher plants have also been used for algae, like acetic–carmine, feulgen, haematoxylin, brazilin and methyl green pyronin.

The acetic-carmine stain is the one most widely applied following the iron alum-acetic-carmine method developed by Godward (1948). In Chlorophyceae, the fixation is followed by mordanting in aqueous iron alum solution, used in different dilutions, depending on the material, for a period not exceeding 30 s. The material is washed repeatedly. Super-saturated carmine solution in 45 per cent acetic acid is added and the preparation boiled to dissolve the starch. Filamentous forms are held with forceps and passed through the solutions, spores are handled as settled on slides or cover slips, and unicellular cultures in centrifuge tubes. The carmine-stained preparations are passed as usual through acetic acid-ethanol (1:3), 95 per cent and absolute ethanol grades before mounting in euparal. Acetic-orcein has been used instead of carmine but gives a paler stain. In a method followed by Thomas (1940), a few drops of super-saturated solution of ferric acetate in 45 per cent acetic acid is added to the fixative (acetic-ethanol 1:3), This addition of iron salt has also been found to be very effective in Charophyceae. The material is transferred directly from the fixative to carmine and back to the fixative, followed by 95 per cent euparal essence and euparal.

Members of the Dinophyceae are stained after fixation in a drop of acetic-carmine on a slide. Acetic-orcein is equally effective. For Laminariales, acetic-carmine, containing 1 drop of saturated ferric acetate solution per 25 cm^3, is added to the material *after* squashing; followed by alternate heating and cooling. Large parenchymatous brown algae require fixation in acetic-ethanol (1:3), enriched by a few drops of ferric acetate. The material is washed in 70 per cent ethanol, hand sections are cut and mounted in 6 per cent Na_2CO_3. After squashing, distilled water, and later the stain, are added and the slide gently boiled. The iron-acetic carmine schedule has also been used successfully for the Rhodophyceae after slight modifications (Cole, 1963).

Feulgen staining schedule has been applied in almost all forms of algae. The method of handling of the material is similar to that for acetic-carmine staining, both for members of the Chlorophyceae and Dinophyceae. Gametophyte materials of Laminariales take up bright stain in this method. For large parenchymatous brown algae, fixation in Karpechenko's fluid is followed by washing in running water and bleaching for 3–4 h in 20 per cent aqueous H_2O_2 solution and again washing in running water. The material is heated to 60 °C in distilled water, hydrolysed in N HCl at 60 °C for 7–10 min. Without bleaching, hydrolysis has to be done for 15–30 min. The material is transferred to cold distilled water, washed in running water for 10 min, hand sections are cut and squashed in SO_2 water (N HCl, 5 cm^3; 10 per cent $K_2S_2O_5$, 5 cm^3; water, 100 cm^3). Feulgen staining, however, does not yield consistent results with the red algae.

Heidenhein's haematoxylin schedule and its modifications have been used in the red algae. Haematoxylin in acetic acid stains the nuclei in green algae but it is not permanent.

Brazilin was extensively used by Drew (1934) in red algae. Both the stain and its mordant require a long period of ripening in the dark to give satisfactory preparations. Sections or squash preparations are treated in a

2 per cent ferric ammonium sulphate solution in 70 per cent ethanol for 1 h, washed in 70 per cent ethanol and stained for 12–16 h in 0·5 per cent brazilin in 70 per cent ethanol. Finally they are washed twice in 70 per cent ethanol and dehydrated through ethanol grades. This stain avoids the swelling of chromosomes due to aqueous solutions and the procedure is much shorter than haematoxylin.

Methyl green pyronin staining schedule has been applied in the green algae, using a B.D.H. dye mixture. A very small amount is dissolved in water and the material, after acetic fixation, is mounted in it (Godward, 1966). However, it has not given good results with the red algae (Magne, 1964).

Preparation of permanent slides from temporary ones can be made by any of the schedules followed for higher plants (chapter on mounting).

In addition to the general methods, as outlined above, a few sample schedules used specially for members of the Phaeophyta are given in brief as in this group a special method for softening is necessary.

(a) Acetic-carmine preparations as used in Laminariales (Evans, 1966):

Fix in acetic-ethanol (1:3) for 12–18 h; wash in running water; immerse in 1 M lithium chloride solution for 15 min; keep in water for 15 min; dissect out material and squash under a coverslip; insert a few drops of acetic-carmine solution with a trace of ferric acetate at the side of the coverslip, heat at intervals for 30 s without boiling, squash again and blot excess stain. If over-stained, de-stain by heating in a mixture of acetic-carmine solution and glacial acetic acid (1:5) for 15 s. Treatment in lithium chloride can be omitted for filamentous material, like gametophytes.

In filamentous Phaeophyta, fixation is carried out for 24 h and a few drops of saturated ferric acetate in 45 per cent acetic acid is added to the fixative. After washing in acetic-ethanol (1:3), add stain, cover, boil, squash.

(b) Feulgen schedule for Phaeophyta (Naylor, 1959):

Fix coverslips with growing gametophytes in acetic-ethanol (1:3), wash, hydrolyse in N HCl at 60 °C for 8–10 min, transfer to cold water, and then keep in decolorised Schiff's reagent for 8 h at room temperature, bleach in three changes of SO_2 water; squash in SO_2 water and later dehydrate through ethanol grades and mount in euparal. Variations in hydrolysis and staining periods are adopted for different materials.

(c) For softening and isolating female conceptacles, split the receptacles longitudinally and fix for 12 h, followed by treatment for 20 min in a mixture of saturated ammonium oxalate solution and 20 volumes H_2O_2 (1:1), wash for 15 min before staining. Alternatively, for male material, macerate 30 g for 4 min in 150 ml of fixative in a Waring Blendor; transfer to a graduated cylinder; pipette off middle layer with sex organs and stain as usual (Lewis, 1956).

FUNGI

A group of simple organisms living as parasites or saprophytes and categorised as fungi, hold a unique position in the plant kingdom. Even though belonging to Eucaryota in its nuclear constitution, this group is quite distinctive in its extremely simple thalloid constitution and absence of such pigments that are universal for higher plants. The meagre cytological data so far available have also indicated the occurrence of unusual features in certain groups such as the existence of nuclear membrane during division and the controversial double reduction division during meiosis. Needless to say, from the standpoint of the evolution of the structure and behaviour of chromosomes, there is immense potentiality in the cytological analysis of this assemblage. Unfortunately, it has not attracted the attention of chromosome cytologists to the extent that it should (McIntosh, 1954; Mundkur, 1954; McClary *et al.*, 1957; Roberts, 1957; Bakerspiegel, 1959; Dowding and Weijer, 1960; Somers *et al.*, 1960; Turian and Cautino, 1960; Emerson, 1967).

Chromosome analysis in fungi is not only important from a fundamental standpoint, including biochemical mutagenesis in *Neurospora crassa*, but also because their implications in utilitarian research cannot be ignored. The genetic basis of the fermenting capacity of the *Saccharomyces* complex, the medicinal values of scores of fungi including species of *Aspergillus* and *Penicillium* and the infecting capacity of fungi, need solutions in which chromosome research may play a significant role. An understanding of the cytogenetic basis of these properties would pave the way for their improvement, or prevention, with the aid of physical and chemical agents having established mutachromosomic property.

The methodological data so far obtained, though few, are here outlined:

Fixatives

Helly's modified fluid (mercuric chloride, 5 g; potassium dichromate 3 g; distilled water, 100 cm^3; add 5 cm^3 of formalin just before use) is found to be very successful as a fixative in a wide range of fungi (Heim, 1952, 1954, 1956, 1958; Robinow, 1961). Other fixatives used are osmium tetroxide vapour and acetic-ethanol but their effects are less desirable.

Stains

Different chromosomal stains have been tried out, some of the commoner ones being described here:

Giemsa staining gives effective results for quite a large number of filamentous fungi and yeasts. The preparations can be stained directly or after hydrolysis in N HCl. A stain used for yeast chromosomes contains 16 drops of Gurr's Giemsa R66 dissolved in 10–12 cm^3 of Gurr's Giemsa buffer at pH 6·9.

If hydrolysis is required, fixed cells are usually extracted for $1\frac{1}{2}$ h with 1 per cent NaCl at 60 °C, the time, concentration and temperatures depending on the material (Ganesan, 1959); then treated for 10 min with N HCl at 60 °C, rinsed with tap water and kept in the stain for several hours.

Haematoxylin staining has been advocated by Henderson and Lu (1968) for squash preparations, following fixation in Lu's BAC fixative (1962).

Feulgen staining is utilised according to the usual schedule of acid hydrolysis followed by staining.

Schiff reagent prepared with Diamant Fuchsin is used for yeast chromosomes (Robinow, 1961). The smears can be mounted in water or acetic-carmine.

Representative Schedules

For yeast following Robinow (1961)—Place a loopful of slimy growth from a 2–5 day-old plate culture on a No. 1 coverslip. Place another coverslip on top with corners turned away at an angle of 5 degrees to the former. Allow the drop to spread, pull apart the coverslips so that a smear is formed on each and immerse in Helly's modified fixative for 10 min. Rinse thoroughly in 70 per cent ethanol, transfer to Newcomer's fixative for preservation in the cold (for composition, *see* chapter on fixatives). The film must not be permitted to dry prior to fixation. The yeast culture may be exposed briefly to formalin vapour before the smear is drawn for better preparation. In Giemsa staining, extract fixed cells for $1\frac{1}{2}$ h with 1 per cent NaCl at 60 °C, treat for 10 min in N HCl at 60 °C, rinse in tap water and stain for several hours in Giemsa solution. Differentiate by moving the smears repeatedly for 10–12 s at a time in 40 cm^3 of distilled water to which a few drops of acetic acid have been added (pH 4·2). Observe under a water immersion lens to control extraction of excess stain. When the nuclei are brightly differentiated, mount the coverslip on a slide in a drop of buffer containing 2–3 drops of Giemsa per 10 cm^3. Blot excess medium, squash with uniform pressure, seal and observe. Squashing without pressure may also be done.

In feulgen schedule, hydrolyse the smear for 10 min, rinse with tap water, stain in Schiff's reagent for $3\frac{1}{2}$ h, rinse quickly in 10 changes of SO$_2$ water (tap water 90 cm^3; N HCl 5 cm^3; 10 per cent sodium metabisulphite 5 cm^3), keep in running water for 20 min and mount in water of acetic-carmine.

For mitosis in basidiomycetes (Ward and Ciurysek, 1961)—Centrifuge the culture and suspend in distilled water. Homogenise, spread in a thin film on slide, air dry, fix in acetic-lactic-alcohol (6:1:1) for 10 min, pass through 95 per cent and 70 per cent ethanol and rinse in water. Hydrolyse in N HCl at room temperature for 5 min, then at 60 °C for 6 min, wash thoroughly in distilled water and suspend for 5 min in a phosphate buffer (pH 7·2). Stain for 25 min in Giemsa's stain, rinse successively in water and buffer, drain, flood with Abopon and cover with a coverslip.

For general mitotic preparations (Lu, 1962)—Pour a mixture of BAC fixative (n-butyl alcohol, acetic acid and 10 per cent aqueous chromic acid solution, 9:6:2) over the culture in a Petri dish and store under partial vacuum at 0–6 °C for 1–5 days. Dissect out fruit bodies in a mixture of conc. HCl and 95 per cent ethanol (1:1), heat gently, wash with Carnoy's fixative for 1–2 min, heat in propionic-carmine solution with a trace of iron, apply coverslip, allow to stand, heat to boiling and press the coverslip. Add a drop of 45 per cent acetic acid to all corners, apply a drop of glycerin-acetic acid mixture to one corner, blot and seal.

For members of the Pezizales—Fixation in Carnoy's fluid for 24–72 h followed by either direct squashing in acetic-carmine or hydrolysis in N HCl, staining in Feulgen and squashing in acetic-carmine have been found to yield suitable preparations (Thind and Waraitch, 1968; Sahay and Prasad, 1968).

For basidia of agarics (Lowry, 1963)—Fix bits of hymenial tissue in Newcomer's fixative for 1–12 h, hydrolyse and mordant in aqueous N HCl containing 2 per cent aluminium alum, 2 per cent chrome alum and 2 per cent iodic acid for 5 min at room temperature, then at 60 °C for 10–15 min. Wash in three changes of distilled water and keep in Wittmann's acetic-iron-haematoxylin for 2 h. Mount in a drop of stain, cover with a coverslip, press, heat to just below boiling and seal.

For chromosomes of *Neurospora* (E. G. Barry)—*see* legend of Plate 16 in Plate section.

REFERENCES

Abbas, A. (1963).   *Ph.D. thesis,* London University
Alfert, M. (1954).   *Int. Rev. Cytol.* **3,** 131
Austin, A. P. (1959).   *Stain Tech.* **34,** 69
Austin, A. P. (1960).   *Ann. Bot. Lond.* NS, **24,** 296
Bajer, A. (1955).   *Experientia* **11,** 281
Bajer, A. and, Molé Bajer, J. (1956).   *Chromosoma* **7,** 258
Bakerspiegel, A. (1959).   *Am. J. Bot.* **46,** 180
Baranetsky. (1880).   Referred to in Darlington and La Cour (1960)
Barigozzi, C. (1937).   *Comment. pontif. Acad-Sci.* **1,** 333
Bauer, H. (1935).   *Z. Zellforsch.* **23,** 280
Bhowal, J. G. (1963).   *Canad. J. Gent. Cytol.* **5,** 268
Bishop, C. J. (1949).   *Stain Tech.* **24,** 9
Bradley, M. V. (1948).   *Stain Tech.* **23,** 29
Brandham, P. E. (1964).   *Ph.D. thesis,* London University
Brown, S. W. (1949). *Genetics* **34,** 437
Brown, S. W. (1954).   *Univ. Calif. Publ. Bot.* **27,** 231
Brown, S. W. (1966).   *Science* **151,** 417
Callan, H. G. (1942).   *Proc. Roy. Soc.* **130B,** 324
Callan, H. G. (1963).   *Int. Rev. Cytol.* **15,** 1
Callan, H. G. (1966).   *J. Cell Sci.* **1,** 85
Callan, H. G. and Lloyd, L. (1956).   *Nature* **178,** 355
Callan, H. G. and Lloyd, L. (1960).   In *New approaches in cell biology.* New York; Academic Press
Carothers, E. E. (1936).   *Biol. Bull.* **71,** 469
Caspersson, T. (1940).   *Chromosoma* **1,** 562
Cole, K. (1963).   *Proc. IX Pacif. Sci. Congress* **4,** 313
Coleman, L. C. (1940).   *Am. J. Bot.* **27,** 683
Conger, A. D. and Fairchild, A. M. (1952).   *Proc. Nat. Acad. Sci. Wash.* **38,** 289
Conger, A. D. and Fairchild, A. M. (1953).   *Stain Tech.* **28,** 289
Creighton, H. H. (1938).   *Cytologia* **8,** 497
Darlington, C. D. (1926).   *J. Genet.* **16,** 237
Darlington, C. D. (1929).   *J. Genet.* **21,** 207
Darlington, C. D. (1933).   *Cytologia* **4,** 229
Darlington, C. D. (1939).   *J. Genet.* **37,** 341
Darlington, C. D. and La Cour, L. F. (1942).   *J. Genet.* **40,** 185

Darlington, C. D. and La Cour, L. F. (1960). *The Handling of chromosomes.* London; Allen and Unwin

Darlington, C. D. and La Cour, L. F. (1968). Ibid 5th ed.

De, D. (1958). *Stain Tech.* **33,** 57

Demerec, M. (1950). *Biology of drosophila.* New York; Wiley

Dixon, P. S. (1966). *The Rhodophyceae* in *The chromosomes of the algae.* London; Edward Arnold

Dodge, J. D. (1966). *The Dinophyceae* in *The chromosomes of the algae.* London; Edward Arnold

Dowding, E. S. and Weijer, J. (1960). *Nature* **188,** 338

Drew, K. M. (1934). *Ann. Bot. Lond.* **48,** 549

Drew, K. M. (1945). *Nature* **161,** 223

Drew, K. M. (1956). *Bot. Rev.* **22,** 553

Droop, M. R. (1959). *Proc. XI Int. Bot. Congress,* Montreal

Duryee, W. R. (1941). *Univ. Pa. Bicent. Conf. Cytology, Genetics, Evolution* 129

Dyer, A. F. (1966). *Stain Tech.* **41,** 227

Eigsti, O. J. (1942). *Am. J. Bot.* **29,** 626

Emerson, S. (1967). *Ann. Rev. Genet.* **1,** 201

Estable, C. and Sotelo, J. R. (1952). *Stain Tech.* **27,** 307

Evans, L. V. (1963a). *Phycologia* **2,** 187

Evans, L. V. (1963b). *Nature* **198,** 215

Evans, L. V. (1966). *The phaeophyceae* in *The chromosomes of the algae.* London; Edward Arnold

Ford, L. (1969). *Nucleus* **12,** 93

Fries, L. (1963). *Physiologia Pl.* **16,** 695

Gall, J. G. (1952). *Exp. Cell Res.* Suppl. **2,** 95

Gall, J. G. (1954). *J. Morph.* **94,** 283

Gall, J. G. (1956). *Brookhaven Symp. Biol.* **8,** 17

Gall, J. G. (1966). In *Methods in cell physiology* **2,** 37 New York; Academic Press

Ganesan, A. T. (1959). *Compt.-rind. trav. Lab. Carlsberg* **31,** 149

Godward, M. B. E. (1948). *Nature* **161,** 203

Godward, M. B. E. (1961). *Heredity* **16,** 53

Godward, M. B. E. (1962). In *Physiology and biochemistry of algae.* Academic Press, New York

Godward, M. B. E. (1966). Ed. *The chromosomes of the algae.* London; Edward Arnold

Godward, M. B. E. and Newnham, R. E. (1965). *J. Linn. Soc. (Bot.)* **59,** 99

Gottschalk, J. (1955). *Z. indukt. Abstanam. – u. Vererb Lehre* **87,** 1

Grasse, P.-P., Hollande, A., Cachon, J. and Cachon-Enjumet, M. (1965). *C. R. Acad. Sci. Paris* **260,** 6975

Grell, K. and Schwalbach, . (1965). *Chromosoma* **17,** 230

Haque, A. (1954). *Stain Tech.* **29,** 109

Harris, R. (1932). *Ann. Bot. Lond.* **46,** 893

Heilborn, O. (1937). *Lantbr. Hoqsk. Ann.* **4**

Heim, P. (1952). *Rev. Mycol.* **17,** 3

Heim, P. (1954). Ibid **19,** 201

Heim, P. (1956). Ibid **21,** 93

Heim, P. (1958). Ibid **23,** 273

Heitz, E. (1931). *Planta* **15,** 495

Heitz, E. and Bauer, H. (1933). *Z. Zellforsch.* **17,** 67

Henderson, S. A. and Lu, B. C. (1968). *Stain Tech.* **43,** 233

Hillary, B. B. (1940). *Bot. Gaz.* **102,** 225

Hutner, S. H., Provasoli, L., Shatz, A. and Haskins, C. P. (1950). *Proc. Amer. Phil. Soc.* **94,** 152

Inoh, S. and Hiroe, M. (1956). *La Kromosomo* **27–28,** 942

Inoué, S. and Bajer, A. (1961). *Chromosoma* **12,** 48

Iyengar, M. O. P. and Balakrishnan, M. S. (1950). *Proc. Ind. Acad. Sci.* **31B,** 135

Izawa, M., Allfrey, V. G. and Mirsky, A. E. (1963). *Proc. Natl. Acad. Sci. US* **50,** 81

James, T. W. and Cook, R. (1960). *Exp. Cell Res.* **21,** 585

Jona, R. (1967). *Stain Tech.* **42,** 113

King, G. C. (1960). *New Phytol.* **59,** 65

Kodani, M. (1942). *J. Hered.* **33,** 115

Kostoff, D. (1930). *J. Hered.* **21,** 323

Kumagae, N., Inoh, S. and Nishibayashi, T. (1960). *Biol. J. Okayama Univ.* **6,** 91

Kuwada, Y. and Nakamura, T. (1934). *Cytologia* **5,** 244

Kuwada, Y., Shinke, N. and Oura, G. (1938).   *Z. Wiss. Mikr.* **55,** 8
La Cour, L. F. (1935).   *Stain. Tech.* **10,** 57
La Cour, L. F. and Fabergé, A. C. (1943).   *Stain Tech.* **18,** 196
Lafontaine, J. G. and Ris, H. (1958). *J. Biophys. Biochem. Cytol.* **4,** 99
Leedale, G. F. (1958).   *Nature* **181,** 502
Leedale, G. F. (1962).   *Arch. Mikrobiol.* **42,** 237
Leedale, G. F. (1966).   *The Euglenophyceae* in *The chromosomes of the algae.* London; Edward Arnold
Lettré, R. and Siebs, W. (1961).   *Pathol. Biol. Semaine Hop.* **9,** 819
Levan, A. (1946).   *Hereditas* **32,** 449
Levan, A. (1948).   *Portug. Acta biol.* **2,** 167
Levan, A. (1949).   *Hereditas* **35,** 77
Levan, A. (1952).   *Chromosoma* **5,** 1
Lewis, K. R. (1956).   *Ph.D. thesis,* University of Wales
Lima-de-Faria, A. (1948).   *Portug. acta biol.* **2,** 167
Lima-de-Faria, A. (1949).   *Hereditas* **35,** 77
Lima-de-Faria, A. (1952).   *Chromosoma* **5,** 1
Löve, R. (1962).   *J. Histochem. Cytochem.* **10,** 227
Löve, R and Walsh, R. J. (1963).   *J. Histochem. Cytochem.* **11,** 188
Löve, R., Clark, A. M. and Studzinski, G. P. (1964).   *Nature* **203,** 1384
Lowry, R. J. (1963).   *Stain Tech.* **38,** 199
Lu, B. C. (1962).   *Canad. J. Bot.* **40,** 843
Ma, T. (1967).   *Stain Tech.* **42,** 285
MacGregor, H. C. (1965).   *Quart. J. micros. Sci.* **106,** 215
Magne, F. (1964).   *Cah. Biol. mar.* **5,** 461
Matsuura, H. (1941).   *Cytologia* **8,** 142
McClary, D. O., Williams, M. A., Lindegren, C. C. and Ogur, M. (1957). *J. Bact.* **73,** 360
McClintock, B. (1931).   *Bull. Mo. Agric. exp. Sta.* **163,** 1
McClintock, B. (1933).   *Z. Zellforsch.* **19,** 191
McClintock, B. (1934).   Ibid **21,** 294
McIntosh, D. L. (1954).   *Stain Tech.* **29,** 29
Metz, C. W. (1935).   *J. Hered.* **26,** 177
Metz, C. W. (1941).   *Cold Spring Harb. Symp. quant. Biol.* **9,** 23
Miller, O. L. (1965).   *Natl. Cancer Inst. Monog.* **18,** 79
Morrison, J. W. (1953).   *Canad. J. Agri. Sci.* **33,** 309
Morrison, J. H., Leak, L. V. and Wilson, G. B. (1959).   *Trans. Amer. micr. Soc.* **76,** 368
Mundkur, B. D. (1954).   *J. Bact.* **68,** 514
Myers, J. (1962).   In *Physiology and biochemistry of algae.* New York; Academic Press
Naora, H., Naora, H., Isawa, M., Allfrey, V. G. and Mirsky, A. E. (1962). *Proc. Natl. Acad. Sci. US*
    48, 853
Naylor, M. (1956).   *Ann. Bot. Lond. NS* **20,** 431
Naylor, M. (1957).   *Nature* **180,** 46
Naylor, M. (1958).   *Brit. Phycol. Bull.* **1,** 34
Naylor, M. (1959).   *Nature* **183,** 627
Nebel, B. R. (1934).   Referred to in Darlington and La Cour 1960
Newnham, R. A. (1962).   *M.Sc. thesis,* London University
Nizam, J. (1960).   *Ph.D. thesis,* London University
Norris, R. E. (1957).   *Univ. Calif. Publ. Bot.* **28,** 251
Östergren, G. (1947).   *Bot. Notiser* **2,** 176
Östergren, G. and Bajer, A. (1958).   *Hereditas* **44,** 466
Östergren, G. and Heenan, K. (1962).   *Hereditas* **48,** 332
Oura, G. (1936).   *Z. Wiss. Mikr.* **53,** 36
Painter, T. S. (1933).   *Science* **78,** 585
Painter, T. S. (1934).   *Genetics* **19,** 175
Painter, T. S. (1941).   *Cold Spr. Harb. Symp. Quant. Biol.* **9,** 47
Paolillo, D. J. Jr. (1960).   *Stain Tech* **35,** 152
Patel, R. J. (1961).   *Ph.D. thesis,* London University
Papenfuss, G. F. (1937).   *Symb. bot. Upsaliens* **2,** 1
Papenfuss, G. F. (1946).   *Bull. Torrey Bot. Cl.* **73,** 419
Pavan, C. and Da Cunha, A. B. (1969).   *Proc. Int. seminar on chromosome, Nucleus* **12,** suppl. 183

Pavan, C. and Ficq, A. (1957). *Nature* **180,** 983
Pontecorvo, G. (1944). *Nature* **153,** 365
Poulson, D. F. and Metz, C. W. (1938). *J. Morph.* **63,** 363
Pringsheim, E. G. (1951). In *Manual of Phycology*
Provasoli, L. (1958). *Ann. Rev. Microbiol.* **12,** 279
Provasoli, L. (1958). and Hutner, S. H. (1964). *Ann. Rev. Plant Physiol.* **15,** 37
Provasoli, L. McLaughlin, J. J. A. and Droop, M. R. (1957). *Arch. Mikrobiol.* **25,** 392
Rattenbury, J. A. (1952). *Stain Tech.* **27,** 113
Read, R. W. (1964). *Stain Tech.* **39,** 99
Ris, H. (1945). *Biol. Bull.* **89,** 242
Ris, H. (1962). In *The Interpretation of ultrastructure Symp. Int. Soc. Biol.* Academic Press, New York
Ris, H. and Crouse, H. C. (1945). *Proc. nat. Acad. Sci. Wash.* **31,** 321
Roberts, C. (1957). *Nature* **179,** 1198
Roberts, M. (1966). *The Phaeophyceae* II in The *Chromosomes of the Algae.* London; Edward Arnold
Robinow, C. F. (1961). *J. Biophys. Biochem. Cytol.* **9,** 879
Rogers, M. E. (1965). *J. Cell Biol.* **27,** 88A
Sahay, B. N. and Prasad, A. N. (1968). *Proc. 55th Ind. Sci. Cong.* III Abst., 309
Saito, M. (1961). *J. Protozool.* **8,** 300
Sakamura, T. (1927). *Bot. Mag. Tokyo* **41,** 59
Sarma, Y. S. R. K. (1958). *Ph.D. thesis,* London University
Sarma, Y. S. R. K. (1969). *Proc. Int. Seminar on chromosomes. Nucleus* **12,** Suppl. 128
Sarvella, P., Holmgren, J. B. and Nilan, R. S. (1958). *Nucleus* **1,** 183
Savage, J. R. K. (1957). *Stain Tech.* **32,** 283
Sax, K. and Humphrey, L. M. (1934). *Bot. Gaz.* **96,** 353
Schrader, F. (1939). *Chromosoma* **1,** 230
Schreiber, E. (1930). *Planta* **12,** 331
Semmens, C. J. and Bhaduri, P. N. (1941). *Stain Tech.* **16,** 119
Sharma, A. K. (1951). *Nature* **167,** 441
Sharp, L. W. (1943). *Fundamentals of cytology.* New York; McGraw-Hill
Sinha, J. P. (1963). *Cytologia* **28,** 194
Somers, C. E., Wagner, R. P. and Hsu, T. C. (1960). *Genetics* **45,** 801
Studzinski, G. P., Reidbord, H. E. and Love, R. (1967). *Stain Tech.* **42,** 301
Subrahmanyan, R. (1957). *J. Ind. Bot. Soc.* **36,** 373
Svedelius, N. (1956). *Svensk. bot. Tidskr.* **50,** 1
Swanson, C. P. (1940). *Stain Tech.* **15,** 49
Swift, H. and Rasch, E. M. (1955). *J. Histochem. Cytochem.* **2,** 456
Tamiya, H. (1963). *J. Cell Comp. Physiol.* **62,** 157
Thakur, M. (1964). *Ph.D. thesis,* London University
Thind, K. S. and Waraitch, K. S. (1968). *Proc. 55th Ind. Sci. Cong.* **III,** Abstr., 309
Thomas, P. T. (1940). *Stain Tech.* **15,** 167
Tjio, J. H. and Levan, A. (1950). *Ann. Estac. exp. Aula dei* **2,** 21
Turian, G. and Cautino, E. C. (1960). *Cytologia* **25,** 101
Über, F. M. (1940). *Bot. Rev.* **6,** 204
Upcott, M. B. (1939). *J. Genet.* **37,** 303
Wada, B. (1966). *Cytologia* **30,** Suppl.
Ward, E. W. B. and Ciurysek, K. W. (1961). *Candad. J. Bot.* **39,** 1497
Westbrook, M. A. (1935). *Beih. Bot. Zbl.* A**53,** 564
Wolf, B. E. and Wolf, E. (1969). In *Chromosomes today* **2,** 44. London: Oliver and Boyd
Yabu, H. (1965). *Bull. Fac. Fish. Hokkaido Univ.* **15,** 205
Yabu, H. and Tokida, J. (1963). *Ibid* **7,** 61
Zirkle, C. (1937). *Science* **85,** 528

9 Light Microscope Autoradiography

Becquerel's pioneer attempt in 1896 to detect radioactivity with the aid of a photographic plate has been advantageously used in later years to develop autoradiographic techniques. The purpose of this method is to locate radioactive material in a specimen with the help of photography. Whereas with the Geiger counter, an electronic wave is measured by an electrical device, in autoradiography, the radioactive material is detected by a photographic process of development. The method is based on the principle that if a photographic emulsion is brought into contact with radioactive material, the ionising radiation will so convert the emulsion as to show spots at certain points after being developed.

Specimen and film are brought into contact with each other for a certain period of exposure and there is decay of radioactive atoms. The radiation thus emitted affects the emulsion, activating silver halide crystals. The final result is the formation of a latent image which can be developed to denote the location, intensity and distribution of radioactive material. If the radioactive substance is tagged with a metabolic precursor, the distribution of the grains in the autoradiograph will indicate the distribution of the precursors and thus the metabolic path can be determined with accuracy.

STATUS OF THE PRINCIPAL METHODS

Of all the methods so far presented, Lacassagne's (1924) technique of using X-ray film and making contact between the sensitive pellicle and the tissue is the simplest, but it is not widely applied because errors occur, due to uneven contact between the object and the film (Hamilton, Soley and Eichorn, 1940). Observation at culture level was made possible for the first time in Belanger and Leblond's technique (1946), due to the direct appearance of the autograph on the section. The sensitive gelatin, separated from a photographic plate, is first moistened and kept in position on the tissue. Two disadvantages of this method are the risk of the appearance of bubbles after the emulsion dries up, and the increasing number of background grains which occur on melting the emulsion, thus giving an incorrect picture (Ficq, 1959). Evans (1947), Endicott and Yagoda (1947) and Bourne (1949) presented a method involving mounting sections on a photographic plate or film, but as the sections remained coated with celloidin, difficulties were often encountered with the tissue when photographic baths were in use. There was always the risk of the loss of material while removing the celloidin.

The stripping film method, as adopted by Pelc (1951, 1956), Andresen (1952), Taylor and McMaster (1954) and Kopriwa and Leblond (1962), was considered, for a long time, to be advantageous for the study of chromosome structure, bringing out a sharp contrast between radioactive spots and background tissue. In this technique, the outline of the schedule is to cover the preparation with a sensitive emulsion with a bottom layer of inert gelatin against a glass plate.

Lately, however, application of emulsion in the liquid form has been found to be most suitable for autoradiography. It is now widely used for plant, animal and human tissues and also for their cultures (Prescott, 1964; Prescott and Bender, 1964). This method allows the formation of a monolayer of emulsion on the tissue. In the study of high resolution autoradiographs (*see* chapter on high resolution autoradiography), the liquid emulsion technique has been found to be the most convenient one.

In both the above techniques, the film or the emulsion must always be stored in a cool, dark chamber (18–20 °C). Gude (1957) suggested that film plates be stored at 3–10 °C in a desiccator containing saturated aqueous calcium nitrate solution at a humidity of nearly 51 per cent. Moreover, after the application of the emulsion, air drying of slides and storage should also be carried out under similar conditions.

FACTORS INVOLVED IN THE INCORPORATION OF RADIOACTIVE TRACERS

As the particles are ejected in all directions, the photographic emulsion may become spotted over a much wider area than that covered by the actual location of tracers. This difficulty can, however, be eliminated by the use of particles of a suitable range, as the spreading must be limited within this range. In this respect ^3H, ^{14}C, ^{35}S are convenient tracers since they emit soft β-particles, affording good resolution.

The grain yield and radioactive decay should always be taken into account in making a quantitative assessment of the incorporation of tracers in the tissue. For example, in the emitters of soft β-rays, such as tritium, self-absorption is one of the important factors: half the particles are emitted in the reverse direction of the emulsion and, as such, a correct estimate of the number of particles emitted will be twice the number obtained in emulsion. Quantitative assessment depends on the correspondence between amount of radioactive material and density of the grains or 'track' in this region.

In order to use a radioactive molecule as a tracer, it is necessary to utilise the specific precursors of the molecule, the distribution of which is to be studied within the cell. For this purpose, in the study of chromosome metabolism, labelled uridine is used for the detection of RNA, thymidine for the detection of DNA, and specific amino acids for the detection of proteins. With ^{32}P alone, the localisation of DNA is difficult, as phosphorus is incorporated in different metabolic products of the cell unless such products are extracted or digested. The measurement of radioactivity for quantitative study, however, depends on several factors, such as the amount of radioisotope, its half-life, and the energy of β-rays emitted.

The concentration of the radioactive tracer in the labelled medium, as well as the time interval between application and fixation, are important factors in the study of autoradiography. As the rate of incorporation varies for different labelled compounds, the period of treatment must be fixed with this information in mind.

To obtain an accurate autoradiograph, the tissue should be treated with at least a certain minimum concentration of the labelled substance. Approximately 10 grains per 100 μm^2 of emulsion is just enough, requiring an exposure of $10/\delta$. β-particles (where δ denotes the yield of developed grains per particle hitting the film), which were obtained from the decay of double the number of radioactive atoms at the time of exposure. The half-life should also be considered, because for short-lived isotopes, the exposure of two half-lives results in the decay of three-quarters of the atom. In consideration of the various factors, the minimum concentration c for short-lived and long-lived isotopes has been suggested (Pelc, 1958) as:

1. For short-lived isotopes: (half-life H in days, in 5 μm sections and emulsions, and f denoting the proportion of labelled to unlabelled tissue)

$$C = \frac{11 \cdot 5 f}{H^\delta} \text{ in } \mu\text{Ci/cm}^3$$

2. For long-lived isotopes:

$$C = \frac{12 \cdot 5 f}{d^\delta} \text{ in } \mu\text{Ci/cm}^3$$

The isotope being long-lived in this case, the decay during the time of exposure (d) is negligible.

For fine-grained stripping-film, the value of δ can be taken as 1.

In addition to the factor of minimum concentration, the actual amount of radioactive substance required for feeding the tissue depends on the metabolic role of the metabolite in which it is supposed to be incorporated. For example ^{32}P, when injected into the tissue, mixes with the free phosphate in the organism. To obtain an autoradiograph, therefore, a sufficient quantity of the tracer is needed, but if the tracer is applied labelled with a particular metabolite which is not present in a large amount in the tissue, the quantity of tracer needed will be very low. A specific example is seen in mice, where labelling of DNA requires at least 1 mCi of ^{32}P per mouse while ^{14}C—labelled adenine is required in an amount of 25 μCi. Similarly, in leucocyte culture from peripheral blood, application of tritiated thymidine at 1 μCi/cm^3 for 5–6 h before harvesting is needed for incorporation in chromosomal DNA.

For a detailed consideration of these aspects, the reader is referred to the reviews by Boyd (1955), Pelc (1958), Ficq (1959) and Perry (1964).

AUTORADIOGRAPHY IN THE STUDY OF CHROMOSOME CHEMISTRY

In order to apply the principle of autoradiography in cytochemistry, precise localisation of the radioactive material within the tissue must be secured. Radioactive substances are introduced into the tissue either in a given chemical form or tagged with certain precursors of metabolism, and as at present specific and precise forms of labels, such as, thymidine, uridine, etc. can be obtained, the method can be usefully employed in the study of chromosome replication as well as in the chemical make-up of chromosomes. The radioactive molecule, at the precise site of its occurrence, following treatment, can be identified by several methods, namely, digestion through enzymes, selective staining techniques, extraction of different components and precipitation. For instance, for extracting DNA and RNA, the trichloracetic acid technique (Schneider, 1945) and the perchloric acid method (Ogur and Rosen, 1950) can be followed, and a study of the autoradiograph in the non-extracted and extracted slides will give an exact idea of the location of the radioactive tracer in the specific type of nucleic acid. The same object can be achieved by using specific nucleic acid enzymes, like deoxyribonuclease and ribonuclease (Brachet, 1942). For the localisation of basic and non-basic proteins, pepsin and trypsin can be applied to bring about digestion. For details of the enzyme treatment method, please see chapter on enzymatic methods.

There are, however, certain inherent limitations in the procedure for extraction. For example, perchloric acid may cause desensitisation of the emulsion. Hydrolysis in N HCl at 56 °C for 5 min (Vendrely and Lipardy, 1946) removes not only purine bases of DNA but RNA as well. This method is utilised for the extraction of RNA (Pearse, 1968). Here it is necessary to find out how far the proteins are affected by hydrolysis. For details of extraction procedures, please see chapter on Extraction.

TECHNICAL STEPS IN THE PREPARATION OF AUTORADIOGRAPHS

The different steps followed in autoradiography are:
(*a*) administration of the tracer into the tissue; (*b*) fixation; (*c*) paraffin

embedding or smearing; (*d*) staining; (*e*) application of the photographic emulsion; (*f*) drying; (*g*) exposure, and (*h*) photographic process.

Staining can be done before or after the application of the emulsion. For details of the types of emulsion fluid, developers, exposure time, etc., please see chapter on high resolution autoradiography.

Administration of the Tracer into the Tissue

The radioactive tracer can be obtained in specific salt solutions or tagged with metabolic precursors.

For instance, ^{32}P can be obtained in orthophosphoric acid in dilute HCl solution or as orthophosphate in isotonic saline solution containing phosphate buffer. Similarly, ^{35}S is available as H_2SO_4 in dilute HCl solution or as sulphate in isotonic saline solution. Tagged isotopes are available in the form of tritium (^{3}H) labelled thymidine, ^{14}C-labelled adenine, thymine, uracil, etc., representing nucleic acid bases. Similarly, for proteins ^{35}S-labelled methionine, phenylalanine, etc. are used.

They can be administered in the medium in which the organism grows, or can be injected into the tissue. With animal materials, direct injection of the substance can be made into the tissue concerned, or the tissue can be cultured in an artificial medium containing the radioactive substance at a known concentration. The same procedure is adopted for normal and malignant tissue in human materials. In plants, for somatic cells, intact roots are allowed to grow in the medium. For flower buds, a method has been devised in which the inflorescence is floated in the medium containing the isotope. For culture of excised tissue in artificial media, the principle is the same for all biological objects. For the study of nucleic acid in somatic cells of plants, seedlings can be cultured in water, containing ^{32}P as orthophosphate (Pelc, 1958) or ^{14}C as adenine (Howard and Pelc, 1951; Clowes, 1956) or tritium (^{3}H) labelled thymidine (Taylor, 1956). Marimuthu (1970) studied ^{3}H-thymidine-labelled nuclei in developing pollen of *Tradescantia*. Savage and Wigglesworth (1971) devised a method of direct injection of ^{3}H thymidine through calyx and corolla (0·001–0·002 mCi for 2 days, that is, the concentration needed for flooding the precursor pool as recommended by Cleaver, 1971) in buds for the study of post-meiotic stages followed by the routine method for cytological preparation of autoradiographs (Savage, 1967). For meiotic studies, inflorescences can be cultured for 8–24 h in a medium containing ^{32}P in aqueous solution (0·02 mg KH_2PO_4 and 20 μCi of ^{32}P/cm^3) (Taylor, 1953; Taylor and Taylor, 1953) or in White's medium (Plaut, 1953). Similarly, for the study of proteins of chromosomes, materials can be cultured in a medium containing ^{35}S and Na_2SO_4 (Pelc and Howard, 1956). Cut inflorescences can be kept in media containing 0·036 mg of $MgSO_4·7H_2O$ and 10 μCi of ^{35}S/cm^3 (Taylor and Taylor, 1953). For the study of nucleic acid in mammals, intraperitoneal injections of ^{32}P, ^{14}C, adenine and ^{35}S *dl*–methionine can be administered to mice. In the same way, for human bone marrow cells, mouse prostates etc., culture medium containing the radioactive substance can be used. For peripheral blood cultures too, labelled radioactive compounds can be added to McCoy's or Eagle's medium, together with calf serum. For chinese hamster fibroblast lines, labelled precursors are added to cell cultures. Application in monolayer cultures is necessary for short pulse exposures (Schmid, 1966). In the study of *Drosophila* salivary gland chromosomes, the tracers can be administered in food in the medium.

Fixation

The fixation can be performed either through freeze substitution (Lison, 1953; Freed, 1955; Woods, 1955; Burstone, 1969) or through a number of non-metallic fixatives; however, the fixative should not leach the radioactivity during preparation. Wolman and Behar (1952) suggested a method for slow fixation at $-8\,^{\circ}C$ for 4 h, followed by washing in ethanol. As far as practicable, metallic fixatives should be avoided; ethanol, a mixture of ethanol and acetic acid, neutral formalin, and formol-saline can all be employed for fixation, especially where isotopes are applied for incorporation into nucleic acid or protein of chromosomes, but the most reliable method so far found is freezing. Taylor (1951, 1952) observed that fixation of cells of anthers of *Tradescantia paludosa* in acetic-ethanol mixture, followed by hydrolysis in HCl for 10 min at $10\,^{\circ}C$ for Feulgen staining, resulted in the complete loss of radioactive phosphorus from the tissue, except that in the nuclei, and as such, the remaining phosphorus, being only in DNA, could be specifically localised. The fixative removes the acid-soluble phosphorus. If necessary, phospholipides can be removed by treatment with hot ether-ethanol (1:3) for 5 min (Taylor and Taylor, 1953) after fixation. For bone marrow or fibroblast cell cultures, 50 per cent acetic acid is recommended, though for detachment of fibroblasts, trypsin treatment is necessary (Schmid, 1965). In Chinese hamster fibroblast lines and in other mammals, acetic acid followed by acetic-ethanol (1:3) is recommended. Treatment in hypotonic solution and Hank's medium is necessary prior to fixation (Prescott and Bender, 1964). Frøland (1965), for blood cultures too, followed hypotonic treatment in one containing 25 per cent Hank's solution.

Paraffin Embedding or Smearing

In order to secure a high resolution in paraffin sections, the sections should not exceed 5µm in thickness, should preferably be cut in a freezing mictotome, and should be mounted on slides coated with a film of alum-gelatin (0·5 per cent aqueous solution of gelatin and 0·1 per cent chrome alum) or with egg albumin. Absolute drying for at least 48 h is necessary before mounting the sections. Similar slides should be used in the case of smears, though preparations made with ordinary slides have occasionally yielded good results. In order to soften the tissue for smearing after fixation, the use of pectinase or cellulase is preferable instead of prolonged hydrolysis in

N HCl, which often results in the removal of nucleic acids; the sections or smears should not be uneven and, especially in the latter, uniform pressure must be applied after smearing in dilute acetic acid to obtain a one-layered smear. Heating should be avoided as far as practicable.

Staining

Staining of the preparations can be done prior to, or after, the application of the photographic emulsion, but in the former, the protection of the material by a a very thin layer of impermeable substance, like celloidin, may be necessary (Pelc, 1958; cf. Belanger, 1961). Experience in this laboratory (Chatterjee, A. unpublished) has shown that uncoated tissue, especially in the study of chromosomes, allowed good resolution.

The dyes chosen should have certain prerequisites, namely, a good penetrating capacity, no injurious effect on the tissue containing the radioactive substance, and moderate staining effect. These qualities must be strictly observed in those instances where the photographic emulsion is applied before staining. Strong dyes and high temperatures are to be avoided, as they often attack the silver grains and artefacts may be produced. Leucobasic fuchsin, methyl green-pyronin, toluidine blue and Leishmann-Giemsa are satisfactory stains for autoradiography (Gude *et al.*, 1955; Prescott, 1964).

Bergeron (1958) suggested staining of nuclei with basic fuchsin at pH 3·5–4·0 before film development. Frøland (1965) recommended acetic-orcein staining prior to the application of emulsion for human blood cultures. Benge (1960) carried out staining of autoradiographs at low temperatures.

Fussell (1966) described a method for methyl green-pyronin staining of autoradiographs, involving the gradual bringing down to water through ethanol grades of fully processed autoradiographs. Two changes are given in dilute McIlvaine's buffer (1:10) at pH 4·2 for 5 min each, followed by staining for 15 min in a solution containing 80 cm^3 of diluted buffer and 2 cm^3 of 2 per cent aqueous solution of purified methyl green, to which 0·5 g of pyronin is added. The slides are quickly rinsed in two changes of water, blotted and air-dried.

Application of the Photographic Emulsion

The photographic emulsion can be applied over the section either in the form of a liquid (Belanger and Leblond, 1946; Belanger, 1950) or smear, or as a film pressed on the material (Pelc, 1956; Taylor, 1956), being composed principally of a fine layer of gelatin, containing numerous silver halide crystals. The formation of radioactive spots is based on the principle of the production of ion pairs due to an electronic event, ultimately manifested in the single grain of black silver. Nuclear emulsion of small grains should not be used because of the possibility of their being washed off during washing.

The emulsion should be stored at 4 °C to prevent melting and should be applied at 25–30 °C. Sawicki *et al.* (1968) have, however, shown that both storage and exposure of AR-10 films at 18 °C increases overall efficiency. Even though fogging may occur, it does not exceed the safe values of background. Undesirable background effect and artefacts can be avoided if the operation is performed away from the direct illumination of safelight in the dark room. Different types of emulsion with different grain sizes are available in the market. For Kodak, NTB, NTB2, NTB3 denote decreasing grain sizes, with increasing resolution and sensitivity.

Drying

The tissue, after being coated by emulsion, either in liquid form or else as stripped film, must be dried quickly, preferably in a strong current of air in a cold chamber. Quick drying is necessary to prevent the formation of air bubbles within the tissue and also the production of artefacts in the presence of excess moisture.

Exposure

The period of exposure in a cool dark chamber depends on the type of isotope used and its concentration, and is the period required for a specified number of ionised particles to heat a unit area. The time needed for any emulsion can be increased to secure a satisfactory image with a smaller dose—from a few days to even one or two months' continuous exposure may be necessary in certain cases. The relationship between the exposure time and half-life of the isotope has already been discussed. For special techniques, exposure at even − 18 °C has been recommended (Sawicki *et al.*, 1968).

Photographic Process

The photographic process to secure autoradiographs consists of the following steps: development of the latent image, fixation, washing and drying.

Development—The principle of the first step is to develop the latent image in an aqueous solution of a reducing agent in a dark chamber. The widely used developers are hydroquinone and methyl-*p*-aminophenol, which have the property of reducing the silver halides. The developing solution contains some sulphide salt as well, which helps in the removal of undesirable products, and also potassium bromide, a constituent of the developer, which checks the formation of foggy grains. A number of developers of Kodak and Ilford have been found to be very suitable.

Fixation—The fixation of the intact image involves removal of the silver halide from the emulsion and hardening of the image. It is not possible to wash the silver halide away in water as it is only sparingly soluble; however, it forms a soluble ion complex with the thiosulphate salt, usually of sodium, which is added to remove the silver halide from the solution. In autoradiography the use of an acid–hardener fixing bath becomes necessary for the protection of film or plate; therefore potassium aluminium sulphate (potash alum) or potassium chromium sulphate (chrome alum), which is effective in acid solution, is used in the acid fixing solution, having the added advantage of neutralising any alkaline developer remaining in the solution. The effect of the hardener also prevents swelling and spreading of the gelatin. The authors, however, have noted in the study of chromosomes that a dilute solution of sodium thiosulphate yields quite accurate autoradiographs. The temperature of the chamber should be between 22 and 24 °C, and the humidity should be low. An increase in temperature and humidity causes fading of the image by reaction of metallic silver and sodium thiosulphate.

Washing and drying—Continuous washing in running water is necessary to remove traces of any excess sodium thiosulphate, and a temperature of 15·5–21·0 °C speeds the washing and prevents softening of the emulsion. The slides can then be dried in air or by passing through grades of ethanol, the latter technique being more satisfactory for microscopic work as it does not allow dust particles to accumulate.

Where the experimenter prefers to stain after photographic processing rather than before, the slides can now be stained by the recommended schedules described earlier. Finally the slides may be mounted, the most satisfactory medium being euparal. For observation of unstained slides, mounting can be done either in distilled water or in a special fluid containing 3 g gelátin, 0·2 g chrome alum, 80 cm^3 distilled water and 20 cm^3 glycerol, stored at 4 °C and melted at 50 °C. Messier and Leblond (1957) suggested that, following exposure, the developed emulsion can be protected with a coat of vyrylite and mounted under a cover slip.

SOME IMPORTANT SCHEDULES FOR THE PREPARATION OF AUTORADIOGRAPHS

METHOD OF ADMINISTRATION OF ISOTOPE

For Plants

1. For the study of somatic chromosomes, grow young seedlings of *Vicia faba* in medium containing 2 µCi of ^{32}P (as orthophosphate) per cm^3 in tap water. Fix young healthy roots at different intervals ranging from 2 days to 1 month in acetic–ethanol or chilled 80 per cent ethanol.

For the study of meiosis in plants, 20–75 µCi of ^{32}P in 1 cm^3 of water can be administered (Moses and Taylor, 1955) on flower buds of *Tradescantia paludosa* or *Rhoeo discolor*, or 4 µCi/cm^3 labelled ^{32}P in White's medium, on inflorescence of *Lilium henryii* (Plaut, 1953).

For the study of proteins in plants, roots can be grown in tap water in which 1 µCi of ^{35}S as Na_2SO_4 is present per cm^3. The exposure required is at least 70 days. The details of prophase chromosome can be studied (Pelc and Howard, 1956).

For Animals

1. For the study of nucleic acids in mammalian chromosomes, 40 μCi of ^{32}P/g can be injected intraperitoneally in rodents. Similarly, ^{14}C adenine can be injected in the dose of 0·3 μCi/g and the animal can be killed 24 h after injection. Fixation can be made in acetic-ethanol mixture. For salivary gland chromosomes of *Drosophila*, autoradiographs can be prepared after feeding *D. melanogaster* with 20 μCi/cm^3 of ^{14}C adenine for at least 24 h. In other tissues, for the study of proteins for example, intraperitoneal injection of 2 μCi/g of ^{35}S *dl*-methionine can be given. Exposure time may vary from 20 to 70 days. Animal tissues can also be cultured in a medium containing 2 μCi/cm^3 of ^{35}S *dl*-methionine with air exposure time of 14–28 days.

For Human Material

Human bone marrow cells can be cultured in a medium containing 0·5 μCi/cm^3 of ^{14}C adenine or 1 μCi/cm^3 of ^{32}P (Lajtha, 1952); fixation is performed in methanol. For human peripheral blood culture autoradiograph (Frøland, 1965), preparations can be made following the method of Moorhead and colleagues (1960). Tritiated thymidine (1 μCi/cm^3) may be added 5–6 h before termination of the culture. Colcemid may also be added 2 h before excising the cells. Air-dried preparations can be made after treatment in hypotonic salt solution and staining the chromosomes with 2 per cent acetic-orcein solution. Slides are then covered with liquid emulsion (Ilford K5) for 2 days' exposure. Developing is done in Amidol solution for 4 min, followed by fixing in 30 per cent sodium thiosulphate solution for 7 min. Slides are dried in air and mounted in DPX (Frøland, 1965).

2. For cutting sections or preparing smears, grease-free slides should be used, originally coated with a mixture of photographic gelatin (5 g), chrome alum (0·5 g) and water (1 litre) and dried.

For paraffin blocks, dehydrate by passing through 70 per cent to absolute ethanol grades, followed by ethanol-chloroform grades and the usual process of embedding in paraffin. Cut thin paraffin sections (5 μm thick) and bring the slides down to water after de-paraffinising through xylol by the usual process adopted in microtomy.

For preparing smears, hydrolyse the material in N HC1 for 5 min at 60 °C and wash in water.

Slides from both paraffin block and smear schedules are now ready for autoradiography. If staining is desired before applying emulsion, for the former, slides should be hydrolysed for 10–12 min at 60°C, washed in water and stained in leuco-fuchsin solution according to Feulgen procedure;

for the latter, the period of hydrolysis is extended by 10 min and the same procedure for Feulgen staining is followed. After staining, the slides are brought to water, passing through 45 per cent acetic acid. Further steps are carried out in a cool dark chamber.

Savage (1962) suggested, for plant materials, squashing in a long subbed cover slip under another rectangular cover in 45 per cent acetic acid, followed by removal of cover slips by dry ice technique. Only the cover slip is to be further processed. Jona (1963) recommended squashing under a scotch-tape which can be removed before applying autoradiographic emulsion.

3. For coating the slides with emulsion, either of the two following methods can be used:

Coating with Liquid Emulsion

For emulsion film on glass plate, cut the film with a blade parallel to the four edges at 1·27 cm from the edge. Slowly peel away the emulsion from the glass plate, blowing moist air at the point of bonding; films on 35 mm film-base can also be similarly peeled off. Place the free emulsion strip in distilled water at 18–20 °C, for 10 min, transfer it to a 50 cm^3 beaker on a water bath at 37 °C cover, and allow the emulsion to melt completely without stirring (15 min).

For the bulk emulsions (gels), transfer about 2 cm^3 of the emulsion to a 50 cm^3 beaker with a clean glass spatula.

In both cases, warm the slides and with the warm slide held horizontally between the thumb and the forefinger of one hand, apply 2 drops of emulsion per square inch of slide from a medicine dropper. Spread the drops quickly with a brush (previously warmed) over an area already outlined by a diamond pencil scratch, rotate the slide from side to side so that the emulsion flows evenly over the area, and keep the slide in the warm condition for 30–60 s for uniform spreading of the emulsion, then transfer to cold temperature for 30 min to harden the emulsion.

The other method, which is more convenient, though requiring more emulsion, is to keep the melted emulsion (melted at 42–45 °C) in a long trough, and dip the subbed slide in it for 4–5 s followed by draining off the emulsion.

Coating with Stripping Film

For plates, outline with a blade a square area about 3·80 cm^2 on the film. Keep the plate in total darkness for 10 min, then slip the blade under one edge of the outlined square and pull upward slowly. If the humidity is high, stripping becomes difficult, so the plate should first be kept in a desiccator for 10 min. Invert the stripped film and float it, with emulsion downwards, on distilled water at 22–24 °C for 2–3 min. Slip a slide with specimen upwards into the water beneath the floating film and lift the film out on top of the section. Dry in front of a fan at room temperature.

For 35 mm films, cut a piece 3·80 cm long, rub the cut edge and slowly strip away the film from the film base with thumb and forefinger. The later stages are the same as followed with glass plates.

4. The slides, when dry, are kept in black boxes containing a desiccant, and are sealed with black tape and kept so that the slides are horizontal. Keep in a refrigerator for a period of one month.

5. For processing, use dilute Kodak D-19 developer with distilled water (1:2) and filter it. Develop the autoradiograph for 2–10 min at 14–20 °C, rinse in distilled water at the same temperature for 30 s, and fix in filtered acid–hardener–fixer at the same temperature until clear. Leave the slides in the fixer for a period equalling half the time taken for clearing, and then wash in running water at 17 °C for 30 min, and dry at room temperature in a dust-free box.

6. If not done previously, the slides can now be stained, preferably in Feulgen or pyronin–methyl green, in accordance with the usual schedules.

Note: If necessary, in place of ^{32}P, ^{14}C adenine (200 µCi/1000 cm^3) or ^{3}H-labelled thymidine can be used for the study of DNA in chromosome. In the former case, it is necessary to remove the RNA fraction present, by digestion through ribonuclease at 37 °C for 1–2 h (1 mg in 1 cm^3 of N/10 NaOH). Distribution of RNA can be studied if ^{14}C-uridine is used as the radioactive substance.

7. For restaining, after removal of the cover slip with xylene and rehydration, Frøland (1965) suggested treatment in a mixture of iodine (2 g) and potassium iodine (4 g) in 100 cm^3 of water for 1 h to remove silver grains, followed by rinsing in running water and treatment in 30 per cent sodium thiosulphate solution for 30 min. The slides are finally washed in running water before restaining.

8. For removal of silver grains after taking photographs (carbol fuchsin stained, tritiated human leucocytes), Bianchi *et al.* (1964) recommended that slides should be kept in 7·5 per cent potassium ferricyanide solution for 3 min, followed by keeping in 20 per cent $Na_2S_2O_3$, $5H_2O$ solution for 3–5 min. The slides should then be washed thoroughly, kept at 37 °C for 10 min in 0·2 M Tris buffer (pH 7), water and trypsin mixture (50 cm^3 + 50 cm^3 + 0·19), washed, air dried and observed under Shillaber's immersion oil.

I. Liquid emulsion autoradiography for plant, animal and human materials (Prescott, 1964)

In order to study materials fed with labelled amino acids or nucleosides, acetic–ethanol (70–95 per cent) 1:3 fixative is recommended. Stock solution of the emulsion should be kept at 22–24 °C. Kodak NTB series is used with different grain sizes. With decrease in grain size, higher resolution and sensitivity are expected. The operations (1) to (5) must be performed in a dark room with a covered safe light (15W bulb—Wratten Series 1).

Melt the emulsion at 42–45 °C using a constant temperature water bath. To avoid background effect, it is better to develop a clean dry slide, without any material, as control.

Subbed slide is not necessary if Kodak emulsion is used. Fit two slides with materials back to back and dip in the emulsion, kept in a long trough, for 4–5 s. Drain off the emulsion. After separating the slides, dry in racks against a stream of air and keep in slide boxes, sealed with tape, in a cool dark chamber, for the required period of exposure (2–3 weeks).

After exposure, develop the slides in Kodak Dektol, D-11 or D-19 developer for 2 min.

Fix in Kodak Acid-Fixer for 2–5 min.

Rinse in running water for 20 min. Give a final rinse in distilled water and dry.

For staining, three alternative stains are recommended:

(*a*) 0·25 per cent aqueous solution of toluidine blue (pH 6);

(*b*) methyl green-pyronin;

(*c*) Giemsa stain as adopted by Gude *et al.* (1955).

Wash in 95 per cent ethanol, make air dry preparations and mount in euparal.

II. *Method for mammalian chromosomes from cultures* (Prescott and Bender, 1964); (developed for chinese hamster fibroblast line CHEF/125 but also applicable to other mammals).

Culture cells in Petri dishes and add colchicine (10^{-6M}) a few hours before harvesting.

Decant off the medium and replace it by Hank's phosphate buffered solution.

Replace again by hypotonic saline (10–15 per cent) of Hank's solution, in which the culture may remain even up to 1 h.

After 3–4 min, cells in metaphase are automatically separated from the glass.

Rotate the medium gently so that the cells aggregate in the centre.

Observe under a dissecting microscope and draw out 0·5 µl of medium in a braking pipette.

Withdraw an equal volume of glacial acetic acid in the pipette. As substitute, to minimise loss of material, 0·5 µl of 10 per cent formalin can be taken followed by an equal volume of acetic-ethanol (1:3).

After a few seconds, eject the entire contents of the pipette on a slide on which a drop of acetic-ethanol (1:3) had been spread a few seconds earlier. Add another drop of acetic-ethanol after a few seconds prior to drying.

Wipe off the excess fluid from the margin and air-dry the preparation.

The remaining procedure is the same as schedule I for coating the emulsion. This technique has several advantages, like the absence of serum protein in the culture; availability of a large number of plates; absence of any cytoplasmic layer between the chromosomes and the emulsion and also of background radioactivity due to clearing the cytoplasm. Moreover, if neutral formalin is used, the loss of chromosomal protein can be checked.

III. *Autoradiography of peripheral blood culture for human chromosomes* (Schmid, 1965)
(also applicable to bone marrow and fibroblast cells)

For culture: Allow 10 cm^3 of heparinised blood to stand for 2–3 h at 22–24 °C. Take 2–4 cm^3 of supernatant in a 2 oz flask and make up to 10 cm^3 by adding the medium (McCoy's or Eagle's medium + 20 per cent foetal calf serum). Add 0·1 cm^3 of phytohaemagglutinin M or P.

For continuous labelling of DNA, add ^{3}H-labelled thymidine (1 µCi/cm^3 of medium) 6 h before termination of culture and add colcemid at a final concentration of 0·03 µg/cm^3. For short pulse labelling, monolayer cultures are necessary, cells need to be centrifuged and colcemid at a concentration of 0·06 µg/cm^3 is needed.

(*a*) For squash preparations from leucocyte suspension or fibroblasts, detach cells with 0·2 per cent trypsin and centrifuge in 12 cm^3 conical tubes for 4 min at 500 rev/min and pour off the supernatant. Add 10 cm^3 of 1 per cent sodium citrate and leave for 10 min. Re-centrifuge and add 10 cm^3 of 50 per cent acetic acid without disturbing the cell pellet. Decant off the fixative after 20 min and make air-dried smears. If staining is desired, add 2 per cent acetic-orcein and squash. Follow Conger and Fairchild's (1953) dry ice schedule for permanent preparations, treat in absolute ethanol and store the dried slides.

(*b*) For blood culture, trypsinisation is not necessary and the usual schedule of air-dried preparations can be followed without staining.

In the preparation of autoradiographs, the stripping film coating technique of Kopriwa and Leblond (1962) was followed. All operations should be performed in the dark room in the safelight at 20–22 °C.

(*a*) Cut 40 × 40 mm squares of AR10 film mounted on glass slides and keep the plate for 3 min in 75 per cent ethanol. Transfer to another tray containing absolute ethanol.

(*b*) With forceps, lift a single square of film and float, with emulsion side down, in a tray of distilled water, so that the film automatically spreads out.

(*c*) Bring the glass slides containing the material below the film under water and lift the slides so that the film adheres neatly on the surface of the slide with the material on it. Dry in a stream of air and store in a slide box, sealed with tape, containing silica gel desiccant for the required period in the cold.

(*d*) Prior to developing, paint the reverse side of the slide with a paint and dry. Develop with Kodak developer D-196 for 5 min. Treat with the acid fixer as usual, wash and dry. Detach paint and film with a blade and follow the usual procedure for mounting.

(*e*) For staining, where this has not been performed before, stain with Giemsa (dist. water, 100 cm^3; 0·1 M citric acid, 3 cm^3; 0·2 M Na$_2$HPO$_4$, 3 cm^3; methanol and Giemsa stock solution, 5 cm^3) for 4–7 min. Rinse in distilled water and dry. Detach paint and film and follow the usual procedure for mounting.

PRINCIPAL CONCLUSIONS ON CHROMOSOME STRUCTURE AND
METABOLISM ARRIVED AT THROUGH AUTORADIOGRAPHY

This method has been advantageously employed in interpreting the
controversial issues regarding the structure and metabolism of chromosomes. In this respect, the use of tritium-labelled thymidine, with low
energy β-particles has made possible an analysis of the single chromosome
and its different loci, and has aided in unravelling the sequence of DNA
replication and basic protein metabolism in the macronucleus of *Euplotes*
and also the relationship between RNA and DNA metabolism in the
'puff' of salivary gland chromosomes of *Drosophila* and other organisms.
The precision of this method has been shown further in the analysis by
Levinthal and Thomas (1957) of DNA duplication in single viral particles.

Taylor, Woods and Hughes (1957) have demonstrated clearly, on the
basis of their experiments on the uptake of tritiated thymidine by roots, that
DNA replication follows the pattern outlined by Watson and Crick (1953).
According to the latter authors, the DNA molecule is composed of a double
helix, joined by base pairs, and during duplication, each strand serves as
a template for the synthesis of a sister strand so that the new molecule,
after duplication, is composed of an old and a new strand. Taylor, Woods
and Hughes (1957) showed that both chromatids in the first generation of
treated cells in metaphase exhibited equal labelling, in the second generation
only one of the chromatids was found to be labelled, and in subsequent
generations there was an equational decrease in the number of labelled
chromatids. Evidently the findings confirm the replication of the double
helix as suggested by Watson and Crick (1953). Though this observation has
been questioned on technical grounds by La Cour and Pelc (1958), and on
unequal labelling of chromosomes by Mazia and Plaut (1956), yet other
experimental approaches support the conclusions of Taylor, Woods and
Hughes (1957), Taylor (1962, 1968) and the semiconservative replication of
DNA has been established.

Evidence of the differential nature of euchromatin and heterochromatin
has been secured with the application of autoradiography. Lima-de-Faria
(1959) investigated the uptake of tritium-labelled thymidine in the heterochromatic and euchromatic chromosome segments in *Melanopus* and
Secale and the data show that heterochromatin synthesises DNA later than
euchromatin. The functional differences of even two types of heterochromatin namely α and β have also been claimed (*see* Wolf, 1969). The
results may be taken to imply the specialised nature of heterochromatic
segments as suggested by other cytological data as well (*see* Sharma and
Sharma, 1958, Brown, 1966).

That DNA synthesis takes place in the interphase has been clearly
elucidated through autoradiographic studies involving tritiated thymidine
incorporation (for review, *see* Taylor, 1968). The categorisation of the
interphase into two growth phases, G_1 and G_2, intercalated by the DNA
synthetic 'S' phase has been made possible largely through such incorporation studies.

The study of chromosome structure has also been greatly facilitated through the use of labelled halogenated unusual nucleosides like FUDR and BUDR. Autoradiographs of chromosomes, following their incorporation, have not only demonstrated decisively that DNA forms the fundamental skeleton of the chromosomes but at the same time it has adduced evidence that change or breakage of the DNA molecule may ultimately result in gene mutation or chromosome breakage.

The elucidation of the functional aspects of chromosomes has been the most outstanding contribution of autoradiography in recent years. The very dynamicity of the chromosomes during different stages of development and differentiation has been well illustrated in the autoradiographs of the puffing patterns in dipteran salivary gland chromosomes (Beermann, 1964; Keyl and Hagele, 1966; Pelling, 1964; Swift, 1964). The amounts of RNA and proteins associated with chromosomes have been shown to be variable in different stages of development (Busch, 1965; Busch *et al.*, 1964; Edström, 1964; cf. Moses and Coleman, 1964; Plaut and Nash, 1964; Swift, 1964). But later works showing DNA puffs in Sciarideae and variable replication of nucleolar organiser in amphibia (Miller, 1966; Peacock, 1965) have indicated that even the DNA and histone of chromosomes may vary within an organism. The identification of the nature of metabolic DNA, which is synthesised in excess or independently of chromosome duplication, as termed by Pavan (1965) and Pavan and Da Cunha (1968, 1969), was principally due to this method.

An understanding of the DNA replication pattern of the sex chromosomes of mammals (Evans *et al.*, 1965; Lyon, 1963; Monesi, 1965; Mukherjee, B. B. *et al.*, 1967, 1969; Schmid, 1963) and diptera (Berendes, 1965; Mukherjee, A. S. *et al.*, 1968, 1969; Müller and Kaplan, 1966) has been gained through advances in this incorporation technique. The studies by Mukherjee, B. B. and his colleagues (1967, 1969) confirmed the existing concept of the differential rate of late replication of sex chromosome DNA at the synthetic phase. But at the post-implantation stage, their findings further imply that though the pattern of sex chromosome replication varies in different spermatogonia, yet in the spermatocyte, the replication is either earlier or synchronous with the autosomes.

Mukherjee, A. S. and his colleagues (1968, 1969), on the other hand, working on the problem of dosage compensation, have shown that in *Drosophila*, the single X-chromosome in the salivary gland of the male fly, compensates, by its transcriptive activity and enlargement in size, the functions performed by the X-chromosome in the female. But, they have also recorded that the pattern of transcription of the X-chromosomes, in both sexes of two species of *Drosophila*, having different X-chromosome constitutions, is nearly identical.

In the field of differentiation of plant organs, application of autoradiography is gaining importance. In our own laboratory, it has been shown that in the nuclei of mature organs, which have been induced to divide through hormone treatment, the presence of both diploid and polytenic nuclei has been recorded. As gene action, dependent on transcription, has now been shown to be independent of replication, the existence of polyteny raises interesting speculations. A comparative study of the RNA synthesising activity of the diploid and polytenic strands in the nuclei of mature organs through autoradiography may indicate the *modus operandi* of one aspect of the genetic control of differentiation in higher plants.

In the analysis of chromosome nucleolus relationship, autoradiographic techniques have been of special use. On these studies, it has been claimed that RNA is synthesised independently of the general chromatin in the nucleolus during interphase, even though some of its components might have been ultimately derived from the multiple chromosomal centres at the end of telophase (Brinkley, 1965; Das, 1963; Hsu *et al.*, 1965; Hay, 1969; LaFontaine, 1969; LaFontaine and Lord, 1966; *see* Perry, 1965, and Prescott and Bender, 1962). Bernhard (1966) and Bernhard and Granboulan (1969) have elucidated, to some extent, the metabolic pathway involving nucleus, nucleolus and cytoplasm. Significantly, the presence of a considerable amount of intranucleolar DNA, in contact with the nucleolus-associated chromatin, has been indicated. Active incorporation of tritiated thymidine in the nucleolar DNA of vertebrates has been secured.

In the field of molecular hybridisation, at the chromosomal level, as done extensively by Gall, Pardue and others, the incorporation method has thrown considerable light on the loci of differential gene action.

In summing up, the most significant contribution of autoradiography in recent years may be claimed to be the elucidation of the dynamicity of the structure and behaviour of chromosomes (Gay, 1966; Kaufmann *et al.*, 1960; Sharma, 1968), necessary for controlling differentiation in different phases of development and growth in higher organisms.

REFERENCES

Andresen, C. C. (1952).   *Exp. Cell Res.* **4**, 239
Becquerel, H. A. (1896).   *C. R. Acad. Sci., Paris* **122**, 420
Beermann, H. (1964).   *J. Exptl. Zool.* **157**, 49
Belanger, L. F. (1950).   *Anat. Rec.* **107**, 149
Belanger, L. F. (1961).   *Stain Tech.* **36**, 313
Belanger, L. F. and Leblond, C. P. (1946).   *J. Endocrin.* **39**, 8
Benge, W. P. J. (1960). *Stain Tech.* **35**, 106
Berendes, H. D. (1965).   *Chromosoma* **17**, 35
Bernhard, W. (1966).   *Natl. Cancer Inst. Monog.* **23**, 13
Bernhard, W. and Granboulan, N. (1969).   In *The nucleus, Ultrastructure in biological systems,* **3**, 81
Bergeron, J. A. (1958).   *Stain Tech.* **33**, 221
Bianchi, N., Lima-de-Faria, A. and Jaworska, H. (1964).   *Hereditas* **51**, 207
Bourne, G. H. (1949).   *Nature* **163**, 293
Boyd, G. A. (1955).   *Autoradiography in biology and medicine.* New York; Academic Press
Brachet, J. (1942).   *Embryologie chimique.* Paris; Masson
Brinkley, B. R. (1965).   *J. Cell Biol.* **27**, 411
Brown, S. W. (1966).   *Science* **151**, 41
Burstone, M. S. (1969).   Cryobiology techniques in Histochemistry, in *Physical techniques in biological research* **3C**, 1, New York; Academic Press
Busch, H. (1965).   *Histones and nuclear proteins.* New York; Academic Press
Busch, H., Starbuck, W. C., Singh, E. J. and Ro, T. S. (1964).   In *The role of chromosomes in development* New York; Academic Press
Cleaver, H. H. (1967).   *Thymidine metabolism and cell kinetics.* North-Holland; Amsterdam
Clowes, F. A. H. (1956).   *New Phytol.* **55**, 29
Conger, A. D. and Fairchild, L. M. (1953).   *Stain Tech.* **28**, 289
Das, N. K. (1963).   *Science* **140**, 1231
Edström, J. E. (1964).   In *The role of chromosomes in development.* New York; Academic Press

Edwards, L. C. (1955). *Stain Tech.* **30,** 63
Endicott, K. M. and Yagoda, H. (1947). *Proc. Soc. exp. Biol., N.Y.* **64,** 170
Evans, T. C. (1947). *Proc. Soc. exp. Biol., N.Y.* **64,** 313
Evans, H. J., Ford, C. E., Lyon, M. F. and Gray, J. (1965). *Nature* **206,** 900
Ficq, A. (1959). Autoradiography in *The Cell* **1,** 67, eds. Brachet, J. and Mirsky, A. F. New York; Academic Press
Freed, J. (1955). *Lab. Invest.* **4,** 10
Frøland, A. (1965). *Stain Tech.* **40,** 41
Fussell, C. P. (1966). *Stain Tech.* **41,** 315
Gay, H. (1966). *Carnegie Inst. Year book* p. 581
Gude, W. D. (1957). *Stain Tech.* **32,** 197
Gude, W. D., Upton, A. C. and Oddl, T. T. Jr. (1955). *Stain Tech.* **30,** 161
Hamilton, J. G., Soley, M. H. and Eichorn, K. B. (1940). *Univ. Calif. Publ. Pharm.* **1,** 339
Hay, E. D. (1969). In *The Nucleus, Ultrastructure in Biological Systems* **3,** 2, New York; Academic Press
Howard, A. and Pelc, S. R. (1951). *Exp. Cell Res.* **2,** 178
Hsu, T. C., Arrighi, F. E. and Klevecz, R. R. (1965). *J. Cell Biol.* **26,** 539
Jona, R. (1963). *Stain Tech.* **38,** 91
Kaufman, B. P., Gay, H. and McDonald, M. R. (1960). *Intern. Rev. Cytol.* **9,** 77
Kopriwa, B. M. and Leblond, C. P. (1962). *J. Histochem. Cytochem.* **10,** 269
Keyl, H. G. and Hagele, K. (1966). *Chromosoma* **19,** 223
Lacassagne, A. (1924). *J. Radiol. electrol.* **9,** 506
La Cour, L. F. and Pelc, S. R. (1958). *Nature* **182,** 506
La Fontaine, J. G. (1969). In *The Nucleus, Ultrastructure in Biological Systems* **3,** 152, Academic Press; New York.
La Fontaine, J. G. and Lord, A. (1966). *Natl. Cancer Inst. Monog.* **23,** 67
Lajtha, L. G. (1952). *Exp. Cell Res.* **3,** 696
Levinthal, C. and Thomas, C. A. Jr. (1957). *Biochem. biophys. Acta* **23,** 453
Lima-de-Faria, A. (1959). *J. biophys. biochem. Cytol.* **6**
Lison, L. (1953). *Histochemie et Cytochemie Animales* p.29. Paris; Gauthier-Villars
Lyon, M. F. (1963). *Genet. Res. Camb.* **4,** 93
Marimuthu, K. M. (1970). *Stain Tech.* **45,** 105
Mazia, D. and Plaut, W. S. (1956). *J. biophys. biochem. Cytol.* **2**
Messier, B. and Leblond, C. P. (1957). *Proc. Soc. exp. Biol. Med.* **96,** 7
Miller, O. L. Jr. (1966). *Natl. Cancer Inst. Monog.* **23,** 53
Moorhead, P. S., Nowell, P. C., Mellmann, W. S., Battips, D. M. and Hungerford, D. A. (1960). *Exp. Cell Res.* **20,** 613
Monesi, V. (1965). *Chromosoma* **17,** 11
Moses, M. J. and Taylor, J. H. (1955). *Exp. Cell Res.* **9,** 474
Moses, M. J. and Taylor, J. H. (1964). In *The Role of chromosomes in development.* New York; Academic Press
Müller, H. J. and Kaplan, W. D. (1966). *Genet. Res. Camp.* **8,** 41
Mukherjee, A. S. and Datta Gupta, A. K. (1968). *Proc. XXI Int. Genet. Cong. (Abst.)*
Mukherjee, A. S., Lakhotia, S. C. and Chatterjee, S. (1969). *Proc. Int. Seminar on Chromosomes, Nucleus* Suppl., Calcutta
Mukherjee, B. B., Sinha, A. K., Mann, K. E., Ghosal, S. K. and Wright, W. C. (1957). *Nature* **214,** 710
Mukherjee, B. B. and Ghosal, S. K. (1969). *Exp. Cell Res.* **54,** 101
Ogur, M. and Rosen, G. (1950). *Arch. Biochem. Biophys.* **25,** 262
Pavan, C. (1965). *Brookhaven Symp. Biol.* **18,** 222
Pavan, C. and Da Cunha, A. B. (1968). *Proc. Symp. Nucl. Differentiation,* Belo Horizonte
Pavan, C. and Da Cunha, A. B. (1969). In *Proc. Int. Seminar on Chromosomes Nucleus, Suppl.* **12,** Calcutta
Peacock, W. J. (1965). *Natl. Cancer Inst. Monog.* **18,** 101
Pearse, A. G. E. (1968). *Histochemistry, theoretical and applied.* London; Churchill
Pelc, S. R. (1951). Paper presented in Medical Research Council, Radiotherapeutic Research Unit
Pelc, S. R. (1956). *Nature* **178,** 59
Pelc, S. R. (1958). *Gen. Cytochem. Meth.* **1,** 279
Pelc, S. R. and Howard, A. (1956). *Exp. Cell Res.* **10,** 549
Pelling, C. (1964). *Chromosoma* **15,** 71

Perry, R. P. (1964).   *Quantitative autoradiography* in *Methods in cell physiology* **1,** 305, New York; Academic Press

Perry, R. P. (1965).   *Natl. Cancer Inst. Monog.* **18,** 325

Plaut, W. S. (1953).   *Hereditas, Lund* **39,** 438

Plaut, W. S. and Nash, D. (1964).   In *The role of chromosomes in development.* New York; Academic Press

Prescott, D. M. (1964).   *Autoradiography with liquid emulsion* in *Methods in cell physiology* **1,** 365, New York; Academic Press

Prescott, D. M. and Bender, M. A. (1962).   *Exptl. Cell Res.* **26,** 260

Prescott, D. M. and Bender, M. A. (1964).   *Preparation of mammalian metaphase chromosomes for autoradiography* in *Methods in cell physiology* **1,** 381, New York; Academic Press

Sawicki, W., Ostrowski, K. and Rowinski, J. (1968).   *Stain Tech.* **43,** 35

Savage, J. R. K. (1962).   *J. R. micros. Soc.* **80,** 291

Savage, J. R. K. (1967).   *Sci. Rev.* **48,** 771

Savage, J. R. K. and Wigglesworth, D. J. (1971).   *Stain Tech.* In press

Schmid, W. (1963).   *Cytogenetics* **2,** 175

Schmid, W. (1965).   *Autoradiography of human chromosomes,* in *Human chromosome methodology.* New York; Academic Press

Schneider, W. C. (1945).   *J. biol. Chem.* **161,** 293

Schneider, W. C. (1968).   *Sci. Cult.* **34,** 69

Sharma, A. K. and Sharma, A. (1958).   *Bot. Rev.* **24,** 511

Srinivasan, B. D. and Harding, C. V. (1963).   *Stain Tech.* **38,** 283

Swift, H. (1964).   In *The Nucleohistones,* Holden Day Inc.

Taylor, J. H. (1951).   Paper presented at Genetics Society of America (AIBS) meeting, Minneapolis

Taylor, J. H. (1952).   *Exp. Cell Res.* **4,** 164

Taylor, J. H. (1953).   *Science* **118,** 555

Taylor, J. H. (1956).   *Biol. Res.* **3,** 545

Taylor, J. H. (1968).   The *Replication of DNA in chromosomes* in *molecular genetics.* New York; Academic Press

Taylor, J. H. and McMaster, R. D. (1954).   *Chromosoma* **6,** 489

Taylor, J. H. and Taylor, S. J. (1953).   *J. Hered.* **44,** 129

Taylor, J. H., Woods, P. S. and Hughes, W. L. (1957).   *Proc. nat. Acad. Sci. Wash.* **43,** 122

Vendrely, R. and Lipardy, J. (1946).   *C. R. Acad. Sci., Paris* **223,** 342

Watson, J. D. and Crick, F. H. C. (1953).   *Nature* **171,** 737

Wolf, B. E. (1969).   *Proc. Int. Seminar on Chromosome, Nucleus, Suppl.* **12,** Calcutta

Wolman, M. and Behar, A. (1952).   *Exp. Cell Res.* **3,** 619

Woods, P. S. (1955).   *Stain Tech.* **30,** 123

10 Electron Microscopy

The introduction of electron microscopy to the study of ultrastructure has helped to clarify chromosome structure at the submicroscopic level. The high magnification attained, coupled with good fixation and proper preservation of components *in vivo*, has resolved the inner details of chromosome and other cellular components. As the wave properties of the electron and the focusing properties of the magnetic field were recognised (Busch, 1926), it became possible to design an electron microscope, in pattern with the light microscope. Light microscopes are generally designed with suitably shaped surfaces of solid media, abruptly reflecting or refracting the light rays. In electron microscopes, electrical or magnetic fields are adequately shaped to refract electrons, producing the desired image. The transformation of the electron image to a visible light image is effected on the fluorescent screen. The limit of resolution of a good electron microscope is at present nearly 3 Å and until now very good resolution of biological macromolecules of 5–10 Å has been achieved (Sjöstrand, 1962, 1966). The basic constituents of the electron microscope are assembled in a vertical column. They are, serially: an illuminating system, a specimen chamber, an objective lens, intermediate and projecting lenses, and a viewing chamber with facilities for photographing electron images. In the entire column, vacuum is maintained by oil and mercury diffusion pumps, at specific pressure. The source of the electron providing the illuminating system is generally constructed from a hot tungsten filament, serving as a cathode, and an anode with a surrounding Wehnelt cylinder, provided with a hole for the passage of the electron beam. The resultant image is detected either on a fluorescent screen or on a photographic plate. The majority of electron microscopes have a specimen holder with a cap for fitting the specimen screen. A new plastic handling tool for the RCA EMU3 electron microscope specimen holder has been devised by Chuble and Ellwanger (1967). Magnification is dependent on the proximity of the objective lens to the specimen, thus the orientation should be carefully controlled. In short, to the microscope, which lies in a vacuum, a cathode filament supplies the source of electrons forming the illuminating system, the electrons being focused on the material by an electromagnetic condenser lens. Similarly, electromagnetic objective lenses collect the electrons from the materials and form a magnified image. Through further electromagnetic projection by the eyepiece, the specimen is magnified and projected on to a fluorescent screen, the magnification being controlled by manipulating the current in the projector lens, and focusing by varying the magnetic field of the objective lens. For details of instrumentation and the working of the microscopes, such as, RCA-EMU, Philips EM, HITACHI, Siemens ELMISCOP, different models of BENDIX-AKASHI etc., the reader is referred to Menter (1962), Pease (1960), Pashley (1964), Porter (1964), Siegel (1964), Cosslett (1967), Sjöstrand (1967, 1969) and Toner and Curr (1968). Modifications in the operating method have also been devised so that there has been a reduction in the specimen contamination rate, an increase in the life of the filament as well as a decrease in the load on the condenser lens stabiliser (Meek, 1968). Better reproduction of photographic exposures and strong contrast micrographs have been achieved. In principle, contrast signifies the difference in electron scattering power between the object studied and its surrounding medium.

To allow the passage of electrons ejected from a source, and for good resolution of the object, ultra-thin sectioning in ultramicrotome is essential. The method devised for cutting such fine sections has helped the use of electron microscopy for the clarification of cytological details. For the preparation of materials to be observed under the electron microscope, the following steps should be adopted under strictly controlled conditions: (*a*) fixation, (*b*) dehydration, (*c*) embedding, (*d*) sectioning and (*e*) mounting. The advantages and limitations of the different steps are now considered.

FIXATION

Low Temperature Fixation

With low temperature fixation, the life-like preservation of cells is maintained because of the absence of any chemical compounds used in fixation. Frozen materials, after proper dehydration, either by sublimation of ice as in freeze-drying, or by substitution with organic solvents as in freeze-substitution, have decided advantages in electron microscopy. Cryogenics—which deals with the use of extremely low temperatures—has often been utilised in ultra-structural studies. Cryogenic gases have a boiling point greatly below $-100\,°C$ ($-148\,°F$) and include nitrogen, oxygen, argon, neon, krypton, xenon, helium, hydrogen, methane, fluorine as well as ethylene. Certain workers prefer rapid freezing with the use of certain cryogenic gases, like helium, while others are of the opinion that slow freezing, with the use of compounds like glycerol, ethylene glycol, propylene glycol, etc., protecting against ice crystal formation, is preferable (see Burstone, 1969). Following freezing, the freeze-drying procedure, which includes dehydration from the frozen state in the beginning, involving sublimation of ice *in vacuo*, is generally carried out at −30 to −40 °C. After several days of drying in the cold, the temperature of the material is gradually raised and maintained under vacuum for several hours and finally the ice is allowed to sublimate on its surface. Embedding can be done in vacuum or outside, but with precautions against the absorption of moisture from the air.

In order to eliminate chemical reactions resulting from the use of fixatives, Sjöstrand and Baker (1958) employed the freeze-drying method of fixation (−72 °C dry ice temperature seems sufficient). The specimen is dipped in ethyl ether (pre-cooled with dry ice and ethanol mixture), and later the frozen tissue is transferred to a pre-cooled container which should afterwards be evacuated. Freeze-dried tissue presents difficulties in infiltration with a plastic monomer, and in order to avoid damage to the tissue during polymerisation, it is generally done in a vacuum. Some segments of the freeze-dried tissue often show signs of denaturation if the materials are not properly dehydrated. Refrigeration systems for cryostat and freeze-drying equipments have also been developed (Gersh, 1948; Stowell, 1951; Glick and Bloom, 1956; Clements, 1962; McClintock, 1964; Rutherford *et al.*, 1964; Lotke and Dolan, 1965; *see* Burstone, 1969).

Freeze substitution technique, in which the effects of freezing are separated from those of vacuum drying, involves the substitution of a fluid in the tissue in place of complete vacuum dehydration. This method allows better preservation than freeze-drying. Freeze-drying technique, without the use of chemical fixatives, can no doubt keep the tissue structure intact up to a certain limit (Pease, 1966; Malhotra, 1968), but freeze substitution by fixation may often be necessary for better representation of the cellular structure. Fluids used in freeze substitution are methyl cellosolve, ethanol, diethyl ether, chloroform, butanol, iso-amyl alcohol, isopentane, isobutane, propane, propylene (Rebhun, 1965; Burstone, 1969); acetone and ethanol-acetone, the last two showing the most rapid dehydrating action, the temperature being maintained between −38° to −78 °C during dehydration.

Fernandez-Moran (1959 a, b, 1962) used freeze-substitution to replace fixation. In this method, the tissue can be directly dehydrated and infiltrated with an embedding medium at a low temperature, through cold baths and boxes, and does not require a pumping system, as is needed for freeze-drying. The operation may be carried out at −130 °C to −180 °C, and for freezing, isopentane, Freone 22 chilled with liquid nitrogen or helium can be used. Methyl cellosolve can be employed as a substitute for tissue water and glycerol and finally, methacrylate can be used for infiltration. Infiltration is carried out with methacrylate below −75 °C, followed by u.v. polymerisation at −20 °C. This procedure is supplemented by a cooling specimen support (−10° to −120 °C), with a modified liquid nitrogen stage for observation under an electron microbeam of low intensity. Persijn *et al.* (1964), used liquid air instead of mechanical cooling up to about −130 °C. Bullivant (1965) adopted a schedule in which mouse pancreas were infiltrated with glycerol (3–60 per cent) in isotonic veronal acetate buffer (pH 7·2), followed by freezing in liquid propane cooled by liquid nitrogen. The material was then transferred at $-75\,^{\circ}\mathrm{C}$ in dried ethanol to a refrigerator. Further substitution was carried out after three changes of ethanol at this temperature, during a period of 3 weeks, followed by embedding through butyl methyl methacrylate and 1 per cent Lucidol. Durcupan (Stäubli, 1960) and Epon mixture (Luft, 1961) were also tried. Polymerisation was effected with ultraviolet at $-20\,^{\circ}\mathrm{C}$.

Afzelius (1962) recommended Acrolein as the substitution fluid at low temperature. Since formalin forms a solid polymer in the cold its use in substitution fluid is limited. Anhydrous OsO_4 vapour has been suggested for freeze-drying techniques. Feder and Sidman (1958) and Fernandez-Moran (1961) proposed 1 per cent OsO_4 solution in acetone as the substitution fluid at $-40\,^{\circ}\mathrm{C}$ to $-70\,^{\circ}\mathrm{C}$. Pease (1964) recommended polymerisation in ultraviolet at $-72\,^{\circ}\mathrm{C}$ for freeze-substituted materials. Cope (1968) devised a technique for fixation and embedding at $-20\,^{\circ}\mathrm{C}$, after treatment with an antifreeze designed to prevent ice crystal formation at $0\,^{\circ}\mathrm{C}$. A mixture of 50 per cent V/V of dimethylsulphoxide (DMSO) mixed with glycerol or ethylene glycol serves as the anti-freeze. In this method, tissue fixation is performed with 5 per cent glutaraldehyde in 50 per cent anti-freeze solution for 6–8 h at $-20\,^{\circ}\mathrm{C}$, followed by dehydration in 2-hydroxy-ethyl-methacrylate. Final embedding is done in a mixture of 70 per cent dihydroxy–ethyl–methacrylate and 30 per cent n-butyl methacrylate or styrene. Polymerisation is carried out with the aid of ultraviolet rays at $-10\,^{\circ}\mathrm{C}$.

Freeze-drying and freeze-substitution techniques are generally recommended for the study of enzyme localisation and activity and are not applied to the analysis of chromosome structure. But the potentiality of their use in chromosome study cannot be ignored, especially in research involving the distribution of enzymes, such as alkaline phosphatase in chromosomes.

With Chemical Fixatives

In the preparation of the fixative, the osmolar concentration of the solvent is of special importance and the osmolar concentration of the fixative can be varied widely (Maunsbach, 1966). Of all the metallic fixatives so far tried, osmium tetroxide (OsO_4) is the most widely used, and the buffered OsO_4 solution, with added calcium in certain cases (Huxley and Zubay, 1961; *see* Fawcett, 1964), is considered the most adequate one. That it preserves the cell structure intact, to some extent, is observed by the direct correlation between the spacing of certain structures obtained by birefringence and X-ray on the one hand, and the electron micrograph on the other. As osmium tetroxide is a stain for phospholipids, a sharp contrast is often obtained in osmium-fixed material, but as this contrast imparts brighter staining to non-DNA containing materials, it should be used cautiously in the study of chromosomes. For best results in chromosome preparations with osmium tetroxide as the fixative, both potassium dichromate solution (3 per cent) and lanthanum nitrate solution (2 per cent) are applied in conjunction with 2 per cent osmium tetroxide solution—the lanthanum salt helps in the preservation of nuclear structures. Several workers have tried osmium tetroxide fixation, buffered from pH 4 to 7, and have obtained very satisfactory results. Afzelius (1962) reported that liquid osmium at 4 °C as well as a solution of osmium in 40 per cent carbon tetrachloride, gives good fixation.

The serious drawbacks of osmium fixation are its slow rate of penetration, reduction of enzyme activity (Holt and Hicks, 1962), and its non-reaction with carbohydrates and nucleic acids (Bahr, 1954). In cytochemical procedures, the main limitation of OsO_4 fixation lies in its capacity to render the protein impervious to proteinase digestion (Leduc and Bernhard, 1962). Swift (1962) has further pointed out that osmium fixation causes the protein to become basophilic by combining with amino groups—a drawback which should be avoided. It has been observed that the best contrast with osmium fluid can be obtained if it is used as a post-fixation fluid, after formaldehyde or glutaraldehyde. Gasser (1955) used osmium-dichromate mixture for fixation. Treatment with osmium fluid may vary from 10 min to 4 h (Sjöstrand, 1969) and a brief washing in isotonic salt solution is always desirable. The following osmium-containing fluids have been recommended by Sjöstrand for mammalian and frog tissues:

1. According to Zetterqvist (1956):
   Stock solution A: Sodium acetate, 9·714 g
   Veronal acetate, 14·714 g
   Distilled water to make 500 cm^3.
   Stock solution B: Sodium chloride 40·25 g
   Potassium chloride 2·1 g
   Calcium chloride 0·9 g
   Distilled water to make 500 cm^3.
   Hydrochloric acid 0·1N

The solutions are mixed in the following proportions:

(a) For mammalian tissue: Solution A, 10 cm^3; solution B, 3·4 cm^3; 0·1 N HCl, 11 cm^3.
(b) For frog tissue: Solution A, 7·4 cm^3; solution B, 2·6 cm^3; 0·1 N HCl, 8·1 cm^3.

Distilled water is added to make 50 cm^3 and the pH adjusted to between 7·2–7·4 by adding 0·1 N HCl. Then 0·5 g osmium tetroxide is added to the mixture and stored in the cold in a brown, glass-stoppered bottle.

2. According to Millonig (1962):
   Solution A: 2·26 per cent dibasic sodium phosphate solution,
   Solution B: 2·52 per cent sodium hydroxide solution,
   Solution C: 5·4 per cent glucose,
   Solution D: 41·5 cm^3 of solution A + 8·5 cm^3 of solution B.

A mixture is prepared with 45 cm^3 of solution D, 5 cm^3 of solution C, and 0·5 g of osmium tetroxide and the pH corrected to 7·3–7·6.

Alternatively, formaldehyde may be used as a fixative (Pease, 1962). Solution C is replaced by a new solution C_{form}, containing 40 per cent formaldehyde and 5·4 per cent glucose. Of this solution, 5 cm^3 is mixed with 45 cm^3 of solution D and the pH corrected to 7·3–7·6.

These fixatives have, however, been devised for histochemical procedures, and their efficacy in chromosomal materials is yet to be proved.

Dalton (1953) recommended fixation by perfusion rather than by immersion and Sjöstrand (1969) suggested dripping the fixative on to the surface of an intact organ and perfusion. Potassium dichromate has also been used but it has been claimed that it is incapable of preserving all cellular parts (Gersh, 1959; and *see* Casselman, 1955a, b; Baker, 1966). Other fixatives containing mercuric chloride or acids are considered to result in artefacts, manifested especially as fibrils in the protoplasm.

The importance of maintaining the pH in the range 7·3–7·6 was pointed out by Palade (1952). As regards the medium, certain authors prefer a balanced salt solution whereas others recommend osmotically active substances, such as polyvinyl pyrilidone (de Robertis and Iraldi, 1961), sucrose (Caulfield, 1957), etc. to be added.

Of the non-metallic fixatives, formaldehyde is widely used (Baker and McCrae, 1966). The effects of concentration, duration of treatment, as well as temperature, on its action, have all been worked out in detail. Formalin does not have any damaging effect on polymerisation (Swift, 1962). The rapidity of fixation and maintenance of enzyme activity, even after a long period of fixation, are special advantages of formalin fixation. Holt and Hicks (1962) further noted that formalin-fixed materials not only offer resistance to the effect of cold solvents but even allow strong binding with the dye. Formaldehyde vapour is used as a fixative also in freeze-drying (Falck and Owman, 1965). Ris (1962) observed that neutral fixatives containing osmium tetroxide or formalin preserve the distribution of cellular structures.

Leduc and Bernhard (1962) recommended formalin fixation for specific extraction of nucleic acids and proteins and suggested fixation of rat pancreas in 10 per cent formaldehyde buffered at pH 7·3 with Michaelis veronal acetate, or with Sörensen's phosphate at −3°C. Sjöstrand (1969) confirmed Millonig's (1962) observation that phosphate buffer is a good medium for formaldehyde. Pearse (1968), however, stated that formalin should not be used as a fixative for nucleic acids or nucleoproteins as it blocks a large number of reactive groups, hampering stainability with both acidic and basic dyes. Afzelius (1962) also recorded that formaldehyde-fixed structure is often damaged by electron beams, especially in methacrylate-embedded material, which can be overcome by using other embedding materials like Epon.

In the majority of cases, formaldehyde fixation yields very good results if the materials are post-fixed with osmium tetroxide, and stained in a saturated solution of uranyl acetate. The presence of the heavy metal allows better contrast and sharper staining and, as such, this procedure is often recommended for ultrastructural studies (Fawcett, 1964; Sjöstrand, 1969).

Another aldehyde with wide use in electron microscopy is glutaraldehyde, or more precisely, glutaric di-aldehyde, $(CH_2)_3$ CHO·CHO, with the formula

$$
\begin{array}{c}
\text{H}_2 \\
\text{C} \text{---} \text{CHO} \\
\diagup \\
\text{H}_2\text{C} \\
\diagdown \\
\text{C} \text{---} \text{CHO} \\
\text{H}_2
\end{array}
$$

It was first introduced as a cytochemical fixative by Sabatini, Bensch and Barrnett (1963). Its bifunctional nature, having the property of forming cross linkage, was worked out by Bowes (1963). At lower concentrations it is effective at pH 8·0, whereas pH 6·5 is needed for its optimum activity when given in larger quantities. As the action is very rapid, a dilute solution (generally 1 per cent) is recommended (Pearse, 1968). As with formaldehyde, this compound, if post-fixed with buffered osmium fluids, can be used without any deleterious effect on staining. Sjöstrand (1969) has suggested distillation of the commercial solution for the removal of contaminants. In the case of perfusion fixation of brain, kidney, etc., fixation for approximately 5 min is recommended prior to transfer to osmium fluid. Several formulae for fixatives containing glutaraldehyde (Maunsbach, 1966) have been presented by Sjöstrand (1969) but their efficacy in chromosome study has not yet been tested.

Acrolein, otherwise known as 2-propenal or acrylic aldehyde, is another bifunctional aldehyde having the capacity of forming cross links between end-groups of proteins (Bowes, 1963; Cater, 1963). Under laboratory conditions, it is generally prepared by heating a mixture of anhydrous glycerol, acid potassium sulphate, and potassium sulphate, in the presence of a small amount of hydroquinone, followed by distillation in the dark (Merck Index, 1968). Acrolein, having the formula $CH_2 = CH \cdot CHO$ (mol. wt. 56·06) is unstable at alkaline pH and polymerises, especially under light, forming a plastic solid (disacryl). Watson and Aldridge (1961) used it as a fixative for ultrastructural studies, in which potentially reactive groups other than nucleic acids, were blocked by acetylation. As acrolein is a non-metallic fixative, similar to formalin, the chance of damage due to polymerisation is not significant. Bowes (1963) noted that cross-links produced by acrolein and glutaraldehyde are more resistant to boiling water and acid hydrolysis than formaldehyde. Afzelius (1962) studied fixation of liver tissue in 10 per cent acrolein followed by post-fixation in osmium fluid. Swift (1962) suggested fixation in 10 per cent acrolein in phosphate or tris buffer, with a post-fixation treatment in 1 per cent osmium tetroxide, buffered at pH 7·6, for 2 h. In general, the use of acrolein is gaining importance in studies on ultrastructure.

Ethanol and acetone, in spite of their fixing properties, are not generally used for electron microscopic studies because of their disturbing effect on the morphology of tissue structures, particularly chromatin shrinkage (Ris, 1962). Moreover, Leduc and Bernhard (1962), in relation to their work on the nucleolus, noted that interchromatinic substances are not properly preserved after ethanol fixation.

Another chemical used in electron microscopic techniques is potassium permanganate, and Luft (1956) suggested that the fixing property is due to permanganate and not to potassium. A serious limitation of this fixative is

the presence of black granules in the background which have been attributed to the formation of manganese dioxide granules formed by the
decomposition of the permanganate ions. Feder and Sidman (1958)
recommended permanganate in acetone as a fixative. Afzelius (1962)
observed that different cations differ with regard to the preservation of
different structures and calcium permanganate shows a greater lipid
stabilising activity than potassium permanganate. In chromosome fixation,
potassium permanganate does not prove to be of much use.

DEHYDRATION AND EMBEDDING

Removal of water from the material is generally performed by the usual
procedure of passing through ethanol and acetone grades. For embedding
media for ultrastructural studies, however, certain essential specifications
have to be fulfilled. Pease and Baker (1948) used Parlodion and Carnauba
wax, with slight success, as embedding media and obtained 0·5 µm thick
sections, but collapse of the structure after removal of the wax presented
difficulties in its interpretation. In order to secure ultra thin sections,
embedding in hard cross-linked plastic was found to be necessary. Newman
et al. (1949) introduced an acrylic plastic—the amorphous *n*-butyl poly-
methacrylate as an embedding medium, which for a long time was
considered to be adequate for electron microscopic studies. In fact, most of
the earlier techniques were based on methacrylate-embedded materials. In
general, *n*-butyl and methyl methacrylates, mixed in the proportion of
8:2 (Newman *et al.*, 1950), were considered suitable. But serious limitations
of methacrylate embedding gradually became apparent, particularly
because of its unpredictable hardness and distortion of the tissue during
polymerisation. This problem in relation to the consistency could be
solved to some extent by varying the proportions of the two methacrylates,
while damage due to polymerisation could be minimised by quick polymerisation by the addition of a strong concentration of catalyst, raising the
temperature, and processing in a nitrogenous atmosphere (Moore and
Grimley, 1957). Another serious drawback of methacrylate embedding
involved sublimation of a part of the methacrylate when exposed to an
electron beam, causing appreciable distortion of the material. A carbon or
methacrylate coating of the material, resulting in sandwiching the tissue
between two films, was also adopted (Watson, 1957) but it evidently
further complicated the procedure. However, its susceptibility to thermal
effect was advantageously employed in controlling the degree of hardness
through temperature regulation.

In addition, the difficulty of sectioning methacrylate blocks has been
principally attributed to heavy cross linkage of the plastic, rather than to the
hardness of the block (Luft, 1961). This difficulty was overcome by adding
ethylene glycol dimethacrylate during polymerisation of the monomer.
Klotz (1962) suggested that different water soluble proteins, polysaccharides or synthetic polymers with SH groups may also be used for
embedding. By oxidation in the air, they can give an insoluble cross-linked

matrix. – SH groups can react with phosphotungstate, mercury or gold for contrast. A water soluble polymer at $-4\,°C$, polyvinyloxazolidinone and its ethyl derivative, are insoluble at $4\,°C$. So impregnation in ice and subsequent rendering it insoluble by warming are possible. Similarly, if dimercaptons (water soluble), are followed by doubly substituted mercury derivatives, cross-linked polymers can be obtained.

General Principles of Procedure for Methacrylate Embedding

As methacrylate monomer mixes well with ethanol, the latter is generally employed for dehydration. The process should be carried out in the cold and as rapidly as possible, with a maximum treatment for 5 min in each grade of ethanol, as the materials for electron microscopy are small enough to allow rapid penetration by ethanol. To ensure complete dehydration, several changes in ethanol are necessary. Pease (1960) noted that dehydration is often incomplete with acetone, due, principally to the difficulty of keeping anhydrous acetone.

For embedding, Pease (1960) recommends transfer of the tissue to methacrylate monomer. Gay (1955) observed that satisfactory results can be obtained if a mixture of absolute ethanol and *n*-butyl methacrylate monomer (1:1) is used for at least 1 h. Methyl methacrylate, being cheaper, is generally used, but 10–20 per cent of ethyl methacrylate solution can also be employed. For final embedding and polymerisation, a mixture of pure methacrylate monomer and a catalyst (such as, 2, 4-dichlorobenzol peroxide) is used with several changes for at least 2–3 h (Newman *et al.*, 1949); 1,2-dichlorobenzyl peroxide can also be applied (Pease, 1960).

Small gelatin capsules (size 00), containing partially polymerised methacrylate, may later be used as containers for polymerisation, as this size fits well in the Porter–Blume microtome. For the sake of convenience, the materials are placed on a flat glass plate covered with a gelatin capsule containing methacrylate of syrupy constituency (partially polymerised).

The temperature necessary for polymerisation can be attained either on a warming pan at $47\,°C$, or through ultraviolet radiation secured by a cheap fluorescent lamp (Sun-Lamp bulb 5 TT12). For ultraviolet polymerisation, the amount of catalyst added does not need a critical control, while for heat polymerisation, 2 per cent dichlorobenzoyl peroxide should be mixed with methacrylate. For heat polymerisation, Gay (1955a) recommends treatment for 48 h on a covered warming pan at $47\,°C$, followed by keeping for 24 h at $22\,°C$ for hardening. In case of ultraviolet polymerisation, it is preferable to keep the material 2·54 cm away from the source for at least 48 h. The latter method is superior, according to Pease (1960), as it imparts a rubbery consistency to the material and the polymerisation is uniform. The desired degree of softness of the block can be achieved by regulating the proportion of methacrylate.

As bubbles are formed in the block due to the presence of water, Moore and Grimley (1957) suggested the use of de-gassed methacrylate under dry N_2 atmosphere to replace O_2 by N_2. Due to the transparent nature of methacrylate, allowing the transmission of light, the material can be observed directly under the microscope when the block, on being detached from the glass plate, is fixed on the stage. The desired portion of the material can be delimited with a marker and trimmed for sectioning. Ordinary light microscope photographs can be taken for comparing finally with the electron micrograph.

With the increasing advances in electron microscopic studies, the need for a suitable embedding material with three primary requisites was felt, namely, (a) allowing ultra thin sectioning, (b) not producing artefacts during polymerisation and (c) no sublimation against electron beam. Evaporation during exposure to electrons causes excessive deformation of structure, arising out of extensive distortion and overlapping adpressed lines due to surface tension. In order to meet these requisites, other embedding materials were also tested, leading ultimately to the discovery of two important epoxy and polyester resins: Araldite and Vestopal W. Chemically, an epoxide is a triangular configuration of an oxygen atom joining two carbon ones of an organic molecule, ethylene oxide being the simplest form. Several such epoxies may yield compounds having potentialities of forming resins and plastics. Epoxy resins are often obtained through the use of epichlorohydron and a polyol, which is mostly a bisphenol obtained from acetone and phenol.

Araldite is an aromatic epoxy-resin which has a remarkable stability against the electron beam (Maalöe and Birch-Andersen, 1956; Glauert et al., 1956; Glauert and Glauert, 1958). It permits ultra thin sectioning without polymerisation damage and can be cured without significant contraction. The section need not have any support for mounting, even a simple carbon film may serve the purpose. No doubt, there is a slight lowering of the image contrast, which can be compensated by using metallic staining. However, two serious limitations of this epoxy embedding, the high viscosity of the resin and its low rate of penetration, had to be overcome. One of the short chain mono-epoxides, propylene oxide (1–2 epoxy propane) was therefore used to dilute the resin, which had to be introduced after ethanol dehydration and before and during resin embedding. The low viscosity and the low boiling point of propylene oxide, as well as its ready miscibility with ethanol and resin led to its choice. In Glauert's technique, a hardener, 964B or dodecenyl succinic anhydride (DDSA), is added to Araldite casting resin M, or to Araldite No. 6005 for curing. In order to regulate hardening, a plasticiser, dibutyl phthalate or *n*-butyl phthlate has been found to be useful. To accelerate proper setting, 964 C or trimethamine methyl phenol is also added to the mixture prior to final embedding. Glauert and Glauert (1958) suggested a mixture containing Araldite casting resin M, or Araldite No. 6005, 10 cm^3; hardener 964 B or DDSA, 10 cm^3; di-butyl or *n*-butyl phthalate, 1 cm^3, and accelerator 964 C, 0·5 cm^3. The proportions may be modified to suit different types of tissue. These chemicals are mixed at 60 °C in an incubator. The first three are mixed in a conical flask 15 min before use, stirred for about 30 s, and the last ingredient is added later.

For embedding, half of the ethanol is removed from the specimens dehydrated in ethanol and an equal quantity of the Araldite mixture is added, without the accelerator, in glass-stoppered bottles. The liquids are mixed by shaking the bottles and they are then incubated for 30 min to 1 h at 48 °C. This ethanol mixture is removed from the specimens with a pipette and 5–6 changes are given with pure Araldite mixture at intervals of 1–2 h. The specimens may be kept overnight in the last change at 48 °C. Depending on the type of specimen, the period taken for embedding may range from 24 to 48 h, or even longer. Two to three hours prior to the final embedding, the accelerator is added to the Araldite mixture, during which time the complete medium is changed at least three times.

The specimens are then placed, with a pipette, into dry gelatin capsules, or glass tubes, and the capsule is filled with the complete Araldite mixture, taking care that air bubbles are not included. The capsules are then closed and incubated at 48 °C for 30 h for complete hardening, the period required for setting the Araldite mixture being 6 h.

With American Araldites (Araldite 502 Ciba), the procedure given above did not yield very satisfactory results owing to inadequate penetration and difficulty in sectioning. A variant of this technique was therefore published by Luft (1961, as prepared by Dr. R. L. Wood), involving a three-stage curing programme which results in low heat distortion point of the resin, reflecting low cross linkage. In this method, *n*-butyl phthalate was omitted and DMP-30 (2,4,6-tri (dimethylaminomethyl) phenol) was used as accelerator. The mixture contained, Araldite 502, 27 cm^3; DDSA, 23 cm^3, and DMP-30, 1·5–2 per cent V/V added just before use. The first two ingredients can be mixed and stored for several weeks, but in moisture-free conditions.

The material is dehydrated in ethanol with one change of absolute ethanol and then passed through two changes of propylene oxide, keeping 10–15 min in each. The evaporation must be checked to prevent drying of the specimen and inhalation of the vapour. After the removal of the second change, 1–2 cm^3 of fresh propylene oxide is added. To it, is added an equal volume of the resin mixture. After 1 h, another equal quantity of resin mixture is added and infiltration is allowed to take place for 3–6 h, or even longer. The material is then placed into a gelatin capsule with as little of the propylene oxide–Araldite mixture as possible. The capsule is then filled up with pure resin mixture. The capsule, with material, is kept over-night each successively at 35 °C, 45 °C and 60 °C for complete polymerisa-tion. Alternatively, it can be kept directly at 60 °C for 12 h.

The precautions to be followed are: the tissue must be cut into very small bits, between 0·1–0·2 mm in one dimension, and the glassware used should be expendable since, though liquid Araldite is soluble in ethanol, after setting, it is insoluble in organic solvents.

Anderson and Doane (1967) published a method for epoxy embedding of thin layer materials in commercially available vinyl cups. Embedding of cell cultures, impression smears, and settled cell suspensions are performed in cups, from culture, fixation, and dehydration to embedding.

Epon 812 is another glycerol based aliphatic epoxy resin introduced by Kushida (1959), Finck (1960) and Luft (1961) with many more desirable qualities than Araldite. With epon embedding, the sections show very strong contrast using osmium fixation alone, without any metal staining. It has a very low viscosity, coupled with a very rapid penetrating capacity. In this procedure, another curing agent used in addition to DDSA, is MNA (methyl endomethylene tetrahydrophthalic anhydride), as otherwise the blocks remain soft. Two sets of solutions, using different hardeners with epon 812, are prepared and mixed in the requisite proportions before use, following a flexible schedule. In the original method of Kushida (1959), DDSA was used as the hardener with epon 812 and 815 in different proportions, whereas in Finck's schedule (1960), a plasticiser was added to control the hardness. In general, epon embedding allows a wide range of block hardening and it may be considered as the most widely used embedding medium.

In the method evolved by Luft (1961) for epon embedding, the materials used were: Epon 812 from the Shell Chemical Corporation, San Francisco; dodecenyl succinic anhydride (DDSA) and methyl nadic anhydride (MNA) from the National Aniline Division of the Allied Chemical and Dye Corporation, New York, and trimethyl amino methyl phenol (DMP-30 accelerator) from Rohm and Haas, Philadelphia. Two mixtures were prepared, namely, (*a*) Mixture A containing epon 812, 62 cm^3 and DDSA 100 cm^3, (*b*) Mixture B containing epon 812, 100 cm^3 and MNA 89 cm^3. The former gives soft blocks and the latter hard ones, and they are mixed in different proportions depending upon the degree of hardness required, a proportion of A:B = 2:1 being recommended for general use. The two mixtures are stored separately in the cold; 1·5 per cent V/V of accelerator (DMP-30) is added and A, B and DMP-30 are mixed very thoroughly just before use.

Fixation of the tissue is done as usual in osmium tetroxide, followed by dehydration in ethanol, it is then passed through two successive changes of propylene oxide to accelerate infiltration, keeping to 15–30 min in each case. The tissue is then treated for 1 h in a mixture of resin-accelerator mixture and propylene oxide (1:1); for 20–30 min in 100 per cent resin-accelerator mixture and then finally transferred to gelatin capsules nearly filled with the mixture. The resin is allowed to set either at 60 °C overnight, or kept at 35 °C, 45 °C and 60 °C successively for 12 h in each.

According to Luft (1961), preparation of the mixture is the most crucial factor. Measuring out the mixture in a 10–15 cm^3 conical graduated tube and continuous stirring for 5 min is recommended. After mixing A and B, DMP-30 should be taken in a tuberculin syringe with a long, large-bore needle and then added to the above mixture in the required quantities. Propylene oxide can be handled in a similar syringe. In a modified schedule of Porter (1964), instead of a three-stage incubation programme, vacuum incubation at 60 °C is carried out for 24–36 h, during which time the excess propylene oxide is also evaporated off.

Maraglas-655 was first used by Freeman and Spurlock (1962) as an embedding medium. It is a clear epoxy-resin having a viscosity of 500 Hz at 25 °C in the liquid, uncured state. Curing to a solid state, with a heat distortion point at 87·8 °C is obtained after keeping it for 72 h at 60 °C. It is not miscible with ethanol but mixes readily with acetone, propylene oxide and styrene and can be stored at 22–24 °C. It yields excellent ultra thin sections without chatter and without the granular background obtained with epon-embedded materials. It has a wide range of miscibility, stability against electron beam, and a low viscosity, though its preservation quality is similar to that of epon.

To obtain blocks of suitable consistency, a monoepoxide, Cardulite NC 513 is also used with Maraglas. Since this amber-coloured liquid resin has a viscosity of 50 Hz, it reduces the viscosity of maraglas–cardulite mixture and thus helps penetration. It is compatible with propylene oxide. Storage in the dark (25 °C) is necessary as otherwise there is an accumulation of metallic deposits. Two additional reagents used in this schedule are benzyldimethylamine, as the curing agent, and dibutyl phthalate as plasticiser controlling the hardness of the block.

Spurlock *et al.* (1963) suggested the formula of the resin mixture as: maraglas-655, 68 cm^3; cardulite NC 513, 20 cm^3; dibutyl phthalate, 10 cm^3; benzyl dimethyl amine, 2 cm^3. The schedule for the material includes the following steps:
(a) Dehydration successively in 50, 70 and 95 per cent ethanol.
(b) Two changes of 15 min each in absolute ethanol.
(c) Propylene oxide–resin mixture (1 :1) for 30 min.
(d) Resin mixture—treatment for 1 h.
(e) Transfer to fresh resin mixture in dried gelatin capsules, hardening for 48–72 h at 60 °C.
Staining of the sections is recommended with lead hydroxide, uranyl acetate, etc. Prolonged storage of the capsules causes accumulation of air and softening.

Vestopal W—Ryter and Kellenberger (1958) and Kellenberger *et al.* (1956) used this polyester resin in electron microscopy specially because of its rapid penetrability, resistance against electron bombardment and polymerising capacity. It is generally prepared by esterifying maleic anhydride with glycerol or some other polyhydric alcohol. Since it is not soluble in ethanol, acetone is used for dehydration. It is used in combination with 1 per cent tertiary butyl parabenzoate as initiator and 0·5 per cent cobalt naphthenate as activator. All these compounds require storage in a cool dark place and the initiator and activator should be replenished every 2 months, since their inactivation causes softness of the block. During the process, acetone should be removed completely and passage through acetone–Vestopal W mixtures should be slow and gradual as otherwise the block will be porous and the material will shrink. Polymerisation in vacuum is preferable. The schedule is given below:
(a) Dehydration is carried out successively in 30, 50 and 75 per cent acetone for 15–30 min in each, followed by 30–60 min successively in 90 and 100 per cent (dried over $CuSO_4$) acetone.
(b) The tissue is treated successively, for 30–60 min each, in mixtures of dry acetone and Vestopal W 3:1, 1:1 and 1:3, followed by 12–24 h in Vestopal W—1 per cent initiator and 0·5 per cent activator-mixture, with repeated changes to prevent polymerisation.

(c) The tissue is finally transferred to gelatin capsules filled with the last mixture and allowed to polymerise at 60 °C for 12–24 h.

In preparing the embedding mixture, the initiator is mixed thoroughly with Vestopal W before adding the activator. The mixture can be kept only for a few hours at room temperature.

Water Soluble Embedding Media

Quite a number of water soluble embedding media have been devised, with the advantage that aqueous fixatives meant for cytochemical work can be used for the selective extraction of chromosome components. Otherwise the use of strong solvents may damage the cytochemical and enzymatic patterns. The embedding medium itself may further serve as the dehydrating agent. Such water soluble media include Aquon (Gibbons, 1959, 1960)—the water soluble constituent of epon; glycol methacrylate— the ethylene glycol ester of methacrylic acid (Rosenberg *et al.*, 1960) and Durcupan—an aliphatic polyepoxide (Stäubli, 1960). The chemical structures of these compounds, as studied on a comparative basis by Leduc and Bernhard (1962) are:

Glycol methacrylate   Methyl methacrylate

$CH_2 = C\!-\!COOCH_2\cdot CH_2OH$ $CH_2 = C\!-\!COOCH_3$

$\qquad\quad |$ $|$

$\qquad\quad CH_3$ CH_3

Epoxy resins : General formula

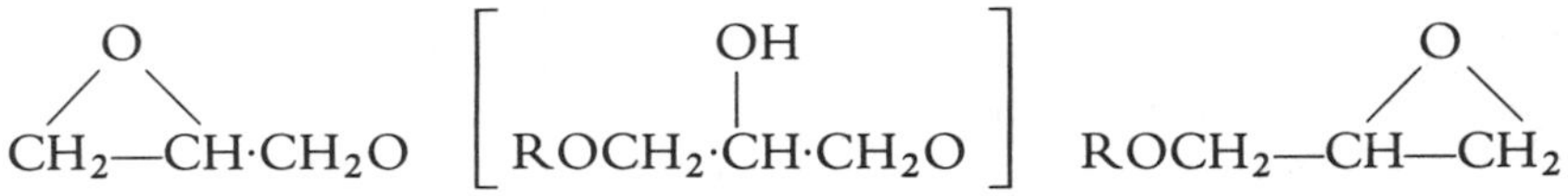

$$\underset{CH_2-CH\cdot CH_2O}{\overset{O}{\triangle}} \left[\underset{ROCH_2\cdot CH\cdot CH_2O}{\overset{OH}{|}} \right] ROCH_2-\underset{}{\overset{O}{\triangle}}CH-CH_2$$

Where R is usually

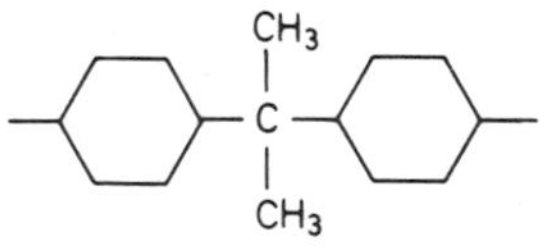

Diphenyl propane

The use of Durcupan is rather limited as it allows only acid hydrolysis and digestion by proteases (Granboulan and Bernhard, 1961; Bernhard, 1966; Bernhard and Granboulan, 1968). Aquon and glycol methacrylate have a comparatively wider use. In addition, 2-hydroxypropyl methacrylate has been used for the same purpose by Leduc and Holt (1965).

Aquon is a colourless resin prepared (Gibbons, 1959) by extraction of epon 812 with two volumes of water, separation through salting with sodium sulphate and finally drying off the residual water in a vacuum desiccator. It has a low viscosity and at a low temperature (15 °C), is soluble in water. Polyepoxides like aquon form intermolecular linkages with proteins and nucleic acids and may to some extent serve the purpose of fixation. Stacey *et al.* (1958) demonstrated that carbonyl, amino and imidazolyl groups of proteins and sulphydryl groups of denatured proteins are reacted upon by epoxides. They further esterify phosphate groups and nitrogenous bases.

For the preparation of the embedding medium, aquon (10 cm^3) is mixed with a hardener—DDSA (25 cm^3), and accelerator benzyldimethylamine (0·35 cm^3). The material is fixed in 10 per cent formaldehyde–veronal acetate buffer (pH 7·3) at 3–4 °C and washed. It is dehydrated through increasing concentrations of aquon in water to pure aquon at 4 °C and finally kept immersed in the embedding mixture for 4 h. Curing is performed by transferring the material to a gelatin capsule with fresh embedding mixture and keeping it at 54–60 °C for 4 days. Ultra thin sections can be cut conveniently with glass knives in a Porter-Blum microtome. Silver interference colour in distilled water can be checked and staining does not require the removal of the embedding medium.

Glycol methacrylate (GMA)—is a hygroscopic, colourless liquid, readily miscible with water, ether and ethanol. Ethylene glycol monomethacrylate is prepared either by re-esterification of methylmethacrylate or directly by esterifying methacrylic acid (Rosenberg *et al.*, 1960). In the former procedure, 0·5 per cent triglycoldimethacrylate is added to glycol methacrylate before polymerisation to secure a three-dimensional polymer. In the latter method, side products ultimately result in the formation of such a polymer. It has a boiling point of 89–92 °C at 7 mmHg, density 1·065 (at 20 °C), viscosity 0·701 poise (20 °C), and refractive index 1·4540.

In cytological procedures, dehydration of fixed or frozen sections can be performed directly by increasing concentrations of GMA, polymerisation being induced by 0·5–1 per cent ammonium persulphate. Leduc and Bernhard (1962) modified the above method slightly for cytochemical procedures, including polymerisation at 37°C like Durcupan and involving extraction of specific cellular components, such as, DNA, RNA, basic and non-basic proteins. The schedule is outlined as follows:

(a) Pancreas of young rats were fixed in 10 per cent formaldehyde buffered at pH 7·3 with Michaelis' veronal acetate buffer at −3 °C for 15 min. The tissue was cut into pieces, not more than 1 mm thick. 10 per cent formol buffered with Sorensen's phosphates also serves as a good fixative.

(b) Embedding is carried out by passing the tissue through (*i*) 20, 40, 60 and 80 per cent glycol methacrylate (GMA) in distilled water—15 min in each; (*ii*) 97 per cent GMA, 30–60 min; (*iii*) 97 per cent GMA and embedding mixture, mixed in the proportion of 1:1—30 min. The tissue is then transferred to a dry gelatin capsule containing the embedding mixture [70 parts 97 per cent GMA plus 0·5 per cent ammonium persulphate, and 30 parts of methacrylate (85 per cent butyl, 0·5 per cent methyl) containing 1 per cent Luperco (benzoyl peroxide)].

(c) Polymerisation takes place at 37 °C.
(d) Sectioning of GMA is possible if the knife is slightly inclined at 3–4 degrees. Sections spread very rapidly in water at the knife edge and change in colour from delicate gold to silver.
(e) Sections are then floated in hydrolysing solution without supporting grids; after incubation, floated in distilled water, picked up on formvar-coated grids and stained for 30 min in a filtered aqueous saturated solution of uranyl acetate (pH 4·0). The thick sections are dried on glass slides and stained by the routine Feulgen method for DNA and 0·25 per cent methylene blue for 2 min for cytoplasmic basophilia.
(f) The hydrolysing solutions contain:
  (i) 0·5 and 0·1 per cent pepsin in 0·1 N HC1 (Nutritional Biochemicals Corpn.)
  (ii) 0·3 per cent trypsin in distilled water, subsequently adjusted to pH 8·0 with 0·01 M NaOH.
  (iii) 1, 0·2 and 0·1 per cent desoxyribonuclease (Worthington Biochem. Corpn.) in 0·003 M, 0·0006 M and 0·0003 M $MgSO_4$ respectively and 2 per cent desoxyribonuclease in distilled water, all subsequently adjusted to pH 6·2 to 6·6 with 0·01 M NaOH (Swift, 1965).
  (iv) 1 and 2 per cent ribonuclease in distilled water adjusted to pH 6·8 with 0·01 M NaOH (Swift, 1959). All incubations are carried out at 37 °C for 1–2 h in a watchglass. Sections may be handled by a moistened narrow wooden stick.

Though both osmium and formaldehyde fixations yielded good results, the latter however, was found to be more suitable.

SECTIONING

Ultramicrotome

For ultra thin sectioning (0·1–0·01 µm), several models of ultramicrotomes are at present available, the earliest attempt in ultra thin sectioning by von Ardenne dating back to 1939. A detailed historical account has been given by Wachtel, Gettner and Ornstein (1966). The first suitable model was manufactured by Ivan Sorvall Inc. in 1953, commonly known as the Porter–Blum microtome (Sorvall—MT-1). In general, the principle of its operation was based on a system of screw thread, lever arm and proper bearings. When necessary, it could be driven by a motor and due to its simplicity of working, did not require a complicated maintenance process. In the slightly improved model, the specimen holder, which is of a collet type screwed at the free end of the aluminium rod, is moved vertically across a glass or diamond knife. On the return stroke, the specimen end of the aluminium rod is allowed to follow the trajectory of a parallelogram. Coarse and fine adjustment screws are provided for thick sectioning, while for ultra thin sections, there is a mechanical device to advance the block towards the knife. To induce thermal expansion of the aluminium rod, an electric lamp is used. In a later model (Sorvall—MT-2), the whole operation is much more compact; the knife stage allows controlled motion of the knife, permitting proper trimming for ultra thin sections. Two controls are provided to adjust the thickness of sections between 100 Å to 4 μm.

In the Huxley model (1959, Cambridge Instrument Co.), steel leaves are employed for hinging the arm. It is based on a mechanical advance with a double leaf spring suspension system. Here, though the gravitational pull is responsible for section cutting, one oil-filled dashpot controls the rate of downward movement of the cutting arm (*see* Wachtel *et al.*, 1966).

In the ultramicrotome of LKB-Producer AB-Stockholm, fluctuation in section thickness is eliminated to a significant extent, the principle being based on a thermal advance system. A cantilever arm is the principal moving part, one end of the arm holds the specimen block, the other end is attached to a leaf spring joined to the base of the microtome. This spring causes the up and down motion of the bar. Thermal control of the cutting arm guides the advance of the block against the knife. The gravitational force controls the cutting stroke and a motor regulates the motion and the upward movement. An electromagnetic force which acts during the return stroke, causes the flexing of the base below the knife holder necessary to ensure the by-pass of the cutting surface and knife edge during the return stroke.

Other thermal advance microtomes have been designed by Sjöstrand (1953), Philpott (1955), Fernandez–Moran (1956) and Hellström (1960), a recent one being the Reichert Om U$_2$ model. The working principles of all the different ultramicrotomes have been discussed in detail in an excellent review by Wachtel *et al.* (1966).

Knife

Of the different types of knives used, including hard steel (Ekholm *et al.*, 1955), diamond (Fernandez–Moran, 1956) and glass (Latta and Hartmann, 1950), the latter is the most convenient for ultra thin sectioning and is, therefore, widely used. Due to the short life of the glass knife, a diamond one is sometimes recommended: this is often useful for epon embedded materials. But the problem of resharpening and the disadvantage of securing a satisfactory meniscus to receive sections, because of its hydrophobic property, have resulted in only a limited application of the diamond knife. A common method for preparing the glass knife, as developed by Porter (1964) is outlined as follows.

Make a 1·27 cm score mark with a sharp cutter on a clean 20 × 20 cm sheet of plate glass, at right angles to the base of the glass plate. Position the scored edge of the plate to overlap the edge of the working surface by about 0·63 cm. Keep the edge of the glass parallel to the edge of the table. Take a pair of glass breaking pliers with wide parallel jaws. (A narrow strip of adhesive tape is placed on the inner surface, from the cutting edge to halfway to the middle of the bottom jaw. Two lateral strips are placed at the edges of the inner surface of the top jaw. Inner surfaces of both jaws are then covered with wide pieces of adhesive tape, smoothly.) Keep the jaws open and with the central piece of tape of the bottom jaw centred beneath the score mark on the glass, push the face of the bottom jaw flush against the table. Gently squeeze the pliers to produce a slow, even and straight break, with two smooth new surfaces, which will be free of artefacts except for the short line where the initial score was made. Turn the two pieces of glass through a right angle so that the smooth edges are away from the table edge. Score one piece of glass in the centre of the old long edge and repeat the procedure to have two 10×10 cm plates. Repeat the process till $2·54 \times 2·54$ cm squares are obtained, each with at least two smooth edges meeting at a 90 degree angle. Choose the best adjacent edges for the final break. Start a diagonal score 1 mm or so from the apex of the angle where the faces meet and extend to bisect the opposite corner. Carefully centre the pliers halfway along this line and gently increase pressure till the glass breaks to give a triangular knife. The good knife should have an even and straight cutting edge, an absolutely flat front surface and a back face with either a right or left-handed configuration when viewed from above. The part of the knife edge closest to the top of the arc formed by the back surface is best for thin sectioning. A good 45 degree angle knife is usually suitable for cutting tissue embedded in media of average hardness.

Trough

In order to receive sections after cutting in an ultramicrotome, troughs are prepared in various ways (Claude, 1948; Gettner and Hiller, 1950). It is necessary because after the sections are cut, ultra thin sections have a tendency to adhere to the dry knife and collection becomes difficult. The trough is an integral part of the section-cutting equipment for diamond knives. With a glass knife, the usual procedure is to prepare a trough with adhesive-backed cloth or paper tape, which is disposable. The exposed adhesive surface of the trough is coated with paraffin to prevent contamination with the trough liquid. It is sealed to the glass with melted paraffin.

Certain prerequisites are necessary in the liquid in the trough. It should be able to detach the sections from the knife, eliminate all electrostatic charges, spread the sections through solvent action, and should have an adequate surface tension to penetrate the layer between section and knife facet. Gettner and Hillier (1950) first introduced a method of trough and flotation fluid, the section being floated in a fluid which wets the glass without wetting the methacrylate. Different concentrations of acetone, preferably 40 per cent, may be used as the fluid. In order to stretch the sections properly, methacrylate blocks could be softened with a proper solvent, such as xylene, applied with a camel-hair brush to the sections as they float in the trough.

In general, several surface tension reducing mixtures serve as good trough fluids. Acetone in 10–40 per cent concentration may be used. It allows rapid relaxation of compression after sectioning, especially in the case of polymethacrylates. High concentration is avoided as it makes the methacrylate soft. Water may be used but relaxation takes a little longer. Ethanol and dioxane, in 20–60 per cent concentrations, are sometimes used. Ethylene glycol, silicone fluids and glycerine solutions are the other chemicals utilised for this purpose. The higher surface tension fluids are useful for epoxy and polyester embedded materials as well, after the knife has been moistened with water. Pease (1965) recommended the use of glycol for this purpose for hydroxypropyl methacrylate embedded materials. In epon-embedded materials, trichloroethylene has been found to be suitable (Porter, 1964). The fluid must have a slightly convex meniscus and the level in the trough should be just below that of the cutting edge, so that the specimen block is not wetted.

Trimming and Sectioning

In the actual procedure, the cutting edge of the knife, the embedding material, the cutting face of the block and operating speed are the principal controlling factors.

In case of hard epoxy-embedded blocks, fine files and jeweller's saws are required for trimming, followed by a final finish with an acetone or chloroform-washed razor blade to ensure that it is free from oil. To prepare the block for sectioning, the side walls of the portion delimited from the tissue should be trimmed—a surface area about 0·3 mm × 0·08 mm is usually desirable. The block should be oriented in the microtome with the long side parallel to the knife edge. After trimming, the final shape of the block should be that of a truncated pyramid or, in the case of larger materials, like a roof-top. Normally the cutting face should be square but some people prefer a trapezoidal block, having asymmetry, with the longer axis oriented towards the cutting face, the upper and lower edges being parallel (Porter, 1964). In general, one side of the pyramid, or the long face of the roof top-shaped top is adjusted parallel to the knife edge.

Pease (1960) suggests the provision of an added support with a superficial layer of hard wax to protect the tissue from direct contact with the knife during practice. The blocks are dipped in a filtered mixture of Carnauba wax and paraffin (1:2) and kept at 80 °C. The tissue specimen can be oriented by mounting the block on a holder made of a wooden dowel rod 5–16 mm in diameter, which fits well in a Porter-Blum microtome. At times, the orientation of the block in relation to the sectioning plane must be precise. In such cases, after appropriate trimming, the block can be stuck with liquid epoxy against a second block and the microtome chuck can be adjusted slightly.

After the block is fitted in the chuck, the front edges of the jaws should clamp it strongly and the projecting portion alone should not be more than 3–4 mm. The knife should be tilted so as to have a 1–3 degree clearance angle and a rake (knife) angle of about 30 degrees (Sjöstrand, 1969). The factor of the rake angle is important for the prevention of chatter, which in addition to a small knife angle, is also accelerated by a very thin block tip, and by rapid movement of the cutting face. The 'chatter' is caused by the vibration in the block, microtome arm or knife edge. In any case, the experienced worker, when using the microtome, can easily devise his own methods for preventing chatter caused by any of the above factors.

The entire process of sectioning should be performed very gently at uniform operational speed. To ensure proper sectioning, it is preferable to check the entire sectioning operation with a block of pure plastic, without the material.

When the sections have been cut, the ribbon can be detached from the knife edge and transferred to a trough in which the level of the liquid is controlled with a hypodermic syringe fitted at the base with a plastic tube (Gay and Anderson, 1954). The level of the fluid is generally maintained over the knife edge, forming a well rounded meniscus, and the ribbon can be detached with a fine-hair brush. In order to estimate the thickness of the section correctly, it is always preferable to use reflection of interference colours while the sections are floating in the trough (Porter and Blum, 1953). The light should be adjusted to allow total reflection on the liquid surface. Peachey (1958) published a detailed account of the thickness of the sections and the corresponding interference colours, the former ranging from 600 to 3200 Å and the latter from grey to yellow. A satisfactory method of observing sections by reflected light is by a fluorescent lamp.

Table 10   INTERFERENCE COLOURS AND CORRESPONDING SECTION THICKNESS (ACCORDING TO WACHTEL *et al.*, 1966

Section thickness in μm	Interference colours		
	n-butyl polymethacrylate		*Vestopal W Walter (1961)*
	Martin and Johnson (1951)	*Peachey (1958)*	
0·013	Iron grey		Grey
0·033	Lavender grey	Grey	Silver grey
0·053	Bluish grey		White
0·073	Clearer grey		
0·079	Greenish white		
0·087	White	Silver	
0·090	Yellowish white		
0·094	Straw yellow		Yellow
0·103	Light yellow		
0·111	Bright yellow	Gold	
0·144	Brownish yellow		
0·169	Reddish orange		
0·180	Red		
0·185	Deep red	Purple	Copper
0·190	Purple		
0·198	Indigo		
0·223	Sky blue	Blue	
0·244	Greenish blue		
0·277	Light green	Green	
0·285	Yellow green		
0·305	Yellow	Yellow	
0·317	Orange		
0·370	Purplish red		
0·380	Bluish violet		

Sjöstrand (1969) observed that ultra thin sections suitable for high resolution work generally appear dark grey in reflected light while for lower resolution work, a section thickness of 700 Å with a silvery shine in reflected light is desirable.

MOUNTING

The removal of the ribbon from the fluid needs special care. For examination under the electron microscope, the ribbon must be mounted on a specimen grid with a backing film (Parlodion and Formvar are the common films used for this purpose because they provide good supporting media, being composed of light atoms). The essential requirements are satisfied as they dissove quickly and become tough when the solvent evaporates. Parlodion is the trade name of nitrocellulose plastic (prepared by Mallinck Rodt Chemical Works, St. Louis). Polyvinyl formal plastic of Shawinigan Products Co., New York is called Formvar. Gay and Anderson's method (1954) appears to be suitable for serial sectioning. The principal implement is a thin film of Formvar supported by a small wire, and these Formvar-coated loops can be inserted in the liquid of the trough in a tilted position. By suitable adjustment, the ribbon can be centred across the diameter, and when the loop is raised, the sections adhere to the Formvar, after which they can be directly transferred to the supporting grids for examination. The grids are placed on a combination of transparent plastic discs, fitted on the top of an adjustable condenser in a standard microscope, and by lowering the condensers, the grid can be kept below the stage and the ribbon can be suitably arranged. Contact is achieved by lifting the condenser. Pease (1960) states, however, that the one disadavantage of this method is that it requires a heavy supporting grid.

Sections can also be mounted directly on large arc hole grids as shown by Sjöstrand (1958, 1969), in a modified version of Gay and Anderson's (1954) method. Another method was developed by Galey and Nilsson (1966) (also see the chapter on high resolution autoradiography). Epon embedded materials can be mounted on 300 mesh copper grids (Afzelius, 1962). For araldite-embedded materials, the sections do not require any support for mounting; even a carbon film may serve the purpose.

STAINING

Significant improvement in securing adequate contrast in ultra thin sections has been achieved by the use of suitable staining procedures (Trump *et al.*, 1961). The tissue is exposed to heavy metal salts to form metal ion complexes with nuclear components. The formation of such complexes increases the image density in electron micrographs. Formation of cross links by oxygen bridges and also —CH_2 bridges with fixed materials has been suggested as the basis of staining (Afzelius, 1962). Positive staining technique involves treatment with components which increase the weight density, whereas in negative staining, the material is surrounded with a structureless material of high weight density. Valentine and Horne (1962), from an assessment of the negative staining techniques, demonstrated that good negative staining can be obtained with sodium tungstate, uranium nitrate or disodium hydrogen phosphate.

The number of staining methods available is increasing gradually and several lead salts, which give a stable staining, such as hydroxide (Watson, 1958), cacodylate (Karnovsky, 1961), tartarate (Millonig, 1961), citrate (Reynolds, 1963) as well as uranyl acetate followed by lead citrate (Reynolds, 1963; Venable and Coggshell, 1965), etc. are applied for securing adequate contrast. Uranyl acetate has been claimed to alter the properties of the molecules so that they bind better with lead citrate. The period of lead staining is generally short, to prevent the formation of lead deposits.

Staining with uranyl nitrate may be performed, using a filtered aqueous saturated solution at pH 4·0. For combined staining, the section on the grid may first be moistened with a drop of distilled water, followed by staining (in an inverted position) in 7·5 per cent uranyl acetate solution for 20 min at 45 °C. The sections are dried on filter paper, again moistened and finally stained with 0·2 per cent aqueous lead citrate solution for 10–60 s (Reynolds, 1963; Venable and Coggshell, 1965). In Salpeter's method (1966), aqueous uranyl acetate or uranyl nitrate staining may be followed by a few drops of lead citrate, before floating off the excess stain with water. The stains are usually selected for specific intracellular components, depending on their stability and non-formation of deposits.

APPLICATIONS OF ELECTRON MICROSCOPY

It is hard to over-estimate the contributions of electron microscopy, as in the study of the chromosome, its ultrastructural details could only be clarified through this method, aided by autoradiography. Yasuzumi and colleagues (1951) studied the origin of the threadlike chromonema isolated from the metabolic nucleus of *Drosophila* and reported that interband fibrils have a periodicity of 300 and 600 Å. In human leucocyte nuclei, double-stranded helices were obtained in threads (*see* Denues, 1953). Holfman-Berling and Kausche (1951) claimed that interphase chromosomes of chicken erythrocytes are polytenic in nature.

Kaufmann (1960), on the basis of observations on ultra-structural details, claimed that the chromosomes of higher plants and animals are multistranded in nature. Fine fibrillar elements can be observed even in the half-chromatids. The diameters of the fibrils vary from 500 Å to 30 or 40 Å. Sixty four fibrillae have been counted in the chromosomes of *Steatococcus* spermatozoon (Nebel, 1959) and staminate hair cells of *Tradescantia* (Kaufmann and De, 1956). Ris (1957) showed that meiotic prophase chromosomes of different insects and plants, as well as lampbrush chromosomes of *Triturus*, etc., consist of a bundle of coiled microfibrils 500 Å thick (*see* also Ris, 1957). Gay (1955b) has shown that the salivary gland chromosome fibril is 200–500 Å thick. Mirsky and Ris (1951) have suggested that sub-units in chromosomes may vary. Bernstein and Mazia (1953, and *see* Mazia, 1954) isolated nucleoprotein from sea urchin sperm and found that the fibres were 200–300 Å thick. Shinke (1959) observed elementary chromosome fibrils, 100–200 Å thick, in the chromonemata of metabolic nuclei of several higher plants.

Ris (1962) considered 100 Å nucleohistone fibrils as the main constituent of chromosomes. These molecules are claimed to be made up of two 40 Å thick nucleohistone macromolecules, held together by histone linkers. In sperms, where the histone is replaced by other basic proteins, the 40 Å fibrils can be observed individually. Indirect evidence has been brought forward to show that nucleohistone fibrils are linked end to end through a non-histone protein. In *Escherichia coli*, the semi-conservative replication of DNA has been demonstrated and the entire bacterial chromosome replication has been worked out (Cairns, 1963; Person and Osborn, 1964; *see* Bleecken *et al.*, 1966).

Interesting evidence has been obtained in the Dinoflagellates regarding the evolution of chromosome structure (*see* Ris, 1962). The members of this group are claimed to have an intermediate structure between viruses and bacteria. Sausage-shaped chromosomes in the interphase nucleus contain fibrils 30–80 Å thick instead of 100 Å fibrils. The observations indicate that chromosomes are formed of 25 Å thick fibrils surrounded by a dense coarsely granular substance. The essential structure of the chromosomes contains DNA and is devoid of any basic protein. Similarly, in the blue-green alga, *Anabaena* sp., fibrils about 25–30 Å in diameter, corresponding to DNA macromolecules, have been found in nucleoplasmic areas (Leak, 1965). Significant data are being gathered through ultra-structural studies, throwing light on the evolution of the complexity in gene-bearing structures (Swartzen-druber and Hanna, 1965; Gaudecker, 1966; Wolstenholme, 1966; Nougarede and Bronchart, 1967 and *see* Schultze, 1969). The structure of the mitotic apparatus has also been clarified through electron microscopy, and is now considered to represent a system of filaments embedded in the compact vesicular mass, possibly including the ribosomes but excluding mitochondria and others (Harris and Mazia, 1962 and *see* Wada, 1966).

In addition to the chromosome details, considerable insight has been gathered regarding the mechanism of gene action as controlled by chromosomes through nucleo-cytoplasmic transfer (Kaufmann and Gay, 1958). In the field of chromosomal control of protein synthesis, RNA synthesis has been localised in the extended state of chromatin, whereas the condensed stage has been shown to represent inactivity in isolated calf thymus nuclei (Littau *et al.*, 1964), in mouse hepatic cells (Noorduyn and de Man, 1966) and cultured monkey kidney cells (Granboulan and Granboulan, 1964). The dynamic role of the nuclear membrane in affecting the nucleo cytoplasmic transfer has been clearly revealed. Electron microscopy offers immense scope, not only in the study of chromosomes, but also in providing a method of understanding the mechanism of the action of genes—its constituent units.

Lastly, in the study of cell fractionation and the identification of isolated units, the·contributions of electron microscopy are immense. It is now possible to identify subcellular particles in thin sections of centrifuged pellets, which could not have been monitored by previous methods (*see* Haggis, 1967). In spite of the phenomenal progress in the study of ultra-structural details, there is still ample scope for further improvements. Even now, the progress in solving the problem of lens aberrations has not been very satisfactory (Cosslett, 1967), though the resolving power may gradually be raised to 1 Å after necessary corrections for mechanical and electrical stability. Refinements in specimen preparation need to be developed to utilise fully the 1 Å resolution through which the nucleic acid bases may ultimately be identified. If the rate of progress is an index, success in this line in the near future is assured.

REFERENCES

Afzelius, B. A. (1962).   In *The interpretation of ultrastructure, Sym. Int. Soc. Exp. Biol.* **1,** New York; Academic Press
Anderson, N. and Doane, F. W. (1967).   *Stain Tech.* **42,** 169
Bahr, G. F. (1954).   *Exp. Cell Res.* **7,** 457
Baker, J. R. (1966).   *Cytological technique,* 5th ed. London; Methuen
Baker, J. R. and McCrae, J. M. (1966).   *J. Roy. Microscop. Soc.* **85,** 391
Bernhard, W. (1966).   *Natl. Cancer Inst. Monograph* **23,** 13
Bernhard, W. and Granboulan, N. (1968).   In *The Nucleus, Ultrastructure in biological systems* **3,** 80, New York; Academic Press
Bernstein, M. H. and Mazia, D. (1953).   *Biochem. biophys. Acta* **10,** 600
Bleecken, S., Strohbach, G. and Sarfert, E. (1966).   *Z. Allgem. Mikrobiol.* **6,** 121
Bowes, J. H. (1963).   *A fundamental study of the mechanism of deterioration of leather fibres,* Brit. Leather Manuf. Res. Assoc. Rep.
Bullivant, S. (1965).   *Lab. Invest.* **14,** 1178
Burstone, M. S. (1969).   In *Physical techniques in biological research* **3C,** 1, New York; Academic Press
Busch, H. (1926).   *Ann. Physik.* **81,** 974
Cairns, J. (1963).   *J. Mol. Biol.* **6,** 208
Casselman, W. G. B. (1955*a*).   *Quart. J. micr. Sci.* **96,** 203
Casselman, W. G. B. (1955*b*).   *Quart. J. micr. Sci.* **96,** 223
Cater, C. W. (1963).   *J. Soc. Leath. Trades. Chem.* **47,** 259
Caulfield, J. B. (1957).   *J. Biophys. Biochem. Cytol.* **3,** 827
Chuble, G. T. and Ellwanger, P. W. (1967).   *Stain Tech.* **42,** 213
Claude, A. (1948).   *Harvey Lectures Ser.* **43,** 121
Clements, R. L. (1962).   *Anal. Biochem.* **3,** 87
Cope, G. H. (1968).   *J. Roy. Microscop. Soc.* **88,** 235
Cosslett, V. E. (1967).   *J. Roy. Microscop. Soc.* **87,** 53
Dalton, A. J. (1953).   *Int. Rev. Cytol.* **2,** 203
Denues, A. R. T. (1953).   *Exp. Cell Res.* **4,** 343
De Robertis, E. and Iraldi, P. A. (1961).   *J. Biophys. Biochem. Cytol.* **10,** 361
Ekholm, R., Hallén, O. and Zelander, T. (1955).   *Experientia* **11,** 361
Falck, B. and Owman, C. (1965).   *Acta Univ. Lund Sect.* II, No. 7
Fawcett, D. W. (1964).   In *Histology and cytology in modern developments in electron microscopy.* New York; Academic Press
Feder, N. and Sidman, R. L. (1958).   *J. Histochem. Cytochem.* **6,** 401; *J. Biophys. Biochem. Cytol.* **4,** 593
Fernandéz- Moran, H. (1956).   *J. Biophys. Biochem. Cytol.* **2,** Suppl. 29
Fernandéz-Moran, H. (1959*a*).   *J. Appl. Phys.* **30,** 2038
Fernandéz-Moran, H. (1959*b*).   *Science* **129,** 184
Fernandéz-Moran, H. (1961).   In *Macromolecular complexes.* New York; Ronald Press
Fernandéz-Moran, H. (1962).   In *Symposia of the International Society for Cell Biology* **1,** New York; Academic Press
Finck, H. (1960).   *J. Biophys. Biochem. Cytol.* **7,** 27
Freeman, J. A. and Spurlock, B. O. (1962).   *J. Cell Biol.* **13,** 437

Galey, F. and Nilsson, S. E. G. (1966). *J. Ultrastruct. Res.* **14,** 405
Gasser, H. S. (1955). *J. Gen. Physiol.* **38,** 709
Gaudecker, von B. (1966). *Z. Zellforsch. Mikroskop. Anat.* **72,** 281
Gay. H. (1955*a*). *Stain Tech.* **30,** 239
Gay, H. (1955*b*). *Ann. Arbor. Publ.* **11,** 407
Gay, H. and Anderson, T. F. (1954). *Science* **120,** 1071
Gersh, I. (1948). *Bull. int. Ass. med. Mus.* **28,** 179
Gersh, I. (1959). *J. Biophys. Biochem. Cytol.* **2,** Suppl. 37
Gettner, M. E. and Hillier, J. (1950). *J. Appl. Phys.* **21,** 68
Gibbons, I. R. (1959). *Nature* **184,** 375
Gibbons, I. R. (1960). *Proc. IV Int. Conf. Electron Microscopy* **2,** 65
Glauert, A. M. and Glauert, R. H. (1958). *J. Biophys. Biochem. Cytol.* **4,** 191
Glauert, A. M., Rogers, G. E. and Glauert, R. H. (1956). *Nature* **178,** 803
Glick, D. and Bloom, D. (1956). *Exp. Cell Res.* **10,** 687
Granboulan, N. and Bernhard, W. (1961). *Compt. Rend. Soc. Biol.* **155,** 1767
Granboulan, N. and Granboulan, P. (1964). *Proc. III European Reg. Conf. Electron Microscopy, Prague,*
   p. 31
Haggis, G. H. (1967). *The electron microscope in molecular biology.* New York; John Wiley
Harris, P. and Mazia, D. (1962). In *The interpretation of ultrastructure.* New York; Academic Press
Hellström, B. (1960). *Sci. Tools* **7,** 10
Holfman-Berling, V. H. and Kausche, G. A. (1951). *Z. Watentersch* **66,** 63
Holt, S. J. and Hicks, R. M. (1962). In *The interpretation of ultrastructure.* New York; Academic Press
Huxley, A. F. (1959). Cambridge Instrument Co. Technical Brochure
Huxley, H. E. and Zubay, G. (1961). *J. Biophys. Biochem. Cytol.* **11,** 273
Karnovsky, M. J. (1961). *J. Biophys. Biochem. Cytol.* **11,** 729
Kaufmann, B. P. (1960). *The cell nucleus,* p. 251, London; Butterworths
Kaufmann, B. P. and De, D. N. (1956). *J. Biophys. Biochem. Cytol.* **2** (Suppl.), 419
Kaufmann, B. P. and Gay, H. (1958). *Nucleus* **1,** 57
Kellenberger, E., Schwab, W. and Ryter, A. (1956). *Experientia* **12,** 421
Koltz, I. M. (1962). Referred to in Afzelius (1962)
Kushida, H. (1959). *Electron microscopy* **8,** 72
Latta, H. and Hartmann, J. F. (1950). *Proc. Soc. Exptl. Biol. Med.* **74,** 436
Leak, L. V. (1965). *J. Ultrastruct. Res.* **12,** 135
Leduc, E. H. and Bernhard, W. (1962). In *Symposia of the International Society for Cell Biology,* 1 New
   York; Academic Press
Leduc, E. H. and Holt, S. J. (1965). *J. Cell Biol.* **26,** 137
Littau, V. C., Allfrey, V. G., Frenster, J. H. and Mirsky, A. E. (1964). *Proc. Natl. Acad. Sci. U.S.* **52,** 93
Lotke, P. A. and Dolan, M. F. (1965). *Cryobiology* **1,** 289
Luft, J. H. (1956). *J. Biophys. Biochem. Cytol.* **2,** 799
Luft, J. H. (1961). *J. Biophys. Biochem. Cytol.* **9,** 409
Maalöe, O. and Birch-Andersen, A. (1956). *Symp. Soc. Gen. Microbiol.* **6,** 261
Malhotra, S. K. (1968). In *Cell structure and interpretation,* p. 11, London; Edward Arnold
Maunsbach, A. B. (1966). *J. Ultrastruct. Res.* **15,** 242, 283
Martin, L. C. and Johnson, B. K. (1951). *Practical Microscopy.* New York; Chem. Publ. Co.
Mazia, D. (1954). *Proc. nat. Acad. Sci., Wash.* **40,** 521
McClintock, M. (1964). *Cryogenics.* New York; Reinhold
Meek, G. A. (1968). *J. Roy. Microscop. Soc.* **88,** 419
Menter, J. W. (1962). *Proc. Roy. Inst. Chem.* **86,** 415
Merck Index (1968). 8th ed., p. 17, Merck & Co. Inc., Rathway, USA
Millonig, G. (1961). *J. Biophys. Biochem. Cytol.* **11,** 736
Millonig, G. (1962). *Int. Congr. Electron Microscopy V Philadelphia* **2,** 8
Mirsky, A. E. and Ris, H. (1951). *J. gen. Physiol.* **34,** 451
Moore, D. H. and Grimley, P. M. (1957). *J. Biophys. Biochem. Cytol.* **3,** 255
Nebel, B. R. (1959). *Rad. Res.* **1** (Suppl.), 431
Newman, S. B., Borysko, E. and Swerdlow, M. (1949). *Science* **110,** 66
Newman, S. B., Borysko, E. and Swerdlow, M. (1950). *J. Appl. Phys.* **21,** 67
Nougarède, A. and Bronchart, R. (1967). *Compt. Rend. Acad. Sci.* D **264,** 1844
Noorduyn, N. J. A. and de Man, J. C. H. (1966). *J. Cell Biol.* **30,** 655
Palade, G. E. (1952). *J. Exptl. Med.* **95,** 285 and *Anat. Rec.* **114,** 427

Pashley, D. W. (1964).   *Thin metal specimens* in *Modern developments in electron microscopy*. New York; Academic Press
Peachey, L. D. (1958).   *J. Biophys. Biochem. Cytol.* **4,** 233
Pearse, A. G. E. (1968).   *Histochemistry-theoretical and applied.* London; Churchill
Pease, D. C. (1960).   *Histological techniques for electron microscopy.* New York; Academic Press
Pease, D. C. (1962).   *Anat. Rec.* **142,** 342
Pease, D. C. (1964).   *Histological techniques for electron microscopy.* New York; Academic Press
Pease, D. C. (1965).   *J. Appl. Phys.*
Pease, D. C. (1966).   *J. Ultrastruct. Res.* **14,** 356
Pease, D. C. and Baker, R. F. (1948).   *Proc. Soc. Exptl. Biol. Med.* **67,** 470
Persijn, J. P., De Vries, G. and Daems, W. T. (1964).   *Histochemie* **4,** 35
Person, S. and Osborn, M. (1964).   *Science* **143,** 44
Philpott, D. E. (1955).   *Exptl. Med. Surg.* **13,** 189
Porter, K. R. (1964).   *Ultramicrotomy* in *Modern developments in electron microscopy,* New York; Academic Press
Porter, K. R. and Blum, J. (1953).   *Anat. Rec.* **117,** 683
Rebhun, L. I. (1965).   *Fed. Proc.* **24,** S217
Rosenberg, M., Bartl, P. and Lěsko, J. (1960).   *J. Ultrastruct. Res.* **4,** 298
Reynolds, E. A. (1963).   *J. Cell Biol.* **17,** 208
Ris, H. (1957).   In *Symp. on the Chemical basis of Heredity,* p. 23, Baltimore; John Hopkins Press
Ris, H. (1962).   In *The interpretation of ultrastructure, Symp. Intern. Soc. Cell Biol.* New York; Academic Press
Rutherford, T., Hardy, W. S. and Isherwood, P. A. (1964).   *Stain Tech.* **39,** 185
Ryter, A. and Kellenberger, E. (1958).   *J. Ultrastruct. Res.* **2,** 200
Sabatini, D. D., Bensch, K. G. and Barrnett, R. J. (1963).   *J. Cell Biol.* **17,** 19
Salpeter, M. M. (1966).   In *Methods in cell physiology* **2,** 229, New York; Academic Press
Schultze, B. (1969).   Autoradiography at the cellular level. In *Physical techniques in biological research* **3B,** New York; Academic Press
Shinke, N. (1959).   *Nucleus* **2,** 161
Siegel, B. M. (1964).   ed. *Modern developments in electron microscopy.* New York; Academic Press
Sjöstrand, F. S. (1953).   *Nature* **171,** 30
Sjöstrand, F. S. (1958).   *J. Ultrastruct. Res.* **2,** 122
Sjöstrand, F. S. (1962).   In *The interpretation of ultrastructure,* **1** New York; Academic Press
Sjöstrand, F. S. (1966).   *Acta Physiol. Scand.* **67** Suppl., 270
Sjöstrand, F. S. (1967).   *Electron microscopy of cells and tissues* **1,** New York; Academic Press
Sjöstrand, F. S. (1969).   *Electron microscopy of cells and tissues* in *Physical techniques in biological research* **3C,** New York; Academic Press
Sjöstrand, F. S. and Baker, R. F. (1958).   *J. Ultrastruct. Res.* **1,** 239
Spurlock, B. O., Kattine, V. C. and Freeman, J. A. (1963).   *J. Cell Biol.* **17,** 203
Stacey, K. A., Cobb, M., Cousens, S. F. and Alexander, P. (1958).   *Ann. N.Y. Acad. Sci.* **68,** 682
Stäubli, W. (1960).   *Compt. Rend. Acad. Sci.* **250,** 1137
Stowell, R. E. (1951).   *Stain Tech.* **26,** 105
Swartzen-druber, D. C. and Hanna, M. G. (1965).   *J. Cell Biol.* **25,** 109
Swift, H. (1955).   *The Nucleic Acids,* **2,** 51, New York; Academic Press
Swift, H. (1959).   *Brookhaven Symp. Biol.* **12 B N.L (C22),** 134
Swift, H. (1962).   In *Interpretation of Ultrastructure, Symp. Intern. Soc. Cell Biol.* **1,** New York; Academic Press
Toner, P. G. and Curr, K. E. (1968).   *An introduction to biological electron microscopy.* London; Livingstone
Trump, B., Smuckler, E. and Benditt, E. (1961).   *J. Ultrastruct. Res.* **5,** 343
Valentine, R. C. and Horne, R. W. (1962).   In *Symposia for the Society of Cell Biology* **1,** New York; Academic Press
Venable, J. H. and Coggshell, R. (1965).   *J. Cell Biol.* **25,** 407
Von Ardenne, M. (1939).   *Z. wiss Mikroskop.* **56,** 8
Wachtel, A. W., Gettner, M. E. and Ornstein, L. (1966).   In *Physical techniques in biological research,* **3A,** 173. New York; Academic Press
Wada, B. (1966).   *Analysis of mitosis, Cytologia* **30,** Suppl.
Walter, F. (1961).   *Leitz-Mitt. Wiss. u. Technik* **1,** 236
Watson, M. L. (1957).   *J. Biophys. Biochem. Cytol.* **3,** 1017

Watson, M. L. (1958). *J. Biophys. Biochem. Cytol.* **4,** 475, 727

Watson, M. L. and Aldridge, W. G. (1961). *J. Biophys. Biochem. Cytol.* **11,** 257

Wolstenholme, D. R. (1966). *Chromosoma* **19,** 449

Yasuzumi, G., Mijao, G., Yamamoto, Y. and Yokohama, J. (1951). *Chromosoma* **4,** 351

Zetterqvist, H. (1956). The ultrastructural organisation of the columnar absorbing cells of the mouse jejunum. *Doctoral thesis,* Karolinska Institute, Stockholm

Bachmann, L. and Salpeter, M. M. (1965).

11 High Resolution Autoradiography

Since the pioneering work of Liquier-Milward (1956) on the combined use of electron microscopy and autoradiography on tumour cells, high resolution autoradiography has proved to be an effective tool in the study of ultrastructure as correlated with function (*see* Haggis, 1967; Toner and Curr, 1968; Schultze, 1969). An insight into the macromolecular pattern of different biological units, varying between 10–20 Å, combined with a knowledge of their strictly delimited functions, has resulted in a better understanding of the processes of life. Such studies, taken in conjunction with cytochemical findings, form a multipronged approach towards an analysis of cell metabolism.

As with autoradiography in general, this aspect of the subject has been much refined through the use of tritiated compounds of high specificity, now available in forms with specific activity even more than 15 Ci/mM— Schwartz Bio Research Inc. They have a very short range of radiation (0·018 MeV–1 μm), and thus eliminate the difficulty of using high energy β particles emitted by ^{14}C (0·155 MeV/40 μm in water) etc. where the range of radiation exceeds the thickness of emulsion. Moreover, the use of fine-grained nuclear emulsions, with grains as small as 0·1 μm (Ilford Nuclear Research Emulsion K5–L4) has facilitated the preparation of emulsion of uniform thickness and rendered the technique more convenient (Pelc *et al.*, 1961; Revel and Hay, 1961; Silk *et al.*, 1961; Granboulan *et al.*, 1962; Caro and Palade, 1963). However, the need for an autoradiographic resolution of less than 1 μm, prompted research on the modification of autoradiographic techniques used in the case of ultra thin sections observed under the electron microscope. This requirement was due to the fact that the size of the grains reaches beyond the limit of resolution of light optics and the thinness of the section does not allow sufficient contrast under light microscopy. Not only has it led to a clearer understanding of autoradiographs but it has also become an effective cytochemical tool for correlating structure with function. A simpler method, replacing the earlier time-consuming one, has now been devised (Stevens, 1966) so that electron microscope autoradiographs of high resolution can be obtained conveniently.

It has, however, been found necessary to observe thicker sections (0·3–0·5 µm) under light microscopy, prior to the study of high resolution ultra thin autoradiographs, principally to get an idea about the approximate exposure time needed for ultra thin autoradiographs and for their comparative assessment. Experience has shown (Caro, 1964) that exposure time needed for ultra thin sections is about 10 times more than that required for comparatively thicker sections. A maximum of 2 weeks exposure in thicker sections is desirable, as more than 5 months exposure in ultra thin sections results in background interference from heat, light, chemical and other effects. The requirement of 2 weeks exposure for thicker sections would imply that thinner sections would require an exposure of nearly 4–5 months. If, in the thicker sections, even after 3–4 weeks exposure, the response is not adequate, it is preferable to change the set-up to secure better incorporation. The period required for exposure is dependent on several variables, such as: the specific activity of the isotope and its amount of incorporation, the rate of synthesis of the labelled compound, the extent of its absorption in the tissue, and the nature of the cell itself. The last three factors are extremely specific and, as such, a knowledge of these properties is essential to formulate the appropriate procedure for securing good high resolution autoradiographs. Radio-sensitivity of the object to be studied is also an important factor, particularly in eliminating the possibility of radiation damage (Thrasher, 1966).

In addition to the need to predict the exposure time for ultra thin sections, light microscopy is required to obtain a demarcated picture of the area to be studied in ultra thin sections. For both these purposes, the desirable thickness of the sections is between 1500 and 5000 Å.

Fixation and Embedding

For the study of thin and thick sections, the same buffered osmium tetroxide fixation is generally preferred (Pease, 1960). In the analysis of chromosome structure, osmium tetroxide solution buffered to pH 7·2–7·4 is often recommended (*see* Wood and Luft, 1968). The addition of divalent cations like calcium (10^{-2M}) in the fixative, checks swelling and helps in maintaining uniformly the packed macromolecular configuration as observed in erythrocyte nuclei. Several workers have shown that un-buffered solution of pH (6·0–6·4) is quite suitable for fixation (Ryter *et al.*, 1958; Claude, 1961; Schreil, 1964). Lafontaine (1965, 1968) obtained very satisfactory results with chromosomes of *Vicia faba*, by fixing in 1 per cent unbuffered (pH 6·0–6·4) solution of OsO_4 in double distilled water, to which varying amounts of calcium chloride were added. Freeze drying methods can also be adopted for dehydration after quick freeze fixation.

For embedding, any of the usual media, such as methacrylate, epoxy resins, araldite (Glauert and Glauert, 1958), Epon (Luft, 1961) or polyester, Vestopal W (Ryter and Kellenberger, 1958), can be employed. Stevens

(1966), however, preferred Epon 812 embedding due to lack of polymerisation damage and stability of the resin (Luft, 1961). The advantage of methacrylate is that on removal of the embedding medium the sections can be observed under a phase contrast microscope.

Section-cutting

The slides have to be kept ready before cutting the sections. For this purpose, clean slides with frosted ends should be dipped in subbing solution (1 g Kodak purified calfskin gelatin is dissolved in 1 l hot distilled water, cooled, 1 g chromium potassium sulphate is added and the solution stored in the cold). The subbed slides are dried in a dustfree chamber and stored in boxes.

To secure thick sections meant for predicting the exposure time needed for ultra thin sections, as well as for comparative assessment, the block is trimmed so as to obtain a much larger face than that needed for ultra thin sections. As a result, more material, greater ease of operation, and serial ribbons can be secured. A glass knife with a smooth cutting edge is used and is adjusted with a metal or a tape boat. A metal boat is preferred since it presents a larger area. After mounting on the microtome, sections of desired thickness (1500–5000 Å) can be cut by setting the section indicator and observing the interference colour by adjustments of the water level and illumination.

When a ribbon with 2–4 sections is cut, the sections are picked up with a damp, fine-haired, clean nylon brush and transferred to a drop of water placed near the edge of a subbed slide. The sections adhere to the end of the brush where the bristles are narrow rather than in the middle, where it would be difficult to separate the sections from the brush. The slides are then dried at 45 °C according to the schedule followed by Caro (1964), and at 60–80 °C according to Stevens (1966).

In order to locate specific regions in ultra thin sections, Stevens (1966) has suggested a modified procedure. The block is trimmed in such a way that ultra thin sections can be obtained. A slightly better quality of glass knife is selected, without a scratch mark. First a thick section (1200 Å) is cut and transferred to the slide by the method given previously, followed by several ultra thin sections (600–1000 Å) which are shifted on the boat, prior to cutting another thick section of the original thickness. The latter is also mounted on the slide and both are observed. If the desired region is present in both the thick sections, the intervening thin sections are mounted on the grid, since the presence of the desired zone or material in the ultra thin sections has been ensured.

Stevens (1966) has adopted a method for handling the thinner and thicker sections alternately, employing a bent nichrome wire loop (4 mm in diameter and arm length 40 mm) with arm wrapped round a piece of glass tubing for handling. After a thick section is cut, it is brought to the centre with the aid of an eyelash, where it is picked up, together with a drop of water, by the loop which had previously been dipped in desiccate and dried. The section is then dropped on a slide by touching the water drop in the loop with a wooden applicator and fixed, and the same procedure repeated for the second thick section, after the ultra thin sections are cut.

When embedding has been done in methacrylate, it can be removed by the application of amyl acetate for a few seconds. The section may be marked with a diamond pencil on the underside of the slide. It is preferable to observe the very thin sections under the dissecting microscope and to mark the locations. Sections 2000–5000 Å thick may be even examined under phase contrast microscope without staining and comparatively thinner sections (1500–2000 Å) may be examined following staining with aniline dyes, like Azure A, which gives very good results with epoxy-embedded preparations. Since aniline dyes activate silver halide crystals, sections which can be spared should be stained prior to autoradiography or otherwise, other stains should be applied. To find out the exposure time needed for ultra thin sections, 2–4 days exposure on 0·4 μm sections gives an approximate indication.

Swift (1962) noted that suitable preparations of rat liver can be obtained if the tissue is fixed in 10 per cent formalin in 0·2 M phosphate buffer (pH 7·4), or 10 per cent acrolein in phosphate or tris buffer which prevents damage due to polymerisation. It also avoids the limitation of osmium fixation which is that it causes basophilia in proteins, through combination with the amino groups. He advocates freezing after fixation and cutting sections 50 μm or so thick in a freezing microtome, followed by treatment with buffered osmium tetroxide (pH 7·6) for 2 h before embedding in vestopal or epon. Ultra thin sections, cut in the ultramicrotome in the usual manner, can be mounted in carbon-coated titanium grids and stained in 3 per cent uranyl acetate for up to 4 h, and washed in water to secure contrast between chromosomes and nucleoli.

Stevens (1966) has listed the different stains compatible with the different embedding media generally used.

Coating with Emulsion and Observation

The emulsion, either K5 or L4, is first prepared by taking equal volumes of K5 and distilled water and melting at 45 °C for 10–15 min, stirring and cooling for 30 min. The slide is dipped into it, the excess emulsion should be drained off and the slide clamped vertically against a stream of air. A safelight may be used during the process. Occasionally, on prolonged storage, background grains may develop in the emulsion; these can be eliminated by treatment with H_2O_2 vapour (3 per cent) in a moist chamber for 4–6 h and storage in a dark slide box at 4 °C for the required period. At the time of developing, the slides are transferred to a staining jar, brought down to 20 °C, developed in Kodak D-19 developer (2 min for K5, 4 min for L4), rinsed in 1 per cent acetic acid for a few seconds, fixed in Kodak Rapid Fixer for 5 min and then washed. To check fogging, 0·01 per cent benzotriazole in D 19 may be used. Observations may be made under a phase contrast microscope in glycerine. Stained preparations can be examined in an ordinary microscope. Removal of the gelatin, if required for observations under phase contrast, may be carried out by treatment with very dilute NaOH solution.

Swift (1962) preferred an application of 1 per cent potassium permanganate solution before washing in dilute citric acid and water, for vestopal-embedded materials. He used, for emulsion-coating, a week's exposure to Ilford G5 liquid emulsion at 40 °C (1 : 15 diluted) before developing in Kodak D19 developer.

Coating of Ultra Thin Sections

Sections embedded in methacrylate, epon, araldite or vestopal may be used. The success of the operation principally depends on obtaining a fine compact monolayer of silver halide crystals. With increase in the thickness of the layer, the sensitivity increases, at the cost of resolution. Selection of a proper emulsion is one of the most important factors in high resolution autoradiography. Its sensitivity depends on the extent to which it can register and develop the latent images formed by electrons in their path on silver halide and it is measured by the number of grains developed per unit distance in the track of particles with minimum ionisation. The particle energy and the distance the electron has to traverse through the silver halide, controls the formation of the latent image. Normally, of the isotopes used; ^{32}P, ^{131}I, ^{14}C and ^{35}S have long range ionising particles. With Tritium (^3H), nearly all electrons emitted into the upper hemisphere can be developed. During exposure, oxidation of the latent image severely affects sensitivity. Protection against oxidation becomes essential with smaller crystals and fine-grained development. For β particles, Pelc *et al.* (1961) have suggested a 100–500 Å crystal size as suitable. With tritiated compounds, the problems of sensitivity and resolution are not so severe as the emitted β-particles are heavily scattered within one silver halide crystal, of Ilford L4 emulsion. With finer grained emulsions, such as, Gevaert 307 or Kodak NTE, (crystal size 500 Å), multilayered crystals add to the sensitivity. Caro (1962) has recommended Ilford L4 (crystal size 1200 Å) for electron microscopic and Ilford K5 (crystal size 1800 Å) for light microscopic autoradiography. A close-packed monolayer of silver halide adds to resolution by preventing the spread of electrons from the source.

For mounting the sections, both the collodion film and the sections must be perfectly smooth. Electroplated Athene-type copper grids are used for coating with collodion with a thin carbon layer (Caro, 1962). For extra strength, the grids are generally covered with 0·25 or 0·5 per cent parlodion-carbon and dried (Stevens, 1966). A thin film is spread so that resolution is not hampered and at the same time, breakage is avoided during the procedure. Sections are mounted on the grid which is added at one edge by a piece of scotch tape (double coated) to a slide. Several grids can be placed on one slide.

Various methods have been proposed for applying a uniform layer of emulsion. It may be applied, either by dipping the slide in the emulsion or dropping it on the slide (5 cm^3 of distilled water per 1 g of emulsion, Hay and Revel, 1963; Granboulan, 1963; Koehler *et al.*, 1963; Salpeter and Bachman, 1964; Young and Kopriwa, 1964). A thin layer may be allowed to form on a specially constructed loop before applying on specimen grids (Caro and Van Tubergen, 1962; Moses, 1964). The emulsion may be centrifuged directly on the specimen grids (Dohlman *et al.*, 1964) or may be finely layered on agar before application on sections (Caro, 1964). Salpeter and Bachman (1964) and Salpeter (1966) suggested the preparation of a substrate of uniform property before applying the emulsion, and recommended the formation of the emulsion layer on a fine layer of carbon or silicon monoxide, dried on sections mounted on collodionated glass slides. Caro (1964) suggested two ways of applying the emulsion. In the first one, a 50 per cent solution (5 g/10 cm^3) of Ilford L4 is prepared by stirring in distilled water at 45 °C. The solution is cooled for a few minutes in an ice bath, followed by cooling at 20–24 °C for 30 min. In emulsions like Kodak IVTE, containing too much gelatin, it is better to dissolve 1 g in 10 cm^3 of warm distilled water, followed by centrifuging till a clear supernatant is obtained (about 14 000 g for 10 min). The rotor of the centrifuge should first be heated to help separation of gelatin. The supernatant is decanted and the precipitate chilled. This concentrated emulsion should be dissolved in 1–2 cm^3 of water (Salpeter, 1966). A thin platinum, silver or copper wire may then be dipped in the viscous solution and a thin layer should then be allowed to form round it, which, when touched on the slide, forms a fine layer. If the viscosity is lower or higher than that required, a period of cooling or warming may be necessary. The amount of emulsion in the loop should be adequate, otherwise the distribution of silver halide crystals would be uneven, resulting in different thicknesses in the emulsion layer.

In another method (Caro and Van Tubergen, 1962; Caro, 1964) the use of forming a thin layer on agar has been shown again. A thin layer of agar in a Petri dish may be prepared from a 2 per cent solution of agar in distilled water and stored in a cool place to harden. Small rectangular pieces (2 × 3 cm) are cut and placed on slides, previously dried by slight warming. A solution of 0·2 per cent parlodion in amyl acetate is then applied. The slides can be dried in a vertical position and kept in a dark room where the subsequent operations are performed. A 25 per cent solution 91 for L4 emulsion is prepared in water and cooled at 20–24 °C. A thin film of emulsion, taken on a loop as described previously, may be applied to the agar surface. According to Caro (loc. cit.), this method eliminates the limitations of artefacts due to drying because of the percolation of moisture from emulsion to agar and results in a fine, uniform layer. The whole composite structure would then float in water, the emulsion lying on the upper surface. Sections on grids, but not having the supporting membrane, are then fitted, in a fine-meshed metal screen, in water and brought below the composite membrane. They are carefully lifted out of the water, dried and fixed on slides as usual.

Stevens (1966) has outlined a method for applying emulsion, in which thoroughly clean and dry glass rods with rounded tops are placed in filler blocks. A small piece of adhesive tape is scraped and placed on the top of each rod to hold the grid containing the section, which is then laid on the top.

One grid may be placed on each rod. The rest of the operation is performed in the dark room. Ilford L4 emulsion (2 g in 10 cm^3) is melted at 42 °C in a water bath and then diluted with water added in nearly equal proportions, cooled in an ice bath and kept at 32–34 °C. A wire loop is then coated with emulsion, the excess being wiped off in a kimwipe. After the formation of proper interference colour in the upper half (as mentioned below), the gelled emulsion in the loop is brought on the grid and can then be coated with a monolayer and stored in a sealed dark chamber.

Thickness of Emulsion Layer

The thickness and uniformity of the emulsion layer should be checked by the interference colour, based on the principle that interference colours of emulsion layers in reflected white light depend on their thickness. This can be viewed even in the dark on density differences using a yellow safe light (Filter AO). Interference colour is not only an index of thickness of emulsion but the uniformity of colour also indicates uniformity of the emulsion layer, as against patchy colour indicating unevenness. This observation has been confirmed even at the level of the electron microscope. The refractive indices of collodion and plastic being close to that of glass, the underlying film does not affect the interference colour. Following an inferometric determination of thickness, Salpeter and Bachmann (1964, 1965) and Salpeter (1966) have formulated a table on the basis of which emulsion thickness can be worked out by interference colours (see Bachmann and Salpeter, 1965).

The most appropriate method for ascertaining the thickness needed for quantitative work is to use a developer (Devtol) which does not affect silver halide crystals. After developing, and prior to fixation, the slides can be air-dried and viewed in white light for interference colours. It is always desirable to use Dricrite in the storage chamber to keep the slides dry. The time of exposure has to be deduced from the time required for thicker (0·5 µm or so) sections as mentioned above.

Check Against Background Effect and Loss of Sensitivity

Before photographic processing, it is always necessary to eliminate background effect, if any. With crystals too small to be resolved under light microscopy, such as those of Ilford L4, the background must be checked under the electron microscope. Background grains generally result from latent images being formed by some outside effects, such as, light, heat, radiation, or some chemical agents, and are manifested in the form of fine grains indistinguishable from the effects of β-particles on silver grains. In order to remove the background effect (Yagoda, 1949), immediately when the emulsion is dried, the slides are kept inside a pair of staining dishes lined with filter paper moistened with 3 per cent hydrogen peroxide solution. Care should be taken to see that the slides do not touch the solution and the chamber is kept air-tight. The background effect can be completely removed after about 6 h treatment. Since hydrogen peroxide oxidises all latent images including those given by β-particles, it is necessary that the vapour treatment be performed just at the beginning of exposure. Caro (1964) demonstrated that, following such a procedure, grains developed in preparations incorporated with leucine $-{}^3H$, uridine $-{}^3H$ or thymidine $-{}^3H$ were not affected. Stevens (1966) suggested a control preparation to check against the background effect. This method principally involves a control preparation using Parlodion–carbon–coated grids without sections and developing the coated grids just after application of the emulsion. Only a few reduced grains can be seen in a monolayer of emulsion and they do not increase if kept in an air-tight box even up to 5 months. If the emulsion is not shaken, or exposed to chemicals, heat, and stray radioactivity, and the apparatus is kept scrupulously clean, there is no necessity to check background effect, at least with Ilford L4.

In addition to affecting the background, improper storage may influence sensitivity of the crystals and cause obliteration of the latent image. The disappearance of the latent image is mainly caused by oxidation of the particular silver grains. It has been seen that storage in CO_2, nitrogen (Herz, 1959; Ray and Stevens, 1953) and low temperature (LaPalme and Demers, 1947) minimises these adverse effects. However, storage in CO_2 has been shown not to affect the sensitivity to a significant extent with K5 and L4 (Caro, 1964; Herz, 1959). Storage in air of Kodak emulsion for a prolonged period (2 months) resulted in 60 per cent loss in radiosensitivity, whereas in Helium this loss was not noted (Salpeter, 1966). According to him, Ilford L4 does not show any change in sensitivity either in air or in Helium.

Developing

After adequate exposure, developing of the photographs should be carried out in clean and dustfree conditions. The use of developer is meant to reduce the exposed silver bromide crystals, carrying the latent image, to metallic silver. It is always necessary, by trial, to find out the optimum period of development which would permit the maximum number of grains to be developed with least background effect. The grids must always be kept in absolute ethanol for 3–4 min before developing. This hardening schedule is an essential step as it checks the sudden swelling caused by aqueous developer and the resultant loss of grains. In all the steps followed for taking electron microscopic autoradiographs, the problem of contamination is a serious one. To check against this limitation, it is always necessary to use a small quantity of fresh, filtered, solution for each plate.

Several developers are in vogue and their adequacy depends on the type of emulsion used for coating. Different types of physical and chemical developers are available and the processing is carried out at 20–24 °C. From a look at the developed grains of the latent image, it is rather difficult to locate the exact path of the particles, as in a single crystal of silver halide, at least three images from a single tritium decay have been found. *Chemical developers, which reduce silver halide crystals* result in a coil of silver filament of 0·3–0·4 μm diameter in certain developers like D19, or a long filament as in Microdol − X. Since the latter is comparatively easier to interpret in that the initial image may be considered in the middle of the line, it is often preferred (Caro, 1964). Stevens (1966) has suggested a procedure in which Athene-type grids are placed on filter paper for drying after being lifted from ethanol with forceps. The dried grids should be floated in an inverted position on the convex surface of the developer and kept for 6 min at 22–24 °C. After immediate transference to a watch glass containing distilled water, for a few seconds, the grids should be placed in the fixer with the sections facing upwards. The fixer effectively removes all unexposed silver halide crystals and helps in the later removal of the gelatin. Budd (1964) noted that fixers with hardeners cause the emulsion to become brittle and fibrous. After 10 min, the grids should be rinsed in water for 5 min, washed by flushing with distilled water, dried and kept in a dust free chamber.

With *physical developers*, the highest possible resolution can be obtained. The principle is to *dissolve completely the silver bromide crystals*, only the latent image with silver ions is kept, with the use of 1·0 M sodium sulphite and 0·1 M *p*-phenylene diamine, on silver nitrate in varying proportions. (Lumière *et al.*, 1911; James, 1954; Caro, 1964). Development for even 1–2 min at 20 °C is sufficient. This developer is, however, comparatively unstable. With this method, the latent image can be localised with the least possible error and caution is needed because grain size is very small.

Lastly, the gelatin can be dissolved through proteolytic enzymes, such as pepsin (Hampton and Quastler, 1961; Comer and Skipper, 1954; Przybyl-ski, 1961), alkali (Revel and Hay, 1961) or warm water after fixation at 37 °C for 16 h (Silk *et al.*, 1961).

Staining

The gelatin may or may not necessarily be dissolved for staining after the photographic processing is completed. Staining can be performed even before applying the emulsion. In the former case, the gelatin is dissolved by proteolytic enzymes, like an acidic solution of pepsin (Comer and Skipper, 1951; Hampton and Quastler, 1961; Przybylski, 1961) but this method has an inherent limitation in that undigested grains and autoradiographs often look alike and there is a possibility of the loss of grains. Fixation in warm water at 37 °C for 16 h has been recommended by Silk *et al.* (1961) but according to Stevens (1966), this method is yet to be tested with Ilford L4 emulsion. Another procedure (Hay and Revel, 1963; Revel and Hay, 1961) combines gelatin removal and lead staining (Karnovsky, 1961) and yields very satisfactory results but it should also be applied with caution, to check against grain displacement. In Revel and Hay's method (1961), ^{3}H—thymidine incorporation in interphase chromatin fibres of *Ambly-stoma* larvae was studied.

In order to avoid the limitations caused by removal of the gelatin, Caro

(1964) stained the specimens for 10–15 min in a solution containing 1 per cent uranyl acetate and absolute ethanol, to have a final concentration of 30 per cent ethanol (Gibbons and Grimstone, 1960). However, it was applied in methacrylate embedded bacterial material. A serious drawback of keeping the gelatin intact is the possibility of disruption of the grains through shattering of the gelatin layer by the electron beam. The removal of gelatin may be carried out by sublimation by gradually increasing the intensity of electron beams (Moses, 1964). The accuracy of this method is often minimised by the possibility of artefacts—and the risk involved in sublimating the silver grains (Stevens, 1966). In the case of certain stains, such as, methylene blue, difficulties arise in securing staining intensity, which requires a temperature of 60 °C before applying the emulsion. The result is excessive background effect due to the reduction of the emulsion by the stain. Hendrickson *et al.* (1968) suggested NaOH treatment in addition to HIO_4 (Richardson *et al.*, 1960) prior to applying emulsion. Both treatments enhance stainability. Stevens (1966) followed a modification of Granboulan's technique in which, prior to staining, gelatin was first dissolved by acid hydrolysis.

In this schedule the grids are first floated for 30 min on the convex surface of distilled water, kept at 37 °C and then transferred to 0·5 N acetic acid at 37 °C and kept for 15 min, followed by rinsing in a stream of distilled water and then floated in a second change of distilled water at 22–25 °C for 10 min. The staining is performed by first wetting the sections with a drop of distilled water, staining in an inverted position in 7·5 per cent aqueous uranyl acetate for 20 min at 45 °C, subsequent drying on filter paper, wetting again with distilled water and then further staining with 0·2 per cent lead citrate for 10–60 s to a few minutes (Reynolds, 1963; Venable and Coggeshall, 1965). It was suggested that uranyl acetate changes the properties of the macro-molecules in such a way that they bind better with lead citrate. In order to avoid the possibility of removing or damaging the silver grains, Salpeter (1966) suggested a method of staining prior to the application of the emulsion. In the schedule followed, a few drops of lead stain, or lead citrate staining for 5–30 min preceded by aqueous uranyl acetate or uranyl nitrate (Reynolds, 1963) may be added to the sections on the slide and the excess stain immediately flushed off with a stream of distilled water. Evaporation of a carbon layer (50–100 Å Union Carbide SPK spectroscopic carbon) over the sections is necessary for screening the stained sections, and the emulsion, and also for providing a base for the emulsion layer.

Caution is desirable, as a thick layer of carbon may prevent proper resolution. The period of lead staining requires shortening too as dense deposits may otherwise be formed.

In conclusion, it may be stated that the method of high resolution autoradiography as outlined above, with its limitations and advantages, leaves scope for improvements in every step of its methodology. Even so, its superiority over conventional autoradiography is principally reflected in its extremely high resolving power, capacity for strict identification of macromolecular structures in terms of function, and lastly, exceptional contrast in photographs of silver grains and cell structures. As the advantages far outweigh the limitations, which are on their way to refinement, the potentialities of this aspect of biological research is difficult to overestimate.

REFERENCES

Bachmann, L. and Salpeter, M. M. (1965).　In *Quantitative electron microscopy*. Baltimore; William and Wilkins
Budd, G. C. (1964).　*Stain Tech*. **39,** 295
Caro, L. G. (1962).　*J. Cell Biol*. **15,** 189
Caro, L. G. (1964).　*High-Resolution Autoradiography*, in *Methods in cell physiology*, **1,** New York; Academic Press
Caro, L. G. and Palade, G. E. (1963).　Referred to in Caro (1964)
Caro, L. G. and Van Tubergen, R. P. (1962).　*J. Cell Biol*. **15,** 173
Claude, A. (1961).　*Pathol. Biol. Semaine Hop*. **9,** 933
Comer, J. J. and Skipper, S. J. (1954).　*Science* **119,** 441
Dohlman, G. F., Maunsbach, A. B., Hammerstrom, L. and Applegren, L. E. (1964).　*J. Ultrastruct. Res*. **10,** 293
Gibbons, I. R. and Grimstone, A. V. (1960).　*J. Biophys. Biochem. Cytol*. **7,** 697
Glauert, A. M. and Glauert, R. H. (1958).　*J. Biophys. Biochem. Cytol*. **4,** 191
Granboulan, P. (1963).　*J. Roy. Microscop. Soc*. **81,** 165
Granboulan, P., Granboulan, N. and Bernhard, W. (1962).　*J. Microscop*. **1,** 75
Haggis, G. H. (1967).　*Electron microscope in molecular biology*. New York; John Wiley
Hampton, J. H. and Quastler, H. (1961).　*J. Biophys. Biochem. Cytol*. **10,** 140
Hay, E. D. and Revel, J. P. (1963).　*Develop. Biol*. **7,** 152
Hendrickson, A., Kunz, S. and Kelly, D. E. (1968).　*Stain Tech*. **43,** 175
Herz, R. H. (1959).　*Lab. Invest*. **8,** 71
James, T. H. (1954).　In *The theory of the photographic process* New York; MacMillan
Karnovsky, M. J. (1961).　*J. Biophys. Biochem. Cytol*. **11,** 729
Koehler, J. K., Mühlethaler, K. and Frey-Wyssling, A. (1963).　*J. Cell. Biol*. **16,** 73
Lafontaine, J. G. (1965).　*J. Cell. Biol*. **26,** 1
Lafontaine, J. G. (1968).　*Structural components of the nucleus in mitotic plant cells* in *Ultrastructure in Biological Systems, The Nucleus* **3,** New York; Academic Press
LaPalme, J. and Demers, P. (1947).　*Physiol. Rev*. **72,** 536
Liquier-Milward, J. (1956).　*Nature* **177,** 619
Luft, J. H. (1961).　*J. Biophys. Biochem. Cytol*. **9,** 409
Lumière, A., Lumière, L. and Seyewez, A. (1911).　*Compt. Rend*. **153,** 102
Moses, M. J. (1964).　*J. Histochem. Cytochem*. **12,** 115
Pease, D. C. (1960).　*Histological techniques for electron microscopy*. New York; Academic Press
Pelc, S. R., Coombes, J. D. and Budd, G. C. (1961).　*Exptl. Cell Res*. **24,** 192
Przybylski, R. J. (1961).　*Exptl. Cell Res*. **24,** 181
Ray, R. C. and Stevens, G. W. W. (1953).　*Brit. J. Radiol*. **26,** 362
Revel, J. P. and Hay, E. D. (1961).　*Exptl. Cell Res*. **25,** 474
Reynolds, E. S. (1963).　*J. Cell Biol*. **17,** 208
Richardson, K. C., Jarett, L. and Finke, F. H. (1960).　*Stain Tech*. **35,** 313
Ryter, A. and Kellenberger, E. (1958).　*J. Ultrastruct. Res*. **2,** 200
Ryter, A., Kellenberger, E., Birch-Andersen, A. and Maalöe, O. (1958).　*Z. Naturforsch*. **13B,** 597
Salpeter, M. M. (1966).　*General area of autoradiography at the electron microscope level* in *Methods in cell physiology* **2,** New York; Academic Press
Salpeter, M. M. and Bachmann, L. (1964).　*J. Cell Biol*. **22,** 469
Salpeter, M. M. and Bachmann, L. (1965).　*Symp. Intern. Soc. Cell Biol*. **4,** 23
Schreil, W. H. (1964).　*J. Cell Biol*. **22,** 1
Schultze, B. (1969).　In *Physical techniques in biological research* **3B,** New York; Academic Press
Silk, M. H., Hawtrey, A. O., Spence, I. M. and Gear, J. H. S. (1961).　*J. Biophys. Biochem. Cytol*. **10,** 577
Stevens, A. R. (1966).　*High resolution autoradiography* in *Methods in cell physiology*. **2,** New York; Academic Press

Swift, H. (1962). *Nucleoprotein localisation in electron micrographs* in *The interpretation of ultrastructure* **1,** New York; Academic Press

Thrasher, J. D. (1966). *Analysis of renewing epithelial cell populations* in *Methods in cell physiology* **2,** New York; Academic Press

Toner, P. G. and Curr, K. E. (1968). *An Introduction to biological electron microscopy,* Edinburgh; E. & S. Livingstone Ltd

Venable, J. H. and Goggeshall, R. (1965). *J. Cell Biol.* **25,** 407

Wood, R. L. and Luft, J. H. (1968). *J. Ultrastruct. Res.* **12,** 22

Yagoda, H. (1949). *Radioactive measurements with nuclear emulsions.* New York; John Wiley

Young, B. A. and Kopriwa, B. M. (1964). *J. Histochem. Cytochem.* **12,** 438

12 Microspectrophotometry

I. UNDER ULTRAVIOLET LIGHT

Principle and Instrumentation

Ultraviolet radiation was applied, in Köhler's (1904) experiments, to utilise fully the accelerated resolution of numerical aperture due to decreased wavelength. Ultraviolet light, in place of visible light, has the unique advantage of clarifying unstained living cells, due to the strong ultraviolet absorption by nucleoprotein (Caspersson, 1936; 1950). Moreover, it aids the quantitative estimation of the cell nucleoprotein owing to the characteristic absorption of purine and pyrimidine components of nucleic acid at 2650 Å (Walker and Yates, 1952; Rudkin et al. 1955, Rudkin, 1966). A linear relationship between absorption and section thickness and the concentration of DNA has been clearly demonstrated in nuclei (Greenwood and Berlyn, 1968). U.V. absorption spectra of cytological objects generally show absorption between 200 and 400 Å (nm); nucleic acids at 260 nm, and proteins free from nucleic acids at about 280 nm.

The absorption of cellular components is principally attributed to covalent unsaturated groups such as $C = O$, $C = N$ and $N = O$ in organic compounds, i.e. purines and pyrimidines, indole groups of tryptophane, benzene and imidazole rings of tyrosine and histidine, respectively.

The principal difference between an ordinary light microscope and an ultraviolet microscope lies in the fact that in the latter, transparent fused quartz lenses are used in place of optical glasses, which are opaque to shorter ultraviolet wavelengths. The source of u.v.–rays is generally the mercury vapour lamp (British Thomson Houston MB/D or the American AH/4). The slides and cover glasses are made of quartz and in order to secure a monochromatic beam, a quartz monochromator (Baird Atomic Inc., Cambridge Mass; Barr and Stroud Ltd., Scotland, etc.) is fixed between the source and the microscope. The photographic image is obtained by using a photographic plate and a photoelectric cell. Focusing is generally performed in visible light or on a fluorescent screen (Loeser and West, 1962; Freed and Benner, 1964). Computation of the image can also be performed (Ledley, 1964; Mendelsohn *et al.*, 1964).

Within the last 10 years, the availability of excellent photographic devices, such as quartz fluorite refracting achromatic lenses, ultrafluors, powerful lamps, good monochromators, strong photomultiplier tubes, etc. has been responsible for outstanding developments in u.v.–microspectrophotometric studies. Photometric observation of microscopic objects, even after extraction through a micromanipulator, has been possible (Edström, 1964). A television pick-up system has also been used, which is a very sensitive recording device.

As illuminator, the commonly accepted method is to have the source through the exit slit of the monochromator. The aperture of the condenser is generally kept at about 0·3 or less for good photometric work. Modern discharge lamps, such as hydrogen and deuterium lamps and low pressure mercury lamps, with good achromatic objectives, have replaced the original rotating electrode resonant metallic arc (Köhler, 1904; Caspersson, 1936). Such lamps give a wide band at 2540 Å and can be used with suitable monochromators or interference filters.

Most modern equipment, however, takes advantage of Xenon compact arcs, which combine intensity with output through the u.v. range at· 2600 Å. In the monochromators meant for selecting the particular wavelengths, a band of energy is emitted whose wavelength distribution is controlled by the dispersion of the elements showing diffraction and refraction and the size of the slits. Several microspectrophotometers are equipped with grating monochromators where the change in wavelength is done by grating rotation. A number of interference filters have been developed, which serve not only as protective filters but also as wavelength selectors. Photography may be adopted for the integration of absorption of objects and the negatives may further be scanned through densitometry. Photoelectric recording is utilised for a study of the series of absorption spectra needed for each wavelength. The most convenient method is to allow the light to pass through a selected area for final recording in a photomultiplier tube. The intensities may be recorded at different wavelengths and the background intensity measured by removing the object and allowing the light to pass through the empty space. Good photomultiplier tubes (Herrmann, 1965) of high efficiency (quantum efficiency 0·3) are available (EMI Electronics Ltd.)

In principle, to measure changes in the quantity of nucleic acid and protein, the microscope is generally used as a spectrophotometer. The monochromatic beam of ultraviolet light may be split into two beams, in the split beam device, one falling directly on the photoelectric cell (the blank) and the other passing through the ultraviolet microscope to another photoelectric cell. The sample to be measured is placed in the path of the beam passing through the microscope. The light passing through the material is reduced in intensity; this is calculated by counting the difference in the photoelectric current yielded by the two beams, as indicated by a galvanometer.

The entire measurement of absorption is based principally on Lambert–Beer's law which may be stated as follows:

$$I_x = I_o \times 10^{-kcd}$$

where I_x is the changed intensity of the beam of I_o; I_o the incident intensity, after the ray passes d; d the thickness (in cm) of C; C the concentration (in $g/100$ cm^3) of the absorbing molecules; and K the extinction coefficient. As in cytomicrospectrophotometry, relative amounts are obtained, the constant K is ignored, since absolute values are not necessary. The value of K, when necessary, can be worked out from biochemical data.

In the method mentioned above, the values of I_x and I_o can be obtained without changing the position of the material. With the aid of these values, the percentage of transmission (T) can be worked out (I_x/I_o) and the presence of the components per arbitrary units, showing ultraviolet absorption, can be computed. A limitation of ultraviolet microscopy is that the ray may have some deleterious effect on the absorbing material, but in the above method the period of exposure to ultraviolet is very much reduced.

Zeiss–Caspersson ultramicrospectrophotometer UMSP–1 is now being commercially produced, based on Caspersson's model (Caspersson, 1965). The other types are, the Edinburgh microspectrophotometer with dual microscope, double beam (Walter *et al.*, 1963), Wagenar–Grand Instrument with single microscope double beam (Wagenar and Grand, 1963) and lastly the Leitz microspectrograph (Thaer, 1965), in which the absorption spectrum is recorded through image formation with heterochromatic u.v. radiation, and light is then dispersed in a spectrograph, following the microscope, from a small region of the plane of the image. For details of ultra-violet microspectrophotometry and its modifications, the reader is referred to Taylor (1950), Scott and Sinsheimer (1950), Wyckoff (1952, 1959), Stowell (1952), Patau and Swift (1953), Nurnberger (1955), Pollister and Ornstein (1959), Leuchtenberger (1958), Walker (1958), Mendelsohn (1966), Wied (1966) and Freed (1969).

Methods

The fixing fluids for objects meant for spectrophotometric analysis require certain prerequisites in addition to the prevention of any loss or re-patterning of structure. Such fluids should not affect the light scattering properties of the specimen and must not contain compounds which undergo deposition under u.v.-rays. For the study of DNA, 45 per cent acetic acid solution (Rudkin *et al.*, 1955), ethanol–acetone mixture (Zetterberg, 1966) or acid or *neutral* formalin fixation (Sandritter and Hartlieb, 1955) have been found to be suitable. In the case of RNA, however, most of the aqueous fixatives cause extraction of soluble RNA (Swift, 1966) and freeze-substitution method with ethanol–potassium acetate has been shown to retain soluble RNA (Woods and Zubay, 1966). Zetterberg and others (Killander and Zetterberg, 1965; Zetterberg, 1966) have demonstrated that nucleoproteins can best be studied in cultured cells either by ethanol-freezing substitution or by chemical fixation with ethanol–acetone (1:1) at 4 °C for 24 h.

For a correct assessment of the absorbance data, it is desirable to follow the extraction of specific cellular components simultaneously. Such procedures, in addition to aiding identification of the absorption of particular chemical constituents, may also serve as controls. Further, the most significant use of extraction is to secure a *blank*, so that non-specific light loss and light scatter can be corrected. These purposes are served through digestion with proteases or nucleases, since extraction with acids may alter the absorptive properties (*see* Swift, 1966). In mounting the specimens in u.v. microspectrophotometric work, media like pure glycerine, glycerine–water mixture, 45 per cent zinc chloride and paraffin oil possess the essential prerequisite for checking the non-specific light loss (Caspersson, 1950; Rudkin and Corlette, 1957 and *see* Freed, 1969).

Lastly, during the application of the u.v. rays, caution is recommended, since continuous exposure for even 10 min at 257 nm in a Köhler microscope may result in the loss of absorption capacity by the chromosomes treated with acidic fixatives.

Techniques for the u.v. microscopy of cells in culture have been developed to study conditions and changes *in vivo*, for which perfusion chambers with quartz cover slip windows are generally used (Freed, 1963; Freed and Benner, 1964; Petriconi, 1964). Modifications of Eagle's medium have been used for cell culture, which have to be replaced later by a salt solution transparent to u.v.-rays. In the scanning procedure, improvements have lately been devised (Kamentsky *et al.*, 1965; Kamentsky and Melamed, 1967), in which large suspended cell populations are quickly analysed with the rapid cell spectrophotometer but the procedure does not allow intra-chromosal analysis.

Photometric observation through u.v.-rays have also been extended to chromosomes, or chromosome segments, extracted out of the cell with the micromanipulator (Edström and Beermann, 1962; Slagel and Edström, 1967). Micrurgical extraction of polytene chromosome has been dealt with in the chapter on micrurgical techniques. Extraction of nucleic acid through nucleases and the analysis of a drop of the extract has been made through microspectrophotometry. Methods have even been developed for the determination of base constituents in nucleic acid, extracted from incised segments of chromosomes. The extract is applied to a treated cellular fibre, through which the discharge of electric current results in separation of different u.v.-absorbing bands. The photomicrographs of these bands, following densitometric analysis, yield the quantitative values of the substances.

For the identification and quantitation of constituents in the chemical make-up of chromosomes, reliance on microspectrophotometry is continuously increasing and it is hoped that refinements in this last named method will ultimately lead to an understanding of the qualitative and quantitative differentiation of chromosome segments, even at the molecular level.

II. UNDER VISIBLE LIGHT

Visible light, like ultra-violet rays, is utilised for microphotometric work. The principle involved in such studies is to secure absorption curves of coloured substances to help in the chemical identification, localisation and quantitative measurements of chromosomal constituents. The method of analysis is fundamentally based on Beer-Lambert's law, as in u.v.-microspectrophotometry, except that the instrument is not fitted with quartz achromatic optics meant for u.v.-rays. Absorption measurements are carried out on materials stained with Feulgen solution, methyl green pyronin, azure B, Millon dye and such other compounds capable of staining specific cell constituents (Pollister, 1952; Vendrely, 1955; Leuchtenberger, 1958; Pollister and Ornstein 1959; Mendelsohn, 1966; Swift, 1966; Wied, 1966; Dutt, 1967; Welch and Debault, 1968; Pollister *et al*, 1969).

For the fixation of materials, Carnoy's fluid or 10 per cent neutral formalin are widely used to measure chromosomal DNA. Greenwood and Berlyn (1968) studied the effects of formol-acetic-ethanol, Carnoy's fluid, Craf III, formaldehyde and glutaraldehyde fixation on cytophotometric studies and observed that Craf and glutaraldehyde had a depressing effect on the absorption peak and interfered with DNA extraction. Glutaraldehyde caused binding of the dye with the cytoplasm. Neutral 10 per cent formalin has been found to be very suitable, though digestion and extraction procedures were affected, unlike Carnoy's fluid.

Staining may be carried out in Feulgen solution or in pyronin–methyl green mixture or in any other specific dye for binding with nucleic acids or proteins. Mitchell (1967) has shown that dinitrofluorobenzene treatment may be combined with Feulgen procedure for measuring protein and DNA of the same cell. However, the fixation used, as well as the composition of the DNFB solution, and temperature, affect the coloration.

Microscope and accessories

The appliances needed for visible light photometric work are quite simple, including (a) microscope with apochromat lenses, compensating oculars (fitted with a diaphragm, aperture 1–3 mm for eliminating stray light) and condenser with low numerical aperture (0·2–0·4) or more, if necessary; (b) a photometer unit with microammeter, power supply and photomultiplier tubes (Photomultiplier-American Instrument Co., Oster, 1953), 931A phototube or IPZ1, suitable at low and high magnifications; (c) rotating plate type photometer mount; (d) standard tungsten 100 W coil projection light source with a condensing lens and diaphragm; and (e) a suitable monochromator (Parker-Elmer model 83, 33–86–40 or 33–86–45, as mentioned under u.v.-microspectrophotometry).

Adjustment of the Instrument

The procedure for accurately adjusting the instrument is outlined below (*see* Pollister *et al*., 1969). White cardboard is placed about a foot in front of the lamp and the image of the filament is focused on it. The lens, bulb and diaphragm are adjusted till the colour fringes are symmetrical even under the smallest aperture of the diaphragm. The filament image is finally focused on the monochromator entrance slit and all light entering the monochromator is focused on the prism or grating. The alignment of the monochromator elements and the wavelength scale can be checked by a mercury arc lamp and with the aid of a telescope, the centring of the condenser aperture is done vertically and horizontally in relation to the slit image. The microscope is focused on a slide under low power and the light from the monochromator is allowed to enter the condenser. To test the proper alignment of the optical units, a focusing telescope can be inserted in place of the ocular lens at the side viewing tube. Then the low power lens is replaced by the oil immersion one. On final adjustment, the light from the monochromator is a central spot 10–50 μm in diameter, showing no lateral shift with wavelength, in an otherwise dark field, so that an area about 5 μm away from the lighted spot has less than 2 per cent of its intensity. The condenser can also be aligned by looking down the microscope after removing the ocular. The ocular is then replaced and the condenser and mirror adjusted so that all the fringes present concentric appearance, even when the condenser is moved vertically and the diaphragm opened. The first step in adjustment should be the alignment of the lamp and monochromator, followed by testing for condenser alignment. Sheet metal strips or wooden blocks may be used to adjust the optical axis in a vertical plane in the first step.

Method of Analysis

As in u.v. microspectrophotometry, in taking measurements, the data required are the intensity of the background light (I_o) and the reduced intensity of light after absorption by the objective (I_s), the *transmission* being calculated as I_s/I_o. In a completely transparent object, this value is 1 and there is a logarithmic decrease in transmission with increase in absorbing molecule, due to concentration or thickness. Therefore the optical density, or extinction,

$$E = \log_{10} \frac{1}{T} = \log_{10} \frac{I_o}{I_s}$$

This method of absorption analysis can be successfully applied with homogeneous samples, such as the interphase, but in cases with very irregular dye binding, such as the chromatin distribution in chromosomes at different phases of division like prophase, metaphase, etc., it is necessary to compute data at two wavelengths and to work out the ratios between extinctions (Patau, 1952; Patau and Swift, 1953). Distributional error is indicated by dissimilar ratios.

The prerequisite of the two-wavelength technique of absorption analysis is the uniform illumination of the area with a monochromatic source and absence of light scatter through the use of proper mounting medium. In Feulgen-stained sections, it is desirable to carry out the analysis on the same slide, since acid hydrolysis—an essential feature of Feulgen staining—affects the reaction to a significant extent. The method of analysis is as follows (Pollister *et al.*, 1969):

If the two wavelengths selected are λ_1 and λ_2, the extinction (E_1) at the former should be half the extinction (E_2) at the latter, so that $E_2 = 2E_1$. As mentioned above, $E_1 = \log \frac{I_o}{I_s}$ at λ_1, and $E_2 = \log \frac{I_o}{I_s}$ at λ_2. Choice of the proper wavelengths is important. The total amount of absorbing material (M) in a measured area (A) is, $M = KAL_1D$, where K is constant; ($K = \frac{1}{e_1}$, e being the extinction coefficient in λ_1, K may be disregarded for relative determination in cell microspectrophotometry); L_1 is respective light loss and D is the correction factor for distributional error. Area (A) is measured as πr^2 where π may be omitted (Pollister *et al.*, 1969). From transmissions T_1 and T_2, at λ_1 and λ_2, the degree of light loss may be worked out as follows:

$$L_1 = 1 - T_1 \text{ and } L_2 = 1 - T_2$$

With the ratio L_2/L_1 at hand, the value of D can be worked out from the table given by Garcia (1962), given with detailed principles. The following table taken from Pollister *et al.* (1969) gives the value of D corresponding to each L_2/L_1 ratio:

Table: VALUES OF D FOR DIFFERENT VALUES OF L_2/L_1

L_2/L_1	0·00	0·01	0·02	0·03	0·04	0·05	0·06	0·07	0·08	0·09	
1·0 ·	—	4·033	3·461	3·134	2·907	2·734	2·595	2·479	2·380	2·294	1·0
1·1	2·218	2·150	2·089	2·033	1·982	1·935	1·892	1·851	1·813	1·777	1·1
1·2	1·744	1·712	1·683	1·655	1·628	1·602	1·578	1·555	1·533	1·511	1·2
1·3	1·491	1·471	1·453	1·435	1·418	1·400	1·384	1·368	1·353	1·339	1·3
1·4	1·324	1·310	1·297	1·284	1·271	1·259	1·247	1·235	1·224	1·213	1·4
1·5	1·202	1·191	1·181	1·171	1·162	1·152	1·143	1·133	1·124	1·116	1·5
1·6	1·107	1·098	1·091	1·083	1·075	1·067	1·059	1·053	1·045	1·038	1·6
1·7	1·031	1·024	1·017	1·011	1·004	0·998	0·991	0·985	0·979	0·973	1·7
1·8	0·968	0·962	0·956	0·950	0·945	0·940	0·934	0·928	0·923	0·918	1·8
1·9	0·914	0·909	0·903	0·899	0·894	0·890	0·884	0·880	0·876	0·871	1·9
2·0	0·867										

In Mendelsohn's method, the value of $L\alpha\,C$ can be worked out from any two transmissions and the value of $L_1\,D$ can be obtained by dividing it with 0·868, since $D = 0·868C$, where the values for C have been tabulated by Patau (1952). The two-wavelength method should always be followed for studying the distribution of DNA in non-homogeneous materials.

The photometric technique with visible light has proved to be a very effective and inexpensive tool in working out the quantitative correlation of DNA with the number of genomes present and also the distribution of this genic material at different phases of development and growth. This method will gradually have a much wider application, with the progressive invention of other dyes having stoichiometric specificity for different chromosome constituents.

REFERENCES

Caspersson, T. (1936).   *Skand. Arch. Physiol.* **73,** Suppl. 8, 1
Caspersson, T. (1950).   *Cell growth and cell function.* New York; Norton
Caspersson, T. (1965).   *Acta Histochem. Suppl.* **6,** 21
Dutt, M. K. (1967).   *Nucleus* **10,** 168
Edström, J. E. (1964).   In *Methods in cell physiology* **1,** 417, New York; Academic Press
Edström, J. E. and Beermann, W. (1962).   *J. Cell Biol.* **14,** 371
Freed, J. J. (1963).   *Science* **140,** 1334
Freed, J. J. (1969).   In *Physical techniques in biological research* **3C,** 95, New York; Academic Press
Freed, J. J. and Benner, J. A. Jr. (1964).   *J. Roy. Microscop. Soc.* **83,** 79
Garcia, A. M. (1962).   *Histochemie* **3,** 178
Greenwood, M. S. and Berlyn, G. P. (1968).   *Stain Tech.* **43,** 111
Herrmann, R. (1965).   *Acta Histochem. Suppl.* **6,** 189
Kamentsky, L. A. and Melamed, M. R. (1967).   *Science* **156,** 1364
Kamentsky, L. A., Melamed, M. R. and Derman, H. (1965).   *Science* **150,** 630
Killander, D. and Zetterberg, A. (1965).   *Exptl. Cell Res.* **38,** 272
Köhler, A. (1904).   *Z. Wiss. Mikroskopie* **21,** 129, 275
Ledley, R. S. (1964).   *Science* **146,** 216
Leuchtenberger, C. (1958).   In *General cytochemical methods* **1,** 219, New York; Academic Press
Loeser, C. N. and West, S. S. (1962).   *Ann. N.Y. Acad. Sci.* **97,** 346
Mendelsohn, M. L. (1966).   In *Quantitative cytochemistry.* New York; Academic Press
Mendelsohn, M. L., Kolman, W. A. and Bostrom, R. C. (1964).   *Ann. N.Y. Acad. Sci.* **115,** 998
Mitchell, J. P. (1967).   *J. Roy. Microscop. Soc.* **87,** 375
Nurnberger, J. (1955).   In *Analytical cytology,* **4,** 1, New York; McGraw-Hill
Patau, K. (1952).   *Chromosoma* **5,** 341
Patau, K. and Swift, H. (1953).   *Chromosoma* **6,** 149
Petriconi, V. (1964).   *Z. Wiss. Mikroskopie* **66,** 213
Pollister, A. W. (1952).   *Lab. Invest.* **1,** 106
Pollister, A. W. and Ornstein, L. (1959).   In *Analytical cytology,* New York; McGraw-Hill
Pollister, A. W., Swift, H. and Rasch, E. (1969).   In *Physical techniques for biological research* **3C,** 201, New York; Academic Press
Rudkin, G. T. (1966).   In *Introduction to quantitative cytochemistry,* 387, New York: Academic Press
Rudkin, G. T., Aronson, J. F., Hungerford, D. A. and Schultz, J. (1955).   *Exptl. Cell Res.* **9,** 193
Rudkin, G. T. and Corlette, S. L. (1957).   *J. Biophys. Biochem. Cytol.* **3,** 821
Sandritter, W. and Hartlieb, J. (1955).   *Experientia* **11,** 313
Scott, J. F. and Sinsheimer, R. L. (1950).   *Medical physics* **2,** 537, Chicago; Yearbook publishers
Slagel, D. E. and Edström, J. E. (1967).   *J. Cell Biol.* **34,** 395
Stowell, R. E. (1952).   *Lab. Invest.* **1,** 129
Swift, H. (1966).   In *Introduction to quantitative cytochemistry,* p. 1, New York; Academic Press
Taylor, E. W. (1950).   *Proc. Roy. Soc.* **137B,** 332
Thaer, A. (1965).   *Acta Histochem. Suppl.* **6,** 103
Vendrely, R. (1955).   In *The nucleic acids.* New York; Academic Press
Wagener, G. N. and Grand, C. G. (1963).   *Rev. Sci. Instr.* **34,** 540
Walker, P. M. B. (1958).   In *General cytochemical methods,* **1,** 164, New York; Academic Press
Walker, P. M. B. and Yates, H. B. (1952).   *Proc. Roy. Soc.* **140B,** 274
Walker, P. M. B., Leonard, J., Gibb, D. and Chamberlain, P. J. (1963).   *J. Sci. Instr.* **40,** 166
Welch, R. M. and Debault, L. E. (1968).   *J. Roy. Microscop. Soc.* **88,** 85
Wied, G. (1966).   ed. *Introduction to quantitative cytochemistry.* New York; Academic Press
Woods, P. S. and Zubay, G. (1966).   *Proc. Natl. Acad. Sci. US* **54,** 1705
Wyckoff, H. (1952).   *Lab. Invest.* **1,** 115
Wyckoff, H. (1959).   In *The cell* **1,** 1, New York; Academic Press
Zetterberg, A. (1966).   *Exptl. Cell Res.* **42,** 500

13 Microscopy

1. PHASE AND INTERFERENCE MICROSCOPY

In principle, these two types of microscopy are identical in the sense that the purpose is to bring about visible change in intensity from an undetectable phase change. The basic resemblance between the phase contrast and the interference has been elaborated diagrammatically and mathematically by Zernike (1952), Osterberg (1956) and Barer (1966). However, in spite of their fundamental similarity, considerably greater attention has been given to developing interference methods than to standard phase contrast. Interference systems, being more plastic, allow variable phase changes, and no doubt the variable phase contrast system can also be planned, although for quantitative measurement it is of little use. One can say that the phase contrast system is just an imperfect form of interference.

The phase change is represented as $\varphi = (n_p - n_m)t$, where n_p and n_m are the refractive indices of the object and the immersion medium respectively, and t is the thickness of the object. The formula indicates that with increase in the value of t, there will be a decrease in the detectable difference between the refractive indices.

The advantages of phase microscopy are:
(a) Simple and easily adjustable arrangement.
(b) Low cost of the apparatus.
(c) Insensitivity to slight variations in slide and coverslip.
(d) Internal details are often better resolved through 'zone of action' effect.

The *principal limitation* of the phase system, however, is that it is not possible to carry out quantitative measurements conveniently and the presence of a 'halo' prevents proper resolution to some extent. This limitation is inherent in the very principle of phase contrast microscopy. The direct light is allowed to fall on a conjugate area (annulus) here, while the diffracted light is separated and falls on the entire phase plate. Consequently,

276

the conjugate area also receives some diffracted light, which is responsible for the formation of a 'halo' around the object.

In addition to the attachments of an ordinary microscope, the phase contrast microscope is fitted with: (a) a sub-stage annular diaphragm to produce a narrow cone of light for illuminating the object, and (b) a 'diffraction' plate fitted on the rear focal plane of the objective, where the deviated and undeviated light rays are separated after emerging from the object. A layer of phase retarding material is present on that portion of the diffraction plate which is covered by either of the two rays, and it helps in changing the relative phase of the two rays. Two types of phase contrast microscopes are generally available, namely, (a) the instruments based on the principle of negative phase contrast, and (b) those depending on positive phase contrast. In the positive phase contrast system, the slightly retarding object details appear brighter against a lighter background, thus resembling visually stained preparations; while in the negative system, the object is decidedly lighter than the background. For routine use in cytology, the former is usually preferred due to the excellent contrast of living cells in aqueous media, where details of the retarding object appear sharper than the background, simulating stained preparations.

The importance of phase contrast microscopy has been widely appreciated in recent years and in cytological laboratories it has become a routine method for the study of various aspects of the structure and movement of chromosomes in the living cell. All good research microscopes can now be fitted with phase contrast equipment, including proper objectives and a sub-stage condenser. In view of its capacity to resolve the phase difference, it is effectively employed for the study of the three-dimensional nature of the objects and the phases of chromosome movement in a living cell. In the study of bacterial nuclei, phase contrast microscopy has helped considerably in resolving the details (Hewitt, 1951; Clifton and Ehrhard, 1952; *see* Barer 1956, 1966).

For measuring the refractive index and solid concentration of cell structures, the method of 'immersion refractometry' has been applied in phase contrast microscopy. This method is based on the principle that, if an object is immersed in a medium having a refractive index equal to its own, it affords the least contrast to the viewer. In addition, in order to have a choice of media for watching, a large number of fluids (isotonic solutions with different refractive indices) should be used. In the case of living materials, however, the choice of the immersion medium must be restricted to compounds which show no toxicity and which lack the ability to penetrate the cell or deform the cellular structure. The cell is initially observed in physiological saline or body fluid by positive phase contrast. Later, the observation is carried out in isotonic protein solution. If the refractive index of the medium is more than that of the cytoplasm, the latter presents a bright appearance, and vice versa. After several trials with different strengths of the medium, an optimum stage can be obtained, where the two will be nearly identical. The concentration of the solid can be worked out in g per 100 cm^3 of protoplasm, on the basis of the formula (Barer, 1966):

$$C = (n_p - n_s)/(\alpha \cdot n_s)$$

where n_s is the refractive index of the solvent, which is either water or dilute salt solution. For protoplasm, the value of α is 0·0018. In terms of wet weight, the gramme concentration of solid per 100 g of protoplasm, is

$\dfrac{C}{1+C/400}$. Following immersion refractometry, the change in refractive

index during cell division has been observed and it has been contended that there is a decrease in cytoplasmic concentration, which reaches minimum density during diplotene (Ross, 1954) or early metaphase (Barer and Joseph, 1957; Joseph, 1963). This method allows a study of the changes in the refractive index of the cellular constituents during cell division.

The convenience of securing quantitative results is the principal reason for the development of interference microscopy. The basis is empirical in determining the mass of single and different tissue elements. The actual optical path difference, or phase change, imparts a quantitative aspect to the measurement by interference microscopy (*see* Barer 1956, 1964 and Davies, 1958). In interference microscopy, the light is split into two beams by a beam-splitting mirror, one beam being transmitted through the object and the other passing some distance to the side of it. The interference is produced by the two beams combining at the semi-reflecting mirror.

The advantages of interference microscopy are (Barer, 1966):

1. Phase changes of the material can be measured.
2. Bright colour effect can be secured.
3. The contrast can be varied, allowing a proper type of contrast to be selected with respect to the object, to secure the intracellular details. The system is elastic; variable phase contrast can be obtained.
4. 'Halo' and 'Zone' effects are absent.
5. Mass per unit area of the cell can be conveniently measured.

The flatness of the image obtained and glare are serious disadvantages of the interference system.

Microscope interferometry, in conjunction with immersion refractometry, permits the determination of the refractive index, the solid concentration, the thickness of structures and the dry mass. With the aid of microscope interferometry, the changes in the amount of protein and DNA in the nuclei of mouse fibroblasts, as well as in ascites cells, have been measured (Richards and Davies, 1958), and other cancerous tissues studied (Longwell, 1961). In cytology, the most successful application has been made by Mellors and others (*see* Mellors, 1959); they claim to have measured the total protein in the chromosomes, interphase nuclei, and sperm heads in mice, and have concluded that in the prophase of germ cells (haploid) the chromosome set is equal in mass to the sperm head; similarly, the double set of chromosomes equals in mass the resting somatic nucleus of mouse liver. The implication is, therefore, that all the protein of the nucleus, at least in prophase, is located in the chromosomes. These conclusions have, however, been questioned by other workers (Richards and Davies, 1958). In view of the results obtained by different workers, Davies has suggested that the protein associated with the haploid set of chromosomes is definitely more than the amount associated with the DNA in the sperm head.

An additional use of interference microscopy has been in the investigations on isolated nuclei. The total protein content has been found to be more reliable as deduced from interferometry than by the u.v.-light absorption technique. The results of Hale and Kay (1956), on the measurement of dry mass of nucleoprotein of calf thymus, isolated in three media (namely, citric acid, sucrose calcium chloride and non-aqueous solution) are nearly identical with those obtained by ordinary weighing procedures. Cell division has been filmed by using the interference system with excellent colour effects (Ambrose and Bajer, 1960; Ambrose, 1963).

Interference microscopes, though sometimes applied for standard observation, are chiefly measuring instruments, whereas for routine work phase contrast is definitely adequate and until now irreplaceable. The details of the two-beam interference microscope have been dealt with exhaustively by several workers (Davies, 1958; Hale, 1958; Barer, 1959, 1966; Francon, 1961; Krug *et al.*, 1964).

II. POLARISATION MICROSCOPY

Polarisation microscopy is based on the principle of birefringence against polarised light exhibited by certain objects of complex molecular arrangement. This gives an indication of the physical heterogeneity of the structure at the molecular level. There is no doubt that with the development of electron microscopy, the importance of this method has been minimised, but even so, birefringence and dichroism can reveal the molecular orientation of structures beyond the resolution of electron microscopy. When a beam of polarised light is received by such a structure, the ray is split into two rays polarised in mutually perpendicular lines. Of these, the ray which obeys the ordinary laws of refraction is known as the 'ordinary' ray, and the other, whose velocity through the object is different, is known as the 'extraordinary' ray. The difference in the refractive indices $(\mu_e - \mu_o)$ associated with the process gives the value of birefringence. The two polarised rays, after emerging from the object, recombine but, because of different velocities through the object, one shows retardation as compared with the other. The value of 'retardation' T, which is based on the birefringent property, is counted by multiplying birefringence $(\mu_e - \mu_o)$ with the thickness of this object t, i.e., $T = (\mu_e - \mu_o) \, t$ and is expressed in terms of wavelength in Ångström units.

Birefringent or anisotropic objects possess one, or several, optical axes. In general, the optical axis of the fibre of a biological object coincides with its lengthwise or perpendicular direction. 'Positive birefringence' signifies that the refractive index for light vibrating parallel to the long axis is greater than the one perpendicular to it. Similarly, 'negative birefringence' denotes the reverse property. Nucleic acid and nucleoprotein fibres are negatively birefringent, whereas mere protein fibres are positively birefringent. 'Isotropic' objects are those in which, owing to their complexity and heterogeneity of structure, both properties are present and one neutralises the other. 'Intrinsic birefringence' is due to regularity in the pattern of molecules whereas 'form birefringence' in elements is caused by preferred orientation of asymmetrical particles.

Polarisation microscopes are more or less identical in principle to ordinary microscopes, but fitted with polarising elements. Instead of the ordinary light, a low intensity carbon arc or a high pressure mercury arc is used as the source of illumination. Two principal polarising elements fitted to the microscope are the 'polariser' and 'analyser'. A sheet of 'polaroid' film (iodine polyvinyl alcohol polarising film) fitted below the substage condenser serves as the polariser; it is better than the prism polarisers used previously, and the analyser is fitted in the body tube of the microscope, above the objective lens, and can be rotated to obtain alternate bright and dark appearances at every 180 degrees turn. When the axis of transmission of the analyser is parallel to that of the polariser, the maximum amount of light transmission can be obtained, while in the case of no transmission, otherwise known as extinction, the positions of polariser and analyser are crosswise. In addition to these two, the 'compensator' is another necessary component, being formed of birefringent plates or crystals, and as its 'retardation' value is known, it helps in determining the slow and fast axes of transmission of an object. A weak birefringent object when viewed between the crossed polariser and analyser assumes a faintly bright appearance against a dark background. Maximum brightness can be observed when the object axis lies at 45 degrees with respect to the analyser. At this stage, if the compensator is inserted below the analyser, the object will brighten when the compensator is rotated in one direction, and lighten when it is rotated in the opposite direction. When the object shows maximum brightness, it is implied that the directions of the slow ray of the compensator and the object are identical. The brightness is therefore due to the joint refraction of the two.

The selection of lenses is an important factor in polarisation microscopy and as such, a lens is selected so that it does not contain components (fluorite) which often show anomalous birefringence. Moreover, in order to reduce scattering of light through air–glass surfaces, it becomes necessary to 'gloom' the lens surface which involves the deposition of a thin transparent film of desired refractive index and thickness. For details on polarisation microscopy, the reader is referred to Bennett (1950), Frey-Wyssling (1953) and Mellors (1959).

In cytology, polarisation microscopy has been particularly helpful in studying the structure of the mitotic spindle and the changes it undergoes during cell division (Barer, 1955; Inoué, 1959; *see* Mazia, 1961; Harris and Mazia, 1962). Because of the birefringent nature of the specimen, phase contrast microscopy is not suitable. On the basis of his observations on the birefringence of fibrils in centrifuged and compressed eggs of sea urchins, Inoué (1951) suggested the presence of chromosome fibres at least (between pole and chromosome) and spindle fibres from pole to pole (*see* Swann, 1951). Birefringence apparently decreases at anaphase, and Swann has interpreted this as being due to the release of some active substances from the chromosomes but Inoué interprets the inactivation of the spindle by colchicine as due to the disorganisation of micelles by depolymerisation.

Birefringence in the sperm head has been claimed to be due to the DNA molecules, oriented parallel to the longitudinal axis of the chromosomes (Schmidt, 1937). Such an arrangement of the DNA molecules has also been suggested for the salivary gland chromosomes of diptera. Frey-Wyssling (1943), on the other hand, presented evidence against the assumption of this parallel arrangement in living chromosomes. Birefringence has also been attributed to the effects of ethanol fixation (Ruch, 1945). Inoué and Sato (1962) suggested the orientation of DNA molecules in the form of a 2000 Å thick coil—a constituent of a supercoil of 8000 Å thickness in the living sperms of *Ceutrophilous nigricans*. In living endosperm cells, Inoué and Bajer (1961) noted birefringence in the centromere.

Though birefringence still presents a number of difficulties in the interpretation of structures, it provides a good approach to understanding the orientation of birefringent objects, and should be supplemented with other methods to study the property of the oriented molecules.

As compared with birefringence, dichroism in ultraviolet light, which is specific at least in relation to certain compounds, has been found to be more useful in the study of molecular orientation. Ambrose and Gopal-Ayengar (1952), on the basis of observations on chromosomes vitally stained with neutral red, suggested the folded nature of the DNA molecule, the chains in these folds lying parallel to one another, in dipteran salivary gland chromosomes. Caspersson's (1940) method of analysing dichroism in ultraviolet rays, and the differential absorption of nucleic acids (2600 Å) and proteins (2800–2900 Å) has been employed to work out the molecular relationship between the two in chromosomes (Ruch, 1951, 1955; Wilkins, 1951; Seeds, 1953). High absorption has been considered to be an index of molecular direction. Parallel arrangement of DNA molecules has been indicated to some extent. Ruch (1966), with the aid of an apparatus meant for measuring weak anisotropic objects, studied the orientation of DNA and protein in dipteran salivary gland chromosomes fixed in 50 per cent acetic acid with 1 per cent lanthanum acetate. Slight longitudinal orientation of DNA molecules has been shown and comparative analysis under fixed and living conditions has been recommended. Undoubtedly the study of dichroism in u.v.-rays is emerging as an additional tool in studying molecular inter-relationships of cellular constituents, but at the present level of refinement, its use in the study of chromosome structure is rather limited.

III. X-RAY MICROSCOPY

The direct study of biological materials at the molecular level is principally by electron microscopy, although it generally does not allow the resolution of structures below 10 Å. In polypeptide chains, where the distance between the atoms is of much lower magnitude (1·5 Å), resolution by this method is not possible. On the other hand, x-rays are scattered due to diffraction by all forms of matter and from the diffraction pattern, which is dependent on the position of atoms, one can get an idea of the molecular orientation. This method even allows the study of molecular patterns from gels and concentrated solutions.

In studying the chemical and physical nature of chromosomes, different properties of x-rays, such as diffraction, emission and absorption are advantageously employed. For analysis of the x-ray diffraction pattern, a narrow beam of x-rays is passed through the object, and a scattering of rays occurs because of diffraction by the atoms. On emerging from the object, the rays diverge, following the pattern of diffraction which is recorded on a photographic plate placed in the vicinity of the object (0·06–0·1 cm), the plate is developed later, using an amber glass safe lamp. The symmetry and pattern of the structure can be determined on the basis of the angle and intensity of the diffraction patterns. In the study of bio-molecular structures when repeating units are present, the distance between each unit can also be worked out by the diffraction pattern of x-rays of known wavelength from the fact that the angle of diffraction and the distance between the units are inversely proportional (Astbury, 1945, 1947; Furberg, 1950; Oster, 1950, 1951; Bragg, Kendrew and Perutz, 1950). In the study of chromosomes, outstanding achievements have been made with the aid of this technique and the establishment of the double helix configuration of the DNA molecule (Watson and Crick, 1953; Wilkins and colleagues, 1953) and the pattern of arrangement of different components of DNA has been made possible.

For details on x-ray microscopy, the reader is referred to Oster (1955), Engström (1962) and Wilson and Morrison (1961).

For the application of analytical principles based on the use of x-rays, it is necessary to have an x-ray imaging system. On the basis of x-ray absorption characteristics, several qualitative and quantitative assessments at the cellular level can be carried out. For dehydrated biological specimens and thin sections of soft biological tissues, up to 1–10 μm thick with up to 35 per cent dry mass, it is preferable to generate the spectrum at 500–2000 V. Engström (1966) has devised a number of models for microradiography with ultra-soft x-rays, incorporating vacuum system, x-ray tube and high voltage source (Combée and Engström, 1954; Lindström, 1955; Henke *et al.*, 1957, etc.). Point projection x-ray microscopes have also been evolved (Cosslett and Nixon, 1953, 1960). Projection microscopes have been utilised for obtaining representation of chromosomes, clearly delineating the mass and water distribution (Engström and Lindström, 1950; Engström, 1966). In the preparation of microradiographs in connection with dry weight determination of objects, such as ascites tumour cells, salivary gland chromosomes, etc., it is preferable to adopt the smearing technique and freeze-drying of the smears for examination in high vacuum x-ray tubes.

For mass determination, in cells from solid tissues, the best procedure is to use thin frozen sections and freeze-dry in the sample holders. Fixation with chemicals may cause distortion of the structure and is not recommended as such (Engström, 1966).

Because the molecular interaction between chromosomes and x-radiation follows certain fixed principles, x-ray microscopy is proving to be an extra method for the qualitative and quantitative assessment of the structure and behaviour of chromosomes.

IV. FLUORESCENCE MICROSCOPY

The utilisation of the fluorescence shown by some of the cell constituents, as well as of some special dyes, against ultraviolet light, forms the basic principle of fluorescence microscopy. In this sytem, intracellular constituents are detected either through their own property of autofluorescence, or by secondary fluorescence due to the adherence of labelled or fluorescent dye.

To detect an intracellular constituent on the basis of its own fluorescence, short wavelength ultraviolet light transmitted from a quartz condenser fitted with a microscope is required. Several compounds can be identified on the basis of their own fluorescence, provided an understanding is obtained as to the exact wavelength which excites this, as well as the wavelength of the excited substance. One of the serious limitations of this technique is that an apparatus fitted with an ultraviolet monochromator and ultraviolet transmitting quartz condenser is difficult to obtain.

The general practice in the majority of laboratories is to utilise the property of secondary fluorescence obtained by fluorochrome preparations, but such dyes become effective only in long blue or near ultraviolet light (Cowden, 1960). Like other dyes, these agents are also either acidic, basic or amphoteric in nature and can be used in accordance with their application in cytological practice, but a special technique is necessary for the detection of secondary fluorescence of specific compounds within the cell. The method involves the observation of frozen sections or squashes against ultraviolet light and through a standard microscope fitted with adequate condensing lenses and filters. A special advantage of this technique is that unfixed materials can be studied, thus eliminating the artefacts which often arise due to fixation.

Conjugated planar dye molecules, in solution, form complexes having properties different from those in monodispersed form. The metachromatin dye polymer complexes allow observation of fluorescence and absorption spectra. One of the planar dye molecules is acridine orange, dissociating near pH 7 (Strugger, 1949; Albert, 1951; Zanker, 1952). Two equivalent resonating structures are,

and

This dye, as a monomer at a pH 6·0, has fluorescence and absorption peaks at 535 and 490 nm, and as a polymer the peaks are at 660 and 455 nm respectively. It is difficult to distinguish spectroscopically between acridine orange—DNA and acridine orange—RNA complexes (Loeser *et al.*, 1960; Wolf and Aronson, 1961; Steiner and Beers, 1961; Ranadive and Korgaonkar, 1960; West, 1969). However, in fixed preparations (Armstrong and Niven, 1957; Mayer, 1963), it may be possible to distinguish DNA and RNA on the basis of metachromasia of acridine orange, a red colour specifying RNA, and the yellowish-green colour indicating DNA. But under conditions *in vivo*, no such differentiation could be recorded.

On the basis of the assumption that amino groups in the 3, 6 positions of acridine react with phosphoric groups of nucleic acids, Lerman (1963, 1964) suggested internucleotide binding of planar dye molecules. Dye-polymer complexes and their optical properties have also been studied (Bradley, 1961; Stone, 1964, 1967; Stone and Moss, 1967).

Puchtler *et al.* (1967) worked out the relationship between structure and fluorescence of azodyes. Several azo dyes become fluorescent after combining with tissues. Aromatic compounds in the dye control its fluorescent property. In other classes of dyes, ring closure, coplanarity of chromophores, accumulation of ring system and low dye concentrations exert an inhibiting influence. There are also other factors, which control the fluorescence of triphenyl methane, anthraquinone and quinone-imine dyes.

Quantitative measurements can be carried out by fluorescence microspectroscopy. Long exposure time for photographs should be avoided as the intensity may fade with continuous exciting irradiation of the fluorochrome-stained cell, in certain cases (*see* West, 1969).

In preparing tissue sections meant for observation under the fluorescent microscope, chemical fixatives are avoided as far as practicable, freeze-dried or chilled preparations of unfixed materials being preferable for study by this method. In cases where natural fluorescence is absent, the choice of proper and specific dyes, with fluorescent properties, is the most critical step in the procedure, as the dye imparts fluorescence to its specific substrate which can be detected *in situ*.

Frozen sections of unfixed tissue can be cut in a microtome and maintained in a refrigerated cabinet at $-20\,°C$. Freeze-dying, during which the temperature of the tissue should be maintained at about $-40\,°C$ by means of a slush of diethyl oxalate, can also be employed. In principle, this involves molecular distillation, and in order to secure full molecular distillation, the cold spot, that is, the liquid nitrogen water trap ($-194\,°C$), is placed within a short distance of the tissue block, the pressure being kept at $2–5 \times 10^{-4}$ mm/mg. This adjustment within a short distance is necessary so that the mean free path of the water vapour molecules, at the pressure employed, will be more than this distance. Within this procedure, freeze-drying can be completed in 7 h. Solid carbon dioxide can also be used as the drying agent, when drying may take 2–3 days (Danielli, 1953). Freeze-dried sections should always be floated on non-aqueous medium or adequate fixative. Glycerine, isobutyl methacrylate, methyl salicylate, etc. are good mounting media.

The most adequate light source, emitting ultraviolet rays, is the carbon arc, preferably with a direct current and fitted with an electromagnetic field. Among other sources, high pressure mercury vapour arcs may be mentioned. Coons (1958) recommended the use of A-H6 (General Electric Co.), HBO 200,109 (Osram), ME/126 (Mazda) mercury vapour arcs.

Suitable condensing lenses of single bispheric type are generally fitted in front of the light source so that the back lens of the dark-field condenser gets the image of the horizontal carbon crater. The filter is prepared in the following manner (Coons, 1958; *see* also Richards, 1955).

A Pyrex water cell, containing copper sulphate solution, is fitted in front of the light source. The cell is fitted with a cooling device by means of a glass coil immersed in the $CuSO_4$ solution inside which running water circulates, as without this cooling device, the $CuSO_4$ solution may boil within a short period of turning on the carbon arc. A filter to remove the middle range of the spectrum is attached to the side of the cell away from the arc.

The light, passing through the $CuSO_4$ solution and the filter, ultimately falls on an ordinary glass mirror. The microscope receiving the light is a standard one fitted with a chromatic objective and dark-field condenser. A protective Wratten filter must be used in the eyepiece to remove undesirable wavelengths, such filters generally being colourless, as colouring may cause difficulty in detecting fluorescence of the same colour. Monocular lenses are always preferred over binocular ones in order to reduce scattering of light. Image intensity in the microscope is generally directly proportional to the square of the objective NA.

For photography, 35 mm films, with exposure time necessary for fast films, are usually employed; the negatives are generally thin and strong contrast printing is desirable.

The advantages of fluorescence microscopy were first realised in attempts to detect specific antigens. Richards (1955) and Coons (1958) have adequately reviewed this aspect of the use of fluorescence microscopy. In principle, the method involves the forced production of antibodies by the injection of antigens labelled with fluorescin isocyanate into the tissue. The conjugation of the gamma globulin of the antibody with fluorescin isocyanate in the antigen, makes visible the site of the antibodies within the tissue.

On the basis of this principle, originally applied to antigens, modifications of the technique have been evolved for the detection of proteins, as well as DPN and TPN, in cytological practice (*see* Glick, 1959). A measurement of the amount of fluorescence also allows a quantitative estimation of the substance transmitting secondary fluorescence. As with visible light and u.v.-photometric methods, fluorescence emission and absorption analyses are also being carried out quantitatively according to the same principle. This analytical method has a wide application. The details of modifications in instrumentation for quantitative fluorometric study have been given by West (1969). Both absorption and emission spectra are used for quantitation. Methods have been developed for photographic microspectroscopy and for photoelectric microspectrophotometry. The fluorometric technique developed by Holter and Marshall (1954) is interesting. With the photo-multiplier tube attached to a fluorescent microscope, on the stage, the sample is fitted in a capillary tube.

The method of Lowry, Roberts and Kapphalm (1957) is convenient for the study of DPN and DPNH. DPN does not show any fluorescence as such, but in a strong alkaline medium it develops this property, while DPNH, although it is itself fluorescent, shows an increase of fluorescence if it is oxidised to DPN and treated with a strong alkali. These properties have been made use of in detecting their presence, or absence, as well as their amount in the tissue. Krooth *et al.* (1961) demonstrated that mammalian chromosomes react with several human sera, including those of diseased patients. The method involves an initial exposure of the chromosome preparation to the serum, followed by exposure to horse anti-human globulin, conjugated with fluorescent dye (Holborow *et al.*, 1957). Fluorescence has been observed in all the chromosomes, though several diseased sera did not show any reaction. The fluorescence is based on the principle that gamma globulin of the sera reacts with the nuclei.

Further, anti-nuclear globulins from patients with collagen diseases have been shown to bind with various nuclear constituents in the interphase and metaphase (Razavi, 1968). In different sera, the specificity and titer of anti-nuclear factors have been found to differ, though in the majority of cases at least two or more cell constituents are reacted upon. The nuclear constituents which react with serum differ in distribution and concentration in various types of cell, and according to the phase of cell metabolism. As fixation is also an important factor, difficulties are often encountered in the interpretation of fluorescence data. With the aid of fluorescin-labelled anti-DNA serum, the distribution of single stranded DNA in lymphocyte chromosomes has been demonstrated (Razavi, 1968). As gene activity has been found to be continuous throughout the chromosome cycle, it has been suggested that the DNA molecules may just serve as the coding device, whereas gene expression is controlled by changeable cytoplasmic and nuclear factors.

For the study of the effect of chemicals on the nucleus and cytoplasm, fluorescence microscopy is often recommended. In the study of cellular nucleoproteins and nucleic acids, diaminoacridine dyes, such as acriflavine, acridine orange and acridine yellow have been employed because of their affinities (Strugger, 1949). When applied to living cells, as mentioned before, they react principally with nuclear DNA and RNA—as such, they are showing increased application in chromosome study (De Bruyn *et al.* 1950, 1953). With acridine orange, RNA and DNA can be differentially stained, as the former appears red and the latter looks green under a fluorescence microscope. This coloration and its intensity show alterations in diseased conditions, or in injury. Therefore, for the detection of the effects of ionising radiations as well, this method has been applied (Meisel *et al.*, 1961; Seydel and Lawson, 1966). The effect of aromatic diaminidines in nuclei and cytoplasm has also been measured through fluorescence microscopy by Snapper *et al.* (1951).

Bushong *et al.* (1968) demonstrated that the intensity of fluorescence in acridine orange-stained preparations varies at different wavelengths. Maximum effect was observed at 525 nm and a progressive shifting towards shorter wavelengths was noted. At 580 nm, with increase in illumination time, there was a decrease in intensity. At levels lower than 580 nm, initial brightness was followed by a gradual decrease in intensity.

In the detection of diseases, such as, malignancies (Bertalanffy, L. 1958; Bertalanffy, F. D., 1960; Sherif, 1963), by fluorescence microscopy, the wavelength and intensity, as previously mentioned, are of special importance.

Porro *et al.* (1963) have published a list of biological dyes with detailed information on their fluorescent property and absorption spectra, to which the interested reader is referred. For further details of instrumentation, Richards (1950, 1955), Mellors (1959) and West (1969) should be consulted.

For identification and quantitative estimation of cell constituents, the importance of methods based on absorption and emission of fluorescent compounds is gradually being realised (*see* Passwater, 1970). Television fluorescence microspectrophotometry (West, 1965) is now allowing the detection of even extremely minute quantities of substances in the living cell. Fluorescent antibody technique has been a powerful aid in detecting chemical differences (Goldman, 1968). In the field of chromosome study, there is ample potentiality in the combined analytical technique based on chromosome isolation and their fluorescence emission analysis as well as specific fluorescence of chromosome segments. With the gradual invention of specific antisera, their application on isolated chromosomes may ultimately lead to chemical identification of specific segments of chromosomes, based on their differential fluorescence.

SCHEDULE FOR THE STUDY OF CHROMOSOME FLUORESCENCE FOLLOWING TREATMENT WITH ANTI-DNA SERUM (RAZAVI, 1968):

Cell Preparations

(a) Use as substrate, intact white cells from mixed short-term unsynchronised primary cultures, washed in globulin-free serum for preliminary experiments. Smear the cells on a slide, applying spirally with a glass rod. Fix the slide in cold acetone for 5 s. Wash in phosphate-buffered saline (pH 7·2, 0·15 M—'PBS') for 5 min at room temperature, and stain.

(b) Prepare osmotically isolated interphase and metaphase nuclei of lymphocytes from partially synchronised phytohaemagglutinin-stimulated cultures at 45–50 h (*see* chapter on human and mammalian chromosomes). Add required amount of distilled water to dilute cultures to 100 m mol, determined for each batch of medium with the help of an advanced instrument freezing point osmometer. After 20 min, centrifuge at 500 rev/min for 5 min and decant supernatant. Add 2 cm^3 of pre-cooled acetic acid–methanol (1:3) mixture to the tube drop by drop, while gently agitating the cells in an iced water-bath. Keep overnight at 4 °C. Change fixative to acetic acid–methanol mixture (1:2). Suspend by pipetting. Centrifuge and remove fixative, keeping only 0·2 cm^3. Add 0·2 cm^3 of 50 per cent acetic acid, resuspend the cells. Prepare air-dry smears by spreading a drop on a slide at 55 °C and drying for 30 s. Wash and stain.

Nuclease Treatment

To one set of unstained slides add desoxyribonuclease solution ($0 \cdot 1$ mg/cm^3) at 37 °C, in PBS containing $0 \cdot 05$ mg MgCl$_2$/cm^3. Treat controls in enzyme-free solutions, wash all slides in PBS and stain.

Antisera

Conjugate 'reticular' anti-DNA serum procured from microbiological laboratories at a fluor: protein ratio of $1 : 100$. Remove the free dye by DEAE-sephadex chromatography, using saline (buffered at pH $7 \cdot 0$ by $0 \cdot 0175$ M phosphate) as eluant, according to Tokumaru's technique (1962). Concentrate to the original volume by dialysis against 25 per cent pyrrolidone and store at 4 °C with $1 : 10 000$ merthiolate as preservative.

Fluorescent Antibody Staining

Treat slides, in Petri dishes lined with damp filter paper, with serum for 1 h at 37 °C. Wash in 500 cm^3 PBS for 15 min with gentle stirring. In preliminary experiments, see section (*a*), a two-layer technique with unlabelled LE serum and conjugated horse anti-human globulin serum (Sylvana, Millburn, N.J.) may be followed.

Ultraviolet Microscopy and Photomicrography

Mount cover slips with 10 per cent PBS in glycerol. Illuminate by a sylvania L50 mercury vapour lamp attached to a microscope with a dark-field condenser. Observe through apochromatic objectives, using UG2 (4 mm) exciter and No. 47 barrier filters. Photograph through BG12 exciter and No. 47 barrier filters on Agfachrome 35 mm film with 3 and 6 min exposures.

Observations

From anti-DNA sera tests on osmotically isolated, and acetic-methanol fixed nuclei, obtained in cultures partially synchronised by overnight treatment at room temperature after 8 h initial incubation at 37 °C, discrete attachment of the antibody is seen in interphase and metaphase nuclei. In the brightly stained metaphase chromosomes, the more frequent attachments are telomeric, or near secondary constrictions. Pre-heating to 80 °C for 5 min intensifies these patterns; treatment with desoxyribonuclease reduces them and brief pre-heating in an open flame destroys them.

Localisation of Differential Segments of Chromosomes

For the localisation of chemically differentiated segments of chromosomes, especially the heterochromatic ones, Caspersson *et al.* (1968, 1969, 1971) utilised the different methods of binding of fluorescent DNA reagents, such as, propyl quinacrine and quinacrine mustard dihydrochloride. The principle involved in the treatment is that the biological effect of alkylating agents is the alkylation of nucleophilic sites, i.e. reaction with the N-7 atom of guanine. The guanine-rich segments are expected to show the accumulation of the dye. With the use of bifunctional quinacrine mustard, differentiation of the chromosome segments could be observed. For observation and recording, they have devised methods involving ultramicro-spectro-photography (UMSP) computerised sorting and ultramicro-fluorometry (UMFL), and for scanning and integration the ultramicro-interferometer has been used.

Zech (1969) noted that in human spermatozoa, the distal proportion of the Y chromosome shows brighter fluorescence. For other differences in size and surface properties of cells and other applications, the reader is referred to the review by Beatty (1970) and Chapter 15.

Vosa (1970), Pearson *et al.* (1970) and Barlow and Vosa (1970) have developed methods for identifying Y chromosomes in buccal smears of human interphase nuclei as well as for differentiating specific sites in human lymphocyte chromosome preparations, using quinacrine (0·005 per cent quinacrine mustard in deionised water for sperm, or 0·5 per cent aqueous or 1 per cent solution in absolute ethanol). George (1970) also used quinacrine mustard (250–300 µg/ml in glass distilled water) for human chromosomes. In all these methods, except for sperm, the material after treatment with colchicine or colcemid, is fixed in acetic-methanol or ethanol, washed, rinsed in buffer (pH 4·1–5·5), stained for a few minutes and mounted in distilled water, buffer or buffered glycerol. For photography, Kodak-Tri X-Pan film with exposure time of 10–30 s has been recommended. For viewing, the Zeiss photomicroscope with fluorescent illumination (Filter-tr. range– 300–500 nm) is recommended. Conen *et al.* (1970) and Thuline (1971) used the same principle to identify Y chromosomes in blood smears.

V. ORDINARY LIGHT MICROSCOPY

In light microscopy, the underlying principle is to obtain a real, inverted, and enlarged image of the material by means of the objective lens, followed by the formation of a virtual image by means of the eyepiece lens. In the study of chromosomes, where only light microscopes are required, the compound microscope should have at least the following attachments:

1. Apochromatic objective and oil immersion lenses ($\times 100$)—(1·3–1·4 N.A.);
2. Sub-stage aplanatic and achromatic condenser—1·4 N.A.
3. Compensating eyepieces ($\times 10$, $\times 15$, $\times 20$);
4. Fitted mechanical stage.

The tube length for each microscope is fixed for the operation of particular objectives. In British instruments, it is generally 160 mm (Haskell and Wills, 1968). Details of the instrumentation in microscopes are available in several textbooks; the designs are based on two original types of instrument, named after their inventors, the Huygens and Ramsdens. For convenient reference, Corrington (1941) and Oliver (1947) may be mentioned. The principles of the use of special types of lenses are here outlined.

Special lenses are used principally to eliminate two types of aberration: spherical and chromatic. Spherical aberration is inherent in lenses with spherical surfaces. Here the rays passing through the periphery of the lens focus at a different point to those passing through the centre or close to the axis. In *aplanatic* lenses, which are compound and constituted of different kinds of glass, all the rays are brought to a common focus by suitable corrections.

Chromatic aberrations imply that ordinary sources of illumination, being composed of light of different wavelengths, or in other words of different colours, focus at different points because of a variation in the path followed by the rays. Colour fringes sometimes appear. In suitably constructed lenses, these aberrations are eliminated as far as is practicable, so that preferred rays are made to focus at a common point. In *achromatic* objectives the chromatic aberration is corrected for two colours, or more precisely, two wavelengths, and the spherical aberration for one colour. In *semi-apochromats* or *fluorite* lenses, a higher degree of correction than the former is achieved. In *apochromatic* lenses, the chromatic aberration is corrected for three colours and spherical aberration for two colours. Such lenses do not allow the formation of any secondary spectrum.

Different colours undergo different degrees of magnification, which may result in a number of images of different colours; because of superimposition, difficulties arise in the proper clarification of the object. This inequality of colour·magnification is corrected by means of *compensating* eyepieces, while apochromatic lenses are used as the objectives. In the case of achromatic lenses, this defect is generally corrected in its own combination.

The separation of finer details in an image is dependent not on the magnification of the lenses but on their resolving power—this power, or resolution, being dependent on the wavelength of the illuminating source as well as on the *numerical aperture* of the lens. The numerical aperture is calculated on two factors; the angular aperture of the lens, and the refractive index of the medium through which the light enters. Resolution is always proportional to the numerical aperture, which is $n \sin \mu$; n being the refractive index of the cover slip, and μ the maximum angle to the optical axis formed by any ray passing through the specimen, before the formation of total internal reflection. With an objective of N.A. 1·4, it has been calculated that good resolution can be achieved if the intervening minimum distance from the specimen is $0·24 \times 0·001$ mm (Haskell and Wills, 1968). Brightness increases with increase in numerical aperture, but decreases with increase in magnification. Visual magnification is obtained by multiplying the magnification of the objective with that of the eyepiece.

Oil immersion objectives are used for critical work where the scattering of light, due to its passage through media of different refractive indices, is to be avoided. If a fluid of refractive index similar to that of glass and canada balsam is used to bridge the gap between the cover slip and objective lens (and if necessary, between slide and condenser), then a homogeneous medium can be achieved for the path of light, avoiding light scattering as far as possible. Cedarwood oil is generally used for the purpose, its refractive index (1·510) being close to that of glass (1·518), and balsam in xylol (1·524), but several synthetic media are now available. Darlington and LaCour (1968) used media with liquid paraffin and α-bromonaphthalene, or olive oil and α-bromonaphthalene, as constituents. For the source of illumination, Pointolite, Ribbon filament, mercury arc, or, if these are not available, even a 100 W ordinary lamp may be used. Proper screens should always be chosen for observation, depending on the colour of the light and stain used. Wratten yellow-green filters for violet-stained preparations and blue filters for red-stained preparations are useful.

Camera lucida, drawing head or drawing prisms (Zeiss, Leitz, etc.) may be used for drawing on a board placed next to the microscope, and for direct measurements ocular micrometers may be used if necessary. The scale in the ocular micrometer is standardised with the stage micrometer in which the value of each division is known. The latter is fitted onto the stage in place of the slide and, by focusing, the ocular and the stage micrometer scales are brought into the same focus. The number of ocular divisions per stage division is worked out in several readings and from the mean, the value of each ocular division can be calculated. The magnification of the drawing can be estimated by drawing a stage division on the paper, with the same combination of lenses and drawing head as used for observation of the material. The length of the drawing is measured on a millimetre scale. The value (in millimetres) of the drawing, divided by the actual value of stage division (which is known) gives the magnification of the drawing.

For photography, a Zeiss (Ikon)—35 mm camera or plate camera may be used. The camera is fitted over the eyepiece and a number of cameras with microphotographic attachments, like Zeiss, Leitz, Olympus, etc. are now available. They are designed as a 35 mm or plate camera without the lenses, as the real image is formed over the eyepiece. These cameras are always fitted with adaptors for fitting in the microscope tube. Elaborate, and even built-in camera attachments can also be obtained with several microscopes, with provision for timing and exposure (*see* Stevens, 1957; Needham, 1958 etc.).

Both slow-and high-speed films are satisfactory, depending on the requirements, the former requiring a longer exposure period. Rapid process panchromatic plates, Kodak, Gavaert or Ilford, are most suitable. The plate size is generally $3\frac{1}{4} \times 4\frac{1}{4}$ in but 9×12 cm plates may also be obtained.

The selection of an adequate period of exposure is a matter of experience and is fixed after some trials. For developing and printing, the procedure for ordinary photography is followed, using the proper developers recommended for particular films. Preparation of lantern slides from negatives with the aid of enlargers and slides from contact prints, can be made following the usual procedure.

REFERENCES

Albert, A. (1951). *The Acridines*. London; Arnold
Ambrose, E. J. (1963). In *Cinemicrography in cell biology*, p. 123, New York; Academic Press
Ambrose, E. J. and Bajer, A. (1960). *Proc. Roy. Soc.* **153B**, 357
Ambrose, E. J. and Gopal-Ayengar, A. R. (1952). *Heredity* **6**, 277 and *Nature* **169**, 652
Armstrong, J. A. and Niven, J. S. F. (1957). *Nature* **180**, 1335
Astbury, W. T. (1945). *Nature* **155**, 501
Astbury, W. T. (1947). *Symp. Soc. exp. Biol.*, **1**, 66
Barer, R. (1955). *Analytical cytology* p. 301, New York; McGraw-Hill
Barer, R. (1956). In *Physical techniques in biological research* **3**, 1st ed., p. 29, New York; Academic Press
Barer, R. (1959). In *Analytical cytology* p. 159, New York; McGraw-Hill
Barer, R. (1964). In *Cytology and cell physiology*, 3rd ed., p. 91, New York; Academic Press
Barer, R. (1966). In *Physical techniques in biological research* **3A**, 1, 2nd ed., New York; Academic Press
Barer, R. and Joseph, S. (1957). *Exptl. Cell Res.* **13**, 438 and *Symp. Soc. Exptl. Biol.* **10**, 160
Barlow, P. and Vosa, C. G. (1970). *Nature* **226**, 961
Beatty, R. A. (1970). *Biol. Rev.* **45**, 73
Bennett, H. S. (1950). *Microscopical techniques*, New York; Hoeber
Bertalanffy, F. D. (1960). *Can. Med. Assoc. J.* **83**, 211
Bertalanffy, L. von (1958). *Cancer* **11**, 873
Bradley, D. F. (1961). *Trans. N.Y. Acad. Sci.* **24**, 64
Bragg, W. F., Kendrew, J. C. and Perutz, M. F. (1950). *Proc. Roy. Soc.* **203A**, 321
Bushong, S. C., Watson, J. A. and Atchison, R. A. (1968). *Stain Tech.* **43**, 273
Caspersson, T. (1940). *Chromosoma* **1**, 605
Caspersson, T. and Zech, L. (1971). *Abstr. IV Int. Congr. Hum. Genet., Paris*, 42
Caspersson, T., Farber, S., Foley, G., Kudynowski, J., Modest, E. J., Simonsson, E. and Waugh, U. (1968). *Exp. Cell Res.* **49**, 219
Caspersson, T., Zech, L., Modest, E. J., Foley, G. E., Waugh, U. and Simonsson, E. (1969). *Exp. Cell Res.* **58**, 128
Clifton, C. E. and Erhard, H. (1952). *J. Bact.* **63**, 537
Combée, B. and Engström, A. (1954). *Biochim. et Biophys. Acta* **14**, 432
Conen, P. E., Lewin, P. and Vaco, D. (1970). *Am. J. Hum. Genet.* **22**, 22
Coons, A. H. (1958). *General cytochemical methods* **1**, 400, New York; Academic Press
Corrington, J. D. (1941). *Working with the microscope*, New York; McGraw-Hill
Cosslett, V. E. and Nixon, W. C. (1953). *J. Appl. Phys.* **24**, 616
Cosslett, V. E. and Nixon, W. C. (1960). *X-ray microscopy*. London; Cambridge Univ. Press
Cowden, R. R. (1960). *Int. Rev. Cytol.* **9**, 369
Danielli, J. F. (1953). *Cytochemistry: a critical approach*. New York; Wiley
Darlington, C. D. and La Cour, L. F. (1968). The *Handling of chromosomes* 5th ed., London; Allen and Unwin
Davies, H. G. (1958). *General cytochemical methods* **1**, 55, New York; Academic Press
DeBruyn, P. P. H., Robertson, R. C. and Farr, R. S. (1950). *Anat. Rec.* **108**, 279
DeBruyn, P. P. H., Farr, R. S., Banks, H. and Northland, F. W. (1953). *Exptl. Cell Res.* **4**, 174
Engström, A. (1962). *X-ray microanalysis in biology and medicine*. Amsterdam; Elsevier
Engström, A. (1966). In *Physical techniques in biological research* **3A**, 87
Engström, A. and Lindström, B. (1950). *Biochim. et Biophys. Acta* **4**, 351
Francon, M. (1961). *Progress in microscopy*, Oxford; Pergamon
Frey-Wyssling, A. (1943). *Chromosoma* **2**, 473
Frey-Wyssling, A. (1953). *Submicroscopic morphology of protoplasm* 2nd ed., Amsterdam; Elsevier
Furberg, S. (1950). *Acta Chem. Scand.* **4**, 751
George, K. P. (1970). *Nature* **226**, 80
Glick, D. (1959). In *The Cell*, **1**, p. 139, New York; Academic Press
Goldman, M. (1968). *Fluorescent antibody methods*, New York; Academic Press
Hale, A. J. (1958). *The interference microscope*, London; Livingstone
Hale, A. J. and Kay, E. R. M. (1956). *J. Biophys. Biochem. Cytol.* **2**, 147
Harris, P. and Mazia, D. (1962). *The interpretation of ultrastructure*, p. 279, New York; Academic Press
Haskell, G. and Paterson, E. B. (1962). *Genetica* **33**, 52
Haskell, G. and Wills, A. B. (1968). *Primer of chromosome practice*, Edinburgh; Oliver and Boyd

Henke, B. L., White, R. and Lundberg, B. (1957).   *J. Appl. Phys.* **28,** 98

Hewitt, L. F. (1951).   *J. gen. Microbiol.* **5,** 287

Holborow, E. J., Weir,   and Johnson, G. D. (1957).   *Brit. Med. J.* **2,** 732

Holter, H. and Marshall, J. M. Jr. (1954).   *C. R. Lab. Carlsberg (Ser. Chim.)* **29,** 7

Inoué, S. (1951).   *Exp. Cell Res.* **2,** 513

Inoué, S. (1959).   *Biophysical Science—a study program,* p. 402, New York; Wiley

Inoué, S. and Bajer, A. (1961).   *Chromosoma* **12,** 48

Inoué, S. and Sato, H. (1962).   *Science* **136,** 1122

Joseph, S. (1963).   Ph.D. thesis, University of Oxford

Krooth, R. S., Tobie, J. E., Tjio, J. H. and Goodman, H. C. (1961).   *Science* **134,** 284

Krug, W., Rienitz, J. and Schultz, G. (1964).   *Contributions to Interference Microscopy,* London; Hilger and Watts

Lerman, L. S. (1963).   *Proc. Natl. Acad. Sci. US* **49,** 94

Lerman, L. S. (1964).   *J. Mol. Biol.* **10,** 367

Lindström, B. (1955).   *Acta Radiol. Suppl.* **125**

Loeser, C. N., West, S. S. and Schoenberg, M. D. (1960).   *Anat. Rec.* **138,** 163

Longwell, A. C. (1961).   *Hereditas* **47,** 647

Lowry, O. H., Roberts, N. R. and Kapphalm, J. I. (1957).   *J. Biol. Chem.* **224,** 1047

Mayer, H. D. (1963).   *Intern. Rev. Exptl. Pathol.* **2,** 1

Mazia, D. (1961).   In *The Cell,* **3,** New York; Academic Press

Meisel, U. N., Bramberg, E. M., Kondratjera, T. M. and Barsky, F. J. (1961).   In *The initial effects of ionizing radiation in cells,* p. 107, New York; Academic Press

Mellors, R. C. (1959).   *Analytical cytology* 2nd ed. New York; McGraw-Hill

Needham, G. H. (1958).   The *Use of microscope including photomicrography,* Springfield, Ill.; Thomas

Oliver, C. W. (1947).   *The intelligent use of the microscope,* London; Chapman and Hall

Oster, G. (1950).   *Progress in Biophysics* **1,** p. 73, London; Butterworth-Springer

Oster, G. (1951).   *C. R. Acad. Sci. Paris* **232,** 1708

Oster, G. (1955).   In *Physical techniques in biological research* **1,** p. 439, 1st ed., New York; Academic Press

Osterberg, H. (1956).   In *Physical techniques in biological research* **1,** 377, New York; Academic Press

Passwater, R. A. (1970).   *Guide to fluorescence literature.* **2,** New York; Plenum

Pearson, P., Barrow, M. and Vosa, C. G. (1970).   *Nature* **226,** 78

Porro, T. J., Dadik, S. P., Green, M. and Morse, H. T. (1963).   *Stain Tech.* **38,** 37

Puchtler, H., Sweat, F. and Gropp, S. (1967).   *J. Roy. Microscop. Soc.* **87,** 309

Ranadive, N. S. and Korgaonkar, K. S. (1960).   *Biochim. Biophys. Acta* **39,** 547

Razavi, L. (1968).   In *Nucleic acids in immunology,* p. 248, Berlin; Springer-Verlag

Richards, B. M. and Davies, H. G. (1958).   *General cytolochemical methods* **1,** New York; Academic Press

Richards, O. W. (1950).   *Medical physics* **2,** p. 530, Chicago; Yearbook Publishers

Richards, O. W. (1955).   In *Analytical cytology,* p. 501, New York; McGraw-Hill

Ross, K. F. A. (1954).   *Quart. J. Microscop. Sci.* **95,** 425

Ruch, F. (1945).   Referred in Ruch (1966)

Ruch, F. (1951).   *Exptl. Cell Res.* **2,** 680

Ruch, F. (1955).   Referred in Ruch (1966)

Ruch, F. (1966).   In *Physical techniques for biological research* **3A,** p. 57, 2nd ed., New York; Academic Press

Schmidt, W. J. (1937).   *Protoplasma Monogr.* **11** and *Protoplasma* **29,** 435

Seeds, W. E. (1953).   *Progr. in Biophys. and Biophys. Chem.* **3,** 27

Seydel, H. G. and Lawson, N. S. (1966).   *Int. J. Rad. Biol.* **10,** 567

Sherif, M. (1963).   *Acta. Abs. Gyn. Scand.* **42,** 181

Snapper, I., Schneid, B., Leiben, F., Gerber, I. and Greenspan, E. (1951).   *J. Lab. Clin. Med.* **37,** 562

Steiner, R. F. and Beers, R. F. Jr. (1961).   *Polynucleotides.* p. 301, Amsterdam; Elsevier

Stevens, G. W. (1957).   *Microphotography at extreme resolution.* New York; Wiley

Stone, A. L. (1964).   *Biopolymers* **2,** 315

Stone, A. L. (1967).   *Biochim. Biophys. Acta* **148,** 193

Stone, A. L. and Moss, H. (1967).   *Biochim. Biophys. Acta* **136,** 56

Strugger, S. (1949).   *Fluoreszenmikroskopie und Mikrobiologie.* Hannover; M. H. Schaper

Swann, M. M. (1951).   *J. exp. Biol.* **28,** 434

Thuline, H. C. (1971).   *J. Paed.* **78,** 875

Tokumaru, T. (1962).   *J. Immunol.* **89,** 195

Vosa, C. G. (1970).   *Chromosoma* **30,** 367

Watson, J. D. and Crick, F. H. C. (1953).   *Nature* **171,** 737

West, S. S. (1965).   In *Methoden und Ergebnisse der Zytophotometrie, Acta Histochem. Suppl.* **6,** 135

West, S. S. (1969).   In *Physical techniques in biological research* **3C,** p. 253, 2nd ed., New York; Academic Press

Wilkins, M. H. F. (1951).   *Pubbl. Staz. Zool. Napoli* **23,** 105

Wilkins, M. H. F., Stokes, A. Z., Seeds, W. E. and Wilson, H. R. (1953).   *Nature* **171,** 738

Wilson, G. B. and Morrison, J. H. (1961).   *Cytology,* New York; Reinhold

Wolf, M. K. and Aronson, S. B. (1961).   *J. Histochem. Cytochem.* **9,** 22

Zech, L. (1969).   *Exp. Cell Res.* **58,** 463

Zernike, F. (1952).   *Physica* **9,** 686 and 974

Zanker, V. (1952).   *Z. Physik. Chem. (Leipzig)* **199,** 225

14 Study of Chromosomes from Cultured Tissue

Evolution from lower to higher organisms has been associated with specialisation in anatomical details and complexities of internal environment. This internal environment has a specific constancy—though it is endowed with a certain degree of flexibility—which does not undergo considerable changes, even when exposed to a variety of external conditions. Constancy of a complex internal environment is the result of specialisation in evolution.

In this complex environment of cells, the application of any physical and chemical agents or the accumulation of certain metabolites *in vivo* initiates a series of reactions and interactions at the level of cells, tissues and organs, and such reactions are difficult to analyse as they must be studied *in vivo*. Tissue culture technique allows the cultivation of cells, tissues and organs *in vitro* in natural medium or in artificial medium, but reproducing as far as is practicable the conditions *in vivo*.

The technique of tissue culture has immense application in the study of biology in general, and cytology and cytochemistry in particular. In the first place, it allows an analysis of the effects of different physical and chemical agents on the cell of the blood or vascular supply, the importance of which cannot be over-estimated as it gives an understanding as to how far the effect is exerted directly on the cell or is influenced during transport. The effect of a physical or chemical agent on an organism is studied under ordinary conditions after the manifestation of a visible expression, but one of the limitations of this method of approach is that it only allows an analysis of the end-product, and no direct understanding can be obtained of the *modus operandi* as the visible expression is, in principle, many steps ahead of the initial reaction. Tissue culture, on the other hand, allows a complete analysis of the sequence of reactions from the initial stage, and also of the chromosomal mechanism involved in the operation.

Secondly, the mechanism of differentiation can be directly understood only through tissue culture. The cultivation of one morphological type of tissue having the capacity of unrestricted growth offers immense scope in

295

the study of differentiation. Differentiation involves the development of different organs from a tissue potentially capable of giving origin to various tissues during its growth and development; consequently, the culture of this type of tissue *in vitro*, and an analysis of the chemical, morphological and chromosomal behaviour associated with the transformation into different organs, is considered to be a reasonable approach towards understanding the exact mechanism involved in differentiation.

In the study of cytogenetics in relation to plant breeding, tissue culture, or more precisely embryo culture, has acquired the status of a routine technique in the laboratory. Its importance lies principally in incompatible crosses where hybrid embryos, though formed, fail to reach maturity. In such cases, nutritional disbalance or deficiency is the principal cause of the abortion of the hybrid embryos, which generally show chromosomal irregularities as well. Culture of excised young embryos in *in vitro* medium, containing all the growth requirements for the normal development and cytological behaviour of an embryo, helps in securing mature plants which can be grown in these conditions.

Lastly, another sphere in which the contribution of tissue culture methods is significant is the study of malignant cells of diverse kinds, including both carcinoma and sarcoma. The requirements for the growth and spread of the cancerous growth can be studied directly *in vitro*. Factors inhibiting the development of malignant growth can similarly be analysed and the importance of this knowledge in the therapy of cancer is obvious. This aspect of investigation has clearly brought out the special requirements for cancerous growth, and has demonstrated that live or dead cells of normal type may often be necessary for the successful culture of malignant cells. An understanding has also been attained with regard to the histological distinction between normal and malignant cells. Only through this line of approach has it been possible to bring out a clear relationship between the processes of differentiation and neoplastic growth, both of which have been shown to be associated with a synchronous and abnormal division, along with high chromosomal irregularities.

COMPARATIVE ADVANTAGES AND DISADVANTAGES OF TISSUE CULTURE OF PLANTS AND ANIMALS

While dealing with the culture of excised tissues for cytological studies, the relative advantages and disadvantages of the cultivation of animal and plant cells should be clearly understood. In view of the extremely divergent and specialised nature of the organs, as well as the differences in their metabolism, the culture of these two groups of biological objects presents special problems. In general, however, the culture of animal cells, for reasons noted below, is comparatively easier than the cultivation of plant cells *in vitro*.

It is obvious that the most congenial medium for the growth of any cell, organ or tissue is the natural medium of the object concerned. Except for stigmatic juice or stylar fluid, which provides the best natural medium for the germination of pollen and growth of its tube, there is no complete extractable natural medium for plant cells. Xylem sap, though containing many inorganic salts, lacks several organic elements and as such is not a complete medium in itself, while phloem sap, which contains nearly all necessary nutrients, is in contact with only a few of the specialised types of cells and as such does not form a natural medium for the culture of all organs. Coconut milk only, to some extent, serves the purpose of a natural medium. Animal cells, on the contrary, provide no difficulty in securing a natural medium of growth. Blood plasma and serum, embryo extract, lymph, body fluid, etc., satisfy all the requirements of a natural medium and different organs can be cultured at ease in them, and thus animal cells are more easily cultured than plant cells.

Regarding the preparation of a synthetic medium, the requirements for plants are quite simple and precisely known, and with a gradual understanding of the growth factors in terms of vitamins, nitrogenous substances, hormones, etc., the preparation of a good synthetic medium for the culture of plant tissues has become quite an easy task. Owing to the plastic and versatile nature of their metabolism, plants can utilise very simple molecules for the synthesis of complex substances, while animals, with their stable and limited metabolic activities restricted to special organs, evidently cannot do so. Thus it is easier to prepare a synthetic medium containing, principally, sucrose, a mixture of vitamins, glycine, etc., in addition to an inorganic salt mixture, for the cultivation of plant cells *in vitro*.

The presence of a rigid cellulose wall in most of the lower and all higher plants provides another difficulty in the culture of plant cells. In the *in vivo* condition, the presence of plasmodesma and other living factors, which cannot necessarily be reproduced exactly *in vitro*, help in the process of transport and uptake of food between cell and cell; therefore under cultural conditions the access of food at intracellular sites is a problem which needs constant attention. Animal cells, on the other hand, being devoid of any cellulose wall do not present this difficulty; moreover, the presence of a tough but mobile pellicle delimiting the organs in animals helps in their excision from the parent body without any difficulty, thus causing the least injury to the tissue concerned.

Lastly, the different type of growth in plant and animal organs is another important factor controlling the success of their cultivation under cultural conditions. In animals, growth is not restricted to any specialised segment and all the different parts, as well as organs, are capable of growth, while in plants, growth is restricted to special segments and unless those segments are maintained intact under cultural conditions, their cultivation *in vitro* may present further problems; therefore this specialised nature of growth makes plant tissues rather inconvenient materials for culture. Animal tissues in this respect are quite convenient to handle.

DEVELOPMENT OF TISSUE CULTURE METHODS

The development of tissue culture methods dates back to 1887 when Arnold first successfully cultivated leucocytes. He soaked thin pieces of alder pith in aqueous humour of the frog, implanted them in the peritoneal cavity of the frog and successfully sub-cultured them in saline or aqueous humour solution.

Earlier, Vöchting (1878) established that the development of the morphogenetic pattern is a function of the organism as a whole. He observed that from a cut portion of stem, the distal region develops the roots and the proximal one develops the leaves; but if the length of the stem is further shortened and the distal and proximal regions are just a few centimetres apart, the same proximal portion forms the leaf and the distal one the root. Evidently polarity is always maintained but the development of the organ is a function of the organism as a whole. In 1898, Ljungren was able to keep a piece of skin alive in ascitic fluid. Jolly (1903) successfully kept leucocytes alive in serum.

The idea of single cell culture was first suggested by Haberlandt (1902) and since then, this technique has been very successfully applied in animal materials on a large scale (Sandford, Earle and Likely, 1948) and in a restricted sense in plants (Northcraft, 1951; Blakely and Steward, 1961; Mitra and Steward, 1961, etc.). Haberlandt, although responsible for the suggestion in the first place, was not successful in his efforts, even after several years' sustained attempts, and White (1934) pointed out that the failure of his experiments was due to the fact that his work was conducted on mature green cells which were incapable of meristematic growth.

Harrison (1907) first cultured tissues successfully from frog neuroblasts in clotted lymph as medium, and introduced the term 'tissue culture' (*see* White, 1954). Since then, considerable developments have been made, especially by Carrel (1912), and a number of media suggested, including embryo juice, as a growth-promoting nutrient. Research has centred principally on exploring the possibilities of different types of cells in tissue culture, and in 1923 Carrel introduced the flask culture, named after him, which replaced to a great extent the 'hanging drop method' followed by previous workers. This flask culture method was pursued, with refinements, by several of his colleagues, including Barski and colleagues (1951) and Harris (1952).

In the field of plant tissue culture, White (1934) first invented the 'organ culture' technique and successfully cultivated excised roots of tomato *in vitro*. Nobecourt (1937, 1939) and Gautheret (1934, 1939) were pioneers in developing methods for unlimited cultures of undifferentiated materials, raising cambial tissues of tobacco and carrot *in vitro*. White (1939) also cultured tissues from tumours or from fleshy organs composed of undifferentiated parenchyma. Gradual development of the tissue culture method has led to the evolution of techniques for culturing hybrid embryos and also for cultivating sterile crown gall tissue of plant cancer. The methods have been developed and refined to such an extent in recent years that even free cells, small cell clusters, special portions of a colony, for example of carrot (Steward, 1958) and *Haplopappus gracilis* (Blakely and Steward, 1961) can all be cultured *in vitro*. Together with advancements in techniques for plant tissue culture, the methods for culturing animal tissue also underwent progressive refinement. Fell (1928, 1931) introduced the 'watch-glass technique' for the culture of animal organs such as bone, teeth, eye, etc., this method being found to be much simpler than the complex 'Carrel flask technique'. This 'watch-glass technique' was followed later by different workers (Carpenter, 1942; Martinovitch, 1950). Gey and Gey (1936) substituted the 'roller tube', which has the advantages of being simpler, less expensive and equally effective for many purposes, over the 'Carrel flask'. Bryant, Earle and Peppers (1953) were successful in evolving a special method of sterilising nutrients without loss of potency—a factor of immense importance in the success of tissue culture.

PRINCIPAL STEPS INVOLVED IN TISSUE CULTURE

So far as animals are concerned, there is not much difficulty in choosing suitable tissue because the worn-out cells are continuously replaced by fresh growth. In this way, cultures of skin, epithelium of the eye, kidney, liver, thyroid, ovary, bone connective tissue, skeletal muscle, lung, heart, different types of embryonic tissue, etc., can be obtained. From the cytologist's point of view, leucocytes, tumours and cancers of different types, ascites fluid, HeLa cells, etc., can all be cultured with ease.

In plants, on the other hand, due to the restricted nature of growth in certain specialised and local regions, not all the tissues and organs form convenient material for cultural studies. The highly active growing points, such as apices of stem and root, buds, lateral cambium, intercalary meristems, leaf meristems, phloem tissues, etc., provide excellent materials for tissue culture. Pollen, embryos and endosperm of higher plants, as well as spores and prothalli of lower groups, do not present much difficulty in culturing in a suitable medium (Mehra, 1961). Even flowers, fruits, excised leaves and seed primordia have been cultivated (White, 1954). Phleom tissue of carrot and tomato root, *Haplopappus* stem and pith callus of tobacco, provide ideal materials for study.

Preparation of the medium is the second important factor on which the success of tissue culture depends, and even with considerable advancements in our knowledge of nutrient requirements, the use of different kinds of biological fluids is the principal basis of the preparation of media for tissue culture.

For human and animal materials, serum is the most important fluid and is generally obtained from the blood of adult humans or placental cord, or from horse or calf. The method of preparation is quite simple. The entire plasma is allowed to coagulate and sinerese and the exudate serum is removed, and a test for sterility before use is always made. In addition, amniotic fluid, especially of bovine type, obtained from a pregnant uterus, ascitic and pleural fluids, as well as aqueous humour, are also satisfactory materials for tissue culture. Ox-eye provides the most easily available source of aqueous humour.

Extracts of tissue have also been used successfully, the most successful being one made from embryo. Two methods for preparing extracts from embryos are practised; in one, less damage is caused to the tissue, while in the other, the preparation of a homogenate, including complete destruction of the tissue is necessary. Simple extracts from intact cells are secured in chick embryos, about 1–2 weeks old, the extraction being carried out by putting a few embryos within a sterile syringe and collecting the juice in a receiver, when the concentrated juice can be diluted with salt solution, and the supernatant liquid isolated by centrifuging.

The preparation of bovine embryo extract is a good example of a homogenate. In this method, amniotic fluid is first removed from the excised pregnant uterus. After washing it carefully and swabbing with iodine, and following an incision, and folding back, the embryo can be removed. Chopped pieces of the embryo are then homogenised in a Waring Blendor with salt solution, and the fluid is centrifuged to isolate the supernatant liquid. Treatment with hyaluronidase has been suggested by Bryant, Earle and Peppers (1953), and ultra-centrifugation for the removal of solid matter becomes necessary.

In addition to these natural biological media, different balanced solutions and different media containing these balanced salt solutions, together with other ingredients, have been suggested by several authors (*see* White, 1954). The balanced solutions suggested by Tyrode (1910), Gey (1945), etc., include different salts like sodium chloride, potassium chloride, calcium chloride, magnesium chloride, magnesium sulphate, several phosphate salts, glucose, different bicarbonates, etc. Makino and Hsu (1954) have, however, suggested that for rat tissues, Gey's medium, without sodium chloride, may be necessary for short duration media. Ringer's medium includes only chlorides of sodium, potassium and calcium. In the synthetic media, as suggested by different authors (Fischer, 1948; White, 1949; Parker and Healy, 1955), different amino acids, vitamins, coenzymes and even antibiotics and nucleic acid derivatives (Parker and Healy, 1955) together with inorganic salts, are added.

Temperature is a major factor while incubating in these media and, as far as practicable, incubation is carried out at temperatures approximating the body temperature of the organism.

For the culture of plant tissues, as already mentioned, natural media, except coconut milk, are not available. In order to supplement xylem sap, inorganic salt mixtures are generally added and sucrose, mixture of vitamins, glycine, etc., supplement the phloem sap. Of the inorganic salts, sulphates, nitrates, chlorides, phosphates, etc., of sodium, potassium, calcium, magnesium and iron are usually given. For the growth of tumorous tissue, White's medium (1954) is generally suitable, whereas Gautheret's medium (1955) is congenial for normal tissues. However, in addition to these basal media applied in suitable proportions with agar, coconut milk, naphthalene acetic acid, casein hydrolysate, yeast extract, etc., are often added to supplement the growth.

Morel and Wetmore (1951) dealt in detail with the specialised processes needed for the tissue culture of monocotyledonous materials. They further observed that spores of *Osmunda cinnamomea*, the fern, can be grown in Knudson's medium with a few additional nutrients, and that cytological study with acetic-carmine staining can be performed on undifferentiated callus tissue. In endosperm of maize cultured in White's medium with the yeast extract, chromosomal irregularities were noted in sectioned material, which were attributed to the presence of yeast extract. Partanen, Sussex and Steeves (1955) obtained tumour growth following spore culture of *Pteridium aquilinum* in agar medium with modification of Knop's (1884) or Knudson's solution (1925) and cytological study was carried out on Carnoy-fixed, squashed material under a 'phase contrast' microscope. An increase in chromosome number through endopolyploidy and associated loss of morphogenetic potentiality was noted in the tumour. Tulecke (1957) was successful in culturing pollen of *Ginkgo biloba*, the gymnosperm, in an agar medium supplemented by coconut milk, tomato juice or yeast extract and indole acetic acid.

The most important factor in tissue culture is the apparatus used for cultivating the tissue *in vitro*. The different techniques adopted for the explants are based on the same principle, the chief differences being the type of container employed for the purpose. They can be classified under three categories, (*a*) hanging drop, (*b*) flask and roller tube, and (*c*) watchglass techniques. All three different methods are practised for different animal and plant materials and each has separate advantages and disadvantages. Modifications of the different methods have been made on the basis of the requirements of long- or short-term growth or the convenience of observation, and in this connection it should be noted that sub-culture from the primary explant is an essential step in tissue culture. The apparent peripheral expansion of a tissue in culture does not necessarily signify growth of the tissue, as the interior cells of the original tissue may often shift towards the periphery, and no new growth occurs; thus sub-culturing at this stage is fruitless. In order to be certain of tissue regeneration, the best method is to weigh the culture at specific intervals, but other alternative methods are also applied. An increase in weight denotes growth.

The necessity of maintaining strictly aseptic conditions in all phases of tissue culture, including sterilisation of the object, medium, etc., is obvious: contamination at any stage may not only spoil the culture but also give erroneous data. Strictly aseptic conditions and regulated temperature and growth are essential for successful tissue culture, and in cases of difficulty in maintaining a well-equipped large laboratory for tissue culture, all the steps necessary can be carried out in small chambers in which temperature, humidity and sterility are strictly controlled. For details of the set-up needed for a tissue culture laboratory, the reader is referred to the treatises by White (1954), Paul (1959) and Willmer (1965) and for principles, to the review by White (1959).

An outline of the different types of culture methods for the cultivation of tissues is now given.

Hanging Drop Method

This technique is otherwise known as the 'slide culture' technique, originally devised by Harrison (1907). In this method, a drop of a suitable medium, such as serum, lymph or salt solution, is generally placed on a cover slip, the tissue is transferred to the medium and the cover slip, with tissue, is inverted over a grooved sterilised slide. The edges of the cover glass are then sealed with vaseline or paraffin wax and the slide kept in the incubator. After the growth of the tissue, cells can be sub-cultured by following a similar procedure in different sets with bits of tissue taken from the original one. After implantation, the cells soon start growing, and on emerging from the tissue, spread on the glass surface, provided the latter has the required properties (Rappaport, 1960). For quick and rapid emergence, older tissues may need pre-treatment with 1 per cent trypsin in physiological salt solution. Saline extract of minced embryo tissues or of brain, cartilage and heart may also serve as growth promoters (Moscona *et al.*, 1965). Certain tissues have distinct but changeable points of contact on the glass surface (Ambrose, 1961). Chick connective tissue shows the rapid emergence of mechanocytes with increase in concentration of the medium. The rate may even be 20 μm/h (Moscona *et al.*, 1965). Serine (Lockart and Eagle, 1959), inositol (Eagle *et al.*, 1957), cepalin (Fujii, 1941), fetuin—the glycoprotein from foetal calf serum (Puck, 1961), as well as several polypeptides (Lieberman and Ove, 1958), help in the growth and spreading of the tissue.

The problem of surface contact in glass cover slips is often minimised by providing a standardised surface by dipping the cover slips in a dilute solution of celloidin, followed by drying in a vertical position. The cells, on being flattened over the glass or celloidin surface, can conveniently be observed under the microscope.

In addition to a fluid medium, solid media like blood plasma can be utilised for culture. Fowl plasma is one of the most commonly used media, as under strictly controlled conditions it does not undergo clotting. By heparinisation, early clotting can be prevented and the plasma shows clotting only after the tissue is implanted. The plasma not only provides nutrition for the cells, but also facilitates the invasion and spreading of tissues because of the digestibility of its fibrin threads by enzymes. However, the use of embryo extract or plasma is gradually being replaced by more synthetic media, in which the ratio of constituents can be controlled. This method, however, has certain limitations. If the medium is not heavy, evaporation may change its concentration, resulting in degeneration of cells, especially at the centre, due to the increased concentration of metabolites and inadequacy of food and oxygen supply. This difficulty can be minimised to some extent by rapid sub-culturing in fresh medium. For prolonged culture, this method is not very effective, but even then the suspension of free cells on the surface of the cover glass and convenience of observation under the microscope make it a very suitable technique for the study of tumour cells in particular. Makino (1953) and Makino and Nakahara (1953) pointed out certain limitations of the original hanging drop method, especially as the drying of the fluid with increase in temperature and the condensation of water at the bottom of the slide hamper visibility. These difficulties can be overcome by putting liquid paraffin in the groove of the slide. Following this method, they made phase contrast observations of single tumour cells obtained from ascites fluid.

Slide Culture Method

In order to overcome the limitations of the hanging drop method, Pomerat (1951) improvised a perfusion chamber, based on the slide culture technique, but having a continuous circulation of the fluid through inlet and outlet tubes surrounding the culture. Tissues cultured in a perfusion chamber are very suitable for cytological studies.

Slide Chamber Method

Several such chambers have been devised by various authors (Mackaness, 1952; Buchsbaum and Kuntz, 1954; Rose, 1954; Pulvertaft *et al.*, 1956; Sykes and Moore, 1960). All these chambers are constructed on the basic principle of continually renewing the medium or even securing a slow continuous perfusion. In one of the devices (Rose, 1954), silicone–rubber rings are used to separate the two cover slips and two steel plates are applied for firmly clamping the cover slips. The rubber allows the introduction of a hypodermic needle through which the gas or medium can be injected. By inserting a bubble and rotating the slide, the stirring needed for oxygenation can be obtained. The tissue may be grown directly on the glass in the fluid medium or in plasma. To eliminate the observational difficulties arising out of the growth of different layers on both the interfaces (Rose, 1962), cells are enclosed and sandwiched between two pieces of sterilised cellophane, or cellophane on one side and glass on the other.

In the Mackaness type (Mackaness, 1952), a slide of Lucite (Perspex) or stainless steel is used, having an aperture bored on both sides for fitting two circular cover slips, sealed with wax. Two plugged lateral holes are used for introducing the medium. In the Pulvertaft type (Pulvertaft *et al.*, 1956), a glass slide is provided with a central pillar of agar, on which the cover slip is placed and its rim sealed. In another device a Lucite slide is used with a central pillar and a moat, and the cover slip on the top is sealed.

Following all these methods, the cells can be grown in a single layer, either between two cellophane sheets, or between glass and cellophane, or between serum agar and glass. Such monolayer growth has decided advantages for observation. Time lapse cinematographic attachment fitted with perfusion systems (Buchsbaum and Kuntz, 1954) has helped in detecting the continuous phase of growth.

Glass rings, similar to the process followed for pollen tube culture, can be used instead of a grooved slide (Gautheret, as mentioned in White, 1954).

Flask and Tube Method

The Carrel flask technique of 1923, which was later modified by Barski and colleagues (1951) and Harris (1952), allows the growth of tissue in the required medium, within a flask, which is flat at the top and bottom and has an oblique neck at one side. This flask has the advantage of holding a large quantity of medium, which can be repeatedly washed so that frequent sub-clotting of the original tissue is not necessary. The provisions for adding a known gas mixture and its renewal are a decided advantage. One serious drawback of the flask method is that the position of the tissue is not convenient for observation, but this difficulty is, to some extent, overcome by sealing cover slips over the perforations made in the body of the flask or, alternatively, the tissues can be cultured on cover slips inside the flask and removed for observation later.

Gey and Gey (1936) devised the 'roller tube technique' which, though a modification of the 'flask method', is yet much simpler, less expensive and equally effective for many purposes. In this case, flasks are substituted by test-tubes, in which the culture can be made either in flat or obliquely placed media, but as difficulty is often encountered in measuring growth due to the oblique nature of the bottom of the tube, cultures are often grown on cover slips placed within the test-tubes (Ehrmann and Gey, 1953). Collagen was used to line the tube, thus facilitating adhesion and spreading (Ehrmann and Gey, 1956).

Makino and Hsu (1954) successfully employed the roller tube culture for cytological studies of embryonic lung, spleen, heart and liver of rats. For proper aeration and constant circulation of the nutrients, it may be necessary to rotate the culture in the tube.

As cells on static surfaces suffer from lack of oxygenation, 'shaker flasks' (Earle, Bryant and Schilling, 1954), for large volumes, are now used. The medium is placed in Erlenmeyer flasks or balloon flasks, with living cells and then placed in a shaker oscillating at approximately 400 rev/min, allowing proper aeration, but not permitting the cells to settle on the glass, and free multiplication of cells in this fluid.

In the case of carrot tissue, Blakely and Steward (1961) maintained stock cultures in large quantities by rotating culture flasks containing 250 cm^3 of medium following the device of Steward and Shantz (1956), or by using flasks of 500 cm^3 capacity to contain 150 cm^3 of the medium and by agitating them on a horizontal shaker. Roller tube drums may be used for rotation, if available. The medium used was White's basal solution with coconut milk and casein hydrolysate. The same method was followed by Mitra and Steward (1961) on *Haplopappus gracilis* and cytological observations were made. Callus tissues were obtained by culturing severed stems of *Haplopappus* sp. These cells, free or in clusters, were smaller than carrot cells. The free cells and cell aggregates received in suspension, were also distributed on agar medium or in Petri dishes to secure further small colonies. Suspension cultures have been dealt with in detail later.

For this purpose, Blakely and Steward (1961) first eliminated the larger clusters by straining the suspension through cheese cloth, and they then further concentrated it by sedimentation, slow speed centrifugation and finally the concentrated suspension was spread either on semi-solid medium or mixed with liquid agar before pouring into the plates. Observations were carried out on Carnoy-fixed callus tissues which were stained as usual with orcein–HC1 mixture and mounted in orcein. By sub-culturing, free cells, small cell aggregates in suspension, as well as peripheral cells of a colony, were all cultivated and their cytological behaviour studied. The aberrant chromosome behaviour noted thereby has a great bearing on the problem of differentiation or morphogenesis.

For plants, 'tube culture' is the most convenient method and other nutrients with agar, stored vertically (Gautheret, 1942) or in a slope (White, 1943), have been tried successfully. The latter method of storage provides greater surface area. Plugging with cotton (Gautheret, 1942) is an effective measure.

Watch-glass Technique

One of the most widely used methods is the 'watch-glass technique', originally introduced by Fell and Robinson (1929). It is a very convenient method for the cultivation of organised tissue, such as bone, teeth, eye, etc., and is also employed for the study of tumours and cancers.

In the original method, the tissues were grown on the surface of a plasma clot in a watch-glass, enclosed within a Petri dish. The bottom of the Petri dish was covered with damp cotton wool, serving the purpose of a moist chamber preventing evaporation. Later modifications (Chen, 1954; Shaffer, 1956) involved growing the cultures attached to lens paper or rayon strips, which were fixed on the plasma clot or suspended in the fluid nutrient. An advantage of this technique is that the culture is not detached during transfer. Gaillard (1948) and Wolff and Haffen (1952) suggested the use of embryological watch-glasses. The former author preferred a medium of saline, serum, plasma and tissue extract, whereas the latter suggested a mixture of agar with saline serum and embryo extract as the medium of growth. Lasnitzki (1958) recommended the watch-glass technique in general for the cultivation of organized tissues, and for prolonged cultures, Fell and Robinson's technique has been recommended.

Moscona (1956) and Wolff (1957) evolved an excellent technique which involves treatment of the organ with trypsin for organ cultures, as well as for the preparation of cell suspensions, allowing the separation of constituent cells without loss of viability. If the trypsinised disaggregated mass is cultured in nutrient media, reaggregation takes place and the formation of new organs is easily achieved.

Cell Isolation and Suspension Culture

The concept of the pattern of pure strain of cells dates back to 1933 when Baker succeeded in culturing chicken monocytes in serum. The method that was gradually developed was to obtain one class of cells from a suspension by differential centrifugation, in suitable solutions of sucrose or albumen, or gradient sedimentation which was found to be specially effective in blood (Weiss and Fawcett, 1953). Doubtless, to some extent growth of one class of cells can be obtained even from a tissue explant, provided it is homogeneous, such as: certain epithelial cells, fibroblasts, etc., but this homogeneity appears in certain cases to be questionable. A method of mechanical isolation for pure strain culture was worked out by Sanford *et al.* (1948), in which fibroblast culture in plasma clot was carried out in small capillary tubes, and later the tubes were broken into pieces in such a way that each broken piece contained a single cell. It was then cultured in fresh medium, but even so, abnormalities in growth developed. It has been assumed that single cells, without undergoing mutation or adaptation, cannot continue further growth except in a few cases, an example being HeLa cells (see chapter on cancer cells). Later several techniques were developed for preparing cell suspensions, some of which have been outlined here:

Mechanical separation—The simplest method of preparing cell suspension is to make a fine paste of the tissue by grinding and to secure the cells from the sedimented suspension. This method, however, causes serious damage and is not suitable for critical analysis, due to heterogenicity. Rinaldini (1958) outlined different schedules for the preparation of cell suspensions.

The general procedure for securing single cell suspension is through the disintegration of tissues or from standard cell cultures. Such disintegration can be brought about by grinding, in a homogeniser and Waring blendor, in which the cells are generally not exposed to chemical treatment. But the damage caused, as well as the denaturing and lytic actions of the disrupted cell products, are serious limitations. Ultrasonic dissociation has also been attempted (Lutz and Lutz-Ostertag, 1959; Bell, 1960) but its consequent effect is yet to be analysed. Trypsinising the tissues, followed by treatment in a vibrating tube (Auerbach and Grobstein, 1958) has been shown to be useful for embryonic tissues. Treatment with chemicals which has no appreciable effect on cell properties, combined with slight mechanical treatment, or enzyme digestion, forms the basis of most of the recent techniques.

Disintegration through Chemicals or Enzymes

Methods for chemical disintegration involve the elimination of divalent cations like Ca or Mg, chiefly the former which is responsible for maintaining the cell contact and for the attachment of the cell to the substratum. Moscona (1952) observed that such treatment should be associated with trypsin digestion, the principles for which will be discussed later. The chelating agent, versene or ethylene diamine tetra-acetate (EDTA), which affects the calcium binding and the alkaline pH, has been effective in dissociating mammalian embryos (Brochart, 1954; Zwilling, 1954), but even then, in most cases, slight mechanical agitation is needed to secure at least appreciably small fragments. The extent to which versene is effective in mature tissue is debated. Anderson (1953) noted that sieving or grinding lead to the dissociation of liver tissue previously treated with versene or citrate.

Dissociation of the tissue through enzyme digestion is widely practised because it causes the least damage to the living tissue. The digestive property of trypsin was first employed for tissue dissociation in chick embryo culture by Willmer (1945) and later by a number of authors (Dulbecco and Vogt, 1954; Melnick *et al.*, 1955; Rinaldini, 1958, etc.). As the action of trypsin is mainly on arginine and lysine, the constituents of basic proteins, it is presumed that these amino acids are involved in linkages responsible for cell binding. The proteolytic activity of trypsin has been shown to be associated with its cell disrupting effect (Easty and Mutolo, 1960; Moscona, 1963). The enzyme may possibly be taken up ultimately by the cells—a possibility requiring serious consideration in cell culture experiments using serum-free media (Moscona *et al.*, 1965). It is due to the fact that in serum-free media, trypsin treatment often results in the appearance of a highly viscous substance around the cells. The nature of this substance has been suggested as DNA (Medawar, 1957; Moscona, 1962) or mucoprotein (Rinaldini, 1958). Serum or pancreatin, having trypsin inhibiting activity, reduces this effect which is possibly caused by the retention of residual trypsin in the cell.

In addition to trypsin, other enzymes have also been tried in cell separation, such as, pancreatin, elastase, elastomucase, papain, aspergillin-o (a proteolytic enzyme isolated from *Aspergillus oryzeae*), etc. (Easty and Mutolo, 1960; Fell, 1961; Sabina *et al.*, 1963) and the choice is left to the worker dealing with specific types of tissue. As collagen fibres are not attacked by trypsin, collagenase is effective in adult tissues rich in collagen (Lasfargues, 1957; Laros and Stickland, 1961; Grover, 1962). The schedule of trypsin treatment for cell separation is outlined below (Moscona, 1961):

(a) Incubate fragments of embryonic tissue in Ca and Mg-free solution (CMF) at 38 °C for 10–15 min under 5 per cent CO_2–air mixture.
(b) Incubate further at 38 °C for 15–20 min or more in a solution containing 0·25–1·0 per cent crystalline trypsin in CMF under CO_2–air mixture.
(c) Rinse thrice in excess CMF solution (pH 7·2). Precaution must be taken so that the cells are not disrupted.
(d) Keep the tissue in culture medium and flush through a pipette for dispersal.
(e) Sample the stock solution for plating on a plasma clot.

Culturing

Cell suspensions, obtained through tissue disintegration, can be induced to undego a monolayer growth if cultured in nutrient medium at the bottom of culture containers like Carrel flasks, T-tubes, Roller tubes, Erlenmeyer flasks, Petri dishes, etc. In order to check the growth rate, it is always preferable to count cells at the time of primary culturing and after a specific period before subculturing. Further counts may be taken after trypsinisation. The number of cells taken in the medium should be adequate for the amount of nutrient medium used, neither less not more. For α-strain of mouse fibroblasts, the adequate number is 100 000 cells in 2 cm^3 of medium containing chick embryo extract, horse serum and balanced salt solution (Earle *et al.*, 1954). In addition to the growth of settled cultures (originally derived from suspensions) in Carrel flasks, Roux bottles, Roller tubes,

Petri dishes, etc., cells can even be grown in suspension by fitting the flask in a horizontal shaker or by rotating the roller tubes vertically through a special device (Owens *et al.*, 1953). Addition of serum, as well as other compounds like hyaluronic acid or methyl cellulose, aid in increasing fluid viscosity and maintaining cell suspension (Earle *et al.*, 1954; Kuchler *et al.*, 1960). 'Darvan' (0·03 per cent)—a sulphonic acid salt polymer, on being added to the medium, checks clumping of cells (Merchant *et al.*, 1960) in some cases, by coating the cells and conferring a negative charge.

Attempts to secure clonal growth from single cells are primarily made with the sole object of obtaining a uniform population of cells. It is rather difficult to grow a uniform population, though in a number of cases, it has been made possible. Carcinomatous HeLa cells can be cultured in this way (for details, see chapter on cancer chromosomes).

With epithelial and fibroblast cells, Puck *et al.* (1956, 1957) successfully obtained such colony cultures by securing first a monolayer of growth from a thick suspension, the mitotic activity of which was checked later through heavy irradiation with x-rays. Normal cell suspensions were then plated on these feeder cells and good growth was secured. Later it was observed that feeder cells were not needed, if the medium was supplemented with embryo extract and serum. From normal human fibroblasts as well, Puck *et al.* (1958) succeeded in obtaining such clones. Apparently, however, growth in isolation is not the natural property of a cell; in successful cases, it should be regarded as an abnormal event (Moscona *et al.*, 1965). In fact, cells in culture reveal a high frequency of chromosomal abnormalities, which may be attributed to this reason (Hsu and Pomerat, 1953; Chu and Giles, 1959; Frédéric and Corin, 1962).

Determination of the Cell Cycle

The term 'cell cycle' implies the sequential occurrence of different phases of the cell, initiated from a division till the completion of the next cell division. In the actual method of analysis, it involves the time span between one point in a cycle to the same point in the next cycle. The cinemato-graphic analysis of cell division in culture to some extent gives a correct estimate of cell cycle, notwithstanding the fact that even between sister cells, as noticed specially in mammalian cultures, the time requirement may vary to a certain extent (Hsu, 1960, 1965; Sisken and Kinosita, 1961). Leaving aside this complicated and time-consuming method of securing a mean data after statistical analysis, the generally accepted principle of counting the *generation time* is to consider the period during which a parti-cular population doubles its number.

At the time of counting in culture, cells are in the logarithmic phase of growth, in which every cell is active. In this type of culture, the cell count at each stage is nearly proportional to the time taken by the cells to complete this phase. More precisely, if the generation time is 15 h and the frequency of dividing cells is 5 per cent, the time required for mitosis is $15 \times 0·05 = 0·75$ h. This method of calculating the generation time also depends on the nature of the medium and the type of the cell (Lajtha, 1957; Siminovitch *et al.*, 1957; Whitmore *et al.*, 1961). However, irregular mitosis may have a different value for cell cycle and the values may also differ from population to population (Hsu, 1955).

Mitotic indices can also be utilised in working out the duration of the cell cycle (Firket, 1965). The technique is based on the fact that the generation time (T) is inversely proportional to the division frequency of cells per unit time. If M is the mitotic index and d the period of duration of mitosis, then the number of cells entering into mitosis per hour is M/d. The practice of considering the generation time as $T = \dfrac{d}{M}$ ignores the fact that there is a continuous increase in the total number of cells during the cell cycle and so there is an error of nearly 30 per cent. The following formula for working out the generation time fits well the observational data (Stanners and Till, 1960; Smith and Dendy, 1962):

$$ T = \log_e 2 \, \frac{d}{M} = 0{\cdot}693 \, \frac{d}{M}. $$

Mitotic index at any given time is,

$$ M(t) = \frac{n(t+d) - n(t)}{n(t)} $$

when n is the number of cells at the time t, and d is the mean duration of mitosis. In the logarithmic phase of cell multiplication, as in cultures, the number of cells at any time t is also $n(t) = n_0 . \, 2^{\frac{t}{T}}$, where n_0 represents the number of cells counted at the beginning of this phase.

The most important change occurring during the cell cycle is the chemical turnover of nucleoprotein. DNA synthetic phase, during which DNA replication takes place, is only a fraction of the interphase (Taylor, 1958; Firket, 1958). The entire mitotic cycle is divided into four phases (Lajtha, 1957), namely, G_1—growth phase, post telophase; S = DNA synthetic phase; G_2—post-synthetic growth phase, and M—rest of the mitotic phase. Species differ with respect to the durations of these different phases, G_1 phase in general being variable (Mendelsohn, 1960) and taking the maximum and G_2 the minimum (except in HeLa cells) period of interphase. The method of calculating the duration of the different phases in culture is based on autoradiographic procedure. After pulse labelling or short treatment with tritiated thymidine and by varying the time between labelling and fixation, the duration of the three different phases can be worked out. If the cells are fixed after a long interval, only those which were at the S phase at the time of treatment show labelling. The evidence of endoreduplication too, i.e., DNA replication without any visible sign of mitotic activity, has been obtained in cultures of several mammalian tissues (Levan and Hauschka, 1953; Levan and Hsu, 1961).

Synchronisation of Division

In a logarithmically growing cell population, several methods of analysis present serious problems due to non-synchrony of cell division. No doubt, cells in certain cases show natural synchronous division, such as, eggs at the time of cleavage, or p.m.c.s. in the anthers of *Lilium*. Induced synchrony becomes essential for accurate analysis of several aspects of the cell cycle. The principle involved in induced synchrony is to allow all the cells to start DNA replication simultaneously.

The different schedules for inducing synchrony fall under two categories —physical and chemical (Zeuthen, 1964). The former has, however, not been found to be very useful. Of the physical methods, chilling at 4 °C for 1 h and subsequent removal to 37 °C has been found to be successful in HeLa cell cultures (Newton and Wildy, 1959 and cf. Miura and Utakoji, 1961). Even irradiation with x-rays has been utilised for synchronisation.

The chemical methods appear to be more promising in inducing synchronisation as specially noted in the case of mammalian cells. With this object in view, methods have been devised to cause thymine deficiencies, thus blocking DNA replication. A folic acid analogue, amethopterin (4-amino—N^1o-methyl folic acid), has yielded successful results. As long as the culture contains this analogue, DNA-replication remains suspended and it can be resumed on the addition of thymidine and the cell population is doubled within a few hours (Rueckert and Mueller, 1960). However, even here, synchronisation cannot be considered as complete, due to the differential susceptibility of the different phases. After treatment, the cells in G_1 remain in the same phase while those in the later phases complete their cycle and are blocked prior to the succeeding S phase. Cells in the S phase, on the other hand, are heterogeneous in the sense that a few per cent of the cells are at the beginning, some at the middle, and some at the end of the S phase. They are all blocked in their respective positions. As soon as the synthetic activity starts after the addition of thymidine, the cells lying at the different stages of the S phase resume the activity from their respective points and become non-synchronous, as such, while all other cells start from the S phase and show synchrony.

In monolayer cultures of HeLa cells, it has been noticed that interphase cells are not attached to the surface as the mitotic cells. As such, the intermitotic cells can be separated out from the mitotic ones and synchrony can be induced in the latter (Terasima and Tolmach, 1963). An antibiotic which also checks DNA replication is mitomycin C (for detailed structure, please see chapter on effect of chemical agents). This compound permits RNA synthesis to continue, checking only the synthesis of DNA. Cells treated with mitomycin C show the two polynucleotide strands being linked together. Because of the cross linkage, replication, which requires separation of the strands, is obstructed (Pricer and Weissbach, 1964).

In cultures, mitomycin C as such is ineffective, but if activated through chemicals or cell extracts, it can be used *in vitro* (Iyer and Szybalski, 1963). The alkylating action of the ethylene amine group of mitomycin C is responsible for linking it to a DNA strand. Its possibility in synchronisation in higher organisms is yet to be thoroughly studied.

Similar inhibiting effects on replication have been obtained with fluorodeoxyuridine (FUDR) (Eidinoff and Rich, 1959). It is also widely applied, though there is a possibility of chromosome breakage through its incorporation in the DNA molecule, as obtained with BUDR (bromo-deoxyuridine) and hydroxylamine (Somers and Hsu, 1962). Hydroxy-urea has been used as a DNA-synthesis–inhibiting agent as well (*see* Kihlman and Hartley, 1968). Synchronisation has not yet been fully perfected and much greater refinement is necessary for accurate and critical analysis.

A FEW SAMPLE TECHNIQUES IN RELATION TO TISSUE CULTURE

PLANT CULTURES (*see* WHITE, 1954)

PREPARATION OF MEDIA

The reagents necessary are: a standard salt solution, an organic accessory solution, stock solutions of calcium pantothenate, biotin and naphthalene acetic acid, agar, sucrose, ferric sulphate and distilled water.

The apparatus needed are: 125 cm^3 Erlenmeyer flasks, lipless 25 mm × 150 mm test-tubes, 100 mm Petri dishes, watch-glasses and battery jars, all sterilised.

The stages in the preparation are:

(a) *For liquid nutrient medium*—(i) Add 100 cm^3 of 0·005 per cent aqueous ferric sulphate solution to 500 cm^3 of 8 per cent aqueous sucrose solution, 200 cm^3 of standard salt solution, 2 cm^3 of organic accessory solution and add distilled water to prepare 1000 cm^3 of *stock* solution. (ii) Further add 1000 cm^3 of distilled water to the stock solution. Distribute in 50 cm^3 portions in flasks and in 10 cm^3 portions in test-tubes. Plug with sterile cotton wool and autoclave at 18 lb pressure for 20 min.

(b) *For agar nutrient medium*—(i) Prepare stock liquid nutrient medium as described in (i) above. Add an equal quantity of 1 per cent hot agar solution in water to the liquid nutrient, mix, keep half of the mixture thus prepared in portions of 15 cm^3 in test-tubes. (ii) To the remaining half, for every 100 cm^3 of agar nutrient, add 1 cm^3 each of calcium pantothenate, biotin and naphthalene acetic acid solutions, mix, divide into test-tubes, plug and autoclave.

Culture of Tomato Roots

Method I

1. Wash and dry a ripe healthy tomato, then cut it into four quarters with a shallow incision with a sterile scalpel and open it to expose the seeds, without touching them.

2. Transfer selected well formed seeds by forceps to a Petri dish on a sterile filter paper moistened with sterile water and germinate in the dark.

3. After germination, remove healthy roots, 2–3 cm long, with a sterile scalpel and transfer each to a flask of nutrient medium.

4. After a week, cut out 1 cm long healthy root tips from the developing root system with a pair of scissors and transfer individually to fresh nutrient. The roots can be grown indefinitely.

Method II

1. Prepare cuttings from a healthy tomato plant, remove the leaves, wash in a mild antiseptic solution, followed by sterile water.

2. Bore holes in a stiff paraffinised cardboard and cut it to a size slightly larger than the mouth of a battery jar. Push each cutting through the holes in two sheets of this cardboard with stem side up. Place the cardboard covers at the mouth of a battery jar previously lined with sterile blotting paper and containing a thin layer of sterile water at the bottom. The bottom end of the stem protrudes about 25 cm through the cardboard into the moist chamber formed by the battery jar. Roots develop on this end. Keep in dark.

3. When roots develop, cut out healthy root tips, 1 cm long, with sterile scissors and transfer to a flask containing nutrient medium. Some roots will develop into healthy clones.

Culture of Carrot Callus

1. Insert long narrow strips of sterilised filter paper folded twice along the breadth for thickness into test-tubes containing liquid nutrient medium.

2. Wash and dry a healthy carrot, about 15 cm long and break in the middle. Remove a series of cores with a sterile cork borer from the middle with the cambium traversing them lengthwise, and put them on a sterile Petri dish. Cut the cores into discs 1 mm thick.

3. Transfer two discs to each test-tube, placing them side by side on the strip of paper. Keep the tube tilted at an angle of 30 degrees, so that the paper is kept moist by dipping in the nutrient medium but the discs are not immersed in it.

4. The discs develop into callus tissue within a fortnight. They can be transferred to agar nutrient medium with naphthalene acetic acid and will develop into clones, which can grow indefinitely.

Culture of Sunflower Secondary Tumours

1. Raise healthy plants of *Helianthus annuus* from seeds in good soil, keeping one plant in each pot.

2. Prepare a 48 h broth culture of *Agrobacterium tumafaciens*. When the stem of the young plant is about 9 cm long above the cotyledons, inject the broth with a hypodermic syringe into the stem just above the cotyledons by a single puncture. Tumours will form at this point. After about 6 weeks,

some of the plants will develop bacteria-free secondary tumours at the bases of petioles of a few leaves just above the point of inoculation.

3. Cut off the branch 15 cm above and below the secondary tumour. Split the stem at the base through the middle. Continue the split across the tumour to its other end, and separate the pieces, exposing the interior of the tumour.

4. With a sterile scalpel, cut out small bits of tissue, about 1 mm or so thick from the exposed interior of the tumour and place them in test-tubes containing unsupplemented agar medium. Some of these tissues grow rapidly into large cultures of disorganised tissue, malignant in nature. They can be transplanted under the bark of healthy sunflower stems and develop into tumours there as well.

ANIMAL CULTURES (*see* WHITE, 1954)

The reagents necessary are: chick plasma, chick or horse serum, chick embryo extract, a nutrient salt solution, a synthetic nutrient solution of the white type.

The apparatus required are: depression slides, Carrel flasks, cover slips, 16 mm × 150 mm test-tubes, miniature watch-glasses, pipettes.

Culture of Chick Heart by the Hanging Drop Method

1. Incubate fresh hen's eggs for 6 days. Break open the eggs and place the embryonic hearts in a watch-glass containing a nutrient salt solution. Change the solution twice to wash off the blood. Transfer the tissue to a dry sterile watch-glass or depression slide and cut it into small pieces.

2. Place a sterile 22 mm × 22 mm cover-glass flat on the table. Put a drop of embryo extract at the centre of the cover-glass with a pipette or a tuberculin syringe with a No. 26 needle. Transfer a portion of heart tissue to the drop, add a drop of plasma, stir quickly and allow the mixture to clot; then add a drop of nutrient salt solution. Smear the edges of a grooved or depression slide with vaseline and invert it over the cover slip so that the drop with the tissue lies within the depression. Invert the slide again so that the cover slip is on top and the drop hangs downwards in the depression, and seal the edges with molten paraffin wax or paraffin–vaseline mixture. Incubate at 37 °C and observe daily. For prolonged keeping, the nutrient solutions should be changed after intervals of 3 days and the tissue should be cut and sub-cultured as soon as the culture exceeds 5 mm in diameter.

3. For obtaining several cultures at one time, place several drops at equal distances inside a sterile Petri dish, add the tissue, plasma and nutrient salt solution, cover with the other dish of the pair, seal the edges, invert and incubate. Several cultures may develop within the same Petri dish from several drops.

Culture of Mouse Skin by Roller Tube

1. Kill a female mouse, about 3 weeks' pregnant. Dissect out the uterus intact on a sterile Petri dish. Cut out the uterus. Place the embryo in a sterile Petri dish in salt solution to keep it moist.

2. Remove thin strips of skin and cut into squares of 1–2 mm length. Place 12 narrow cover slips across a large slide on drops of water. Place 6 drops of embryo extract on each cover slip. Transfer a bit of skin to each drop and add a drop of plasma. Stir and allow the mixture to clot.

3. Put 1–2 cm^3 of nutrient (60 per cent salt solution: 2 per cent serum: 20 per cent embryo extract) in each test-tube. Place 2 cover slips back to back and put them inside a test-tube and stopper the tube with cotton wool. Place the tubes in a roller tube rotor at 37 °C. If a rotor is not available, keep them at a slight slope, tissue down, in any incubator. Examine daily and renew the nutrient twice weekly.

MACROPHAGE OR LYMPHOCYTE CULTURE

1. Mix 8·4 cm^3 of 35 per cent bovine albumin with 1·6 cm^3 of salt dextrose solution. Transfer the mixture by hypodermic syringe into two centrifuge tubes.

2. Layer fresh heparinised blood (10 cm^3) of either chick or human beings over the albumin in the two tubes, 5 cm^3 in each. Plug the tube and centrifuge at 2000 rev/min for 10 min. The leucocyte layer floats above the red blood cells at the serum–albumin interface.

3. Pipette out 1 cm^3 of the supernatant fluid. Combine the materials from the two tubes and re-suspend in 10 cm^3 of 10 per cent chicken serum in salt dextrose solution. Centrifuge at slow speed for 5 min. Wash off the remaining plasma or albumin.

4. Remove the supernatant liquid containing the leucocytes, add to it 10 cm^3 of nutrient medium and pipette into Carrel flasks without any plasma at 0·5 cm^3 per flask.

5. Within 24 h the monocyte layer attaches to the glass. Remove the nutrient, with other dislodged cell types obtained by shaking, and add fresh nutrient. The culture covers the entire bottom of the flask within 72 h. The addition of slight bicarbonate solution (1–2 drops of 1·4 per cent soln.) helps to maintain the proper pH (see White, 1954; Weiss and Fawcett, 1953).

TISSUE CULTURE FOR OBSERVATION UNDER PHASE CONTRAST
(JONES and colleagues, 1960)

The cells are grown in micro-cultures for extended periods in a manner to permit detailed cytological observations.

1. Prepare hybrid tobacco (*Nicotiana tabacum X.N. glutinosa*) single cell

clones, isolated from stem callus and grown in liquid 'tobacco' supplemented by coconut milk (150 cm^3/l), calcium pantothenate (2·5 mg/l), naphthalene acetic acid (0·1 mg/l) and 2,4–dichloro-phenoxyacetic acid (6·0 mg/l), in tubes within a shaker.

2. Place a drop of paraffin oil near each end of a standard microscope slide. Lower a No. 1, 22 mm square cover slip on to each droplet to form risers for a shallow central chamber on the slide. Put a drop of mineral oil in a rectangle on the slide, connecting the two cover slip risers and covering the inner end of each. Place a droplet of liquid medium at the centre of a third square cover slip.

3. Isolate a single cell or a small cluster of cells from a culture-tube with a pair of flattened teasing needles under a dissecting microscope and transplant it in the droplet of liquid medium on the third cover slip.

4. Invert the cover slip over the rectangle of mineral oil on the slide in such a manner that the mineral oil surrounds the liquid medium with its enclosed cells and the ends of the top cover slip lie upon the inner ends of the cover slip risers. The culture thus lies in a liquid medium in a tiny micro-culture chamber filled with liquid paraffin.

5. Observe the micro-cultures directly under the microscope. Keep them in sterile Petri dishes in the dark at 26 °C at controlled humidity.

6. By this method, the different cytological changes during the growth of the culture in living cells can be observed under both ordinary and phase contrast microscopes.

MEDIA

NUTRIENT FOR PLANT TISSUE CULTURE

WHITE'S NUTRIENT SOLUTION (1943) FOR PLANTS

A. Inorganic Salt Solution

Ca(NO$_3$)$_2$	20 g	MnSO$_4$	0·45 g
Na$_2$SO$_4$	20 g	ZnSO$_4$	0·15 g
KCl	8 g	H$_3$BO$_3$	0·15 g
NaH$_2$PO$_4$	16·5 g	KI	0·075 g

Dissolve one at a time in 8000 cm^3 of double distilled water. Dissolve 36 g of MgSO$_4$ separately in 2000 cm^3 of water. Mix the 2 solutions in a 10 litre bottle to form stock solution, 10 times the concentration needed in the nutrient. Store in dark.

B. Vitamin Supplement

Glycine	300 mg
Nicotinic acid	50 mg
Thiamine	10 mg
Pyridoxine	10 mg

Dissolve in 100 cm^3 of water. It is 100 times the concentration needed. Store in cold.

C. Carbohydrate Solution

Dissolve 40 g of sucrose in 100 cm^3 of water. Dissolve 5 mg of $Fe_2(SO_4)_3$ in 50 cm^3 of water and add to sugar solution.

For liquid nutrient—Add to the mixed sugar solution 200 cm^3 of stock salt and 20 cm^3 of stock vitamin solutions and make up to 2000 cm^3 with water.

For semi-solid nutrient—Make up the nutrient solution to half the above volume. Dissolve 10 g of agar in 1000 cm^3 of distilled water and mix in equal proportions with the liquid nutrient.

This formula has been found to be useful for roots, plant tumours, callus and embryos. It was observed in this laboratory that addition of yeast extract and IAA resulted in profuse growth of cells in carrot tissue culture (Roy, 1969).

GAUTHERET'S NUTRIENT MEDIUM FOR PLANTS (1950) AND ITS ALTERNATIVES

A. Inorganic Stock Salt Solution (According to Knop)

$Ca(No_3)_2$	1 g		KH_2PO_4	0·25 g
KNO_3	0·25 g		Distilled	
$MgSO_4$	0·25 g		water	1000 cm^3

Inorganic stock solution with trace elements (according to Berthelot, 1934).

$Fe_2(SO_4)_3$	50 g		$Ti_2(SO_4)_3$	0·2 g
$MnSO_4$	2 g		$NiSO_4$	0·05 g
KI	0·5 g		$CoCl_3$	0·05 g
$ZnSO_4$	0·1 g		$CuSO_4$	0·05 g
H_3BO_3	0·1 g		H_3BO_3	0·05 g
Conc.			Distilled	
H_2SO_4	1 cm^3		water	1000 cm^3

B. Vitamin Stock Solution

Cysteine–HCl: 100 mg; thiamine: 10 mg; water: 100 cm^3. Autoclave.
    Ca–pantothenate: 10 mg; water: 100 cm^3. Autoclave.
    Biotin: 10 mg; water: 10 cm^3. Filter.
    Inositol: 1 g; water: 100 cm^3. Autoclave.
    Naphthalene acetic acid: 10 mg; 30 per cent ethyl alcohol: 100 cm^3.
    In the different modifications of the final mixture, the proportions shown in Table 5 are used.

Table 5

Ingredients	Solutions				
	M_1	M_2	M_3	M_4	M_5
Knop's soln.	100 cm^3	100 cm^3	100 cm^3	100 cm^3	100 cm^3
Berthelot's soln.	1 cm^3	1 cm^3	1 cm^3	1 cm^3	1 cm^3
Dextrose	30 g	50 g	50 g	50 g	50 g
Agar	6 g	6 g	6 g	6 g	6 g
Cysteine–thiamine	10 cm^3	10 cm^3	10 cm^3	10 cm^3	—
Ca–pantothenate	1 cm^3	—	—	—	1 cm^3
Biotin	1 cm^3	—	—	—	1 cm^3
Inositol	10 cm^3	—	—	—	—
Naphthalene acetic acid	3 cm^3	1 cm^3	—	0·5 cm^3	1 cm^3
Water	875 cm^3	886 cm^3	887 cm^3	886 cm^3	840 cm^3

The final mixture is adjusted with 0·1 N NaOH. Additional ingredients are:

1 cm^3 of each of $CuSO_4$ and MoO_3 solution (at 1·0 mg/l) is added to 1 litre of nutrient (according to Boll and Street, 1951); 30 cm^3 of coconut milk per 1000 cm^3 of nutrient is a beneficial supplement (according to Caplin and Steward, 1948). Boiled extract of 100 mg of yeast in 500 cm^3 of water can be added to the nutrient as a supplement.

NUTRIENTS FOR ANIMAL TISSUE CULTURE (*see* WHITE, 1954; PAUL, 1959)

A. BALANCED SALT SOLUTIONS (g/1000 CM^3) FOR ANIMALS

The ingredients, given in Table 6, are all dissolved successively in distilled water. $NaHCO_3$ and phenol red are dissolved separately in water, filtered through a selas candle and saturated with CO_2, bubbling through a plugged and sterile pipette. Before use, the two solutions are mixed.

Table 6

Solutions	NaCl	KCl	CaCl$_2$	MgSO$_4$7H$_2$O	MgCl$_2$6H$_2$O	NaH$_2$PO$_4$–H$_2$O
Locke (1895)	9·00	0·42	0·24	—	—	—
Ringer (1886)	9·00	0·42	0·25	—	—	—
Tyrode (1910)	8·00	0·20	0·20	—	0·10	0·05
Glucosol	8·00	0·20	0·20	—	0·10	0·05
Gey and Gey (1936)	7·00	0·37	0·17	0·07	0·21	—
Simms (1941)	8·00	0·20	0·147	—	0·20	—
Earle (1954)	6·70	0·40	0·20	0·10	—	0·125
Gey (1945)	8·00	0·375	0·275	—	0·21	—
Hanks (1946) BSS (Hanks, 1955)	8·00	0·40	0·14	0·10	0·10	—
Holtfreter (1929) (Amphibia and Pisces)	3·50	0·05	0·10	—	—	—
Ringer A (Amphibia)	6·50	0·14	0·12	0·20	—	—
Modified Locke (Insects)	9·00	0·42	0·25	—	—	—
Carlson (Grasshoppers)	7·00	0·20	0·02	—	0·10	0·20
White (1949) (dilute 20 times)	14·00	0·75	—	0·55	{ Ca(NO$_3$)$_2$.H$_2$O { Fe(NO$_3$)$_3$.9H$_2$O	{ 0·42 { 0·011

Table 6 (*continued*)

Solutions	$Na_2HPO_4-2H_2O$	KH_2PO_4	Glucose	Phenol red	$NaHCO_3$	Gas phase
Locke (1895)	—	—	—	—	0·02	Air
Ringer (1886)	—	—	—	—	—	Air
Tyrode (1910)	—	—	1·00	—	1·00	Air
Glucosol	—	—	1·00	—	—	Air
Gey and Gey (1936)	0·15	0·03	1·00	—	2·27	5% CO_2 in air
Simms (1941)	0·21	—	1·00	0·05	1·00	2% CO_2 in air
Earle (1954)	—	—	1·00	0·05	2·20	5% CO_2 in air
Gey (1945)	0·15	0·025	2·00	—	0·25	Air
Hanks BSS (1946)	0·06	0·06	1·00	0·02	0·35	Air
Holtfreter (1929) (Amphibia and Pisces)	—	—	—	—	0·20	
Ringer A (Amphibia)	—	—	—	—	—	
Modified Locke (Insects)	—	—	2·50	—	0·20	
Carlson (Grasshoppers)	—	—	0·80	—	0·05	
White (1949) (dilute 20 times)	0·58	0·104	17·00	0·01	2·20	Water 600 cm^3 Sat. with CO_2

Ca and Mg free Hank's solution (solution A) has no calcium and magnesium salts.

B. SOME STANDARD NUTRIENT SOLUTIONS FOR ANIMAL TISSUE CULTURE

Fischer's Nutrients (1948)

These are used to supplement basic dialysed plasma-dialysed embryo-juice substratum and are ineffective without organic supplementation.
A modified form (Ehrensvärd and colleagues, 1949) contains:

Na Cl	7·5 g	$CaCl_2$	0·2 g
Glucose	2 g	$MgCl_2$	0·1 g
$NaHCO_3$	1 g	Glycine	0·01 g
Na_2HPO_4	0·05 g	Arginine	0·004 g
Aminoethyl phosphate	0·2 g	Tryptophane	0·004 g
Glutamine	0·2 g	Cystine	0·01 g
KCl	0·2 g	Water	1000 cm^3

Eagle's Medium (1955)
Synthetic (mg/1000 cm^3)

	mg		mg
Penicillin	0·50	Pantothenic acid	1·0
1-Arginine	17·4	Pyridoxal	1·0
1-Cystine	6·0	Riboflavine	0·1
1-Histidine	3·2	Thiamine	1·0
1-Isoleucine	26·2	Inositol	1·0
1-Leucine	13·1	Biotin	1·0
1-Lysine	18·2	Folic acid	1·0
1-Methionine	7·5	Glucose	2000·0
1-Phenylalanine	8·3	NaCl	8000·0
1-Threonine	11·9	KCl	400·0
1-Tryptophane	2·0	$CaCl_2$	140·0
L-Tyrosine	18·0	$MgSO_4.7H_2O$	100·0
1-Valine	11·7	$MgCl_2.6H_2O$	100·0
1-Glutamine	146·0	$Na_2HPO_4.2H_2O$	60·0
Choline	1·0	KH_2PO_4	60·0
Phenol red	20·0	$NaHCO_3$	350·0
Nicotinic acid	1·0		

Parker and Healy's Medium (1955)
Amino acids

	mg		mg
1-Arginine	70·0	1-Leucine	120·0
1-Histidine	20·0	1-Isoleucine	40·0
1-Lysine	70·0	1-Valine	50·0
1-Tyrosine	40·0	1-Glutamic acid	150·0
1-Tryptophane	20·0	1-Aspartic acid	60·0
1-Phenylalanine	50·0	1-Alanine	50·0
1-Cystine	20·0	1-Proline	40·0
1-Methionine	30·0	1-Hydroxyproline	10·0
1-Serine	50·0	Glycine	50·0
1-Threonine	60·0	1-Cysteine	260·0

Synthetic (mg/1000 cm^3)
Vitamins

	mg		mg
Pyridoxine	0·025	p-Aminobenzoic acid	0·05
Pyridoxal	0·025	Vitamin A	0·10
Biotin	0·01	Ascorbic acid	50·00
Folic acid	0·01	Calciferol	0·10
Choline	0·50	Tocopherol phosphate	0·01
Inositol	0·05	Menadione	0·01

Coenzymes

95% DPN	7·0
80% TPN	1·0
75% COA	2·5
88% TPP	1·0
60% FAD	1·0
90% UTP	1·0
100% Glutathione	10·0

Lipid sources

Tween 80 (oleic acid)	5·0
Cholesterol	0·2

Nucleic acid derivatives

Adenine deoxyriboside	10·0
Guanine deoxyriboside	10·0
Cytosine deoxyriboside	10·0
5-Methylcytidine	0·1
Thymidine	10·0

Miscellaneous

Sodium acetate	50·0
d-Glucuronic acid	3·6
1-Glutamine	100·0
d-Glucose	1000·0
Phenol red	20·0
Ethyl alcohol	16·0

Inorganic salts

NaCl	6800·0
KCl	400·0
$CaCl_2$	200·0
$MgSO_4.7H_2O$	200·0
$NaH_2PO_4.H_2O$	140·0
$NaHCO_3$	2200·0
$Fe(NO_3)_2$	0·1

Antibiotics

Sodium penicillin G (just before use)	1·0
Dihydrostreptomycin sulphate	100·0
n-Butyl para-hydroxybenzoate	0·2
No organic supplement is needed	

White's Nutrient Medium (White, 1954)

It contains 70 cm^3 of sterile water and 15 cm^3 of the mixture of White's inorganic salt solution, $Fe(NO_3)_3$ solution and sugar buffer given in Table 6. To this is added in succession 5 cm^3 each of amino acids, stock, AC stock, B stock and B_{12} stock to prepare the final nutrient with pH 7·4 and a pale red colour.

The ingredients are:
 (i) Inorganic salt solution, $Fe(NO_3)_3$ soln. and sugar buffer are described in Table 6.
(ii) Amino acid stock, containing (mg/80 cm^3 of water):

1-Lysine–HCl	312 mg	dl-Isoleucine	208 mg
dl-Methionine	260 mg	dl-Phenylalanine	100 mg
dl-Threonine	260 mg	1-Leucine	312 mg
dl-Valine	260 mg	1-Tryptophane	80 mg
1-Arginine–HCl	156 mg	1-Glutamic acid	280 mg
1-Histidine–HCl	52 mg	1-Aspartic acid	120 mg
1-Proline	100 mg	1-Cystine	30 mg
Glycine	200 mg	0·01% phenol red	10 cm^3

(iii) AC stock, containing:

Group A	Carotene	10 mg
	Vitamin A	10 mg
	Ethyl alcohol	100 cm^3
Group B	Ascorbic acid	10 mg
	Glutathione	20 mg
	Cysteine HCl	20 mg
	Water	100 cm^3. Filter.
Group C	0·01% aq. phenol red	100 cm^3

Mix 2 cm^3 of A, 10 cm^3 of B, 2 cm^3 of C and 86 cm^3 of water and maintain at pH 7·4 by adding 0·1 N NaOH.

(iv) B stock, containing:

Group B		Group FA	
Thiamine HCl	10 mg	Folic acid	10 mg
Riboflavin	10 mg	NaHCO₃	10 mg
Ca-pantothenate	10 mg	Water	100 cm³
d-Biotin	10 mg		
Pyridoxin HCl	10 mg	Sterilise and filter	
Nicotinic acid	10 mg		
Inositol	10 mg		
l-Alanine	10 mg		
Choline	100 mg		
Water	100 cm³		

(v) B_{12} stock containing 0·015 per cent aqueous vitamin B_{12} solution, sterilised.

SOME NUTRIENT SOLUTIONS FOR EMBRYO CULTURE

The Randolph–Cox Solution (Randolph and Randolph, 1955)

Solution A		Solution B	
Calcium nitrate	23·6 g	Ferrous sulphate	0·2 g
Potassium nitrate	8·5 g	Calgon (NaPO₃)ₙ	1 g
Potassium chloride	6·5 g	Magnesium sulphate	3·6 g
Distilled water	500 cm³	Distilled water	500 cm³

The two solutions are stored separately. For preparing nutrient medium, heat 7 g of agar in 1000 cm³ of distilled water. Add to it 20 g of sucrose and 5 cm³ of each of solutions A and B. Distribute in sterilised bottles.

Knudson's Solution (Referred to in Randolph and Randolph, 1955)

Calcium nitrate	1 g	Ferrous sulphate	0·025 g
Ammonium sulphate	0·5 g	Manganese	
Magnesium		sulphate	0·0075 g
sulphate	0·25 g	Sucrose	20 g
Potassium		Agar	15 g
phosphate	0·25 g	Distilled water	1000 cm³

Concentrated stock solutions of the different chemicals are prepared, except $FeSO_4$. Add proper amounts of each to distilled water in which agar has been dispersed by heating.

REFERENCES

Ambrose, E. J. (1961). *Exp. Cell Res. Suppl.* **8,** 54
Anderson, N. (1953). *Science* **117,** 627
Auerbach, R. and Grobstein, C. (1958). *Exp. Cell Res.* **15,** 384
Baker, L. E. (1933). *J. exp. Med.* **58,** 575
Barski, G., Maurin, J., Wielgosz, G. and Lepine, P. (1951). *Ann. Inst. Pasteur* **81,** 9
Bell, E. (1960). *Exp. Cell Res.* **20,** 378
Berthelot, A. (1934). *Bull. Soc. Chim. biol. Paris* **16,** 1553
Blakely, L. M. and Steward, F. C. (1961). *Amer. J. Bot.* **48,** 351
Boll, W. G. and Street, H. E. (1951). *New Phytol.* **50,** 52

Braun, A. C. (1947). *Amer. J. Bot.* **34,** 234
Brochart, M. (1954). *Nature* **173,** 160
Bryant, J. C., Earle, W. R. and Peppers, E. V. (1953). *J. nat. Cancer Inst.* **14,** 189
Buchsbaum, R. and Kuntz, J. A. (1954). *Ann. N.Y. Acad. Sci.* **58,** 1303
Caplin, S. M. and Steward, F. C. (1948). *Nature* **163,** 920
Carpenter, E. (1942). *J. exp. Zool.* **89,** 407
Carrel, A. (1912). *J. exp. Med.* **15,** 516
Carrel, A. (1923). Referred in Paul, 1959
Chen, J. M. (1954). *Exp. Cell Res.* **7,** 518
Chu, E. H. Y. and Giles, N. H. (1959). *Amer. J. hum. Genet.* **11,** 63
Dulbecco, R. and Vogt, M. (1954). *J. exp. Med.* **99,** 167
Eagle, H. (1955). *J. exp. Med.* **102,** 595 and *Science* **122,** 501
Eagle, H., Oyama, V. I., Levy, M. and Freeman, A. E. (1957). *J. biol. Chem.* **226,** 191
Earle, W. R., Bryant, J. C. and Schilling, E. L. (1954). *Ann. N.Y. Acad. Sci.* **58,** 1000
Easty, G. C. and Mutolo, V. (1960). *Exp. Cell Res.* **21,** 374
Ehrensvärd, G., Fischer, A. and Sjerholm, R. (1949). *Acta physiol.* **18,** 218
Ehrmann, R. L. and Gey, G. O. (1953). *J. nat. Cancer Inst.* **13,** 1099
Ehrmann, R. L. and Gey, G. O. (1956). *J. nat. Cancer Inst.* **16,** 1375
Eidinoff, M. L. and Rich, M. A. (1959). *Cancer Res.* **19,** 521
Fell, H. B. (1928). *Arch. exp. Zellforsch.* **7,** 69 and 390
Fell, H. B. (1931). *Arch. exp. Zellforsch.* **11,** 245
Fell, H. B. (1961). In *La Culture Organotypique* Paris; Coll. Int. C.N.R.S.
Fell, H. B. and Robinson, R. (1929). *Biochem. J.* **23,** 767
Firket, H. (1958). *Nature* **182,** 399
Firket, H. (1965). In *Cells and tissues in Culture* **1,** p. 203, New York; Academic Press
Fischer, A. (1948). *Biochem. J.* **43,** 491
Frédéric, J. and Corin, J. (1962). *C. R. Acad. Sci. Paris* **254,** 357
Fujii, T. (1941). *J. Fac. Sci. Tokyo* **5,** 355
Gaillard, P. T. (1948). *Sym. Soc. exp. Biol.* **2,** 139
Gautheret, R. J. (1934). *C. R. Acad. Sci. URSS* **198,** 2195
Gautheret, R. J. (1939). *C. R. Acad. Sci. URSS* **208,** 118
Gautheret, R. J. (1942). *Manual Technique de Culture des Vegeteux,* Paris; Masson et Cie
Gautheret, R. J. (1950). *C. R. Soc. Biol. Paris* **144,** 173
Gautheret, R. J. (1955). *Ann. Rev. Pl. Physiol.* **6,** 433
Gey, G. O. (1933). *Amer. J. Cancer* **17,** 752
Gey, G. O. (1945). Referred in Paul, 1959
Gey, G. O. and Gey, M. K. (1936). *Amer. J. Cancer* **27,** 45
Grover, J. W. (1962). *Exp. Cell Res.* **26,** 344
Haberlandt, G. (1902). *Sber. Akad. Wiss. Wien* **111,** 69
Hanks, J. H. (1946). Referred in Paul 1959
Hanks, J. H. (1955). In *An Introduction to cell and tissue culture.* Minneapolis; Burgess Publishing Co.
Harris, M. (1952). *J. Cell comp. Physiol.* **40,** 279
Harrison, R. G. (1907). *Proc. Soc. exp. Biol.* **4,** 140
Holtfreter, J. (1929). *Arch. Entw. Mech. Org.* **117,** 221
Hsu, T. C. (1955). *J. nat. Cancer Inst.* **16,** 691
Hsu, T. C. (1960). *Tex. Rep. Biol. Med.* **18,** 31
Hsu, T. C. (1965). In *Cells and tissues in culture* **1,** p. 397, New York; Academic Press
Hsu, T. C. amd Pomerat, C. M. (1953). *J. Hered.* **44,** 23 and *J. Morph.* **93,** 301
Iyer, V. N. and Szybalski, W. (1963). *Proc. nat. Acad. Sci. Wash.* **50,** 355
Jolly, J. (1903). *C. R. Soc. Biol. Paris* **55,** 1266
Jones, L. E., Hildebrandt, A. C., Riker, A. J. and Wu, J. H. (1960). *Amer. J. Bot.* **47,** 468
Kihlman, B. A. and Hartley, B. (1968). *Hereditas* **59,** 439
Knop, W. (1884). *Landw. Versuchsw.* **30,** 292
Knudson (1925). Referred in Randolph and Randolph, 1955.
Kuchler, R. J., Marlowe, M. L. and Merchant, D. J. (1960). *Exp. Cell Res.* **20,** 428
Lajtha, L. G. (1957). *Physiol. Rev.* **37,** 50
Lasfargues, E. Y. (1957). *Exp. Cell Res.* **13,** 553
Laws, J. O. and Stickland, L. H. (1961). *Exp. Cell Res.* **24,** 240
Levan, A. and Hauschka, T. S. (1953). *J. nat. Cancer Inst.* **14,** 1

Levan, A. and Hsu, T. C. (1961). *Hereditas* **47,** 69
Lieberman, I. and Ove, P. (1958). *J. biol. Chem.* **233,** 634
Ljunggren, C. A. (1898). *Deutsch Z. Chir.* **47,** 608
Locke, F. S. (1895). *Boston med. surg. J.* **134,** 173
Lockart, R. Z. and Eagle, H. (1959). *Science* **129,** 252
Lutz, H. and Lutz-Ostertag, Y. (1959). *C. R. 'Acad. Sci. Paris* **249,** 2122
Mackaness, G. B. (1952). *J. Path. Bact.* **64,** 429
Makino, S. (1953). *Cytologia* **18,** 129
Makino, S. and Hsu, T. C. (1954). *Cytologia* **19,** 23
Makino, S. and Nakahara, H. (1953). *Cytologia* **18,** 128
Martinovitch, P. N. (1950). *Nature* **165,** 33
Medawar, P. B. (1957). *The Uniqueness of the individual.* London; Methuen
Mehra, P. N. (1961). Present address, Botany Section, *Proc. 48th Ind. Sci. Cong.,* pt III, 130
Melnick, J. L., Rappaport, G., Banker, D. and Bhatt, P. (1955). *Proc. Soc. exp. Biol. N.Y.* **88,** 676
Mendelsohn, M. L. (1960). *J. nat. Cancer Inst.* **25,** 477 and 485
Merchant, D. J., Kahn, R. H. and Murphy, W. H. (1960). *Handbook of cell and organic culture,* Minneapolis; Burgess Publ. Co.
Miura, T. and Utakoji, T. (1961). *Exp. Cell Res.* **23,** 452
Mitra, J. and Steward, F. C. (1961). *Amer. J. Bot.* **48,** 358
Morel, G. and Wetmore, R. H. (1951). *Amer. J. Bot.* **38,** 141
Moscona, A. (1952). *Exp. Cell Res.* **3,** 535
Moscona, A. (1956). *Proc. Soc. exp. Biol., N.Y.* **92,** 410
Moscona, A. (1961). *Nature* **190,** 408 and *Exp. Cell Res.* **22,** 455
Moscona, A. (1962). *J. Cell. comp. Physiol.* Suppl. 1, **60,** 65
Moscona, A. (1963). *Nature* **199,** 379 and *Proc. nat. Acad. Sci. Wash.* **49,** 742
Moscona, A., Trowell, O. A. and Willmer, E. N. (1965). In *Cells and tissues in culture* **1,** 19, New York; Academic Press
Newton, A. A. and Wildy, P. (1959). *Exp. Cell Res.* **16,** 624
Nitsch, J. (1951). *Amer. J. Bot.* **38,** 566
Nobecourt, P. (1937). *C. R. Acad. Sci. URSS* **205,** 521
Nobecourt, P. (1939). *C. R. Soc. Biol. Paris* **130,** 1270
Northcraft, R. D. (1951). *Science* **113,** 407
Owens, O. von H., Gey, G. O. and Gey, M. K. (1953). *Proc. Amer. Ass. Cancer Res.* **1,** 41
Parker, R. C. and Healy, G. (1955). Referred in Paul, 1959.
Partanen, C. R., Sussex, I. M. and Steeves, T. A. (1955). *Amer. J. Bot.* 42, 245
Paul, J. (1959). *Cell and tissue culture.* Edinburgh and London; Livingstone
Pomerat, C. M. (1951). *J. Nerv. Treat. Dis.* **114,** 430
Pricer, W. E. Jr. and Weissbach, A. (1964). *Biochem. biophys. Res. Commun.* **14,** 91
Puck, T. T. (1961). *Harvey Lect. Ser.* **55,** 1
Puck, T. T., Marcus, P. I. and Cieciura, S. J. (1956). *J. exp. Med.* **103,** 273
Puck, T. T., Cieciura, S. J. and Fisher, H. W. (1957). *J. exp. Med.* **106,** 145
Puck, T. T., Cieciura, S. J. and Robinson, A. (1958). *J. exp. Med.* **108,** 945
Pulvertaft, R. J. V., Haynes, J. A. and Groves, J. T. (1956). *Exp. Cell Res.* **11,** 99
Randolph, L. F. and Randolph, F. R. (1955). *Bull Amer. Iris Soc.* **193,** 2
Rappaport, C. (1960). *Exp. Cell Res.* **20,** 465
Rinaldini, L. M. (1958). *Int. Rev. Cytol.* **7,** 587
Ringer, S. (1886). *J. Physiol.* **7,** 241
Rose, G. (1954). *Tex. Rep. Biol. Med.* **12,** 1074
Rose, G. (1962). *J. Cell Biol.* **12,** 153
Roy, S. (1969). *Proc. Int. Seminar on chromosome, Nucleus* **12,** 223, Suppl., Calcutta
Rueckert, R. R. and Mueller, G. C. (1960). *Cancer Res.* **20,** 1584
Sabina, L. R., Tosoni, A. L. and Parker, R. C. (1963). *Proc. Soc. exp. Biol. Med.* **114,** 13
Sanford, K. K., Earle, W. R. and Likely, G. D. (1948). *J. nat. Cancer Inst.* **9,** 229
Shaffer, B. M. (1956). *Exp. Cell Res.* **11,** 244
Siminovitch, L., Graham, A. F. Lesley, S. M. and Nevill, A. (1957). *Exp. Cell Res.,* **12,** 299
Simms, H. S. (1941). Referred in Paul 1959
Sisken, J. E. and Kinosita, R. (1961). *J. biophys. biochem. Cytol.* **9,** 509
Smith, C. L. and Dendy, P. P. (1962). *Nature* **193,** 555
Somers, C. E. and Hsu, T. C. (1962). *Proc. nat. Acad. Sci. Wash.* **48,** 937

Stanners, C. P. and Till, J. E. (1960).   *Biochem. biophys. Acta* **37,** 406

Steward, F. C. (1958).   *Amer. J. Bot.* **45,** 709

Steward, F. C. and Shantz, E. M. (1956).   *The Chemistry and mode of action of plant growth substances.* London; Butterworths

Sykes, J. and Moore, E. B. (1960).   *Tex. Rep. biol. Med.* **18,** 288

Taylor, J. H. (1958).   *Exp. Cell Res.* **15,** 350

Terasima, T. and Tolmach, L. J. (1963).   *Exp. Cell Res.* **30,** 344

Tulecke, W. (1957).   *Amer. J. Bot.* **44,** 602

Tyrode, M. V. (1910).   *Arch. int. Pharmocodyn.* **20,** 205

Vöchting, H. (1878).   *Über Organbildung im Pflanzenreich.* Bonn; Cohen

Weiss, L. P. and Fawcett, D. W. (1953).   *J. Histochem. Cytochem.* **1,** 47

White, P. R. (1934).   *Plant Physiol.* **9,** 585

White, P. R. (1939).   *Amer. J. Bot.* **26,** 59

White, P. R. (1943).   *A Handbook of Plant tissue culture.* Lancaster, Pa.; Jacques Cattell

White, P. R. (1949).   *J. Cell comp. Physiol.* **24,** 311

White, P. R. (1954).   *The cultivation of animal and plant cells.* New York; Ronald Press

White, P. R. (1959). In *The cell* **1,** 291, New York; Academic Press

Whitmore, G. F., Stanners, C. P., Till, J. E. and Gulyas, S. (1961). *Biochem. biophys. Acta* **47,** 66

Willmer, E. N. (1945).   In *Essays on growth and form.* New York; Oxford University Press

Willmer, E. N. (1965).   *Cells and tissues in culture* **1,** 143, New York; Academic Press

Wolff, E. (1957).   *J. nat. Cancer Inst.* **19,** 597

Wolff, E. and Haffen, K. (1952).   *J. exp. Zool.* **119,** 381

Zwilling, E. (1954).   *Science* **120,** 219

Zeuthen, E. (1964).   *Synchrony in Cell Division.* New York; John Wiley

15 Chromosome Studies from Mammals with Special Reference to Human Chromosomes

The remarkable progress in the study of cytogenetics in lower species encouraged a comparable study of mammalian, and later, of human cytogenetics, which was unfruitful until recently due to inadequate techniques. One of the less studied branches of cytogenetics was mammalian karyology, and progress in this has been made possible by technical advances during the last 15 years. What was almost a completely unexplored terrain has now attracted a large number of investigators from all over the world. There are now many techniques for studying mammalian chromosomes, particularly human ones, each laboratory having evolved its own variant of different published methods. Most of them have been devised in quick succession, as a result of improvements like, development of the tissue culture schedule to obtain cells *in vitro*, either suspended or forming a monolayer; pre-treatment by colchicine or its derivatives to accumulate a large number of mitotic figures; hypotonic solution treatment causing swelling of the cells to aid chromosome scattering and air-drying to force the chromosomes to lie in one plane. The techniques, in general, aid in the study of the karyotype, of the meiosis, and of the sex chromatin. A suitable method for karyotype study must fulfil three demands: (*i*) adequate scattering of chromosomes, (*ii*) minimum distortion and (*iii*) flattening of the cells. The three principal schedules, with variants, utilised in karyotype analysis, are the short term peripheral blood culture, the direct bone marrow technique, and the long term fibroblast culture. For meiotic studies, methods have been developed for use on gonadal tissues. Sex chromatin studies are done from buccal smears, tissue sections and peripheral blood neutrophiles.

As the methods were first developed either for other mammalian materials and then applied to human chromosomes, or vice versa, and because the method employed is decided by the nature of the tissue used, the different schedules will be discussed in relation to the tissue utilised, irrespective of their mammalian or human origin.

The principle of the use of colchicine or other mitostatic chemicals for causing metaphase arrest in both meiotic and mitotic preparations has been discussed in the chapter dealing with pre-treatment. However, since colchicine has been found to produce chromosomal anomalies (Amarose, 1959) and disturbance in DNA synthesis (Lima-de-Faria and Bose, 1962), Turpin and Lejeune (1969) suggest that it should be avoided unless absolutely necessary. The dispersal of chromosomes by hypotonic shock, as discovered by Hsu (1952), led to the evolution of most of the methods on mammalian chromosomes, but its possible drawback is the loss of certain information, like the probable association of chromosomes with satellites between them, the somatic pairing of homologues and their spatial arrangement in the equator (Barton and David, 1962, 1963; Barton *et al.*, 1963). Similarly, the squash schedule results in possible deformation and displacement of the chromosomes. The simple air-drying technique, as devised by Rothfels and Siminovitch (1958a), gives very satisfactory flattening of the chromosomes and is a common feature of most techniques.

I. STUDY OF MEIOSIS

Meiotic configurations from mammalian gonads were obtained by Makino and Nishimura (1952) and later by Darlington and Haque (1955), leading to the evolution of various schedules. The kind of pre-treatment depends on the developmental stage of the germ cells to be studied but the procedure for fixing, squashing and staining is similar for all stages.

Gonadal tissues are usually obtained through biopsy and cut into small cubes. For mitotic prophase and telophase figures of germ cells, repeated suspension in a cold isotonic solution like Medium 858, or Hank's basal salt solution, at 3 °C is adequate. It is also suitable for first meiotic prophase figures of male germ cells, obtained by testicular biopsy. In female cells, the divisional figures are numerous in foetal ovaries between the 4th and 8th months. The ovary is immersed in isotonic solution, dissected, incubated at 37 °C in 10 per cent trypsin solution for 30 min, followed by three 10 min rinsings in isotonic solution. Swelling needed for mitotic metaphase and anaphase stages can be acquired by immersion for 10 min in neutral (pH 7) double distilled water, at room temperature, two or three times.

In the male, meiotic metaphase figures are greatly improved by hypotonic pre-treatment. Individual seminiferous tubules are dissected out from the tissue immersed in isotonic solution, transferred to double distilled water at room temperature and kept for 30 min prior to fixation. In the human female, first and second meiotic metaphase figures are difficult to obtain, since they occur only in the mature follicle in the ovary of a woman in her reproductive span, on the day of ovulation. The mature follicle is punctured in other mammals and the ovum dissected out and incubated at 37 °C. It is then transferred by means of a pipette to the fixative (50 per cent acetic acid), kept for 15 min and covered gently to separate the bivalents (*see* Ohno, 1965 and also Edwards, R. G. 1962).

The period of fixation ranges from 15–45 min, a cube of material (2 × 2 × 2 mm) being placed on a slide in 1 cm^3 of fixative and tapped to release the free cells. It is covered with a cover slip and squashed with straight, uniform pressure. Later, the slide is immersed for 1 min in a mixture of dry ice and methanol, dried in air, treated in methanol for 15 min to remove fatty substances, dried, washed in water, hydrolysed in N HCl at 60 °C for 15 min, and then stained, dried and mounted in synthetic balsam. Staining for 3 h with Feulgen reagent or 5 min with Giemsa or 1 min with 0·25 per cent basic fuchsin solution gives good results (Ohno, 1965). Welshons *et al.* (1962) proposed 2 per cent acetic-lactic-orcein (orcein 2 g, acetic acid 50 cm^3, 85 per cent lactic acid 42·5 cm^3, water 7·5 cm^3) for tubules after fixation in 50 per cent acetic acid added to hypotonic sodium citrate solution. Two sample schedules are given.

SCHEDULE FOR HUMAN CHROMOSOMES

Devised by Ford (1961) and recommended by Turpin and Lejeune (1969):
 1. Place tissue in hypotonic solution (pH 7·0; 50 per cent potassium glycerophosphate, 93·02 cm^3; 20 per cent glycerophosphoric acid, 24·92 cm^3 and distilled water made up to 200 cm^3) and tease out the tubules. Keep at room temperature for 3 min.
 2. Fix in glacial acetic acid–ethanol (1:3) for 1 h. Treat in 30 per cent ethanol and distilled water successively for 3–5 min for hydration.
 3. Hydrolyse in N HCl at 60 °C for 8 min.
 4. Stain in Feulgen reagent for 1 h.
 5. Rinse successively in SO$_2$ water and cold 45 per cent acetic acid, 3–5 min in each.
 6. Transfer tubules to 45 per cent acetic acid and squash under a siliconed cover slip without any sideways motion. If necessary, the tubules can be stored in 45 per cent acetic acid at −12° to −15 °C before squashing.
 7. Papain treatment is sometimes adopted to improve spreading of chromosomes, probably acting through digestion of the tubule membrane. Treat tubules in one per cent aqueous papain solution for 10 min before squashing. Rinse, keep in 60 per cent acetic acid for 5 min, cover and squash. Shrinkage of the cells in papain is followed by swelling to greater than original size.
 8. The preparations are made permanent by dry ice technique or dehydration through ethanol dioxane grades, and mounted in euparal or canada balsam.

SCHEDULE FOR MAMMALIAN TESTES

Followed by Evans *et al.* (1964) and its modifications: Remove testis into 2·2 per cent isotonic sodium citrate solution, swirl. Cut out a portion of the tubule, transfer to 1·125 per cent KCl, mince and tease out the contents. Decant cloudy supernatant and centrifuge at 160 *g* for 5 min. Discard supernatant, add acetic-methanol (1:3), disperse cells and keep overnight at 4 °C. Centrifuge and discard supernatant. Disperse cells in fresh fixative.

Place a drop on a pre-cooled clean slide, tilt and allow fixative to evaporate and air dry. Stain with acetic-orcein or lactic-acetic-orcein, dehydrate through ethanol–xylol grades and mount.

With minor modifications, this method is widely used for mammals (Meredith, 1969) and reptiles. Clendenin (1969) introduced two variations, injection of 4 mg/kg of 1 per cent colchicine solution in Hank's medium intraperitoneally in living animals 1–2 h before killing, and use of 0·563 per cent KCl for swelling the cells. Earlier, Benirschke and Brownhill (1963) had used a variant of this technique in marmoset monkeys by injecting colchicine in the living animals. Peterson *et al* (1967) had emphasised the use of colcemide pre-treatment for meiotic studies and found that hypotonic KCl gave better details of meiotic chromosomes than other chemicals. In a modified schedule (Eicher, 1966), tubules are treated in a 0·7 per cent sodium citrate for 20–30 min, 5 cm^3 of glacial acetic acid is added, mixed and kept for 30 min and the supernatant is removed after centrifugation, 3 cm^3 of 3 M gluconic acid is added to the tissue, it is treated for 3 h, the acid is then removed and the cells are suspended in acetic–ethanol (1:1). Washing in fixative is done by centrifugation and air-dried smears are prepared.

Staining prior to maceration has been suggested by Gardner and Punnett (1964). After swelling in 0·3 per cent sodium citrate for 1–6 h, the tissue is softened for 2 h with 3 M glucono–delta-lactone, stained overnight with acetic- or propionic-carmine, washed in 70 per cent ethanol, macerated in acetic–ethanol (1:1), filtered through gauze, centrifuged and a drop of the supernatant is squashed in Hoyer's medium.

Tarkowski's air-drying method (1966), described in the chapter on processing, was used by him for pre-implantation stages of mouse egg and was adapted for rat eggs as well (Dyban and Udalova, 1967). In an earlier publication, a simpler method of dissecting out the eggs in saline solution, fixing in Carnoy's fluid and floating them in a drop of acetic-carmine between two paraffin ridges on a slide has been described (Spalding and Wellnitz, 1956).

II. KARYOTYPE ANALYSIS

This can be carried out from tissues showing mitotic figures. After Tjio and Levan (1956) definitely established the normal somatic chromosome number in human cells to be 46, their use of colchicine and hypotonic solution on cells *in vitro* was immediately followed by other workers. The biological and clinical significance of this study was established by the demonstration of a specific chromosomal anomaly, trisomy of chromosome G, in mongolism by Lejeune *et al.* (1959). Since then, a vast array of data on chromosomal patterns associated with a variety of clinical disorders has been pouring in from centres all over the world. Screening studies suggest, in fact, that approximately 0·6–0·8 per cent of all live-born individuals carry a chromosomal abnormality (report of WHO, 1969). Karyotype analysis has been the principal means for detecting these anomalies and has thus gained increasing importance. Chromosome studies on human materials are also being used in evaluating homotransplantation barriers; in the study of dosage compensation, of drug effect and radiation on mitosis, of chromosome mapping and of mammalian cell genetics in tissue culture (*see* also, Yunis, 1965).

Depending upon the type of tissue available, different methods have been developed and, with variations, have been applied to other mammals as well.

PREPARATIONS FROM BONE MARROW CELLS

Several schedules are available, which can be divided into two groups: *direct preparations* and *culture techniques*. Direct processing enables identification of the cell line from which the analysed cells are derived and is useful in the study of leukaemia, since leukaemic cells do not grow in the *in vitro* conditions of marrow tissue culture (Sandberg *et al.*, 1962). Two methods, one involving squashing and the other the air-drying schedule of Rothfels and Siminovitch (1958a) have been evolved by Tjio and Whang (1962, 1965). Other techniques include short-term and medium-term cultures and immediate aspiration of bone marrow cells from individuals treated with colchicine.

A. *Short-term culture technique* (Ford and Hamerton, 1956a, b, c; Ford *et al.*, 1958):

Withdraw 1 cm^3 bone marrow from the individual, disperse in Ringer's solution containing 1:20 000 heparin. Centrifuge at 500–1000 rev/min and discard supernatant. Re-suspend in isotonic glucose–saline mixed in equal parts with human AB serum (or serum of the subject). Transfer aliquots of 2–4 cm^3 to McCartney bottles, keep at room temperature overnight. Incubate at 37 °C for 5 h. Add 0·2–0·4 cm^3 of 0·04 per cent isotonic saline solution of colchicine and keep at 37 °C for 2 h. Dilute four times by adding 0·37 per cent sodium citrate solution and keep for 10 min. Centrifuge, discard supernatant, fix and stain by acetic–orcein schedule.

Modifications of this schedule involve elimination of the culture process. Three of them are described below:

(a) *Direct squash preparations*—(Tjio and Whang, 1962)—Aspirate 0·5 cm^3 of bone marrow and treat in two changes of 2–3 cm^3 colchicine solution (0·85 per cent NaCl solution with $6·6 \times 10^{-3}$ M phosphate, pH 7, to which is added 0·3 g/cm^3 colchicine or colcemide). Keep for 1–2 h at 20–30 °C. Transfer to a watch glass with a few drops of 2 per cent acetic-orcein: NHCl (9:1) mixture, heat gently. Squash in a drop of 2 per cent acetic-orcein as usual without sideways movement. Seal and store. It can be made permanent by dry-ice freezing schedule and mounted in permount.

(b) *Direct air-dried preparations*—The preliminary steps are similar for colchicine treatment as the previous schedule. Centrifuge at 400 rev/min for 4–5 min. Remove supernatant, add 2–3 cm^3 of hypotonic aqueous

1 per cent sodium citrate solution, shake and keep for 30 min. Centrifuge, remove supernatant, add acetic-ethanol (1:3), re-suspend by shaking, keep 2–5 min and repeat this procedure twice. With a pipette, transfer a droplet of the suspension to a slide. Blow on each droplet as soon as the cells are attached to the glass surface, to assist spreading. Allow to dry. Immerse 10–20 min in 2 per cent acetic-orcein solution, dehydrate through ethanol–xylol grades and mount in permount. Feulgen reagent, Giemsa stain and crystal violet can be used as alternative stains.

(c) *Immediate examination* (as developed by Bottura and Farrari (1960)) proposes intravenous injection of colchicine into the subject 2 h before sternal puncture, followed by direct treatment in hypotonic solution, fixation and staining. Though used extensively in animals, this method is, for obvious reasons, not recommended for human material.

B. Medium Term Culture Schedules

(a) *Liquid medium technique* (Fraccaro et al., 1960a, b; Hirschhorn and Cooper, 1961) Aspirate out 1 cm^3 of bone marrow, suspend in 3 cm^3 of culture medium: AB serum, 35 per cent; TC 199 medium, 60 per cent; embryonic extract, 5 per cent. Distribute in Petri dishes containing cover slips, incubate at 37 °C for 24–72 h in an atmosphere containing 5 per cent CO_2. Remove the cover slips and treat them in 0·7 per cent aqueous sodium citrate solution, fix and stain following the acetic-orcein schedule.

(b) *Solid medium technique* (Turpin and Lejeune, 1969)—Aspirate 1–2 drops of bone marrow in a heparinised syringe, place on a disc in a Leighton tube covered with a film of chick plasma, as described in study of biopsy tissues. Coagulate with one drop of embryonic extract. Add 5 drops of nutrient medium (TC 199, 5; AB serum or patient's serum, 5, and embryonic extract, 2). Seal and incubate at 37 °C for 3–5 days. Replenish culture after observing under the microscope and disperse; fix and stain as described later under fibroblast culture schedules.

The bone marrow schedule, when applied to vertebrates other than man, generally includes injection of an anti-mitotic agent into the animal 1–4 h before killing it (Young et al., 1960). Several modifications are available. De Vries and van Went (1964) advocated injection of 4 cm^3 of human O-plasma from citrated blood intraperitoneally into a rat 16 h before injecting colchicine. The marrow cells can be washed out from the medullary cavity (Patton, 1967) or expelled from the bone shaft with a needle or a probe (Nadler and Block, 1962). An alternative schedule devised by Lee (1969) is given in outline:

Administer a mitotic inhibitor (0·004 per cent vincaleucoblastine) intraperitoneally into the animal at 0·02 cm^3/g body weight for animals up to 50 g, and 0·01 cm^3/g for those above 50 g. Kill the animal, and extract the bone marrow into 1 per cent sodium citrate solution. Keep at 37 °C for 5–15 min. Suspend the cells by inverting the tube. Decant suspension. Centrifuge twice at 500–700 rev/min for 2–3 min, discard the supernatant. Fix in 6–8 cm^3 of acetic-methanol (1:3) at 4–6 °C for 2–8 h. Re-suspend

and remove debris at the bottom. Wash 2–3 times in fixative by centrifugation. Re-suspend in 0·5–3·0 cm^3 of fixative. Place 4–6 drops on damp, pre-
cooled slides and spread by blaze-drying as described in section on peripheral
blood technique. Any of the usual chromosomal stains can be used, like
acetic-orcein; Feulgen (Battaglia, 1959); Giemsa, haematoxylin, etc.,
following their respective schedules.

The entire process can also be carried out on a slide (Bohorfoush, 1964).
Aspirate 1 cm^3 of bone marrow into a syringe with 3 drops of 15 per cent
K$_2$–EDTA, eject on to a slide with 1–2 extra drops of EDTA. Tilt the slide
to remove solution. Bring marrow to the edge and smear over another
slide. Air-dry the smear, heat at 120–125 °C for 2 min, treat for 30–45 s
with undiluted Wright's stain, dilute with Na$_2$S$_2$O$_3$ solution (0·1–0·2 g/l of
water) and keep 10–13 min.

PERIPHERAL BLOOD CULTURE TECHNIQUES

These are now almost universally practised and their use has been extended
to practically all vertebrates. These methods are based on the finding that
under the influence of phytohaemagglutinin (PHA) and certain other more
specific antigens, the lymphocyte in the peripheral blood is morphologically
changed into a blast-like cell that divides in culture. The first analysis of
human karyotype with this method was performed by Hungerford *et al.*
(1959). A modification developed by Moorhead *et al.* (1960), combining
the air-drying method of Rothfels and Siminovitch (1958a), with that of the
peripheral blood culture, gave an excellent technique for obtaining well-
scattered somatic metaphase plates. This technique has two major advantages: first, the ease with which it can be carried out on large numbers of
individuals with a high success rate, and secondly, almost all the cells
studied in metaphase are in their first division in culture, if suitably controlled.

In PHA-stimulated cultures, small lymphocytes increase in size and start
to synthesise DNA after 24 h and then divide. Court-Brown (1967)
observed abundant cells in metaphase after 40 h in culture: virtually all in
the first division, as seen from studies with H^3-labelled thymidine. The
study of the lymphocyte has gained importance since experiments indicate
that it is an immunologically competent cell, that it has the capacity of
re-circulation and that lymphocyte populations may survive for long
periods *in vivo*.

In the absence of PHA, lymphocytes may also divide in culture in response
to stimulation by an antigen, provided they are suitably sensitised. The
blood culture technique, though excellent for most studies on karyotype
analysis, carries certain limitations regarding the types of abnormality as yet
recognisable. It can also be used effectively in the study of chromosome
damaging agents, particularly ionising radiations.

The first step is to prevent coagulation of the blood obtained for the
culture. Heparin is found to be ideally suitable and heparinised blood can be
stored, without loss of mitotic activity, for 12–24 h in the cold before

culturing. An alternative anticoagulant is acid–citrate–dextrose in which successful storage has been carried out for 2 weeks (Petrakis and Politis, 1962). Separated leucocytes (WBC) can be stored for 96 h at 5 °C, retaining their mitotic potential (Mellman *et al.*, 1962). If culture medium is added, they can even survive at room temperature (Arakaki and Sparkes, 1963).

Leucocytes are separated out from red blood cells (RBC) by centrifugation or gravity sedimentation. Centrifugation at a slow speed (25 g) for 10–15 min is required for fresh blood, and 5–10 min for blood pre-incubated with PHA (Moorhead *et al.*, 1960). High-speed centrifugation has also been found to be effective (Bender and Prescott, 1962). Gravity sedimentation is a useful tool for separating RBC, the optimum temperature being 25–37 °C. More efficient methods are through the application of fibrogen and dextran sedimentation (Skoog and Beck, 1956), these chemicals interfere with the quality of staining, but this may be avoided by washing at the time of harvesting the culture. These two latter methods are most efficacious for leucocyte cultures from small animals (Nichols and Levan, 1962). The presence of a moderate amount of RBC in the leucocyte suspensions does not appear to interfere with mitosis (Mellman, 1965). Tips *et al.* (1963) devised a blood culture method including the whole blood, thus utilising the entire WBC in a given volume. Polymorphonuclear leucocytes (PMN) can be removed by storing the WBC suspensions at 5 °C for 48 h, when the former degenerate. Alternatively, the culture flask is incubated on its side with the cells exposed to a large glass surface for 30–60 min and then stored upright. PMN are eliminated due to their property of adhering to glass, while the other cells settle to the bottom for the remaining period of incubation (Moorhead, 1964). Hastings *et al.* (1961) used differential centrifugation and iron, to promote the magnetic removal of PMN following their phagocytosis of the iron.

The size of the cell inoculum depends on the different variable factors in the culture conditions. Different workers have laid stress on different aspects, ranging from the number of nutrients, and pH, to the maximum log phase required in the culture (see Mellman, 1965). Nowell and Hungerford (1963) recommended determining the inoculum size on the basis of the total white cell concentration in the supernatant plasma.

Mitosis is initiated in blood cultures of normal non-leukaemic individuals by mitogenic agents after a time-lag of 2–3 days. Leukaemic cells, however, divide in culture immediately, without the aid of such chemicals. The principal mitogenic agent used in blood cultures is phytohaemagglutinin (PHA). It is a mucoprotein isolated from seeds of *Phaseolus vulgaris* or *P. communis* by salt extraction (Li and Osgood, 1949; Rigas and Osgood, 1955). At a low pH, two fractions can be separated, a protein haemagglutinin and an inactive polysaccharide. Its capacity of selectively agglutinating and sedimenting mature erythrocytes and of inducing division in leucocytes, has made it an invaluable tool in chromosome analysis from blood culture (Hungerford *et al.*, 1959; Moorhead *et al.*, 1960; Nowell, 1960). Nowell (1960) suggested that the principal action of PHA is on the WBC cell membrane, which on being altered, allows certain substances of the culture medium to penetrate the cell and stimulate mitotic activity. Elves and

Wilkinson (1962) considered the mitogenic effect of PHA to be due to the rejuvenation of the lymphocytes, while Cooper *et al.* (1961) suggested that it might be caused by the induction and progress of DNA synthesis. Hirschhorn *et al.* (1963) held that PHA aggregates the WBC which are converted into dividing cells through close association. This hypothesis is objected to by the fact that WBC divide even while being agitated in spinner cultures (Nowell, 1960). The correlation between mitogenic activity and leucoagglutination has been demonstrated by Hastings *et al.* (1961) and Kolodny and Hirschhorn (1964). Mitogenic activity was tested by them in two separate sets, in one of which haemagglutination was avoided by repeated passage through RBC, while in the other leucoagglutination was similarly eliminated by continuous passage through WBC. In the latter case, mitogenic activity was found to be absent. Byrd *et al.* (1967) demonstrated that the mitogenic capacity of PHA can be checked by an antiserum.

Experiments were also conducted to find out whether agglutination is caused by actual combination with the compound, or by the alteration of the membrane through enzymic action. The latter was considered unlikely and all types of erythrocytes, including O and Rh negative, were seen to be equally affected. Beckman (1962) showed that it can also precipitate α^2 globulin, plant protein, and α and γ globulins. Jaffe (1959) separated phaseolotoxin A and fraction B from *Phaseolus vulgaris* and noted that phaseolotoxin A contains 13·5–14·5 per cent nitrogen, while B has three principal constituents. Bacto-phytohaemagglutinin is available in two forms, M and P (DIFCO code 0528 and 3110). The former one (M) is a stable desiccated muco-phytohaemagglutinin with no toxicity and is the form usually recommended for securing viable leucocyte preparations and nucleated erythrocytes from blood and bone marrow suspensions. The latter (P) is a sterile, highly purified form, from which the polysaccharide has been removed. It is a much more potent form than M in its mitogenic and agglutinating properties. About 0·1 cm^3 of M-form to 5 cm^3 of heparinised blood, is adequate for its optimum action, while the P-form is needed only in 1/10th to 1/50th concentration of the former for differential separation. After the addition of PHA, the container is kept undisturbed for 30 min, followed by centrifugation (500 rev/min) for 1–2 min, and separation of the supernatant containing leucocytes.

Attempts to separate the two distinct cell principles of PHA, one of which has the property of agglutinating RBC and the other of initiating mitosis, are not yet completely successful (Punnett and Punnett, 1963). Unfractionated PHA has been observed to agglutinate RBC preferentially in whole blood but may agglutinate WBC as well when added in larger quantities to WBC suspensions. Therefore, according to Barkhan and Ballas (1963), the non-mitogenic factor in PHA possibly agglutinates RBC and it is different from the mitogenic factor agglutinating WBC. Tuberculin has been found to be an effective substitute for PHA, in human lymphocyte cultures from individuals sensitive to tuberculin (Pearmain *et al.*, 1963), showing a possible solution to the problem of the nature of PHA activity. Downing *et al.* (1968) found that other plant agglutinins commonly agglutinate both RBC and WBC but they seldom also induce mitosis of lymphocytes in tissue culture. This suggests that mitogenic activity in some of them is independent of their agglutinating property.

The basal culture media for leucocyte culture contains mixtures of amino acids, vitamins and buffered salts, the commoner ones being TC 199, NCTC 109, Parker's, Waymouth's, BME Spinner and Eagle's ME media. All of them require serum proteins for successful blood cultures, in a proportion of 10–40 per cent. Calf serum and both autologous and homologous human sera can be used. Mellman (1965) has advocated adding half the volume of the serum to be used as WBC-containing plasma and an equal volume of homologous AB serum. Commercially available foetal calf serum gives better growth. Air-tight bottles are recommended to prevent escape of the CO_2 produced, thus increasing the alkalinity, which has an adverse effect on growth. Alternatively, open vessels can be incubated in an atmosphere of 5 per cent CO_2, or the pH of the medium can be maintained at 7·2 to 7·4 by adding HCl or $NaHCO_3$ daily. Temperature should be kept between 36–37 °C to have peak mitotic activity between 60–72 h, or strictly at 38 °C for optimum activity at 48 h. Bacterial contamination is prevented by the addition of penicillin and streptomycin in human blood cultures. Since certain levels of antibiotics like streptomycin and chloramphenical have been found to be deleterious to mammalian cell cultures (Metzgar and Moskowitz, 1963; Ambrose and Coons, 1963), their use can be omitted if sterile precautions are adopted. The major step in the preparation of metaphase plates for karyotype study from leucocyte culture involves metaphase arrest, hypotonic treatment, fixation, preparation and staining.

For the arrest of mitosis at the metaphase, colchicine or its analogue deacetylmethyl colchicine (Colcemide, CIBA) is most frequently used. The hypotonic treatment for swelling the cell and dispersing the chromosomes, first devised by Hsu (1952) and Hsu and Pomerat (1953) can be carried out by diluting the balanced salt solution used to wash the harvested cells with distilled water and then incubating the cells in it. A solution containing one part human serum to five parts distilled water gives good results in human fibroblast (Lejeune et al., 1959a, b; Lejeune, 1960) and leucocyte cultures (Hungerford and Nowell, 1963). To remove the water present in the culture as much as is practicable, the cells are centrifuged into a small button and the supernatant discarded. The button can be treated in the fixative, with or without shaking, for 30 min. Best fixation results are given by a mixture of glacial acetic acid and methanol (1:3). Two or more changes of the fixative may be necessary to remove completely the coating substances of the chromosomes. Both air-drying and squash schedules can be used for the preparation of the slides, the former being more satisfactory. The stains include most of the common stains for chromosomes, like Feulgen, acetic-orcein, Giemsa, Unna's blue and methylene blue, and the method of staining is similar to that followed for bone marrow techniques.

REPRESENTATIVE SCHEDULES FOR PERIPHERAL BLOOD LEUCOCYTE CULTURE

A. *Macromethod*

Draw in and eject 1 cm^3 of solution of 5000 I.U. heparin per cubic centimetre in a syringe, thus coating its walls completely with heparin. Draw 20 cm^3 of venous blood into the heparinised syringe.

Decant in a test-tube inclined at 45 degrees and permit the RBC to sediment. The blood can also be kept for sedimentation in the syringe itself, with the needle held upward, at room temperature. After 30–50 min, remove supernatant containing exclusively WBC—5 to 10^6 per cm^3. This decantation procedure of Edwards and Young (1961) avoids an alternative method of slow speed centrifugation (200 rev/min) for separating the leucocytes.

Study an aliquot of the supernatant in a Malassez counting cell to obtain finally a concentration of 1–1·5 × 10^6 cells per cm^3, in a medium containing 30–35 per cent of the individual's serum, and 65–70 per cent of the basal medium, by adding required amounts of this medium.

Add 0·2 cm^3 of bacto-phytohaemagglutinin per 10 cm^3 of mixture. Distribute in test tubes, filling them to one-third of their volume. If required, inject a mixture of 5 per cent CO$_2$ and air before sealing the tube. Incubate at 37 °C for 72 h.

Add, 2 h before harvesting, 2 drops of isotonic solution of colchicine at a concentration of 0·04 per cent per cubic centimetre of the medium.

Decant into a centrifuge tube, spin down at 800 rev/min for 5 min. Discard supernatant and add hypotonic solution (0·93 per cent aqueous sodium citrate) and treat for 10 min at 37 °C.

Centrifuge to a button at 800 rev/min. Remove supernatant and add fixative (1:3 acetic-methanol or acetic-ethanol) slowly, suspending the cells. Leave for 30 min and repeat the process using only 2–3 drops of the fixative. Re-suspend cells by pipetting to prevent bubble formation.

Cool several absolutely clean, grease-free slides (or cover slips), by placing on a paper put on an ice block, or on CO$_2$ snow, so that they are covered with a fine mist. Place a drop of cell suspension over a cover slip which immediately spreads. Hold over a flame, ignite (Scherz, 1962) or place in a 60 °C cabinet to cause evaporation of the fixative, flattening the preparation. Alternatively, tilt slide on its long edge, touching absorbent paper, and blow directly on slide.

Dry in air. Store, or stain immediately if necessary. Modifications generally include the number of centrifugations required and the stains used. Antibiotics can be added if necessary, a proposed incubation medium being; autologous plasma, AB human serum or foetal calf serum, 1 cm^3; basal medium TC 199, 6 cm^3; bacto-phytohaemagglutinin, 0·03–0·05 cm^3; penicillin (100 000 units/cm^3) and streptomycin (100 mg/cm^3) solution, 0·02 cm^3. It can be prepared and stored at 20 °C. If, in the dried slide, cells are scarce, concentrate the cell suspension, after adding fixative, by centrifugation and removal of part of the fixative. For chromosomes insufficiently spread or shredded, the number of times of suspension in fixative is increased and the flaming schedule is followed for spreading. If chromosomes are overspread, use a small drop of suspension and do not evaporate quickly.

B. *Semi Micromethods*

According to Mellman (1965)

Add 0·1 cm^3 of aqueous heparin (1000 units/cm^3) and 1 cm^3 of serum to a 2 cm^3 plastic disposable syringe. Fill to the 2 cm^3 mark with venous blood. Mix the contents. Empty the needle by aspiration. Stand the syringe and allow the RBC to sediment at room temperature, as in schedule A, till the clear supernatant is about 1 cm^3 (30–60 min).

Bend the needle and inject contents directly in culture vessel without doing WBC count. The vessel contains the complete culture medium described in schedule A at room temperature. The later steps are similar to schedule A.

Adopted by Nowell et al., (1958) and Hungerford and Nowell (1963)

Draw 10–20 cm^3 of blood in a syringe containing 0·2 cm^3 heparin. Add PHA (M form) to blood (0·2 cm^3 per 10 cm^3), mix and keep in the cold for 30–45 min. Centrifuge at 25 × g for 10 min at 5 °C, remove supernatant and count leucocytes. Plant leucocytes (10^7) in a medium containing: plasma, 3 cm^3; medium 3 cm^3, and antibiotics. Maintain a pH of 7·2–7·4 with HCl or NaHCO$_3$. Incubate at 37 °C for 72 h. The remaining procedure is as above, the hypotonic used being 1:5 dilution of serum. The method should strictly be called a macromethod owing to the quantity of blood needed.

C. *Micromethods*—were developed by several workers using only a few drops of blood (Edwards, 1962; Frøland, 1962; Arakaki and Sparkes, 1963; Grouchy *et al.*, 1964; Reitalu, 1964). The one used by Turpin and Lejeune (1969) is described as follows:

Fill round-bottomed centrifuge tubes (capacity 41 cm^3) with 5 cm^3 human serum, 15 cm^3 TC 199, 4 drops of PHA (mixture of equal parts of phytohaemagglutinin Difco M and P, or phytohaemagglutinin Wellcome and 4 drops of Liquemin (Roche, or equivalent of 5 mg crystalline heparin).

Disinfect skin of index finger or thumb carefully. Incise with a vaccino-style and remove 4–6 drops of blood with a pipette. Transfer directly to the tube containing the medium. Rinse out pipette with medium to suspend the blood. Alternatively, use a few drops of blood drawn from a vein.

Seal and incubate at 37 °C for 48–72 h.

Add 2 cm^3 of 0·04 per cent isotonic colchicine solution. Re-suspend and incubate for 2 h.

Re-suspend, transfer to a conical centrifuge tube.

Centrifuge for 5 min of 800 rev/min, discard supernatant.

Fill up to two-thirds of tube with a mixture of animal serum (1); distilled water (5) and hyaluronidase 2·5 I.U. per cm^3 of mixture. Re-suspend and incubate for 7 min.

Centrifuge for 5 min at 800 rev/min, keep 2 min, discard supernatant hypotonic and add Carnoy's fixative (acetic–chloroform–ethanol 1:3:6). Re-suspend and keep for 45 min.

Centrifuge at 800 rev/min for 5 min, discard supernatant and add acetic-ethanol (1:3). It can be sealed and stored in the cold.

Centrifuge at 800 rev/min, discard supernatant and add 5–6 drops of fixative. Re-suspend by pipetting.

Same as steps (8) and (9) for schedule A.

After air-drying, hydrolyse for 7·5 min in N HCl at 60 °C, rinse in iced water and stain for 10 min in a solution of 1 part Unna's blue and 4 parts neutral water. Rinse in water and dry.

Pass through xylol or toluene grades and mount in canada balsam or directly in permount.

A schedule developed by Hungerford (1965) uses a culture medium composed of: Eagle's basal amino-acids and vitamins, at double strength, in Earle's balanced salt solution (BSS), adjusted to pH 7·0 with 7·5 per cent NaHCO$_3$, supplemented with glutamine 2m M, penicillin 100 units(cm^3, streptomycin 100 µg/cm^3, phenol red 7 µg/cm^3 Foetal agammaglobulin bovine serum and phytohaemagglutinin M (Difco) are added to make 15 and 2 per cent, respectively, of the final volume and also 20 000 USP units of heparin sodium per litre of complete medium. After the usual schedule of inoculation, incubation and treatment with colchicine, the cells are separated by centrifugation and the medium replaced with 0·075 M KC1, 16 USP units/cm^3 heparin sodium, and incubated for 10 min. Subsequently, KCl is removed and the material fixed in two changes of acetic-methanol (1:3), air-dried and stained in 1 per cent orcein in 60 per cent acetic acid. It has the advantage of using a very small quantity of blood and the medium is a modification of an earlier one developed by Hayflick and Moorhead (1961). Wittmann's (1965) acetic–iron–haematoxylin and cresyl violet acetate (Humason and Sanders, 1963) can be used as alternative stains. Vincaleucoblastine (0·15 cm^3 of a stock solution of 0·5 g/cm^3 per 10 cm^3 of culture) was used instead of colchicine by Kolodny and Hirschhorn (1964).

Application in other Human Tissues

Lymph tissue, after maceration, has been cultured, following the micro-method for peripheral blood, giving good chromosome preparations (Baker and Atkin, 1963). Chromosome analysis from capillary blood has been carried out by leucocyte culture also (Robinson *et al.*, 1964). Permanent lymphocytoid cell lines have been established from patients with XYY and XXY chromosome constitutions by cultures following leukaphoresis (Moore *et al.*, 1966; 1969). An important aspect of this technique is the study of foetal blood for pre-natal chromosome analysis in order to check possible chromosomal anomalies in the foetus. A small amount (0.5 cm^3) of blood can be obtained from an umbilical cord, or some such foetal vessel, even by a tiny puncture of the foetal skin. Pre-natal chromosome analysis is usually carried out on cells from the amniotic fluid (drawn out by a syringe) which are concentrated by centrifugation and cultured on tissue culture medium, followed by colcemide treatment, and harvesting as in the schedules described before. This form of analysis could be of value in cases listed for sex chromatin where more data are required; where parents are unaffected carriers of a chromosomal abnormality or have been exposed to mutagens, radiation or virus attacks, and in the rare families having known very high rates of chromosomally abnormal offspring. The method available through amniocentesis usually has several drawbacks including contamination with maternal cells, unreliability of results, and in difficult cases even foetal death by incompatibility, through a small quantity of foetal blood passing into the maternal circulation (Wang *et al.*, 1967). Therefore, extreme caution has to be exercised in such studies (*see* Klinger and Miller, 1968).

Application in other Vertebrates

Modifications of the peripheral blood culture technique have been applied successfully in a large number of mammals, like gorilla (Hamerton *et al.*, 1961); domestic pig (McConnell *et al.*, 1963; Stone, 1963; Srivastava and Lasley, 1968); macaca (Sanders and Humason, 1964); sheep (McFee *et al.*, 1965); dog (Ford, L., 1965); spider monkey (Eide, 1963); ox (Biggers and McFeely, 1963) and the primates (Egozcue and Egozcue, 1966). Ohnuki *et al.*, (1962), however, suggest that a comparison of the leucocytes of monkey and man, in culture, gives poorer results in monkey cells with an identical method. The schedule has also been successfully adapted for use in snakes, fishes and birds (Newcomer and Donelly, 1963; Manna, unpublished).

The major variations involve the type of medium used, its constituents, the relative amounts of PHA and colchicine needed and the periods of treatment and incubation.

PREPARATIONS FROM FIBROBLAST AND OTHER CULTURES

Cultures

This method is a time-consuming one, when compared with the peripheral blood and bone marrow culture schedules. However, fibroblast culture methods can be effectively used for checking unusual karyotypes observed in the blood cells, determining mosaicism by studying chromosomes of different tissues, and for other experimental work where cytogenetic studies are correlated with biochemical, virological or other studies.

In general, any tissue aseptically removed, can give usable cultures. The different techniques have evolved essentially as a result of the manner in which the cells are transferred: *en bloc* in the state in which they are (explant technique, Turpin and Lejeune, 1969), or after trypsinisation (Harnden, 1960; Harnden and Brunton, 1965).

The basic precautions to be adopted during tissue culture are: (*a*) use of scrupulously clean glassware (*b*) adoption of sterilisation measures throughout the process. Though antibiotics have lessened the need for this precaution, yet an excess of antibiotics may result in chromosomal aberrations; (*c*) controlled pH at 7·2, maintained usually by a bicarbonate buffer.

The first step in the technique is a biopsy, which can be obtained from any part of the body by surgical means, the most common one being skin culture. The biopsy should be 2–4 mm in diameter and should include the connective tissue. The surface should be cleaned thoroughly and the use of local anaesthesia is optional. Tissue should preferably be set up in culture immediately, but storage at room temperature overnight is also possible.

The next step is the setting up of the primary cultures and there are several techniques. One way is to digest the tissue with trypsin and plate out the cells thus suspended (Puck *et al.*, 1958). This method, however, is not always recommended, owing to the delicate care required in handling and the uncertain success. The more common method is to cut up the tissue into small bits before placing in culture. Usually the bits are fixed in position by using a plasma clot (Harnden, 1960; Lejeune *et al.*, 1959 a, b); or holding them down with perforated cellophane (Hsu and Kellogg, 1960), or rat tail collagen (Swanson and McKee, 1964), or by placing them on a metal grid (Koprowski *et al.*, 1962), or by pinning under a cover slip (De Mars, 1963 in Harnden and Brunton, 1965), but cultures can also be carried out without holding down the tissue (Davidson *et al.*, 1963).

The tissue is grown in a liquid medium composed principally of a synthetic culture medium to which serum and a growth stimulant like chick or beef embryo extract have been added. Commercially available TC 199 and Eagle's media give good results. Composition of some of the commoner media is given in the chapter on tissue culture. Human AB serum, pooled human serum, calf serum and foetal calf serum can be used, depending on the material. Subcultures should be set up only after adequate growth, after 10–14 days from setting up the culture. The medium should be replenished after 2–3 days to maintain a constant pH.

To start a subculture, either the original tissue is cut free from the outgrowth and replanted in a fresh plasma clot, or trypsin is used to digest the cells free from the clot or the vessel. The cells are washed and placed in a new culture vessel, where they adhere to the glass as a monolayer. The medium is replaced at intervals and new subcultures are made whenever the monolayer becomes too heavy. If a 1 in 10 dilution is used in inoculation, subcultures should be made once a week. Puck *et al.*, (1958) recorded the growth of cultures for a year without any deleterious effects, but Hayflick and Moorhead (1962) and Moorhead and Saksela (1963) observed gradual loss of cell multiplication and aneuploidy after a particular period in cultures.

For chromosome preparations, attempts are made to obtain partially synchronous divisions by changing the medium or subculture. The culture is allowed to become acidic and the pH restored to 7·4 by adding $NaHCO_3$

and embryo extract, resulting in a large number of mitosis after 16 hours. In most techniques, however, colchicine is added to the culture to obtain metaphase configurations. The next stage is the use of a hypotonic solution, ranging from distilled water to dilute sodium citrate, or diluted salt solution or serum.

For obtaining suitable plates, two methods are available. Either the cells are grown on cover slips and processed while attached to them (Harnden, 1960), or the cells are suspended through trypsin digestion and then processed in any of the methods available for peripheral blood culture. The second method gives more satisfactory results. In fixation, acetic-ethanol or acetic-methanol (1:3) is commonly employed. Fixation can be carried out without breaking up the pellet, or by suspending the cells prior to fixation. The cells can be spread by the air-drying schedule originally devised by Rothfels and Siminovitch (1958a) used for blood cells. For better spreading, they can be transferred to 75 per cent acetic acid immediately before drying. Squash preparations are not very successful. Aceticorcein, Feulgen and Unna's blue are the frequently used stains. The length of time from culturing to chromosome smears may range between 1 and 3 weeks, depending on the method and the material used.

Constituents used in the Techniques

Cockerel plasma (CP)—can be stored at 4 °C for several weeks in wax- or silicone-coated glassware. It is prepared by drawing 20 cm^3 of blood from the wing vein of a young bird, centrifuging at 1500 rev/min for 30 min at 4 °C and storing in the cold.

Trypsin (0·25 per cent)—is prepared by dissolving 2·5 g of Bactotrypsin 1–300 in a few cubic centimetres of Hank's solution without Ca and Mg and adjusting the pH to 8·0 with NaOH. The solution is made up to 100 cm^3 with Hank's solution and later diluted with it 10 times before use.

Sodium bicarbonate—(1·4 per cent) contains NaHCO$_3$. 3·5 g; phenol red (0·2 per cent) 2·5 cm^3, and neutral water 247·5 cm^3. 5 cm^3 is added to 200 cm^3 of Hank's solution before use.

Colchicine—Stock solution contains 0·5 per cent colchicine in neutral water. Working solution of 0·005 per cent colchicine is prepared in BSS. 0·5 cm^3 of this solution is added to each 10 cm^3 of culture to have a final concentration of 0·0025 per cent.

Antibiotics—Kanamycin, mycostatin, neomycin, penicillin and streptomycin are all dissoved in neutral water to obtain stock solutions and later diluted with culture medium for the required concentrations (for details, *see* Harnden and Brunton, 1965).

Stain—8 cm^3 of Giemsa stock solution in 192 cm^3 of distilled water.

Chick embryo extract (CEE)—can be obtained commercially. For large quantities, embryos from hen's eggs incubated for 10 days are collected and homogenised by forcing through a syringe. Hank's BSS medium is added (1·25 vol.), the mixture centrifuged at 2500 rev/min for 30 min at 4 °C. The supernatant is decanted, mixed with 2 mg hyaluronidase/100 cm^3 extract, incubated at 37 °C for 1 h, ultracentrifuged at 25 000 rev/min for 1 h and filtered through No. 03 porosity Selas candle filter under pressure of 8–10 lb/in^2.

Serum—AB serum can be obtained by allowing a pint of human AB venous blood to coagulate at 4 °C, drawing out the serum with a syringe, centrifuging at 1500 rev/min for 30 min and filtering through a No. 03 porosity Selas candle filter.

The constitution of the different media is given in the chapter on tissue culture.

Growth medium—used in almost all cultures, contains Eagle's medium with antibiotics, 70 parts; human AB serum 20 parts and CEE 10 parts.

SCHEDULES

Two representative schedules are described below:

A. *Trypsin-digestion Culture Method* (Harnden and Brunton, 1965):

1. *Biopsy*—Clean skin with spirit, inject local anaesthetic, make an incision with scalpel and cut off V-shaped skin (2 mm). Place in about 4 cm^3 Eagle's medium in a small screw-capped bottle. Cover the wound.

2. *Primary culture*—Transfer tissue to glass dish and cut into small pieces. Place equal amount of CEE and CP in two Petri dishes. Transfer one bit of tissue to a third Petri dish with the help of a pipette and suck out the excess medium. Mix one drop each of CEE and CP and draw them and the piece of tissue into the pipette. Transfer to a culture flask spreading out the CEE/CP mixture in a thin layer. Treat other bits of tissue similarly so that a culture flask with 4 cm diameter has 5 pieces of tissue. Allow the plasma to clot for a few minutes. Add 5 cm^3 of growth medium. Add 5 per cent CO$_2$ in air to flask and close with siliconed stopper and incubate at 37 °C.

3. *Subculture*—For first subculture, remove growth medium, add 10 cm^3 of 0·25 per cent pre-warmed trypsin solution, incubate at 37 °C for 15 min, transfer to centrifuge tube and centrifuge at 500 rev/min for 5–10 min. Discard supernatant and re-suspend cells in 1 cm^3 of growth medium. Transfer to fresh culture flask containing 9 cm^3 fresh growth medium, seal and incubate at 37 °C. For subsequent subcultures, use trypsin for 5 min only and after suspending cells in 1 cm^3 of growth medium, use only 0·1 cm^3 for each subculture, to give a 1 in 10 dilution of cells. For maintenance, check cultures every 2 days for growth, sterility and pH, and replace growth medium. Adjust the pH with 5 per cent CO$_2$ or NaHCO$_3$ solution (1·4 per cent) when required.

4. *Processing*—To a healthy culture, add 0·5 cm^3 of 0·005 per cent pre-warmed colchicine solution, incubate for 2–4 h, prepare cell suspension with trypsin as given for first subculture. Centrifuge at 500 rev/min for 10 min, discard supernatant and re-suspend cells in pre-warmed Hank's BSS. Centrifuge at 500 rev/min for 10 min, discard supernatant, re-suspend cells in pre-warmed 0·95 per cent sodium citrate solution and keep 20 min

in 37 °C. Again centrifuge at 500 rev/min for 10 min, discard supernatant, re-suspend cells in small quantity of fluid left, add acetic-ethanol (1 : 3) drop by drop with agitation. Keep in excess of fixative from 30 min to overnight.

5. *Preparation of slides*—Centrifuge at 500 rev/min for 10 min, discard supernatant, add fixative and again centrifuge at 500 rev/min for 10 min. Discard supernatant and re-suspend in a few drops of 75 per cent acetic acid. Place a few drops on a wet, pre-cooled slide, heat over a flame to dry, cool and stain in 2 per cent acetic-orcein for 2–3 h at 37 °C. Dehydrate in cellosolve, treat in euparal essence for 2 min, and mount in euparal.

This schedule, with slight modifications, has been used by Fox and Zeiss (1961).

B. *Plasma Clot Culture Method* (Turpin and Lejeune, 1969):

1. *Biopsy*—For surgical cases, remove under general anaesthesia, a tissue about $4 \times 4 \times 4$ mm in measurement. Wrap in a sterile square of gauze, place in a wide-necked flask 4 cm in diameter. Add 5 cm^3 of sterile physiological saline, seal the flask and store for 24–36 h at room temperature, if necessary. For skin biopsy, after thorough cleansing with soap and sterile water, and sterilisation in ethanol (twice) and ether, pinch the skin between the jaws of a 'Coprostase' clamp, keeping a piece protruding above it. Anaesthetise locally with ether, remove with a scalpel a piece 3–4 mm long, place in a sterile square of gauze and treat as for surgical biopsy. Disinfect and close the suture with band aid.

2. *Explant* (Lejeune et al., 1960)—Wash the tissue in physiological saline and cut into pieces 1–2 mm square. Place a sterilised cover slip in the distal depression of a Leighton tube. Spread a drop of CP over the cover slip. Place the bits of tissue (2–3 per cover slip) on the plasma. Add a drop of CEE to coagulate the plasma and fix the tissues to the glass. Close the tube with a rubber stopper, incubate for several hours or overnight at 37 °C. Add culture medium containing per tube: human AB serum, 5 drops; Hanks' solution with 200 µm/cm^3 penicillin, 50 µg/cm^3 streptomycin and 5 5 µg/cm^3 chloramphenical, 5 drops, and embryonic extract, 2 drops. Incubate.

3. *Subculture*—After development of a crown of fibroblasts around the explants in 4–6 days, remove the explants and transfer to other tubes following the method described in step 2. Replace the medium in the tubes with cover slips with fresh medium. Incubate.

4. *Processing*—After 36 h incubation, add 3 drops of CEE to the medium in each tube. Incubate again for 16 h to obtain a large number of mitosis due to change in the substrate. Remove the cover slip with a curved tip pipette and place it, with the cells uppermost, in a mixture of sterile human or mammalian serum, 1 part; neutral water 5 parts and hyaluronidase 2·5 per cm^3 (used to hasten absorption) of mixture. Incubate at 37 °C for 35 min. Transfer the cover slips to Carnoy's fixative (chloroform-acetic-ethanol, 3 : 1 : 6) and treat for 45 min.

5. *Preparation of slides*—Take out the cover slip and dry in open air.

Hydrolyse in N HC1 for $7\frac{1}{2}$ min at 60 °C. Rinse in neutral water, stain in Unna's blue for 10 min. (1 part Unna's blue solution, 4 parts neutral water). Rinse in neutral water, dry for 5–10 min in open air, pass through toluene and mount in canada balsam.

C. *Liquid Medium Method* (Turpin and Lejeune, 1969). This has been applied to surgical biopsies, mainly of tumours.

1. Biopsy is similar to schedule (B).
2. Cut into small bits, wash three times in Dulbecco PBS solution (with calcium). Transfer to Erlenmeyer flask, add trypsin solution in PBS buffer (25 mg/1), seal and agitate in a magnetic shaker at 37 °C from 10–20 min. Replace trypsin solution every 10–15 min with continued agitation, continuing up to 1–4 h according to the tumour studied. Transfer each change of trypsin solution to a centrifuge tube kept in cold. Centrifuge these solutions for 5 min at 800 rev/min. Wash three times in a solution of casein hydrolysate or a culture medium. Re-suspend sediment in 1–2 cm³ of casein hydrolysate or the culture medium. Place the culture in test tubes with cover slips containing: casein hydrolysate or synthetic culture medium (like 199), 2 parts and human AB serum (1 part to have a concentration of 5×10^5 cells per cm³). Incubate at 37 °C after adjusting pH. Replace the medium every 3 days. Cell multiplication depends upon the nature of the tumour.
3. Processing and preparation of the slides are similar to the schedule given for blood cultures.

Several alternative schedules are in practice, combining or altering different stages in the representative ones described here. For metaphase arrest in cultures grown on cover slips, they can be treated directly with colchicine in culture medium (0·1 g/cm³) for 18–24 h (Axelrad and Mc-Culloch, 1958) or colchicine can be added to the culture medium to give a concentration of 0·0025 per cent (Swanson and McKee, 1964). A saturated solution of Abopon in 0·2 M phosphate buffer at pH 7·0 can be used as a suitable mounting medium after acetic–orcein staining and rinsing in water (Hrushovetz and Harder, 1962).

Application in Other Human Materials

Variations of the above schedules have been developed for cytogenetic investigations on human abortus material (Tjio and Puck, 1958; Basrur *et al.*, 1963; Makino *et al.*, 1963; WHO group, 1966). Some of the commoner ones are outlined in brief:

Method A—Cut out 8 explants from material, approximately 3 mm in diameter. Place in 4 oz medical flats and 5 cm³ of Waymouth's medium enriched with 20 per cent calf serum; place in sloping position so that half of the specimen is in the medium. Incubate at 37 °C. After 4–7 days, primary outgrowths of histocytes and spindle cells are seen, these are followed by round cells which are in turn replaced by fibroblasts. Trypsinise the primary outgrowths for 9–14 days. Subcultures and change of medium are maintained as described in previous schedules. Sufficient cells are obtained by the third or fourth subculture. Thirty six hours after subculturing, add 1 cm^3 colcemide (80 µg/cm^3) and keep for 8 h. Trypsinise cells till freed (5 min), add TC 199, centrifuge, re-suspend in 1·12 per cent sodium citrate, incubate at 37 °C for 7 min. Centrifuge and fix in acetic-ethanol (1:4) and stain as usual.

Method B—Expose tissue to 0·25 per cent trypsin solution for 30 min and shake in magnetic stirrer at low speed. Centrifuge resultant cell suspension at 600 rev/min for 10 min, remove supernatant, re-suspend button in Waymouth's medium and seed on to cover slips or Petri dishes. Incubate at 37 °C in CO_2 enriched medium.

Method C—Sandwich explants 2 mm in diameter between cover slips and incubate in Petri dishes at 37 °C in CO_2. In both methods B and C, transfer explants to fresh cover slips and grow for 48 h. Add 0·25 cm^3 colcemide (80 µg/cm^3 solution) and keep for 4 h. Remove medium, add hypotonic solution, incubate at 37 °C for 12 min and then fix and stain as usual. The workers claim a successful analysis of 20 out of 45 specimens studied.

The tissue culture technique can also be used for prenatal chromosome analysis from chorionic tissue obtained by chorion biopsy. A tiny piece of foetal skin or placental amnion also has possibilities (*see* Klinger and Miller, 1968).

Application in Other Animals

The schedules, with suitable variants, have been employed for the study of tissues from different parts of the body of different animals, as well as human subjects (Penrose and Delhanty, 1961). Chromosomes have been studied successfully in cell cultures from kidney tissue of primates (Chu and Giles, 1957; Bender and Mettler, 1958; Rothfels and Siminovitch, 1958b; cf. Bender, 1965). Chu and Swomley (1961) obtained satisfactory preparations from skin cultures of lemurine lemurs. In an agar-fixation procedure, developed by Pacha and Kingsbury (1962), trypsinised cells from the culture are suspended in Hanks' medium and then transferred to an agar surface containing 1·5 per cent agar and 0·5 per cent sucrose for 10–15 min. An agar square is cut out, processed and stained as usual. Both change of medium, and addition of colchicine, were adopted to obtain a large number of metaphase plates in opossum, followed by double staining in 2 per cent acetic-orcein and Harris's haematoxylin (Shaver, 1962).

DIRECT PREPARATIONS

Originally developed for karyotype study, are modifications of the different procedures discussed previously. They usually involve breaking up the tissue after biopsy and studying it after a series of treatments, including hypotonic solution, colchicine, fixation and staining, but omitting the culture procedure. It can be used for materials which already contain a good number of cells undergoing mitosis, like foetal cells, cornea, etc. A schedule developed for foetal mammalian tissue is outlined (Ford, E. H. R. and Woollam, 1963). Inject 0.3 cm^3 of 0.025 per cent colcemid into 14 days pregnant mice. Kill after 1 h. Break up livers of the foetuses in 0.1 per cent colcemid in phosphate-buffered 0.85 per cent NaC1 and treat for 1 h. Centrifuge, suspend in one per cent sodium citrate for 20 min, centrifuge and fix in acetic-ethanol (1:3) for 30 min, re-suspend twice in 45 per cent acetic acid, air-dry and stain in lactic-acetic-orcein.

Regenerating rat liver can be suspended in 1.12 per cent sodium citrate solution by aspirating, followed by high-speed centrifugation and squashing in 2 per cent acetic-orcein solution (Sparano, 1961). The corneal epithelium of mammals yields good mitotic figures. The entire eye can be fixed in an orcein-acetic-ethanol mixture followed by dissection and mounting (Gay and Kaufmann, 1950). The anterior portion may be fixed in acetic-ethanol (1:3) for 24 h followed by hydrolysis and staining by Feulgen. The epithelium is dissected out and mounted in glychrogel (Howard, 1952). Otherwise, the whole eye can be kept in 0.7 per cent sodium citrate solution at 37 °C for 30 min, fixed in 50 per cent acetic acid and N HC1 (9:1) for 5 min and stained in 2 per cent acetic-orcein for 2 min. Some of the corneal epithelium is scraped off and squashed (Fredga, 1964).

Chromosome preparations from mouse embryos during early organogenesis have been made as follows (Wroblewska and Dyban, 1969): Treat in a glass vessel in 1 per cent sodium citrate at 37 °C for 20–50 min, depending on age. Fix from 3 h to overnight at 4 °C in acetic-ethanol (1:3), stain in 2 per cent acetic-orcein for $\frac{1}{2}$–1 h. Disperse in 1–3 drops of glacial acetic and 50 per cent lactic acid mixture (1:2 to 3:2) for several minutes to 1 h, till cloudiness is produced. Place a small drop of the suspension on a slide, followed by a larger drop of the fixative. Air-dry and stain in lactic-acetic-orcein. It is suitable for embryos between the 7–11th day of pregnancy.

The study of pachytene chromosomes, though dealing with meiotic configurations, has nevertheless been utilised for karyotype analysis in human material (Ford, L. *et al.*, 1968, 1969). It has, therefore, been included in this section. Testicular materials obtained by biopsy are cut into bits (Rowley and Heller, 1966) and kept in distilled water for 30 min. An equal volume of acetic acid is added and the material retained in it for several hours. Then a piece of tubule is squashed in a drop of fixative on a slide under a mechanical press. The cover slip is then floated out in 45 per cent acetic acid and the slide is treated in Carnoy's fixative for 20 min, followed by air-drying and staining in lactic-acetic-orcein solution (Welshons *et al.*, 1962) and passing through ethanol and xylol grades to mount in permount. This method can be recommended for other mammalian materials as well.

Methods for Observation—Observation of mammalian and, more specifically, human chromosomes, are similar to those for other materials. Both light and electron microscopic studies have been carried out. Karyotype and meiotic observations are made following the usual procedure under the light microscope. Divisional configurations are drawn with the aid of a camera lucida, or on a piece of cellophane paper fixed on the ground glass of a projector. These figures are correlated with their corresponding micro-photographs, magnified to a suitable size. Pachytene chromosomes are matched by cutting out from the photographs and mounting prints of each member of a bivalent side by side (Ford L., 1969).

III SEX CHROMATIN STUDIES

The sex chromatin, as seen in the interphase nuclei, has been used in the detection of errors of sex development, to determine the number of X chromosomes in the complement of an individual, and in studies requiring the indication of the sex chromosome complex of a particular tissue, especially in cases of homologous tissue transplants where donor and recipient do not have the same sex. The different sex chromatin techniques are based on the fact that the X chromosome may show the property of heteropycnosis in interphase nuclei and may form a distinctive chromatin mass or chromocentre. It is a female characteristic in all mammals, except the opossum, which shows such masses in the cell nuclei in both sexes. The presence of a female-specific chromocenter was first demonstrated in the nerve cells of the cat by Barr and Bertram (1949), leading to the term 'Barr body'. A long series of publications has established sex chromatin analysis as an effective diagnostic tool for sex chromosome studies, according to the rule that the maximum number of sex chromatin masses is one less than the number of X chromosomes present in the individual studied. The evidence obtained for genetic inertness of the X chromosome that forms the sex chromatin and the possibility of a mosaicism in mammalian females, the paternal X being active in some cells, and the maternal X in others, has opened new lines of research (Ohno, 1961; Lyon, 1961, 1971; Beutler et al., 1962; Grumbach and Morishima, 1962).

The proportion of nuclei with demonstrable sex chromatin masses varies with the type of preparation and the tissue, ranging from less than 60 per cent in buccal smears to almost 100 per cent in thick sections of nervous tissue. The sex chromatin mass shows a feulgen-positive reaction due to its DNA content and stains with chromosomal dyes like cresyl violet, fuchsin, gallocyanin, haematoxylin, thionine and particularly with orcein. It has an affinity for methyl green in pyronin methyl green staining and persists after mild acid or ribonuclease treatment (*see* Barr, 1965). It may be studied from buccal smears, sections of tissues and peripheral blood neutrophiles.

(1) *The Buccal smear* technique for studying sex chromatin mass from buccal mucosa is the simplest one and is most widely used for clinical studies of sex determination. A simple schedule is given below:

Label clean sides with the subject's reference number and side of the body. Draw the edge of a metal spatula firmly over the buccal mucosa. Discard the material; scrape the mucosa gently a second time to obtain healthy epithelial cells from a deeper layer. Spread over a small area of an albuminised slide, using separate slides for smears from right and left sides.

Do not spread too thinly. Immerse inversely in fixative (95 per cent ethanol) for 15–30 min, treat successively in absolute ethanol for 3 min, and 2 min in a 0·2 per cent solution of Parlodion in ethanol–ether mixture (1:1) to attach the cells firmly. Dry in air for 15 s, pass through 70 per cent ethanol (5 min), two changes of distilled water (5 min each) and stain. Alternative fixation can be carried out in ethanol–ether mixture (1:1), immediately after smearing, for periods ranging from 12 h to 2 weeks, followed by gradual hydration.

Staining may be done in 1 per cent aqueous cresyl violet solution or acetic-orcein solution (Sanderson, 1960) to get successful preparations. Barr (1965) suggests staining 5–10 min in working solution of carbol fuchsin followed by differentiation for 1 min in 95 per cent and absolute ethanol successively. (Stock solution of carbol fuchsin: 3 g basic fuchsin in 100 cm^3 of 70 per cent ethanol. Working solution: stock solution 10 cm^3; 5 per cent carbolic acid in dist. water, 90 cm^3; glacial acetic acid, 10 cm^3; 37 per cent formaldehyde, 10 cm^3. Keep for 24 h before use.) Double staining, with biebrich scarlet as chromatin stain and fast green as counterstain, has been suggested by Guard (1959). Dehydration and clearing are carried out as for other squash schedules.

In interpreting the results of buccal smears, certain observations are significant, such as that the nuclei with sex chromatin masses in newborn females reach their normal number only on the fourth day after birth and that the size of the mass has been found to decrease after oral administration of certain antibiotics (Sohval and Casselman, 1961; Taylor, 1963).

(2) *Sex chromatin studies from tissues* have a much more limited application, being restricted mainly to tissues obtained during operation or post mortem (*see* Barr, 1965). A recommended fixative is: 37 per cent formalin, 20 parts; 95 per cent ethanol, 35 parts; glacial acetic acid, 10 parts, and dist. water, 30 parts. After 24 h, fixation tissues are transferred to 70 per cent ethanol. Paraffin blocks are prepared after dehydration through ethanol and xylol grades, sections 5 μm thick are cut. Staining is carried out in Harris's haematoxylin, counterstained with eosin, or in other chromosomal stains like Feulgen, thionin or gallocyanin, following the usual procedures.

Sex chromatin can also be studied from a monolayer of cells growing *in vitro*, or skin biopsies (Moore *et al.*, 1953), or leucocyte cultures, by omitting treatment with hypotonic solution during processing.

Prenatal sex chromatin analysis is useful in detecting male conceptuses with sex-linked recessive hereditary disorders when pedigree data are available. It can be carried out by concentrating cells from fluid samples by centrifugation, fixing on a slide and staining by thionin or Feulgen.

(3) *Studies from peripheral blood neutrophiles*: A drumstick-shaped nuclear appendage, which stains with Wright's, Giemsa or haematoxylin schedules, is observed in occasional neutrophile leucocytes in normal females, but never in the male, and is thought to contain sex chromatin. The number of neutrophiles per drumstick varies in different females. They can be studied from relatively thick blood smears, drawn on cover slips or slides, and followed by staining. Both buccal smear and neutrophile methods should be used simultaneously for the detection of sex chromosome mosaics.

Other Methods for the Study of Differential Chromosome Segments

In certain recently developed methods, observations under the light micro-scope after Giemsa staining have been correlated with banding patterns of chromosomes obtained with Caspersson's fluorescence technique. The method evolved by Arrighi and Hsu (1971; Arrighi et al., 1971), including denaturation of DNA *in situ* in NaOH and subsequent re-annealing of DNA strands with sodium chloride or citrate buffer, followed by Giemsa staining, can distinguish specific heterochromatic segments of chromosomes resembling the polythene banding pattern (Craig and Shaw, 1971). These chromosomes, on being destained and treated with quinacrine dihydro-chloride (Atebrin) for fluorescence microscopy, showed correspondence between the Giemsa-stained regions and the fluorescent banding pattern (Buckton et al., 1971; Evans et al., 1971). Fluorescent distribution curves of chromosomes may also be scanned in Beckman Analytrol (Van der Hagen and Berg, 1970). Ammoniacal silver nitrate staining has been used for differential staining of chromosomes as well (Bartalos and Rainer, 1971). Yunis et al. (1971) described a schedule involving denaturation of flame-dried bone marrow preparations for 10 min at 85–100 °C in 0·06 M phosphate buffer (pH 6·8), fast-cooling at 0 °C, and reassociating through incubation at 65 °C for different periods before Giemsa staining. Ridler (1971) followed a simplified technique of Bobrow, M.—in which lymphocyte cultures, after treatment in colchemide, were treated with 0·3 per cent KCl and 0·5 per cent trisodium citrate (1:1) for 10 min, fixed in acetic-ethanol (2:3), air-dried, treated at 65 °C for 90 min in 0·3 N NaCl + 0·03 M trisodium citrate, washed twice in 70 per cent ethanol and stained in Giemsa (1/20—pH 7 buffer for 5 min. In all these schedules, differential staining of segments were corroborated from fluorescence banding patterns. Zakharov et al. (1971) reported clarification of late-replicating segments of lymphocyte chromosomes by treating with 5-BUDR (200 µm/ml for 5–7 h) or B-mercaptoethanol ($3·6 \times 10^{-3}$, $6·0 \times 10^{-3}$ M for 2–3 min) before fixation, later confirmed through autoradiography with thymidine labelling.

Lately, extensive work has been done on somatic cell fusion techniques, allowing genes to be mapped on the human chromosomes (see next chapter). In the *molecular hybridisation* method, to localise chromosome segments, e.g. satellite DNA on the basis of differential gene action, the principle of hybridisation at the molecular level in *situ* has been applied successfully by Pardue and Gall (1970); (for technique, please see legend of Plate 11 at the end of the book).

CONCLUSION

An outline of the techniques, their application and advantages in human and other mammalian chromosome methodology, as here depicted, is a clear index of the tremendous enthusiasm evinced in this branch of chromosome science. In spite of this explosion of research within the last few years, there is very little reason for complacence in view of the intricacies of the problems still unsolved. Glaring inadequacies are still present in the technology. For example, most of the outstanding achievements are based on culture technique—a method with certain inherent disadvantages. Scepticism has often been expressed regarding the accuracy of the *in vitro* culture data and the extent to which they represent *in vivo* conditions. In plants, it has been demonstrated that even treatment in distilled water exerts an appreciable effect on chromosomes (Sharma and Sen, 1954; Sharma and Sharma, 1960). The elaborate composition of the human chromosome culture medium calls for a simplification of the procedure.

With relation to the initiation of leucocyte division in the peripheral blood culture, the discovery of the plant product, phytohaemagglutin (PHA), should no doubt be considered as a landmark in the evolution of technology. But even then, to all workers in human chromosome analysis, the failure of certain batches of PHA to act is a common experience. It is often attributed to contamination, and in the absence of the latter, the cause is cited to be the physiological set-up of the tissue. It is not unlikely that the genetic constitution of the individual concerned controls its response to PHA, as in the case of leukaemic cells, which do not require this compound for the initiation of division. In order to meet the requirements arising out of the diversities in genetic and physiological make-up, many more effective chemicals like PHA are needed, to provide a wider choice. Exploration of this property in extracts from other beans, allied to *Phaseolus vulgaris*, undertaken in our own laboratory but not so far published, shows considerable promise. Moreover, the application of organic compounds like kinetin, gibberilin, ascorbic acid and other hormones—some of which are known to initiate division in nuclei in other biological objects—on human and mammalian tissue, may give valuable results in this direction.

In addition to leucocyte culture, it is necessary that for every pathogenic tissue, highly simplified and rapid schedules should be evolved, which would make human chromosome analysis a routine clinical practice.

The necessity for further advancement in the study of the details of chromosome morphology cannot be over-emphasised (Hsu and Mead, 1969). Though to some extent, initial work on the details of chromosome structure from pachytene, specific fluorescence banding and karyotype analysis of human chromosomes has been started, yet the progress compares unfavourably with that achieved in other organisms. Extensive use of a range of pre-treatment agents of established value may solve this problem to an appreciable extent. An exhaustive analysis of the chromosome morphology, from both normal and abnormal individuals, and localisation of genes by somatic cell hybridisation technique (see next chapter), would bring complete mapping of the genes on human chromosomes within the realms of possibility. Such a genetic mapping would pave the way, ultimately, for the prevention of congenital disorders.

These possibilities of further research on human chromosome methodology in no way minimise the tremendous achievements in this direction made within the past few years. With further refinements in technology as envisaged above, the judicious handling of the human chromosome, if not the genes, from the pre-natal to the adult phases, may perhaps successfully aim at the betterment of the human race.

REFERENCES

Amarose, A. P. (1959).   *Nature* **183,** 975
Ambrose, C. T. and Coons, A. H. (1963).   *J. Exptl. Med.* **117,** 1075
Arakaki, D. T. and Sparkes, R. S. (1963).   *Cytogenetics* **2,** 57
Arrighi, F. E., Getz, M. J., Saunders, G. F., Saunders, P. and Hsu, T. C. (1971).   *Abstr. IV Int. Congr. Hum. Genet.,* Paris, 18
Axelrad, A. A. and McCulloch, E. A. (1958).   *Stain Tech.* **33,** 67
Baker, M. C. and Atkin, N. B. (1963).   *Lancet* **i,** 1164
Barkhan, P. and Ballas, A. (1963).   *Nature* **200,** 141
Barr, M. L. (1965).   *Sex chromatin techniques.* In *Human chromosome methodology.* N.Y.; Acad. P.
Barr. M. L. and Bertram, E. G. (1949).   *Nature* **163,** 676
Bartales, M. and Rainer, G. D. (1971).   *Abstr. IV Int. Congr. Hum. Genet.* 22
Barton, D. E. and David, F. N. (1962).   *Ann. Hum. Genet. London* **25,** 323
Barton, D. E. and David, F. N. (1963).   *Ann. Hum. Genet. London* **26,** 347
Barton, D. E., David, F. N. and Merrington, M. (1963).   *Ann. Hum. Genet. London* **26,** 349
Basrur, P. K., Basrur, U. R. and Gillman, J. P. W. (1963).   *Exp. Cell Res.* **30,** 229
Battaglia, E. (1959).   *Caryologia* **12,** 186
Beckman, L. (1962).   *Nature* **195,** 582
Bender, M. A. (1965). *Arq. Brasil. Endocrin. Met.*
Bender, M. A. and Mettler, L. E. (1958).   *Science* **128,** 186
Bender, M. A. and Prescott, D. M. (1962).   *Exptl. Cell Res.* **27,** 221
Benirschke, K. and Brownhill, L. E. (1963).   *Cytogenetics* **2,** 331
Beutler, E., Yeh, M. and Fairbanks, V. F. (1962).   *Proc. Natl. Acad. Sci. US* **48,** 9
Biggers, J. D. and McFeely, R. A. (1963).   *Nature* **199,** 718
Bohorfoush, J. G. (1964).   *Stain Tech.* **39,** 339
Bottura, C. and Ferrari, I. (1960).   *Nature* **186,** 904
Buckton, K. E., Evans, H. J., Oriordan, M. and Robinson, J. A. (1971).   *Abstr. IV Int. Congr. Hum. Genet.* Paris 36
Byrd, W. J., Hare, K., Finley, W. H. and Finlay, S. C. (1967).   *Nature* **213,** 622
Chu, E. H. Y. and Giles, N. H. (1957).   *Am. Naturalist* **41,** 273
Chu, E. H. Y. and Swomley, B. A. (1961).   *Science* **133,** 1925
Clendenin, T. M. (1969).   *Stain Tech.* **44,** 63
Cooper, E. H., Barkhan, P. and Hale, A. J. (1961).   *Lancet* **ii,** 210
Court-Brown, W. (1967).   *Human Population Cytogenetics.* North-Holland; Amsterdam
Craig, A. P. and Shaw, M. W. (1971).   *Abstr. IV Int. Congr. Hum. Genet.* Paris, 49
Darlington, C. D. and Haque, A. (1955).   *Nature* **175,** 32
Davidson, R. G., Brusilow, S. W. and Nilowsky, H. M. (1963).   *Nature* **199,** 296
De Vries, G. F. and van Went, J. J. (1964).   *Stain Tech.* **39,** 45
Downing, H. J., Kemp, G. C. M. and Denborough, M. A. (1968).   *Nature* **217,** 654
Dyban, A. P. and Udalova, L. D. (1967).   *Genetika Leningrad* No. 4, 52
Edwards, J. H. (1962).   *Cytogenetics* **1,** 90
Edwards, J. H. and Young, R. B. (1961).   *Lancet* **ii,** 48
Edwards, R. G. (1962).   *Nature* **196,** 446
Egozcue, J. and Egozcue, M. V. de (1966).   *Stain Tech.* **41,** 173
Eicher, E. M. (1966).   *Stain Tech.* **41,** 317
Eide, P. (1963).   Referred to in Sanders and Humason (1964)
Elves, M. W. and Wilkinson, J. F. (1962).   *Nature* **194,** 1257
Evans, E. P., Breckson, G. and Ford, C. E. (1964).   *Cytogenetics* **3,** 289
Evans, H. J., Summer, A. and Buckland, R. (1971).   *Abstr. IV Int. Congr. Hum. Genet.* Paris 64
Ford, C. E. (1961).   Human cytogenetics, *Brit. Med. Bull.* **17,** 179
Ford, C. E. (1961).   Technique described in Turpin and Lejeune (1969)
Ford, C. E. and Hamerton, J. L. (1956a).   *Stain Tech.* **31,** 247
Ford, C. E. and Hamerton, J. L. (1956b).   *Acta Genet.* **6,** 264
Ford, C. E. and Hamerton, J. L. (1956c).   *Nature* **177,** 140 and **178,** 1020
Ford, C. E., Jacobs, P. A. and Lajtha, L. G. (1958).   *Nature* **181,** 1565
Ford, E. H. R. and Woollam, D. H. M. (1963).   *Stain Tech.* **38,** 271
Ford, L. (1965).   *Stain Tech.* **40,** 317
Ford, L. (1969).   *Nucleus* **12,** 93
Ford, L., Cacheiro, N., Norby, D. and Heller, C. G. (1968).   *Nucleus* **11,** 83
Ford, L., Cacheiro, N. and Norby, D. (1969).   *Nucleus* **12,** 1
Fox, M. and Zeiss, I. M. (1961).   *Nature* **192,** 1213

Fraccaro, M., Kaijser, K. and Lindsten, J. (1960a). *Lancet* **i,** 724 and **ii,** 899
Fraccaro, M., Kaijser, K. and Lindsten, J. (1960b). *Ann. Hum. Genet. London* **24,** 45 and 205
Fredga, K. (1964). *Hereditas* **51,** 268
Frøland, A. (1962). *Lancet* **ii,** 1281
Gardner, H. H. and Punnett, H. H. (1964). *Stain Tech.* **39,** 245
Gay, H. and Kaufmann, B. P. (1950) Stain Tech. **25,** 209
Grouchy, J. de, Roubin, M. and Passage, E. (1964). *Ann. Genet. Paris* **7,** 45
Grumbach, M. M. and Morishima, A. (1962). *Acta Cytol.* **6,** 46
Guard, H. R. (1959). *Am. J. Clin. Pathol.* **32,** 145
Hamerton, J. L., Fraccaro, M., de Carli, M., Nuzzo, F., Klinger, H. P., Hulliger, L., Taylor, A. and Lang, E. M. (1961). *Nature* **192,** 225
Harnden, D. G. (1960). *Brit. J. Exptl. Pathol.* **41,** 31
Harnden, D. G. and Brunton, S. (1965). *The skin culture technique* in *Human chromosome methodology*. New York; Academic Press
Hastings, J., Freedman, S., Rendon, O., Cooper, H. L. and Hirschhorn, K. (1961). *Nature* **192,** 1214
Hayflick, L. and Moorhead, P. S. (1961). *Exp. Cell Res.* **25,** 585
Hirschhorn, K. and Cooper, H. (1961). *Amer. J. Med.* **31,** 442
Hirschhorn, K., Kolodny, R. L., Hashem, N. and Bach, F. (1963). *Lancet* **ii,** 305
Howard, A. (1952). *Stain Tech.* **27,** 313
Hrushovetz, S. B. and Harder, C. E. (1962). *Stain Tech.* **37,** 307
Hsu, T. C. (1952). *J. Hered.* **43,** 167
Hsu, T. C. and Kellogg, D. S. (1960). *J. Nat. Canc. Inst.* **25,** 221
Hsu, T. C. and Mead, R. A. (1969). In *Comparative mammalian cytogenetics,* Berlin; Springer
Hsu, T. C. and Pomerat, C. M. (1953). *J. Hered.* **44,** 23
Humason, G. L. and Sanders, P. C. (1963). *Stain Tech.* **38,** 338
Hungerford, D. A. (1965). *Stain Tech.* **40,** 333
Hungerford, D. A. and Nowell, P. C. (1963). Quoted in Mellman (1965)
Hungerford, D. A., Donnelly, A. J., Nowell, P. C. and Beck, S. (1959). *Am. J. Hum. Genet.* **11,** 215
Jaffe, G. (1959). *Nature* **183,** 1329
Klinger, H. P. and Miller, O. J. (1968). In *Diagnosis and treatment of fetal disorders.* Berlin; Springer
Kolodny, R. L. and Hirschhorn, K. (1964). *Nature* **201,** 715
Koprowski, H., Ponten, J. A., Jensen, F., Ravdin, R. G., Moorhead, P. and Saksela, E. (1962). *J. Cell. Comp. Phys.* **59,** 281
Lee, M. R. (1969). *Stain Tech.* **44,** 155
Lejeune, J. (1960). *Ann. de Génét. Paris* **2,** 1
Lejeune, J., Gautier, M. and Turpin, R. (1959a). *C. R. Acad. Sci.* **248,** 602 and 1721
Lejeune, J., Gautier, M. and Turpin, R. (1959b). *Lancet* **i,** 885
Lejeune, J., Turpin, R. and Gautier, M. (1960). *Rev. Franc. Clin. Biol.* **5,** 406
Li, S. G. and Osgood, E. E. (1949). *Blood* **4,** 670
Lima-de-Faria, A. and Bose, S. (1962). *Chromosoma* **13,** 315
Lyon, M. (1961). *Nature* **190,** 372 and (1971) *Nature* **232,** 229
Makino, S. and Nishimura, I. (1952). *Stain Tech.* **27,** 1
Makino, S., Sasaki, M. S. and Fukushima, T. (1963). *Lancet* **ii,** 1273
McConnell, J., Fechheimer, N. S. and Gilmore, L. D. (1963). *J. Animal Sci.* **22,** 374
McFee, A. F., Banner, M. W. and Murphree, R. L. (1965). *J. Animal Sci.* **24,** 551
Mellman, W. J. (1965). *Human peripheral blood leucocyte cultures,* in *Human chromosome methodology.* New York; Academic Press
Mellman, W. J., Klevit, H. D. and Moorhead, P. S. (1962). *Blood* **20,** 103
Meredith, R. (1969). *Chromosoma (Berl.)* **26,** 254
Metzgar, D. P. and Moskowitz, M. (1963). *Expt. Cell Res.* **30,** 379
Moorhead, P. S. (1964). *The blood technique and human chromosomes* in *Symp. Mammalian Tissue Culture Cytol. Sao Paulo,* 1962, New York; Pergamon Press
Moorhead, P. S. and Saksela, E. (1963). *J. Cell Comp. Physiol.* **62,** 57
Moorhead, P. S., Nowell, P. C., Mellman, W. J., Battips, D. M. and Hungerford, D. A. (1960). *Exptl. Cell Res.* **20,** 613
Moore, G. E., Porter, I. H. and Huang, C. C. (1969). *Science* **163,** 1453
Moore, G. E., Grace, J. T. Jr., Citron, P., Gerner, R. E. and Burns, A. (1966). *N.Y. State J. Med.* **21,** 2757
Moore, K. L., Graham, M. A. and Barr, M. L. (1953). *Surg. Gynecol. Obstet.* **96,** 641
Nadler, C. F. and Block, M. H. (1962). *Chromosoma* **13,** 1
Newcomer, E. H. and Donnelly, G. M. (1963). *Stain Tech.* **38,** 54
Nichols, W. W. and Levan, A. (1962). *Blood* **20,** 106

Nowell, P. C. (1960).   *Cancer Res.* **20,** 462
Nowell, P. C. and Hungerford, D. A. (1963).   Quoted in Mellman (1965)
Nowell, P. C., Hungerford, D. A. and Brooks, C. D. (1958).   *Proc. Amer. Ass. Cancer Res.* **2,** 331
Ohno, S. (1965).   *Direct handling of germ cells* In *Human chromosome methodology.* N.Y.; Acad. P.
Ohno, S. (1961).   *Lancet* **ii,** 723
Ohnuki, Y., Awa, A. and Pomerat, C. M. (1962).   *Tech. Doc. Rep. no. SAM-TDR*-62-99, USAF
    Sch. of Aerospace Med., Brooks AFB, Texas
Pacha, R. E. and Kingsbury, D. T. (1962). *Proc. Soc. exp. Biol. Med.* **111,** 710
Pardue, M. L. and Gall, J. G. (1970).   *Proc. III Oxford Chromosome Conf.* London: Oliver & Boyd
Patton, J. L. (1967).   *J. Mammal.* **48,** 27
Pearmain, G. E. Lycette, R. R. and Fitzgerald, P. H. (1963).   *Lancet* **i,** 637
Penrose, L. S. and Delhanty, J. D. A. (1961).   *Lancet* **i,** 1261
Peterson, K. W., Legator, M. S. and Jacobson, C. B. (1967).   *Proc. VI Conf. Mammal. Cytol. & Somatic
    Cell Genet.* Pacific Grove, California
Petrakis, N. L. and Politis, G. (1962).   *New Engl. J. Med.* **267,** 286
Puck, T. T., Cieciura, S. J. and Robinson, A. (1958).   *J. Exp. Med.* **108,** 945
Punnett, T. and Punnett, H. H. (1963).   *Nature* **198,** 1173
Reitalu, J. (1964).   *Hereditas* **52,** 235
Ridler, M. A. C. (1971).   *Lancet* **ii,** 354
Rigas, D. A. and Osgood, E. E. (1955).   *J. Biol. Chem.* **212,** 607
Robinson, J. S., Bishun, N. P., Rashad, M. N. and Marton, W. R. M. (1964).   *Lancet* **i,** 328
Rothfels, K. H. and Siminovitch, L. (1958a). *Stain Tech.* **33,** 73
Rothfels, K. H. and Siminovitch, L. (1958b).   *Chromosoma* **9,** 163
Rowley, M. and Heller, C. G. (1966). *Fertil. Steril.* **17,** 177
Sandberg, A. A., Ishihara, T., Crosswhite, L. H. and Hauschka, T. S. (1962). *Cancer Res.* **22,** 748
Sanders, P. C. and Humason, G. L. (1964).   *Stain Tech.* **39,** 209
Sanderson, A. R. (1960).   *Lancet* **i,** 1252
Scherz, R. G. (1962).   *Stain Tech.* **37,** 386
Sharma, A. K. and Sen, S. (1954).   *Genet. Iber.* **6,** 19
Sharma, A. K. and Sharma, A. (1960).   *Internat. Rev. Cytol.* **10,** 101
Shaver, E. I. (1962).   *Canad. J. Genet. Cytol.* **4,** 62
Skoog, W. A. and Beck, W. S. (1956).   *Blood* **11,** 436
Spalding, J. F. and Wellnitz, J. M. (1956).   *Stain Tech.* **31,** 123
Sohval, A. R. and Casselman, W. G. B. (1961).   *Lancet* **ii,** 1386
Sparano, B. M. (1961).   *Stain Tech.* **36,** 41
Srivastava, P. K. and Lasley, J. F. (1968).   *Stain Tech.* **43,** 187
Stone, L. E. (1963).   *Canad. J. Genet. Cytol.* **5,** 38
Swanson, D. W. and McKee, M. E. (1964).   *Stain Tech.* **39,** 117
Tarkowski, A. K. (1966).   *Cytogenetics* **5,** 394
Taylor, A. I. (1963).   *Lancet* **i,** 912
Tips, R. L., Smith, G. S., Meyer, D. L. and Ushijima, R. N. (1963).   *Texas Rep. Biol. Med.* **21,** 581
Tjio, J. H. and Levan, A. (1956).   *Hereditas* **42,** 1
Tjio, J. H. and Whang, J. (1962).   *Stain Tech.* **37,** 17
Tjio, J. H. and Whang, J. (1965).   *Direct chromosome preparations of bone-marrow cells* in *Human
    chromosome methodology.* New York; Academic Press
Tjio, J. H. and Puck, T. T. (1958).   *J. Exper. Med.* **108,** 25
Turpin, R. and Lejeune, J. (1969).   *Human afflictions and chromosomal aberrations.* Paris; Perg. P.
Van der Hagen, C. B. and Berg, K. (1970).   *Clin. Genet.* **1,** 263
Wang, M. Y. F. W., McCutcheon, E. and Desforges, J. F. (1967).   *Amer. J. Obsret. Gynec.* **97,** 1123
Welshons, W. J., Gibson, B. H. and Scandlyn, B. J. (1962).   *Stain Tech.* **37,** 1
Wittmann, W. (1965).   *Stain Tech.* **40,** 161
WHO Expert Cmtte on Hum. Gen. (1969).   III Rep., WHO Tech. Rep. Ser. No. 416, Geneva
WHO group on the standardisation of procedures for chromosome studies in abortion. (1966)
    Cytogenetics **5,** 361
Wroblewska, J. and Dyban, A. P. (1969).   *Stain Tech.* **44,** 147
Young, W. J., Merz, T., Ferguson-Smith, M. A. and Johnston, A. W. (1960).   *Science* **131,** 1672
Yunis, J. J. (1965).   Ed. *Human chromosome methodology.* New York; Academic Press; also chapter on
    Human chromosomes in disease.
Yunis, J. J., Aldrich, J. E. and Lee, J. L. (1971).   *Abstr. IV Int. Congr. Hum. Genet.* Paris, 193
Zakharov, A. F., Seleznev, J. V., Benjusch, V. A. Baranovskaya, L. I. and Demintseva, V. S. (1971).
    Abstr. IV Int. Congr. Hum. Genet. Paris 193

16 Study of Chromosomes from Malignant Cells

INTRODUCTION

Cancer represents an unchecked, malignant form of rapid growth, perpetuated through several cell generations and probably originating from several causes, both internal and external, including transformation by viruses (*see* Klein, 1966). Hueper and Conway (1964) have made a detailed review of carcinogenesis in man in relation to occupational exposure to chemical substances. The problem has been aggravated further by the discovery of the carcinogenic properties of aflatoxins produced from *Aspergillus flavus* infecting groundnuts and cereals under storage (Roe and Lancaster, 1964 and *see* Raven and Roe, 1967). Irrespective of its mode of origin, cancerous tissue is always characterised by distinct cytological features (Seshachar and Nambiar, 1955; Hansen-Melander *et al.*, 1956; Ishihara *et al.*, 1963; Koller, 1963; Lubs and Clark, 1963; Stich, 1963; Wakonig-Vaartaja, 1963; Atkin, 1964; Makino *et al.*, 1964; Springs, 1964; Wakonig-Vaartaja and Kirkland, 1965; Richart and Wilbanks, 1966; Miles, 1967a, b; Talukder and Sharma, 1968; Lampert, 1971). Although Bauer (1949) claimed that the frequency of mitosis is a symptom but not a specific sign of cancer, chromosomal abnormalities, and especially numerical variation, have been found to be conspicuous features of cancer cells (Koller, 1956; Makino and colleagues, 1959). Variability is, however, low in tumours derived from single cell culture (Hauschka and Levan, 1958) whereas diploid cells are more common in primary tumours (Koller, 1960). Makino (1957) suggests that this irregularity shows that every tumour has its own stemline number which occurs at maximum frequency in that particular tissue. This hypothesis has found wide acceptance, particularly in relation to rat ascites tumours and the later phases of human tumours. Several authors have shown, however, that two or more stemlines may also characterise human neoplastic tumours (Ishihara *et al.*, 1963; Sandberg and Yamada, 1966; Sharma, G.P. *et al.*, 1969). Mitotic instability and chromosomal unbalance, associated with lack of differentiation, are therefore the universal features of cancerous cells. However, no cancer is unquestionably characterised by specific chromosomal changes, except chronic myeloid leukaemia, where a structural alteration in chromosome G (Ph') has been found to be a constant feature. Several, as yet unconfirmed, reports of other such associations are also available (*see* Nowell and Hungerford, 1960; Turpin and Lejeune, 1969). Discovery of the origin of the malignant growth, as well as its prevention, require an intimate knowledge of the cytology of cancer cells. Both growth and its cessation are related with the mechanism of the nuclear division.

The development of techniques for the study of cancer chromosomes and their behaviour was principally stimulated by the discovery that certain human abnormalities or diseases, such as mongoloid idiocy, can be correlated with certain chromosomal irregularities, and since then various techniques have been developed for the study of cancer chromosomes from tissues extracted from individuals, or after growing them in culture (Yunis, 1965; Turpin and Lejeune, 1969). Needless to say, tissue or cell culture has become an important tool in cancer cytology.

Even though the study of cancer chromosomes has been pursued for several years, data on their cytology was mostly nebulous until the discovery of the *ascites tumours* and a method for their observation. The single cells, or groups of cells, suspended in the body fluid are convenient for cytological observation, and the methods of study have undergone so much improvement that solid tumour tissue can now ultimately be converted into ascites tumours for facilitating cytological study.

The importance of tissue culture methods in relation to the study of cancer chromosomes is demonstrated in the discovery of Levan and others. This discovery shows that normal cells, during their transformation into adult cells, show mitotic instability in tissue culture, thus attaining a genetic heterogeneity parallel with the cancer cells. Moreover, such cells, when inoculated into normal tissue, can induce tumour growth. Techniques have been devised to study the comparative cytology of cancerous and normal cells under culture conditions and their behaviour following cell fusion *in situ*.

MATERIALS SUITABLE FOR STUDY

Though cancer is the principal unsolved medical problem in human beings, the latter cannot, for obvious reasons, be utilised for detailed experimental purposes. In addition to blood and bone marrow, only biopsies for studying the chromosomes can be taken from patients suffering from cancer in various organs. In carrying out extensive experiments, using different chemical agents and testing the effect on the development and nuclear behaviour of cancerous tissue under laboratory conditions, inbred strains of white rat, mouse, guinea-pig and rabbit are mainly employed. The incidence of cancer can also be studied in other animals, which can be bred in the laboratory on a standard diet under standard conditions, depending upon their availability.

METHODS OF INDUCTION

In addition to the development of spontaneous cancers, different chemical agents, termed *carcinogens*, can be used for their artificial induction. The different types of carcinogens available in the market belong to the coal–tar derivatives; principally, the hydrocarbons, azodyes and others. A list of some of the common carcinogens, which have been tested for their carcinogenic activity on various animals, is given below. In addition, ultraviolet rays have also been employed to induce neoplasia in animals (Blum, 1950).

Hydrocarbons

1. Derivatives of benzanthracene: methyl 1,2-benzanthracene; dimethyl 1,2-benzanthracene; specially 5,6-dimethyl and 9-10-dimethyl 1,2-benzanthracene.
2. Cholanthrene derivatives: methyl and ethyl cholanthrene.
3. 3,4–Benzpyrene.
4. Dibenzcarbazole.
5. 3,4–Dibenzacridine.
6. Dibenzfluorene.

Azodyes

Derivatives of *o*-aminoazotoluene and dimethyl aminoazotoluene (Butter Yellow): − 2,2′-azonaphthalene.

Mustard derivatives

1. Nitrogen mustard—β, β'—dibromodiethyl sulphide, sulphone-sulphoxide and its dichloro and di-iodo derivatives.
2. Ethyl carbamate or urethane.

Steroids

1. Stilbene derivatives, e.g. 4,4′dihydroxy, α–β–diethylstilbene.
2. Testosterone derivatives.
3. Gonadotrophins.

Others

1. Aerosol, alkaloids, *o*-aminotoluene, calcium gluconate, carbon tetrachloride, chloral hydrate, chloroquinone and benzoquinone chloro-

acetone, nitrosamines—dimethyl and diethyl, *n*-nitrosomorpholine, potassium thiocyanate, sulphonamides.

2. For different viruses which can induce tumours, *see* Dalton and Haguenau (1962), Klein (1966) and Dulbecco (1969).

In plants, crown galls show similarities to animal cancer in the rapid rate of cell growth, mitotic instability, and lack of differentiation. Several species of bacteria, especially *Bacterium tumafaciens* and *Agrobacterium tumafaciens*, can induce crown gall in several plant species belonging to leguminous and other groups. The leaf or other plant surfaces can be abrased with carborandum powder and the inoculum applied in the damaged plant, resulting in the development of gall (Sharma and Nandi, unpublished). Species of *Vicia*, mainly *V. faba*, and *Solanum* are extremely susceptible. Several carcinogens have been tried on plant cells (Nandi, 1969) and certain plant extracts have also been shown to induce tumours in plants (Sharma, A., 1959).

The different methods by which a carcinogen can be administered into tissue to induce cancerous growth in animals are: (i) feeding—applied in a specific dose in the food; (ii) intraperitoneal injection; (iii) subcutaneous injection and (iv) rubbing into the skin. These are applied with different solvents like water, olive oil, sesame oil, lard, glycerine, benzene, acetone, etc. and they are administered in specific doses for a prolonged period, at regular intervals. For example, tumours can be induced with intraperitoneal injections of 3,4-benzpyrene and methyl cholanthrene in rats within 3–6 months. For inducing multicentric hepatoma, *n*-nitrosomorpholine can be administered to inbred rats in drinking water (20 cm^3 per day for 6 days), the dosages being 6 mg per cent solution for up to 34 weeks, 12 mg per cent up to 12 weeks, and 20 mg per cent up to 7 weeks (Druckrey *et al.*, 1961; Bannasch, 1968).

In established mouse fibroblast cell cultures, to study the effect of carcinogens, methyl cholanthrene at 1 $\mu g/cm^3$ of medium can be administered for a prolonged period (one to several weeks). The behaviour of loss of contact inhibition can be detected even after 2–3 days exposure to either methyl cholanthrene or to 3,4-benzpyrene (Berwald and Sachs, 1963). When the treated culture is injected into C_3H mice, maximum tumour production was obtained with 6-day treated cultures.

The two terms *carcinoma* and *sarcoma* are applied, depending on the type of origin. The latter type represents tumorous growth initially originating from the mesoderm, whereas all other types are included within the category of carcinoma. Cancerous growth involving blood cells may result in *leukaemia*. Cancers represented in the small islands of cells in the body fluid are known as *ascites* tumours. Specific names are given to growths in the different organs, based on the organ affected and detailed cytological studies have been carried out on almost all kinds of human cancer, including maxillary, mammary, uterine, pulmonary, hepatic and skin carcinomas, reticulosarcoma, postauricular tumour, etc. to mention a few. After induction of cancer in the body of the subject, cells are maintained in culture and chromosomes can be studied in them through the methods outlined for the study of mammalian chromosomes (*see* chapter 15).

In order to inhibit the cancerous growth, different dosages of x-rays, as well as different types of anticarcinogens, can be applied in the medium. Vincaleucoblastine, podophyllin, α– and β– paltatin, quercelin, etc. are some of the anticarcinogenic agents used. Podophyllin (Makino and Tanaka, 1953) is dissolved in isotonic glucose solution and applied to rats through peritoneal injections at room temperature. Three applications of a concentration up to 0·1 per cent (1 cm^3/100 g of body weight) given every seventh day produced regression of the ascites tumour, associated with a loss of mitotic instability.

STUDY OF CANCER TISSUE IN CULTURE

The rapid advance of research on cancerous materials has been responsible for the development of several techniques for culturing malignant cells *in vitro*. The chromosome studies of these materials form one of the principal sources of our knowledge of the genomic constitution of malignant tissues. To secure growth *in vitro*, methods for the culture of *tissue explants, monolayers* as well as *cell suspensions* have been devised. Most of these schedules have been discussed in detail in the chapters on mammalian and human chromosomes and tissue culture. Only the schedules specially modified for cancerous cells are discussed here.

Explant culture—Most methods for cancer tissue explant culture are identical with those described in the chapters for the study of mammalian chromosomes and tissue culture, including Carrel flask, Roller tube, Hanging Drop and Sandwiching processes (Basrur *et al.*, 1963). Mammalian serum is one of the major constituents of the medium but plasma clot consisting of 50 per cent plasma in balanced salt solution and 50 per cent embryo extract in serum provide the most congenial medium for tissue growth. Walker and Wright (1961) successfully cultured several human tumour tissues, namely, lymphoma, melanoma, carcinoma and tumour of the nervous system in TC 199 containing 20 per cent autologous or pooled human serum, in order to study the effect of vincaleucoblastine on the cultured tissue. Barski and Belehradek (mentioned in Easty, 1967) used a mixture containing 30 parts TC 199 (Morgan *et al.*, 1950) as modified by Barski; 30 parts Hank's medium and 30 parts human serum, without complement, for successful growth of human tumour tissues. For inducing growth of primary human tumour tissues, the medium devised by Evans *et al.* (1956) has been quite effective (Ambrose *et al.*, 1962; Easty and Wylie, 1963) but the addition of insulin and folic acid (Prop, 1961) along with AB serum resulted in profuse growth of mammary gland tissue of mouse. Lumsden (1963) succeeded in obtaining cultures of human brain tumours. In addition to these, several reports have been published of explant cultures of human organs and the different media adopted for the purpose are modifications of those already mentioned.

Monolayer and Suspension Culture from Cell Effusions

Suspension cultures *in vitro* from cell effusions can be obtained most success-fully from ascites tumour cells. De Bruyn (1956), De Bruyn and Hampe (1961) effectively applied the monolayer technique for culturing ascites cells and cell effusions obtained from human pleural and peritoneal fluids. In monolayer culture, it is observed that cancer cells do not exhibit contact inhibition, often manifested in closely adhering normal cells (Abercrombie and Heaysman, 1957; Abercrombie and Ambrose, 1958). Effusion cultures of cells from human ascitic fluid, obtained through centrifugation, were made by Moore and Koike· (1964), who noted that adaptation—the primary requisite for growth in culture—is maximum in the triploid cells where the number of cells in culture reaches as high a count as of 100 000 cells per cm^3. Di Paolo (1964) has also observed growth *in vitro* of several cell effusions. Reports of successful culturing of tumour effusions are continually accumulating (*see* Hsu, 1965, and Ambrose *et al.*, 1967).

Cell Suspensions from Solid Tissue

Of all the techniques evolved for cancer tissue culture, processing of cell suspensions from solid tissue is the most elaborate one. As with normal tissues, disaggregation of cells poses the principal problem, for which mechanical, chemical and enzymic methods have been adopted (for details see chapters on tissue culture and mammalian chromosomes). Of the several mechanical methods available, one (Snell, 1963) involves kneading very small pieces of tumour (1 cm^3 by vol) in the medium with thin-walled rubber tubing and finally forcing the suspension through a metal sieve which gives a good amount of single cell suspensions. In another technique, cells were grown in test tubes with glass beads and after a period of growth, were dislodged by shaking (Leighton, 1958) and dispersed finally by aspirating several times through a pipette. Suspensions of rat liver cells could be obtained by dispersion with a rubber pestle inside a glass tube (Jacob and Bhargava, 1962). Scraping of the layer with a rubber policeman often hampers viability of the cells but yields sufficient material for analysis (Magee *et al.*, 1958).

Of the chemical methods, the most effective is the use of versene or EDTA (see chapter on tissue culture) and shaking in this medium may or may not be supplemented with homogenisation (Di Paolo and Dowd, 1961). Among the different enzymes tried for dispersion, trypsin is un-doubtedly the most useful, both for HeLa cells and other malignant tissues. Trypsin, with a small amount of desoxyribonuclease, has been applied successfully in some cases, but this treatment is limited by the possibility of digestion of DNA (Boyse, 1960; Madden and Burk, 1961). Elastase and collagenese have also been employed—the latter especially in the adult organs. Morgan and Griffiths (1963) used fibrinolysin for disaggregating tissues from cancerous human colon.

The containers used for culturing are the same ones as those required for normal animal tissue cultures, such as, Carrel flasks, T-flasks, Erlenmeyer flasks, Petri dishes, cover slips in test tubes, etc. Pulvertaft (1961) recommended the use of a small culture chamber, 3 mm deep, prepared of polytetrafluoroethylene (nontoxic plastic), the cover slips being fixed with silicon grease and sealed with wax. Ambrose and others (1967) employed small glass tubes with stoppers, and with the aid of an inverted microscope, the monolayer at the bottom of the tube could be analysed.

Both natural extract, like serum and embryo extract, and synthetic media are in use. All laboratory-adapted cell strains in culture grow profusely in Waymouth's (1956) simple medium M.B. 752/1. The optimum growth is generally seen at pH between 7·0 and 7·8, cell disintegration occurring at over 8·0 (Paul, 1952; Taylor, 1962). Lymphocytes survive for long periods in hypotonic media (Trowell, 1963). A buffering system with bicarbonate, employing an atmosphere of 5 per cent CO_2 in air, or with tris-HCl in air, or atmospheres containing 2 per cent carbon dioxide, gives good growth and pH control (Martin, 1964). Easty *et al.* (1964) have demonstrated that tissues differ with respect to their capacity for the uptake of macromolecules of serum proteins. Excess of oxygen is detrimental to proliferative growth (Pace *et al.*, 1962). When normal cells are grown continuously in an atmosphere containing nitrogen, they often become neoplastic (Goldblatt and Cameron, 1953).

Suspension cultures—have been found to be very suitable for malignant cells. For such cultures, continued rotation of the drum containing the culture tubes, steady shaking, silicone-coating of the inner wall of the container to prevent adhesion, continuous stirring and automatic replacement of the medium through cryostat are all applied as in normal tissues (Paul, 1959; Björkland *et al.*, 1961).

After continued and repeated subculturing, *cell lines* can be secured containing a colony of cells showing a rapid rate of proliferation. Cell lines derived from different types of tissue often look alike and lose the morphological characteristics of the original parent line. The cytological basis of the origin of cell lines, that is, whether it is through mutation or selection is not fully known. Just as normal cell lines often show neoplastic growth later, similarly cases have been reported where the malignant cell lines, after a certain period of growth, have become normal (Gey *et al.*, 1954; Foley and Drolet, 1964). An understanding of the mechanism of this transformation into normalcy may have a bearing on the cure for malignancy. In any case, in spite of these limitations, due to the easy method of rapid culturing, several normal and malignant cell lines have now been standardised and are being maintained for use in cytochemical work.

Several schedules for isolated *single cells* have been evolved but in general, single cell cultures of HeLa cells are easier to prepare than cultures of *fibroblast* cells. Methods for obtaining clones from a single cell originally derived from suspensions are also identical with those described for other animal tissues. Aronson and Kessel (1960) devised a method for isolating single cells by making them adhere to wax-soaked glass beads. For HeLa *cells*, the schedule developed by Puck *et al.* (1956), involving the growth of dilute suspensions of single cells on x-ray irradiated feeder layer or on suitable media, has yielded excellent results (Foley *et al.*, 1963). HeLa cells (so named after being derived from carcinoma of the cervix uteri of a patient, Helen Lane), represent an excellent example of adaptation to cultural conditions, and maintained for innumerable generations.

Recently, the potentiality of hybridising mammalian cells in culture for the suppression of malignancy has been elucidated in the work of Harris *et al.* (1969). Hybrid cells in culture were first obtained by Barski and Cornefert (1962), followed by a number of workers who also used polyoma virus to induce malignancy in one of the parent cell lines (Defendi *et al.*, 1964, 1967; Scaletta and Ephrussi, 1965; Silagi, 1967). The fusion of malignant and non-malignant cell lines and the malignant properties of their respective progeny revealed the dominance of malignant characters. Human–mouse somatic cell hybrids have been obtained by Nabholz *et al.* (1969) by the method introduced by Harris and Watkins (1965).

The technique for cell fusion, described later, was first outlined by Harris and Watkins (1965), aided by the Sendai virus through which they show that any mammalian cell can be hybridised in culture. Following this schedule, they have hybridised a non-malignant cell line and different malignant cell lines (Harris *et al.*, 1969; also *see* Ephrussi *et al.*, 1969). The non-malignant cell line chosen was A9 mouse fibroblast—an azaguanine-resistant strain originally obtained from α- cell line of mouse (Littlefield, 1964). Three malignant ascites tumour lines selected were

(a) Ehrlich—originally derived from mouse mammary carcinoma (Hauschka, 1953);
(b) SEWA—obtained initially by subcutaneous injection of polyoma virus into a newborn A-SW mouse (Sjogren, 1964) and
(c) MSWBS—ascites sarcoma, original derivation by methyl cholan-threne injection in a hybrid mouse (A-SW × AF$_1$) (Klein and Klein, 1958).

The detection of these hybrid strains growing in culture did not present any difficulty, as the culture medium (Littlefield, 1964) and the container used did not permit the growth of any of the parent strains. Mixed suspension of parent cells with inactivated sendai virus yielded a large number of hybrid clones. Chromosome analysis of these hybrid cells revealed the sum total of the parent complements, indicated both by the general number and the proportion of marker chromosomes.

Injection of hybrid cells intraperitoneally and subcutaneously in compatible mice was carried out to test the tumerogenic capacity of these hybrid cells. With A9–MSWBS hybrids, no tumours were obtained, whereas with the other two types, reversion to malignancy was noted in only a few cases. In the chromosome complement of such malignant tumours, derived from A9–Ehrlich hybrids, a loss of certain chromosomes was observed; evidently the segregants of the hybrid cells. These data have been claimed to indicate that malignancy can be suppressed by fusion with specific non-malignant cell lines, and also that reversion to malignancy is associated with loss of certain chromosomes. Such results, if later proved to be consistent, bristle with immense possibilities for evolving a cure for malignancy. Moreover, refinements in methods, leading to a further clarification of the structure of chromosomes responsible for tumour inhibition may reveal the ultimate basis of the genetic control of susceptibility to carcinogenic agents.

Lately, excellent work has been carried out on the virus–induced malignancy in animals. Under cultural conditions, oncogenic viruses cause changes in the host cells resulting in tumorous growth of the latter, the process being known as *transformation*. This behaviour is similar to that of neoplastic or malignant growth induced by different carcinogens. Both RNA and DNA viruses can cause transformation (Klein, 1966). The polyoma virus and the simian virus (SV 40) of the latter category have been studied in detail from this aspect (Westphal and Dulbecco, 1968; Dulbecco, 1969). Both of them have a typical circular DNA and a molecular weight of 3×10^6. Three different mammalian cell lines, namely,

(a) BHK—hamster kidney fibroblasts (Macpherson and Stoker, 1962);
(b) BSC–1, kidney line of African green monkey (Hopps *et al.*, 1963) and
(c) 3T3—skin fibroblast of mouse (Todaro and Green, 1963), were used for this investigation on transformation.

Polyoma virus, and SV 40, multiply and cause cell death of 3T3 and BSC–1 cell lines respectively. But BHK and 3T3 lines can be also transformed by both the viruses, though the characteristics of transformation differ, depending on the nature of the specific virus. Transformed cell lines can be distinguished in culture from normal cell lines by their rapid rate of multiplication, low serum requirement, ability to grow in suspension culture in agar, or methocel, and their growth patterns on glass or plastic (Macpherson and Montagnier, 1964; Todaro *et al.*, 1965; Hakomori *et al.*, 1968; Holley and Kiernan, 1968; Stoker, 1968).

In the case of stable transformation, there is complete integration of viral DNA into the DNA of the transformed cells, as specially demonstrated in 3T3 cells. Evidence in support of this has been obtained through

(a) their detection on membrane filters (Gillespie and Spiegelman, 1965);
(b) hybridisation of viral DNA in transformed cells with specific RNA of SV40 (Westphal and Dulbecco, 1968);
(c) restriction of this hybridising capacity to association of the virus with chromosomal DNA and
(d) absence of any free viral DNA in transformed cells. Moreover, this virus can be detached from the chromosomal DNA by fusing the transformed lines with permissive cells by the method of induced cell fusion (Gerber, 1966; Koprowski *et al.*, 1967; Tournier *et al.*, 1967; Watkins and Dulbecco, 1967).

Even though the exact mechanism of this integration of viral genome into the host chromosomes is not fully understood, the importance of this discovery in relation to viral carcinogenesis cannot be overrated. This knowledge of integration, together with the capacity of detachment, may provide essential clues to a solution of the problem of neoplasia and its regulation in animal cells.

SCHEDULES

FOR THE PREPARATION OF TISSUE FOR CYTOLOGICAL STUDY

Direct Biopsies

(a) Biopsies are taken directly from patients suffering from cancer of particular organs. The tissue is directly observed in acetic-orcein, acetic-carmine or acetic-dahlia after staining for a few minutes, preceded by treatment in hypotonic salt solution for a few minutes. If necessary, the fluid may be slightly warmed to facilitate staining and squashing. To secure well-scattered chromosomes, pre-treatment in aqueous coumarin solution for a few minutes, prior to acetic-carmine staining, yields excellent results (Manna, 1954, 1957). Therman and Timonen (1950) observed that fixation in 10 per cent formalin before staining brings out the structure of the spindle quite clearly. They applied acetic-ethanol fixation to biopsies prior to staining.

For solid tumours, slight trypsin treatment (2 per cent at 37 °C for 1 h) may facilitate squashing.

(b) *Ascites tumours*—The peritoneal or pleural fluid can be taken out with the help of a hypodermic syringe or pipette and a drop of this can be stained directly on the slide with any of the above mentioned dyes. The cells may be treated in a hypotonic salt solution (preferably 1·12 per cent sodium citrate) before staining. To obtain a concentrated solution, the fluid is centrifuged. Makino (1957) observed that pre-treatment in a drop of water on the slide for 20–30 min prior to staining with acetic-dahlia yielded well scattered chromosomes. This method has been followed by Takayama and Makino (1961) and others. Propionic-sudan black B is also an effective stain (Tanaka and colleagues, 1955).

For fixing, squashing and staining the fluid directly in acetic-orcein, the material is pressed between specially prepared siliconed surfaces of slide and coverslip (*see* Levan and Hauschka, 1952).

(c) *Direct observation of bone marrow cells*—The techniques for normal cells, as described in the chapter on mammalian and human chromosomes, can be followed.

(d) *Mitotic division of living cells in pleural fluid* according to the culture of ascites tumour, by modified hanging drop method (Makino and Nakahara, 1953)—

1. Fill the depression of a sterile grooved, or depression, slide with liquid paraffin above the level of its edge.

2. Cut open a tumour-bearing rat. Draw out a small amount of tumour ascites fluid from the peritoneal cavity with a pipette. Place the fluid on a sterile dry cover slip without any medium.

3. Invert the cover slip over the depression so that the ascites fluid lies in the liquid paraffin. Apply slight pressure and remove the excess liquid with a blotting paper. Seal and observe under a phase contrast, or light microscope.

(e) *Study from peripheral blood*—The different techniques for studying chromosomes from peripheral blood cells, as described in the chapter on mammalian and human chromosomes, can be effectively used for cancer chromosomes as well.

Tissue Culture in Nutrient Media

Biopsies are taken from solid or soft tumours and can be cultured in media as specified for explant cultures in the chapter on tissue culture. The cytological studies can be carried out as mentioned before for direct biopsies.

Levan and Hsu (1960) observed that if the original explanted tissue, e.g. solid mammary carcinoma of mouse, is trypsinised before transference to McCoy, Maxwell and Cruse's 5A medium (1959) for culture, convenient materials for cytological study can later be obtained. For chromosome analysis, they treated the tissue for 6 h with 0.05 γ/cm^3 colchicine in the medium. Culture of ascitic fluid too, in sterilised pleural fluid can be made similarly *in vitro* after extracting the primary ascitic fluid from injected rats with a pipette. The chromosome studies in cultured cells follow the same methods as those mentioned above.

Another device which is finding great application in the study of tumour cells and the method of their invasion into normal cells involves the provision of a 3-dimensional matrix for growth (Leighton, 1951; Leighton and colleagues, 1956). Cellulose sponge matrix, saturated with serum and embryo extract, has been found to allow the development of 3-dimensional colonies, closely approaching the condition in the body. The disadvantages of this method are the necrosis of the central region and the difficulty of direct observation.

Cloning of HeLa Cells

The dilution technique (Paul, 1959)—The materials necessary are:
1. A strain of HeLa cells growing on a medium of pH 7.4;
2. A balanced salt solution (BSS) without Ca, Mg, or PO_4;
3. 0.05 per cent trypsin dissolved in the BSS;
4. Medium containing 20 per cent human or calf serum and the remaining synthetic nutrient medium;
5. A stainless steel or glass cylinder of approximately 5 mm diameter and 10 mm height;
6. 60 mm Petri dishes;
7. CO_2 incubator;
8. Pasteur pipettes.

The steps are:
1. Prepare a healthy culture of HeLa cells on a clean medium at a pH of 7.4. Examine to identify the strain.
2. Drain off the nutrient medium from the container. Add a quantity of BSS and drain again.
3. Add 0.05 per cent trypsin dissolved in the BSS to the culture and incubate at $37\,^\circ C$ for 5–10 min.

4. Distribute nutrient medium, containing 20 per cent serum and the rest of the synthetic medium, into Petri dishes and test tubes, putting 5 cm^3 into the former and 4·5 cm^3 into the latter. Keep the Petri dishes in a CO_2 incubator.

5. Suspend the cells in the medium by shaking the container gently. Aspirate the suspension in and out of a pipette several times. After all cells have been suspended, add to the suspension half its quantity of growth medium to stop the action of trypsin. With a haemocytometer, determine the number of cells in a known volume of suspension. Add 0·5 cm^3 of the suspension to 4·5 cm^3 of the medium in the first one of the series of test-tubes. Shake well and add 0·5 cm^3 of the fluid from this test-tube to 4·5 cm^3 of the medium in the next one, and so on along the set of test-tubes until, on counting, a suspension yields 1000–2000 cells/cm^3.

6. Pipette out 0·5–1 cm^3 of the final suspension on to one of the Petri dishes containing the medium, to give 100 cells/dish. Treat the other dishes similarly and keep them in the CO_2 incubator at 37 °C for 1 week. For counting the colonies, drain off the medium and stain.

7. For further isolation, outline on the outer glass surface of the Petri dish, with a glass pencil, the location of a particular colony. Coat the bottom edge of a glass cylinder with silicon grease. Drain out the medium and place the glass cylinder on the Petri dish so that it encloses the colony to be isolated and is attached to the Petri dish by the silicon grease. Pour a few drops of BSS into the cylinder, drain and add a few more drops. Drain out this solution, washing the culture within the cylinder. Add a few drops of trypsin solution and incubate at 37 °C for a few minutes. Disperse the cell suspension thoroughly by aspirating in and out with a pipette. Draw out the suspension and inoculate a fresh nutrient in another vessel for subculture.

8. This method is effective for isolating single colonies but it cannot confirm whether the colony has developed from a single cell or a small group of cells.

The isolation technique (Paul, 1959)—The materials needed, in addition to those for the dilution technique, are:

9. Liquid paraffin saturated with the medium, prepared by adding 10 cm^3 of medium to 100 cm^3 of liquid paraffin, shaking, and incubating at 37 °C for some days.

10. Micro-pipettes prepared by drawing out the capillary area of a fine Pasteur pipette on the flame of a micro-burner to form a capillary point with a diameter of approximately 50 μm. It can be coated with silicon if necessary. The micro-burner can be prepared by attaching the rubber tubing of a gas burner to a hypodermic needle. The micro-pipettes are washed in BSS before use.

The different steps of the technique are:

1. Pour a little ether on the bottom half of a Petri dish. Allow it to evaporate. Dry completely by flaming. Fill the dish with liquid paraffin saturated with the medium.

2. Pipette out the medium and put 10 drops around the dish containing the paraffin at approximately equal distances, about 1 cm from the edge. The drops penetrate the paraffin and spread out at the bottom of the dish.

Pick up a few drops of the cell suspension (prepared as in the dilution technique) with a pipette and place in the centre of the medium–liquid paraffin mixture.

3. Place the Petri dish on the stage of a dissecting microscope. Attach a micro-pipette to a mouth tube. Focus the microscope on the drops of cell suspension at the centre of the Petri dish. Take the micro-pipette in the right hand and guide its tip to the cell suspension at the centre of the Petri dish. By observing through the microscope, guide the tip to an isolated cell and draw it into the pipette by sucking cautiously. Now move one of the drops of medium into the field of vision; guide the pipette into it and eject the cell into it gently. Inoculate similarly all the 10 drops, cover the Petri dish and incubate in a CO_2 incubator at 37 °C. Observe after 24–48 h.

4. Each cell develops into a colony within the drop. For transferring the colony add, with the pipette, a drop of trypsin in BSS inside the drop of medium containing the colony and aspirate it out as a suspension with a micro-pipette.

Study of Chromosomes from Subcultures in vivo

To study the behaviour of chromosomes and cell division in different subcultures obtained from the primary tumours, the following methods may be applied:

(a) *Ascitic fluid*—may be taken from an affected rat and injected into the peritoneum of a normal rat. Oksala (1956) inoculated ascitic fluid, suspended in buffered physiological saline (1:9), in a dosage of 0·2 cm³ per mouse. The concentrated filtrate of ascites tumour can be obtained through Seitz EK apparatus (after Hamazaki and colleagues (1953)), if necessary. The behaviour is studied at intervals, taking the peritoneal fluid of the sub-cultured rats (Makino and Tonomura, 1955).

(b) *Single cell sub-culturing from ascitic fluid* (Makino and Kano, 1955; Sasaki and Hishida, 1958; Hansen-Melander, 1958)—is applicable to cases in which the chromosomal behaviour of a cancer cell colony derived from a single cell of a heterogeneous population has to be studied. The inherent heterogeneity in the chromosome behaviour of every cancer cell has been substantiated from such studies.

In this method:

1. Tumour ascitic fluid with tumour cells is first removed from rats.

2. These are then diluted in the diluting fluid in a proportion of 1:20 000, the diluting fluid being prepared by injecting 15–20 cm³ of physiological saline with a sterilised syringe into the peritoneum of rats weighing about 100–120 g. With the same syringe, the peritoneal fluid is taken out from the treated rats after 15 min and centrifuged at 3000 rev/min for several minutes. The supernatant can serve as the diluting fluid provided it is cellfree on microscopical examination.

3. With a sterilised pipette, a drop of the diluting fluid containing tumour cells is taken on a clean cover slip.

4. A microscopic droplet containing a single cell is sucked into the micro-pipette.

5. The droplet is ejected into the peritoneum of a normal rat.

6. Cytological observations of the subculture following the method discussed above can be carried out after a few days.

(c) *Subculturing of solid tumours*—Solid extracted tumours can be mashed in a homogeniser or tissue press and the tissue suspension can be inoculated subcutaneously or intraperitoneally into normal animals. Kikuchi (1960), while dealing with CBA mice mammary tumour, inoculated 0.2 cm^3 into the normal animal. The remaining procedure for cytological study has been described before.

(d) *Ascites developed from solid neoplasms* (Goldie and Felix, 1951)—This technique is very important in the study of cancer cytology, because solid tissues, which are otherwise difficult to observe, can conveniently be converted into free tumour cells suspended in the body fluid. The steps are:

1. Intraperitoneal inoculation of sarcoma or lymphoma cells in mice is performed.

2. Week-old subcutaneous implants of the subsequent tumour are taken and ground in a tissue press and a suspension in 0.85 per cent NaCl solution is prepared.

3. The number of viable tumour cells in the mashed suspension is first determined by separating the tissue fragments and cell clumps from the suspension by sedimentation and decantation; a suspension sample is then placed in a haemocytometer previously coated with 0.02 per cent solution of neutral red in ethanol, and cells coloured with neutral red are considered dead. This step is necessary to find out the amount of dilution required to obtain the requisite number of viable tumour cells per 0.1 mm^3. Mitotic activity can be counted with acetic-orcein staining.

4. The suspension, after proper dilution of $1:10$ or $1:20$, is then inoculated into another mouse intraperitoneally.

5. Growth characteristics, cell division, etc. can be studied from the peritoneal exudate after a few days.

Schedule for somatic fusion between malignant cell lines derived from different mammalian species (Harris and Watkins, 1965).

(a) Materials: Two cell types were obtained in quantity, as suspensions of single cells, namely HeLa cells from suspension cultures and Ehrlich ascites tumour cells from the peritoneal cavity of Swiss mice. The virus selected was a strain of Sendai virus supplied by Dr. H. G. Pareira of the National Institute for Medical Research, Mill Hill. This member of the para-influenza I group of myxoviruses was chosen since one strain of these viruses (HVJ) was found to induce rapid fusion in suspensions of Ehrlich ascites cells *in vitro*.

(b) Propagation of the virus: Dilute infected allantoic fluid (8000 haemagglutinating units/cm^3) with phosphate-buffered saline to 1 in 10^4 dilution. Inject 0.1 cm^3 of the diluted fluid into the allantoic cavity of 10–11 day old fertile hens' eggs and incubate them at 37 °C for 3 days. Transfer to 4 °C, keep for 12 h and draw out the allantoic fluid. Centrifuge the accumulated fluid at 400 g for 10 min and count the titre of haemagglutination in the supernatant. Centrifuge the supernatant at 30 000 g for 30 min. Remove the supernatant and re-suspend the cells at the bottom in one-tenth of the original volume in Hank's solution without glucose and store in 1 cm³ lots at −70 °C. Determine the haemagglutination titre. For the experiments, use the stored solution, diluted in Hanks's medium.

(c) Determination of haemagglutination titre: is carried out in Salk-pattern haemagglutination trays. Prepare doubling dilutions of the virus in 0·5 cm³ of phosphate-buffered saline. Add to each cup 0·05 cm³ of this solution containing about $2·5 \times 10^7$ guinea-pig erythrocytes in suspension. One haemagglutination unit (HUA) is the smallest amount of virus which produces complete haemagglutination after 2 h at room temperature.

(d) Inactivation of virus: Expose 1 cm³ of the concentrated Sendai virus suspension for 3 min (in a watchglass) to ultraviolet light on the surface of the membrane (from a Philips 15 W 18 in germicidal tube, with an intensity of 3000 ergs cm⁻³ s⁻¹). Mix the suspension by pipetting after each minute. To test the infectivity of the irradiated virus, incubate pieces of chorio-allantoic membrane with the virus in TC 199 medium in a tray. After 3 min of u.v.-treatment, a drastic reduction was seen in the ability of the virus to multiply in the membrane, but its capacity to induce cell fusion *in vitro* remains intact.

(e) Cell fusion technique following Okada (1962):

1. Centrifuge and separate out HeLa cells from a suspension culture. Re-suspend in Hanks's solution to a concentration of 2×10^7 cells/cm³. Draw out Ehrlich ascites cells from the peritoneal cavity, wash by centrifugation in Hanks's solution and re-suspend in it at a concentration of 2×10^7 cells/cm³.

2. Draw out 0·5 cm³ of each cell suspension with a pipette and transfer to a chilled inverted T-tube. Add 1 cm³ of the virus, and dilute, if required, with Hanks's solution. The cells clump together and the size of the clumps depends on the quantity of the virus added. Keep the tube with the mixture at 4 °C for 15 min and then shake in a water bath at 37 °C for 20 min at the rate of 100 excursions/min. The cells in the clumps undergo various degrees of fusion during this period.

3. Transfer 1 cm³ of the cell suspension and 5 cm³ of the culture medium, with the help of a pipette, into a 6 cm diameter Petri dish containing 15 cover slips, each 1 cm in diameter. The culture medium contains: 20 per cent calf serum and 1 per cent tryptose, both in TC 199 with 100 units/cm³ penicillin and 100 µg/cm³ streptomycin. Incubate the Petri dishes at 37 °C in a gas mixture of 5 per cent CO_2 in air. Transfer the cover slips to fresh medium after 1 day and again after 4 days. Multinucleate cells are observed adhering to the cover slips 4 h after introducing the suspension into the Petri dishes. Within 24 h, most of them flatten out on the glass. Cell counts show that about 10 per cent of the original single cells in suspension adhere as multinucleate cells to the cover slip. Each multinucleate cell contains about 2 to 20 nuclei of two morphological types.

Modified Schedule for Somatic Fusion Between Malignant and Non-Malignant Cell Lines (Engel *et al.*, 1969)

This method, basically the same as that of Harris and Watkins (1965), was used to secure cell hybridisation between two cell lines, one malignant and the other non-malignant. The materials used were A9 and B82 mutants of Earle's L line of mouse. The first mutant cells were deficient in inosinic acid phosphorylase while the latter lacked thymidine kinase. When a medium containing aminopterin was used, neither of these lines could grow because of blocking of the endogenous biosynthesis of purines, and of thymidylic acid. This principle was followed in developing a medium containing 4×10^{-7} M aminopterin, 3×10^{-6} M glycine, $1 \cdot 6 \times 10^{-5}$ M thymidine and 1×10^{-4} M hypoxanthine (Littlefield, 1964). The parent cell lines were unable to grow in this medium but the hybrid line, derived by cell fusion, could grow in it. The cells of the two mutant lines were fused together by exposure to Sendai virus following the technique described before (Harris and Watkins, 1965) and the hybrid strain was isolated by growing the mixed population in the medium here described. In the experiments by Engel *et al.* (1969), two sub-populations were raised, one (A) from cells fused together by 6000 HAU of inactivated virus and grown in the selective medium as a mixed culture, and another (B) derived from a single colony of hybrid cells isolated 25 days after the primary fusion, using 4000 HAU of the inactivated virus.

In non-malignant man/mouse somatic hybrid cells, where chromosomes are eliminated under normal conditions, Miller *et al.* (1971) recorded specific retention of human chromosome No. 17 in the hybrid cell if grown in HAT selection medium with thymidine kinase.

Successful enzyme study and gene localisation from somatic cell hybrids have also been carried out by different authors (Siniscaleo, 1970; Siniscaleo *et al.*, 1969; *see* Meera Khan, 1971).

A List of the More Common Tumours Induced Artificially in Laboratory Mammals

Hirosaki sarcoma—originated spontaneously in a Japanese stock rat (Usubuchi *et al.*, 1951, 1955).

MTK-sarcoma I and MTK-sarcoma II—first produced in male and female Wistar rats respectively by Tanaka and Kano (1951) through the action of *o*-aminoazotoluene and *p*-dimethylaminoazobenzene.

MTK-sarcoma III—produced in a male Wistar rat by Tanaka (1952, unpublished, referred by Makino 1957), through the action of *p*-dimethyl-aminoazobenzene.

MTK-sarcoma IV—formed by the action of *o*-aminoazotoluene and *p*-dimethylaminoazobenzene on a Wistar rat (Tanaka, 1954 unpublished, referred by Makino, 1957).

Takeda sarcoma—originated spontaneously in a stock rat (Takeda *et al.*, 1952).

Usubuchi sarcoma—first produced in a stock rat by administering methyl-cholanthrene (Usubuchi *et al.*, 1953).

Watanabe ascites hepatoma—first induced in a stock rat by Watanabe and Matsunaga (1954) by repeatedly injecting water at 72 °C into the peritoneal cavity.

Yoshida sarcoma—produced by Yoshida, Muta and Sasaki (1944) in a stock rat through injection of *o*-aminoazotoluene and by painting the skin with potassium arsenite solution.

Ehrlich tumours—originally derived from a mouse mammary carcinoma (Hauschka, 1953).

SEWA—derived initially by subcutaneous injection of polyoma virus into a newborn A-SW mouse (Sjögren, 1964).

MSWBS—ascites sarcoma formed originally through methylcholan-threne into a (ASW × AF$_1$) hybrid mouse (Klein and Klein, 1958).

REFERENCES

Abercrombie, M. and Heaysman, J. E. M. (1957).   *Exp. Cell Res.* **13**, 276
Abercrombie, M. and Ambrose, E. J. (1958).   *Exp. Cell Res.* **15**, 332
Ambrose, E. J., Andrews, R. D., Easty, D. M., Field, E. O. and Wylie, J. A. H. (1962).   *Lancet* **i**, 24
Ambrose, E. J., Easty, D. M. and Wylie, J. A. H. (1967).   *The Cancer cell in vitro.* London; Butterworths
Aronson, M. and Kessel, R. W. I. (1960).   *Science* **131**, 1376
Atkin, N. B. (1964).   *Brit. J. Radiol.* **37**, 213
Bannasch, P. (1968).   *Recent results in cancer research, The cytoplasm of hepatocytes during carcinogenesis.* Berlin; Springer
Barski, G. and Cornefert, F. (1962).   *J. Nat. Cancer Inst.* **28**, 801
Basrur, P. K., Basrur, V. R. and Gilman, J. P. W. (1963).   *Exp. Cell Res.* **30**, 229
Bauer, K. H. (1949).   *Das Krebsproblem*, Berlin; Springer
Berwald, Y. and Sachs, L. (1963).   *Nature* **200**, 1182
Björklund, B., Björklund, V. and Paulsson, J. E. (1961).   *Proc. Soc. exp. Biol. Med.* **108**, 385
Blum, H. F. (1950).   *J. Nat. Cancer Inst.* **11**, 463
Boyse, E. A. (1960).   *Transplant. Bull.* **7**, 100
Dalton, A. J. and Haguenau, F. (1962).   *Tumors induced by viruses: ultrastructural studies.* New York; Academic Press
De Bruyn, W. M. (1956).   *6e Jaarboek van Kankeronderzoek in Kankerbestrijding in Nederland*, 50
De Bruyn, W. M. and Hampe, J. F. (1961).   *11e Jaarboek van Kankeronderzoek in Kankerbestrijding in Nederland* 107
Defendi, V., Ephrussi, B. and Koprowski, H. (1964).   *Nature* **203**, 495
Defendi, V., Ephrussi, B., Koprowski, H. and Yoshida, M. C. (1967).   *Proc. Nat. Acad. Sci. US* **57**, 209
Di Paolo, J. A. (1964).   *Cancer* **17**, 391
Di Paolo, J. A. and Dowd, J. E. (1961).   *J. Nat. Cancer Inst.* **27**, 807
Druckrey, H., Schmähl, D. and Müller, M. (1961).   *Naturwiss.* **49**, 217
Dulbecco, R. (1969).   *Science* **166**, 962
Easty, D. M. (1967).   In *The cancer cell in vitro.* London; Butterworths
Easty, D. M. and Wylie, J. A. H. (1963).   *Brit. med. J.* **1**, 1589
Easty, G. C., Yarnell, M. and Andrews, R. D. (1964).   *Brit. J. Cancer* **18**, 354
Engel, E., McGee, B. J. and Harris, H. (1969).   *Nature* **223**, 152
Ephrussi, B., Davidson, R. L. and Weiss, M. C. (1969).   *Nature* **224**, 1315
Evans, V. J., Bryant, J. C., Fioramonti, M. C., McQuilki, W. T., Sanford, K. K. and Earle, W. R. (1956).   *Cancer Res.* **16**, 77
Foley, G. E. and Drolet, B. P. (1964).   *Cancer Res.* **24**, 1461
Foley, J. F., Kennedy, B. J. and Ross, J. D. (1963).   *Cancer Res.* **23**, 368
Gerber, P. (1966).   *Virology* **28**, 501
Gey, G. O., Bang, F. B. and Gey, M. K. (1954).   *Tex. Rep. Biol. Med.* **12**, 805

Gillespie, D. and Spiegelman, S. (1965). *J. Mol. Biol.* **12,** 829
Goldblatt, H. and Cameron, G. (1953). *J. exp. Med.* **97,** 525
Goldie, H. and Felix, M. D. (1951). *Cancer Res.* **11,** 73
Hakomori, S., Teather, C. and Andrews, H. (1968). *Biochem. Biophys. Res. Commun.* **33,** 563
Hamazaki, Y., Hamazaki, H., Ogawa, K., Mukakami, I., Nakatsuka, H., Ariki, I., Omori, I., Sato, H., Miyake, K., Onishi, N., Kajiyama, Y. and Hayashi, D. (1953). *Gann* **44,** 290
Hansen-Melander, E. (1958). *Hereditas* **44,** 471
Hansen-Melander, E., Kullander, S. and Melander, Y. (1956). *J. Nat. Cancer Inst.* **16,** 1067
Harris, H. and Watkins, J. F. (1965). *Nature* **205,** 640
Harris, H., Miller, O. J., Klein, G., Worst, P. and Tachibana, T. (1969). *Nature* **223,** 363
Hauschka, T. S. (1953). *Trans. N.Y. Acad. Sci.* **16,** 64
Hauschka, T. S. and Levan, A. (1958). *J. Nat. Cancer Inst.* **21,** 77
Holley, R.W. and Kierman, J. A. (1968). *Proc. Nat. Acad. Sci. U.S.* **60,** 300
Hopps, H. E., Bernheim, B. C., Nisalak, A., Tjio, J. H. and Smadel, J. E. (1963). *J. Immunol.* **91,** 416
Hsu, T. C. (1965). In *Cells and tissues in culture* **1,** 397, New York; Academic Press
Heuper, W. C. and Conway, W. D. (1964). *Chemical carcinogenesis and cancers.* Springfield, Illinois; Thomas
Ishihara, T., Kikuchi, Y. and Sandberg, A. A. (1963). *J. Nat. Cancer Inst.* **30,** 1303
Jacob, S. T. and Bhargava, P. M. (1962). *Exp. Cell Res.* **27,** 453
Kikuchi, Y. (1960). *J. Fac. Sci., Hokkaido Univ. VI Zool.* **14,** 463
Klein, G. (1966). *Viruses inducing cancer, implications for therapy,* p. 323, Salt Lake City; Univ. of Utah Press
Klein, G. and Klein, E. (1958). *J. Cell Comp. Physiol.* **52,** 125
Koller, P. C. (1956). *Ann. N.Y. Acad. Aci.* **63,** 793
Koller, P. C. (1960). In *Cell physiology of neoplasia 14th A. Sym. fund. Cancer Res.* Texas; University of Texas Press
Koller, P. C. (1963). *Ciba symposium* **11,** 54
Koprowski, H., Jensen, F. C. and Steplewski, Z. (1967). *Proc. Nat. Acad. Sci. US* **58,** 127
Lampert, F. (1971). In *Advances in cell and molecular biology* **1,** New York: Academic Press
Leighton, J. (1951). *J. Nat. Cancer Inst.* **12,** 545
Leighton, J. (1958). *Lab. Invest.* **7,** 513
Leighton, J., Kline, I. and Orr, H. C. (1956). *Science* **123,** 502
Levan, A. and Hauschka, T. S. (1952). *Hereditas* **38,** 251
Levan, A. and Hsu, T. C. (1960). *Hereditas* **46,** 231
Littlefield, J. W. (1964). *Nature* **203,** 1142 and Science **145,** 709
Lubs, H. A. and Clark, R. (1963). *New Eng. J. Med.* **208,** 907
Lumsden, C. E. (1963). In *Pathology of tumours of the nervous system,* p. 281, 2nd ed. London; Arnold
Macpherson, I. and Stoker, M. G. P. (1962). *Virology* **16,** 147.
Macpherson, I. and Montagnier, L. (1964). *Virology* **23,** 291
Madden, R. E. and Burk, D. (1961). *J. Nat. Cancer Inst.* **27,** 841
Magee, W. E., Sheek, M. R. and Sagik, B. P. (1958). *Proc. Soc. exp. Biol. Med.* **99,** 390
Makino, S. (1957). *Int. Rev. Cytol.* **1,** 25
Makino, S. and Nakahara (1953).
Makino, S., Ishihara, T. and Tonomura, A. (1959). *Z. Krebsforsch.* **63,** 184
Makino, S. and Kano, K. (1955). *J. Nat. Cancer Inst.* **15,** 1165
Makino, S. and Tanaka, T. (1953). *J. Nat. Cancer Inst.* **13,** 1185
Makino, S. and Tonomura, A. (1955). *Z. Krebsforsch.* **60,** 597
Manna, G. K. (1954). *Nature* **173,** 271
Manna, G. K. (1957). *Proc. Zool. Soc. Calcutta, Mookerjee vol.,* 95
Martin, G. M. (1964). *Proc. Soc. epx. Biol. Med.* **116,** 167
McCoy, T. A., Maxwell, M. and Cruse, P. F. (1959). *Proc. Soc. exp. Biol. N.Y.* **100,** 115
Meera Khan, P. (1971). *Doctoral Thesis,* University of Leiden
Miles, C. P. (1967a). *Cancer* **20,** 1253
Miles, C. P. (1967b). *Cancer* **20,** 1274
Miller, O. J., Allderdice, P. W., Miller, D. A., Breg, W. R. and Migeon, B. R. (1971). *Abstr. IV Int. Congr. Hum. Genet.* Paris, 124
Miles, C. P. (1967a). *Cancer* **20,** 1253
Miles, C. P. (1967b). *Cancer* **20,** 1274
Moore, C. E. and Koike, A. (1964). *Cancer* **117,** 11

Morgan, J. F. and Griffiths,   . (1963).   Referred in Ambrose *et al*, 1967
Morgan, J. F., Morton, H. J. and Parker, R. C. (1950).   *Proc. Soc. Exp. Biol. Med.* **73,** 1
Nabholz, M., Miggiano, V. and Bodmer, W. (1969).   *Nature* **223,** 358
Nandi, S. (1969).   *Proc. Int. Seminar on chromosomes, Nucleus* Suppl. 1968, p. 220, Calcutta
Nowell, P. C. and Hungerford, D. A. (1960).   *Science* **132,** 1497
Okada, Y. (1962).   *Exp. Cell Res.* **26,** 98
Oksala, T. (1956).   *Hereditas* **42,** 161
Pace, D. M., Thompson, J. R. and Van Camp, W. A. (1962).   *J. Nat. Cancer Inst.* **28,** 897
Paul, J. (1959).   *Cell and tissue culture.* Edinburgh; Livingstone
Prop, F. J. A. (1961).   *Path. Biol. Paris* **9,** 640
Puck, T. T., Marcus, P. I. and Cieciura, S. J. (1956).   *J. exp. Med.* **103,** 273
Pulvertaft, R. J. V. (1961).   *Rep. Br. Emp. Cancer Campn.* **39,** 312
Raven, R. W. and Roe, F. J. C. (1967).   eds. *The Prevention of cancer.* London; Butterworths
Richart, R. M. and Wilbanks, G. D. (1966).   *Cancer Res.* **26,** 60
Roe, F. J. C. and Lancaster, M. C. (1964).   *Br. med. Bull.* **20,** 127
Sandberg, A. A. and Yamada, K. (1966).   *Cancer* **19,** 1869
Sasaki, Z. and Hishida, Y. (1958).   *Cytologia* **23,** 218
Scaletta, L. J. and Ephrussi, B. (1965).   *Nature* **205,** 1169
Seshachar, B. P. and Nambiar, P. (1955).   *Nature* **176,** 796
Sharma, A. (1959).   *Nature* **184,** 1083
Sharma, G. P., Mittal, O. P. and Sharma, S. D. (1969).   *Proc. Int. Seminar on Chromosome, Nucleus* suppl. p. 268, Calcutta
Sharma, A. K. and Nandi, S. (1970).   Unpublished data.
Silagi, S. (1967).   *Cancer Res.* **27,** 1953
Siniscaleo, M. (1970).   *Proc. III Int. Conf.,* The Hague. ed. Fraser, F. C. and McKusick, V. A., *Exc. Med.* Amsterdam
Siniscaleo, M., Klinger, H. P. and Eagle, H. (1969).   *Proc. Nat. Acad. Sci., U.S.* **62,** 793
Sjögren, H. O. (1964).   *J. Nat. Cancer Inst.* **32,** 361 and 645
Snell, G. D. (1963).   In *Conceptual advances in immunology and oncology,* p. 323, New York; Harper
Springs, A. I. (1964).   *Brit. J. Radiol.* **37,** 210
Stich, H. F. (1963).   *Canad. Cancer Conf.* **5,** 99
Stoker, M. (1968).   *Nature* **218,** 234
Talukder, G. and Sharma, A. K. (1968).   *Ind. J. Exp. Biol.* **6,** 67
Takayama, S. and Makino, S. (1961).   *Z. Krebsforsch.* **64,** 253
Takeda, K., Aizawa, H., Immamura, T., Sasage, S., Matsumoto, K. and Kanchira, S. (1952).   *Gann* **43,** 132
Tanaka, T. and Kano, K. (1951).   *J. Fac. Sci. Hokkaido Univ. VI Zool.* **10,** 289
Tanaka, T., Tonomura, A., Okada, T. A. and Umetani, M. (1955).   *Gann* **46,** 15
Taylor, A. C. (1962).   *J. Cell Biol.* **15,** 201
Therman, E. and Timonen, S. (1950).   *Hereditas* **36,** 393
Todaro, G. J. and Green, H. (1963).   *J. Cell Biol.* **17,** 299
Todaro, G. J., Lazar, G. K. and Green, H. (1965).   *J. Cell Comp. Physiol.* **66,** 325
Tournier, P., Cassingena, R., Wicker, R., Coppey, J. and Suarez, H. (1967).   *Int. J. Cancer* **2,** 117
Trowell, O. A. (1963).   *Expt. Cell Res.* **29,** 220
Turpin, R. and Lejeune, J. (1969).   *Human afflictions and chromosomal aberrations.* Paris; Pergamon Press
Usubuchi, I., Oboshi, S., Iida, T. and Koseki, T. (1951).   *Trans. Soc. Pathol. Japan* **40,** 126
Usubuchi, I., Iida, T., Abe, H., Koseki, T. and Kosugi, S. (1953).   *Gann* **44,** 128
Usubuchi, I., Koseki, T., Terajima, T., Haga, T. and Takeda, T. (1955).   *Gann* **46,** 183
Wakonig-Vaartaja, R. (1963).   *Aust. New Zeal. J. Obstet. Gynae.* **3,** 170
Wakonig-Vaartaja, R. and Kirkland, J. A. (1965).   *Cancer* **18,** 1101
Walker, D. G. and Wright, J. C. (1961).   *Cancer Chemother. Reps.* **i14,** 139
Watanabe, F. and Matsunaga, T. (1954).   *Gann* **45,** 443
Watkins, J. F. and Dulbecco, R. (1967).   *Proc. Nat. Acad. Sci. U.S.* **58,** 1396
Waymouth, C. (1956).   *J. Nat. Cancer Inst.* **17,** 315
Westphal, H. and Dulbecco, R. (1968).   *Proc. Nat. Acad. Sci. U.S.* **59,** 1158
Yoshida, T., Muta, Y. and Sasaki, Z. (1944).   *Proc. imp. Acad. Japan* **20,** 611
Yunis, J. J. (1965).   ed. *Human chromosome methodology.* New York; Academic Press

17 Effect of Physical and Chemical Agents on Chromosomes

The discovery of x-ray-induced mutation in *Drosophila* by Müller in 1927, followed by Stadler in maize in 1928, provided the necessary impetus for research on the effects of outside agencies on chromosomes. The result of this enthusiasm led to the discovery of the mutagenic property of chemicals, by Oehlkers (1943) and Auerbach and Robson (1946), followed by a host of others, and in the meantime, the polyploidising action of colchicine and its effective application on plants had been disclosed through the works of Blakeslee and Avery (1937) and their collaborators. The need for refinements in methods for the intensive investigation of the effects of these and other physical and chemical agents on chromosomes, both from utilitarian and fundamental standpoints, had been realised, and new avenues of research opened up.

In spite of the fact that the initiation of this line of investigation dates back to quite an early period, the standardisation of a method for a systematic attempt to explore the properties of different chemical agents was first made by Levan (1949) and his collaborators. The technique applied by them is known as the Allium test, in which the experimental materials consist of bulbs of *Allium cepa*, the common onion. The advantage of using onion as the experimental material lies in the fact that it is: (*a*) inexpensive, (*b*) easy to handle and (*c*) it yields a fresh crop of roots in tap water, thus providing meristems for study, every two or three days. The chemicals to be tested are prepared in solution and kept in wide-mouthed jars. A bulb with roots intact is then placed over the mouth of the jar so that its roots dip in the solution—the jars may be covered with black paper to allow healthy growth of the roots. After treatment for a desired period, the roots can be excised, fixed, stained and observed directly or kept for recovery in water or nutrient solution under similar conditions before fixation, staining and observation.

The above method is applicable only to root meristems, however, and for meiotic cells or pollen grains the entire inflorescence is generally treated by dipping the stalk in water. But the mode of treatment with these materials may vary and, if necessary, the anthers may be dipped in the fluid directly. For animal materials, the usual technique of application is through feeding or injection, and for post-treatment cultures, the organ can be dissected out or cultured in artificial medium, or the animals can be reared in cages with natural feeding. In tissue cultures, colchicine is generally added to the medium (for details, please see chapters on tissue culture and mammalian chromosomes).

For physical agents such as x-rays, ultra-violet rays, etc., both plant and animal materials are placed in front of the source and the required dosage is applied, the subsequent process of culture being the same as that with chemical agents. The exact purposes for which physical and chemical agents are applied, and an outline of the methods of their application, are given below:

CHEMICAL AGENTS

METAPHASE ARREST

In the majority of plant and animal materials, the nuclear division within the different cells is not synchronous, with the result that the meristematic or the dividing zone represents a heterogeneous mass of cells, in which the nuclei are at different stages of division. The difficulty of such a medium is twofold: (*a*) metaphase stages—the best nuclear phase for chromosome analysis—cannot be obtained in high frequency, and (*b*) with regard to the analysis of an effect, the exact stage affected cannot be ascertained. A large number of cells, showing induced synchrony in division, is therefore necessary. This aspect of induced synchrony has been dealt with in detail by Zeuthen (1964) and Padilla and Cameron (1968).

The most suitable chemical for the purpose of securing a large number of metaphase plates is colchicine as it causes metaphase arrest by inhibiting the operation of the spindle mechanism, the principle underlying spindle arrest being outlined in the chapter on pre-treatment (Chapter 2). To secure metaphase arrest, the organs can be treated directly with aqueous colchicine solution for a required period, being added to the medium in tissue culture or injected into tissue or added to the food, which has been described in the chapter on mammalian chromosomes. Its property of arresting the metaphase has made it an essential pre-requisite for procedures on cell synchronisation. Before harvesting the cells in leucocyte cultures, the addition of colchicine or colcemide is essential for securing a large number of metaphase plates. The concentration needed to secure this effect may vary from a very dilute concentration, 0·01 per cent, to even 2 per cent, and the period of treatment from 10 min to 16 h in some amoeba materials. The mitotic stage, at which colchicine is effective at very low concentrations, is late prophase (Eigsti and Dustin, 1957; also Deysson, 1968; Wagner, 1969). In *Allium cepa*, spindle in root tips can be arrested by just $1\frac{1}{2}$ h treatment in 0·2 per cent colchicine solution, but before fixation and observation, a thorough washing in water for at least 15–20 min is necessary to remove any superficial deposits of this alkaloid, which may hamper staining. In leucocyte cultures, application of colchicine 10–20 h before harvesting, is desirable. The characteristic appearance of metaphase stages, showing clear euchromatid segments, is otherwise known as colchicine-mitosis or c-mitosis.

In addition to colchicine, a number of other compounds such as gammexane, chloral hydrate, acenaphthene, actidione, etc., are all applied for metaphase arrest. Their mode of action and effect on viscosity have also been worked out in detail (Sharma and Chaudhuri, 1961; Sharma and Bhattacharyya, U. C., 1962; Sharma and Sarkar, A. K., 1963; Sharma and Talukdar, 1965; Sharma and Ghosh, S., 1969). The efficacy of colchicine is markedly superior to that of the others. The effective concentrations of these chemicals are given in Table 1 (page 26).

POLYPLOIDY

The importance of polyploidy in agricultural and horticultural practices is well known. The increase in gene dosage resulting from multiplication of chromosome sets brings about giganticism in all characters in general. Enhancement of tolerance and adaptability are also added characteristics of polyploids. Moreover several interspecific and intergeneric crosses have been made fertile through polyploidy. All these facts taken together have made the induction of polyploidy an effective tool in the hands of agriculturists and horticulturists.

The properties of arresting metaphase and the induction of polyploidy are inter-related. Polyploidising chemicals like colchicine inhibit the formation of the spindle with the two poles and confine the chromosomes within one nucleus though their division remains unhampered. Levan (1949) has classified colchicine action under narcosis, as the narcotic action allows the tissue to recover as soon as the influence of the chemical is removed. Evidently polyploid cells, which are formed by colchicine action, divide normally and give rise to polyploid shoots.

As an agent for inducing polyploidy, colchicine is more active in plants than in animals, though reports of animal polyploids, induced through colchicine, are available. Colchicine is applied in the following ways:

On seeds and young seedlings—The seeds may be immersed for 2–48 h in concentrations of colchicine solution varying from 0·02–0·1 per cent before sowing. Just-germinating seeds can be treated with similar concentrations of colchicine solution for 12–48 h with the plumules dipped in the solution, or the entire germinating seedling can be immersed completely in colchicine solution, but the former method is preferable as the root system remains unaffected, allowing the plant to grow normally.

On mature seedlings—(a) Colchicine is added in the form of soaked cotton plugs on the growing shoot, the period of treatment varying from 2–4 h, and the range of concentrations used being the same as given in the previous schedule. Cotton plugs, placed over the growing tip, should, however, be moistened at regular intervals by adding drops of colchicine solution with a brush, and after the treatment, the plug should be removed and the tip washed by brushing with water. The same method can be followed for treating young inflorescences.

(b) In the form of a paste mixed with lanolin or with glycerin. This method has been found to be effective where the growing point lies within the plumules, as in monocotyledonous plants.

On pollen grains and animal tissues in culture—Colchicine may be added in the agar medium meant for pollen tube growth (2 cm^3 of 0·2 per cent colchicine in 8 cm^3 of agar medium or in culture medium for other tissues, as described in the chapter on mammalian chromosomes). Other chemicals such as chloral hydrate, gammexane and acenaphthene, which cause metaphase arrest, can also be applied for the induction of polyploidy. In the authors' laboratory, 0·5 per cent caffeine solution in water has been found to be very effective, especially in leguminous plants, but in the present state of knowledge, for commercial purposes it is always safe to rely on colchicine.

After the induction of polyploidy, especially in seeds and seedlings, there are different methods by which success or failure of the experiment can be detected before the plant reaches maturity and starts blooming. These are :

(a) The chromosome number of young shoots, leaf tips and root tips can be counted, following acetic-orcein or Feulgen squash (for technique, *see* Chapter 7), but it is preferable not to rely too much on the chromosome counts from root tips, as polyploid cells, though present in roots, may be eliminated from the shoot apex.

(b) In case of scarcity of materials, where only one or two samples from their external appearance appear to have been polyploidised, it is desirable not to remove the shoot apex or injure roots for chromosome counting but to study the anatomical characteristics from a portion of the mature leaf to obtain an indication of the success of the experiment. In general, it has been found that polyploidy is associated with an increase in stomatal size and decrease in stomatal frequency per unit area of the leaf, so the lower epidermis of the mature leaf of the polyploid can be peeled off and mounted in 50 per cent glycerin solution. Stomatal size and frequency per unit area can be noted and the result can then be compared with that of diploids obtained following a similar procedure adopted for a control diploid plant. Post-treatment with x-rays of colchicine-treated plants results in better survival of polyploids as x-rays are more effective against diploid cells, causing them to be eliminated in selection.

CHROMOSOME FRAGMENTATION AND OTHER EFFECTS

The capacity of inducing chromosome breakage is a property of several chemical agents. The study of chromosome breakage is beset with immense possibilities. Firstly, fragmentation followed by translocation of some fragments may bring about a new patterning of chromosome segments resulting in heritable phenotypic difference. Its importance in the evolution of new species and desirable varieties is obvious. Secondly, the chromosome breaking property of chemicals has an important bearing on the chemo-therapy of cancer. Incidentally it may be mentioned that the biological basis of the radiation treatment of cancer is the induction of extensive fragmentation of chromosomes which ultimately leads to the cessation of nuclear division. As some chemical agents, which are otherwise known as radiomimetic chemicals (simulating radiation effects), are endowed with the same property, their application in cancer therapy needs no elucidation. The bactericidal effect of u.v. rays is also well known.

Lastly, the study of chromosome fragmentation by chemicals has a special significance in bringing out the differential nature of chromosome segments. Several chemical agents such as, 8-ethoxy caffeine induce chromosome breaks at certain specific loci (Kihlman, 1966). This differential break can be taken as an index of the different chemical nature of susceptible segments from the rest of the chromosome parts (*see* Sharma and Sarkar, A. K., 1963; Sharma and Sharma, 1964).

The modes of action of different chemical agents causing chromosome breaks vary. Some of them affect sulphydryl groups of proteins whereas others act through their influence on hydrogen bonds of nucleic acids. Guanidine cross linkages are held to be involved with mustard compounds. Some agents may affect the oxidation–reduction system within the nucleus. The relationship between deoxyribonucleotide synthesis and chromosome breakage has been discussed in detail by Kihlman (1964, 1971; also *see* Taylor, 1963). Kaufmann, Gay and McDonald (1960) have suggested that localised chromosome breakage in constriction segments of *Vicia faba* by maleic hydrazide (McLeish, 1953) may be due to some specific reaction with RNA. Taylor *et al.* (1962) noted the effect of fluorodeoxyuridine on chromosome breakage and reunion. Maleic hydrazide (1,2, dihydro-3,6-pyridazinedione) may interfere with the RNA ratio to such an extent that breakage occurs in segments involved in its metabolism (*see* also Evans and Bigger, 1961; Evans, 1962). Sharma and Sharma (1960) reviewed the different theories in detail and suggested that the final upset of the nucleic acid metabolism ultimately results in hazards in protein re-duplication causing chromosomes to break at different loci. The effects of different chemical agents and their modes of action have been dealt with in detail by several workers (Sharma *et al.*, 1963; Bell and Wolf, 1964; Sharma and Chatterji, T., 1964; Lawley and Brookes, 1965; Michaelis, Schoneich and Rieger, 1965; Rao and Natarajan, 1965; Sharma and Ghosh, 1965, 1969; Adams *et al.*, 1966; Kihlman, 1966, 1971; Rasmussen and Painter, 1966; Turner *et al.*, 1966; Adams and Lindsay, 1967; Glass and Marquard, 1967; Ito *et al.*, 1967; Young *et al.*, 1967; Zimmerman and Schwaier, 1967; Kihlman and Hartley, 1968; Moutschen *et al.*, 1966, 1968; Wagner *et al.*, 1968; Drake, 1969). From our laboratory, Nandi (1969) studied the effects of several carcinogens on plant cells. Several plant products, such as, pigments (Sharma and Gupta, 1959; Sharma and Chaudhuri, R., 1963; Sharma and Sarkar, A. K., 1967); oils (Swaminathan and Natarajan, 1957) as well as alkaloids (*see* Deysson, 1968), which may even add to the soil mutagenicity, have been shown to influence chromosome breakage. Viral and bacterial infections have also been observed to affect chromosome breakage and other irregularities (*see* Halkka, 1967).

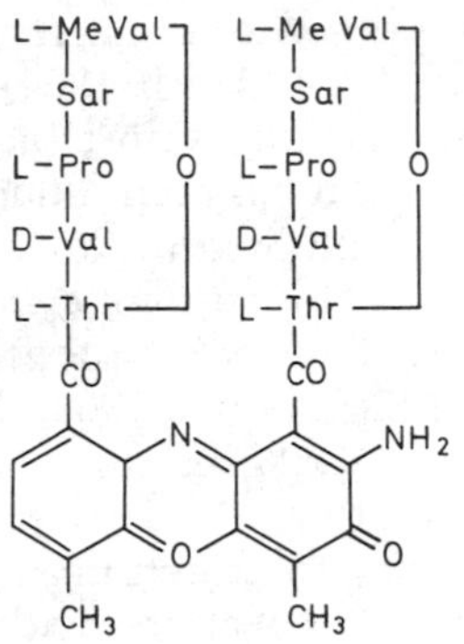

Streptomycin

Actinomycin D

L- Me Val: *N*-methyl-L-valine
Sar : Sarcosine
L-Pro: L-proline
D-Val: D-Valine
L -Thr: L -Threonine

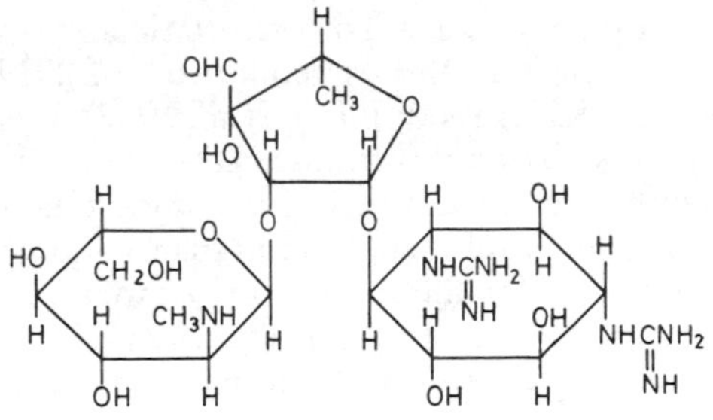

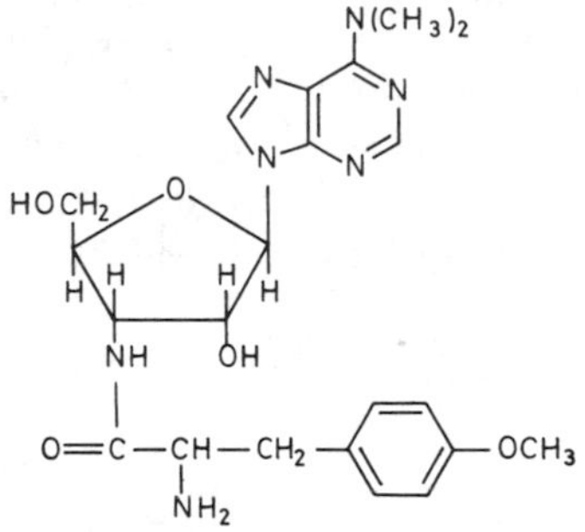

Puromycin

Lately, a number of antibiotics, with a profound influence on the replication of DNA, transcription and protein synthesis (*see* Collins, 1965), have been widely used for affecting the structure and behaviour of chromosomes. These compounds were originally tried out on bacteria and the knowledge gained thereby has frequently been applied on higher organisms, especially in cancer research. In view of their wide applicability, a short account of their properties is given here. Several other antibiotics have also been applied but their mode of action is not yet clear (Sharma and Bhattacharyya, G. N., 1967).

(a) *Mitomycin C*: Hata *et al.* (1956) studied its anti-tumour action. It inhibits DNA replication but transcription is not affected for a certain period (Smith–Kielland, 1964). It acts possibly by linking two polynucleotide strands of DNA (Iyer and Szybalski, 1963; Pricer and Weissbach, 1964). On activation, a mitomycin C molecule loses a methanol (Schwartz, 1962) and the ethylamine group acts as an alkylating agent, linking it with one of the DNA strands.

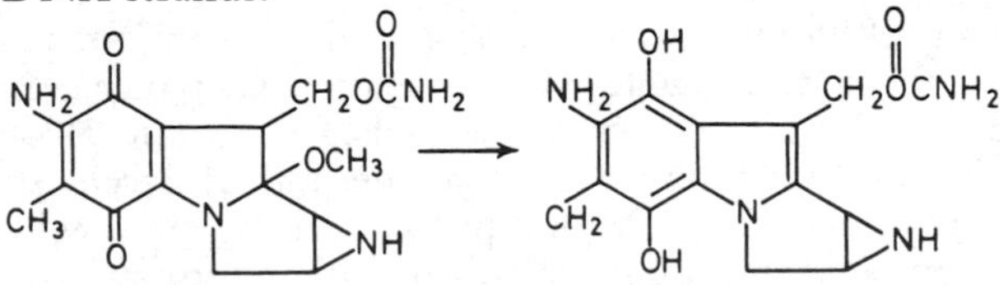

Mitomycin C   Activated form

(b) *Actinomycin D:* This cytostatic and anti-tumour compound checks transcription but allows replication of DNA to continue for a certain period (*see* Kirk, 1960; Reich *et al.*, 1962). It is supposed to form a reversible complex with double-stranded DNA (Haselkorn, 1964; Reich, 1964). Of all the antibiotics affecting chromosome behaviour, actinomycin D has been most widely employed, both in plants and animals. It has been found to be effective even for the suppression of meiosis (Jain and Singh, 1967).

(c) *Streptomycin:* Protein synthesis is checked by this antibiotic, possibly by hampering the transfer of amino acids from transfer RNA to the growing polypeptides (Mager *et al.*, 1962; *see* Kogut and Lightbown, 1964). Davies *et al.* (1964) indicated that cell lethality occurs because of gross misreading of the messenger RNA, resulting in non-functional proteins.

(d) *Puromycin:* This chemical, which is similar to streptomycin, also hampers protein synthesis and interferes in the transfer of amino acids to the growing polypeptide. It is considered to act as an analogue of amino-acyl t-RNA, due to the similarity between the two (Yarmolinsky and Haba, 1959 and *see* Darken, 1964). In fact, it has been detected attached to the released polypeptide in *E. coli* cells treated with puromycin (Nathans, 1964).

$$O_2N-\langle\!\bigcirc\!\rangle-CH-CH-CH_2$$

$$\begin{matrix} & & & NHC\cdot CHCl_2 \\ & & & | \\ O_2N-\bigcirc- & CH- & CH- & CH_2 \\ & | & & | \\ & OH- & - & - & -OH \end{matrix}$$

Chloramphenicol

(e) *Chloramphenicol:* The inhibition of protein synthesis by this compound is affected through action at the ribosomal level. It has been claimed that the binding may be at the same location as puromycin but it checks the binding of the messenger RNA and not that of the transfer RNA (Rendi and Ochoa, 1962; Gale, 1963; Brock, 1964; Jardeztky and Julian, 1964; Kucan and Lipman, 1964; Traut and Monro, 1964; Vazkuez, 1964).

Chemical agents that can cause chromosome breaks have been listed in Table 9. Two methods are outlined below, one for the demonstration of random breakage and the other for localised breakage.

SCHEDULE OF TREATMENT

For Random Breakage

(a) Place a healthy bulb of *Allium cepa*, with root tip intact, on top of a jar containing 0·005 mol/l solution of pyrogallol. Keep the jar in a temperature of 25–30 °C. (b) After 4 h, cut a few roots, fix in acetic–alcohol mixture (1 : 2) for 30 min, and follow the usual method of orcein squashing or Feulgen staining for root tips. Mount in 1 per cent acetic-orcein solution or 45 per cent acetic acid and count the number of random fragments in metaphase and anaphase stages.

(c) Continue the treatment of roots in pyrogallol solution up to 24 h and observe at regular intervals to study the increase or decrease in the frequency of fragments.

For Localised Breakage

(a) Place germinated seeds of *Vicia faba* on a sieve fixed over a jar containing 0·0075 mol/l. 8-ethoxycaffeine solution, in such a way that the roots passing through the sieve remain dipped in EOC solution. Continue the treatment for 6 h at 10 °C.

(b) Allow the roots to recover in tap water at 20 °C for 24 and 48 h.

(c) Cut the root tips and follow the same schedule for observation as in (b) for random breakage.

In the study of the effect of chemical agents on chromosomes, the technique employed for observation should be taken special care of, due to the fact that chromosome breaks may result during slightly prolonged heating with acetic-orcein–HCl mixture (Sharma and Roy, 1956), which is an essential step in the procedure for orcein squashing. Similarly, chromosome breaks have also been observed through mere water treatment by Sharma and Sen (1954*a*). The schedules for the demonstration of chromosome breakage by acetic-orcein and water are outlined below (*see* also chapter on stain).

Schedule for Demonstration of Orcein Breakage

(a) Treat excised young root tips of onion in 0·002 M soln. of 8-oxy-quinoline for 2 h at 16–18 °C.

(b) Fix in acetic–alcohol (1:2) for 30 min.

(c) Heat the root tips gently over a flame in a mixture of 2 per cent acetic-orcein and normal hydrochloric acid mixed in the proportion of 9:1 for 30 s.

(d) After a few minutes, mount and squash in 1 per cent acetic-orcein and observe.

Fragments can be observed in metaphase and anaphase stages, but if the heating is prolonged for a few seconds more, the frequency of fragments shows an increase. High frequency of fragments can also be obtained if acetic–alcohol fixation is omitted. In view of the above results, it is always desirable to keep a check on the period of heating, not exceeding 10 s in the procedure for orcein staining.

This schedule may be applied in animal tissues as well.

Schedule for the Demonstration of Chromosome Breakage Induced by Water Treatment

(a) Treat young healthy roots of *Crinum asiaticum* in tap water at 30 °C for 2 h.

(b) Follow the usual procedure of fixation and orcein staining. Fragments can be detected both in metaphase and anaphase.

In the study of the effect of chemical agents on chromosomes, a control experiment should be set up with water treatment alone in cases where water is used as the solvent of the chemical agent.

DIVISION IN DIFFERENTIATED NUCLEI

Division in adult nuclei, when otherwise the nuclei have ceased to undergo apparent division, can also be induced with the aid of chemical agents. The importance of this line of investigation was realised after the demonstration of the endopolyploid constitution of differentiated nuclei by Huskins (1947), Huskins and Steinitz (1948), and Geitler (1948). Until these works were published, the mechanism of differentiation which is principally operative in adult cells was supposed to be obscure. It is well known that gene action is controlled by specific enzymes, the synthesis of which is again dependent on genes. Manifestation of a trait is the ultimate result of a series of chemical reactions, set up initially at the gene level, and every step in this metabolic path requires the presence of enzymes. If differentiation, which involves the manifestation of diverse characters, is a process controlled by genes, then the latter, it was thought, must reduplicate in adult nuclei for the synthesis of specific enzymes. However, as the nuclei and, consequently, chromosomes of the adult cells do not show any apparent division, the extent to which genes influence the process of differentiation remained questionable. The above-mentioned authors implied that gene action and duplication are interrelated.

The above authors first claimed that the adult nuclei, though apparently non-dividing, undergo endomitotic re-duplication of chromonemata, thus exerting control over the process of differentiation: normally they lie in a polytenic condition. Huskins and Steinitz (1948), with the aid of a special technique using hormones, as outlined later, induced division in the adult nuclei, thus permitting the chromosomes to complete the nuclear cycle; during this division their polytenic and polyploid constitution was revealed. Afterwards other chemicals were worked out having the same property (D'Amato, 1952). Sharma and Sen (1954b) induced division in such cells by nucleic acid treatment, while Sharma and Mookerjea (1954) demonstrated that of the nucleic acid, the sugar constituent alone can induce division, and claimed that the polytenic condition of the adult nuclei is due to deficiency in nucleic acid. Later, other compounds were also shown to induce division (Sharma, and Bhattacharyya, B., 1956; Sharma and Datta, 1956). Even the effect of irradiation on adult nuclei has been studied (Sharma and Mukherjee, R. 1956). Torrey (1961) used a different method which involves the use of kinetin (6-furfuryl-amino-purine) for the induction of division in endomitotic plant cells. A considerable amount of work has been carried out in our own laboratory on the induction of division in adult nuclei. With the aid of 2,4-dichlorophenoxyacetate as well as indolylacetic acid, division could be induced even in the vascular zone

(Sen, 1970). Both diploid and polyploid nuclei have been recorded within the differentiated zone. The occurrence of diploid chromosome numbers in adult nuclei in the differentiated zone can easily be explained on the basis of the fact that transcription of RNA responsible for gene action does not depend on replication of DNA, and one strand of DNA is active in this respect. But the occurrence of polyteny may possibly suggest that there is a limit for continued transcription of a strand after which replication is necessary for the production of a fresh strand to be used in transcription.

The chromosomal control of differentiation has been well illustrated in the puffing pattern in salivary gland chromosomes of diptera at different phases of development. Excellent reviews on this aspect of differentiation have been published by different authors (Pavan, 1965; Schultz, 1965; Stebbins, 1965; Clever, 1965; Nescovic, 1968; Padilla and Cameron, 1968; Pavan and Da Cunha, 1969).

Schedule for the Induction of Division in Differentiated Nuclei

1. Take cuttings or seedlings of *Rhoeo discolor* and remove all the roots.
2. Allow them to grow in jars containing the following culture solution:

	mg/l		mg/l
$Ca(NO_3)_2.4H_2O$	95	Mn	0·5
NH_4NO_3	129	Cl	1·9
$MgSO_4.7H_2O$	180	B	0·5
KH_2PO_4	133·5	Cu	0·02
$K_2HPO_4.3H_2O$	7·35	Zn	0·01
$Fe_2(C_4H_4O_6).H_2O$	5·0		

Use distilled water and adjust the pH to 5·7. Fix the cuttings in such a way that a portion of the stem dips in water. Cover the jar with black paper and keep in a temperature of 22–24 °C. Keep the materials in the jar until freshly generated roots, 3–10 cm long, are obtained.

3. Transfer the materials in a similar jar containing 50–100 p.p.m. indole-3–acetic acid and keep for 24–72 h.

4. Re-transfer the materials in fresh culture as in 2 and keep for a maximum of 72 h.

5. Take out the materials after every 24 h. Remove about 4 mm from the root tip, which is meristematic, then cut 1–2 mm.

6. Follow the usual procedure for fixation in acetic–alcohol and orcein squash technique for root tips.

Polyploid cells can be seen in metaphase in the differentiated region.

Control preparations should be set up allowing the roots to grow in culture solution, in which no division will be found in the similar zone of the root tip.

The properties of different cytokinins have been dealt with by Srivastava (1967).

SOMATIC REDUCTION

The possibility of the occurrence of reductional separation of chromosomes (Metz, 1926; Hughes-Schrader, 1927; Berger, 1938; Christoff and
Christoff, 1948) and its experimental induction (Nemec, 1904) in tissues,
other than the gonadal ones, though pointed out by different authors, did
not receive serious consideration until Huskins (1947) demonstrated the
process on slides prepared by his associates and by Kodani. He claimed that
by treatment with sodium nucleate, reductional separation of chromosomes can be initiated in root-tip cells. The significance of this finding from
both the technical and theoretical points of view was immense because of
its direct bearing on the Precocity Theory of Darlington (1937), in which
synapsis and reductional separation of chromosomes are held to be initiated
because of the single-thread nature of chromosomes in early prophase of
meiosis.

The objection to the technique of Huskins that observations based on
squash preparation may show artificial separation of chromosomes has also
been overruled, as similar behaviour has been observed in sectioned
materials as well. Since the work of Huskins was published, reports of
somatic reduction in plants have emanated from different centres, including
the author's laboratory, where both induced and spontaneous occurrence
has been observed (Sharma and Bhattacharjee, 1953). Spontaneous occurrence has also been reported in sea urchin embryo by Lindahl (1953).
However, there is no distinct evidence yet available that somatic reduction
in plants, though occurring rarely, is associated also with interchange of segments. In *Haplopappus* species in culture, such pairing has been recorded
(Mitra and Steward, 1961).

Somatic reduction, through treatment with sodium salts of ribose nucleic
acid, evidently suggests that the balance of the nucleic acid within the cell is
at least one of the principal controlling factors of mitosis and meiosis. The
induction may therefore be considered as based on the principle that increase
in the proportion of RNA, by its direct application or by application of
certain factors which may enhance its synthesis, may cause somatic reduction. Huskins and Cheng (1950) induced reduction by low temperature
treatment, and though no method has yet been standardised for this, the
original method is now given, which may help in its further refinements.

Schedule of Treatment for the Induction of Somatic Reduction

1. Take a healthy young bulb of *Allium cepa* (onion), denude it of roots
and let it grow in a jar containing tap water until a fresh crop of roots
germinates.

2. When the roots are just 2–3 mm long, fit the bulb on top of a jar
containing 0·1–0·2 per cent sodium nucleate solution in water. Sodium
nucleate supplied by Schwartz Laboratories, Inc., New York (S.N.4509),
yields good results. Place the bulbs in such a way that the roots dip into the
sodium nucleate solution. Treat the bulbs for 6–12 h.

3. Cut the tip portion of the root (meristematic region) and follow the usual method of fixation and Feulgen or orcein squashing.

Reductional separation of chromosomes can be observed in a few per cent of the cells. Each chromosome unit with two chromatids intact moves as it is to either of the poles, so that the distribution of 16 chromosomes is equally divided between the two poles, i.e. 8 at each (Huskins and Cheng, 1950). For inducing somatic reduction through temperature treatments, onion bulbs may be kept growing in tap water at 5–6 °C for 5–64 days. At every 7 or 8 day interval, healthy young root tips may be fixed, stained and observed after the usual procedure. In all cases of reductional grouping in somatic cells, it is preferable to confirm the observation from materials sectioned from paraffin blocks.

PHYSICAL AGENTS

PRINCIPLE

X-rays are at present employed for effectively altering chromosome structure, both at microscopic and ultramicroscopic levels. In addition, other types of radiations are used, such as α-rays, β-rays, γ-rays, fast neutrons, ultraviolet and infrared rays. The biological action of these agents depends on their ionising and non-ionising properties, and as such they are classified under two categories, ionising and non-ionising rays, The term 'ionisation' implies the conversion of an atom into an ion. An atom consists, as is well known, of a positively charged nucleus surrounded by negatively charged electrons, the charges being so balanced as to maintain an electrically neutral state. Ionising radiations dissipate energy during their passage through matter, by ejecting electrons from the atom through which they pass. The ionised atom, losing the negative balance, becomes positively charged and is known as an ion. The result of ionisation is a chemical change of the molecule concerned. Whenever a binding electron between the two molecules is affected, serious after-effects ensue.

Non-ionising radiations, such as ultraviolet rays, infrared rays, etc., cause dissipation of energy within the tissue by molecular excitation, the principal biological action being attributed to the absorption of energy by particular cellular constituents, the most important one being nucleic acid. The recent works on the chemical basis of u.v. action have been discussed later.

EFFECT AND SCOPE

The production of gene mutation is no doubt one of the most important uses of the physical agents. In Darlington's words, 'One can compress a millennium of variation within a few seconds.' In addition to gene mutation, fragmentation of chromosomes is another outstanding effect of these agents, and as such breakage may involve both the chromatids of the chromosome or a single chromatid, depending on the stage in which the nucleus has received the radiation. When the chromosome is already split, either one or both chromatids may break, depending on the path of the rays (for details, *see* Lea, 1955), the breaks being either immediate or delayed (Davidson, 1958). The subsequent effects of chromosome breakage are translocation, deletion, inversion, rejoining as well as stickiness, pyknosis and polyploidy, and can all be studied in treated materials.

Radiation in the production of useful mutations in plants was reviewed elaborately by Gustafsson (1954), Singleton and colleagues (1956) and Smith (1958). A prominent case of translocation in barley, induced by x-rays, was demonstrated by Tjio and Hågberg (1951). The comparative effects of radiation and hybridisation in plant breeding have been assessed by Gregory (1956), Gaul (1967) and Swaminathan (1969).

Another important use of x-rays in agriculture and horticulture is the experimental induction of parthenogenesis. If a species is pollinated with irradiated pollen, haploid plants may result through parthenogenesis. The best examples are *Triticum, Nicotiana*, etc. (Kihara and Yamashita, 1938; Smith, 1946). Effects of cosmic rays in plants were studied by Nizam (1969).

The importance of irradiation, either through x-rays or other agents, in working out the time of chromosome reproduction as well as the structure of chromosomes, has been amply realised (*see* review by Taylor, 1962). Results of radiation breaks are considered as good indices of timing of chromosome reproduction. Treatment, followed by a study of the successive cycles and the scoring of chromosome or half-chromatid aberrations, provides significant clues in this direction and aids the understanding of the effects occurring at G_1, S and G_2 phases.

In animals and human beings the most important use of x-ray is its application in the treatment of cancer. The production of excessive chromosome breaks, followed by the loss of acentric fragments into the cytoplasm, thus causing a disturbance leading to the death of the cell, forms the biological basis of radiation treatment of tumours. The smear technique in cytology has considerably helped not only the diagnosis of malignancy in tumours but also the discovery of the actual dose of x-rays needed for its cure or prevention.

To meet all the above problems, methods have been devised for the application of different physical agents under different conditions on the chromosomes, but before undertaking any work on the effects of physical agents on chromosomes certain factors, which control their sensitivity, should be considered.

Regarding the different factors, it has been found that x-irradiation is more dependent on environmental factors than other types of radiation, especially the thermal-neutron radiation. Four important factors of the organism itself have to be taken into account, namely, its genotype, age of the tissue, divisional stage receiving the radiation and number of chromosomes. Species, or even strains of the same species, vary with regard to their reponse to irradiation (Lamprecht, 1956; *see* Smith, 1958; Sharma and Chatterji, A., 1962; Sharma and Chatterjee, T., 1963). Aged seeds are more susceptible (Nilan, 1956). Evidence so far gathered also indicates that meiotic cells are more sensitive than mitotic ones (Sparrow and Singleton, 1953). Sparrow (1951) observed that late prophase to metaphase is the most susceptible stage with regard to chromosome breakage. Fox (1966) studied the effect of x-rays on embryos. Polyploids are more resistant to a majority of physical agents as compared with diploids (MacKay, 1954). Heterochromatic segments may be more susceptible to x-rays as compared to euchromatin (Keyl, 1956; Bhattacharyya, 1958). Randomness of x-ray breaks has been noted by different authors (*see* Heddle, 1965). With regard to environmental factors, the most important ones are moisture content, oxygen concentration and temperature, and in most cases, an increase in moisture content and oxygen concentration accelerates the response (Nilan, 1956; Sharma and Talukdar, 1964). Similarly, an increase in temperature has an accelerating effect on chromosome breakage during irradiation (Nybom and colleagues, 1953; *see* Evans and Sparrow, 1961; Konzak *et al.*, 1961; Riley and Miller, 1966; Wolff, 1967). The comparative action of different rays and chemicals has also been dealt with by different authors (*see* Ghosh, S. and Sharma, 1968). The importance of nucleic acid and protein in the manifestation of chromosome breaks has also been studied (Sharma and Sharma, 1960, 1961, 1962, 1964).

MODE OF ACTION

Regarding the mode of action of the physical agents, particularly the x-rays, opinions differ, but on the basis of the target theory, the x-rays affect the chromosomes directly, without the intervention of any intermediaries (Lea, 1955). Upholders of the chemical theory hold that the x-ray effect on chromosomes is indirect and is principally conveyed through the cytoplasm (Koller, 1948). Some evidence, including the similar effects of radiomimetic chemicals, oxygen, etc., points to the plausibility of the chemical theory. The dissociation of water molecules into H and—OH ions by x-rays and the later formation of HO_2 or H_2O_2 are considered to be the main effects of x-rays, these chemicals ultimately bringing about chromosome breakage (Nilan, 1956). X-rays have also been thought to affect the hydrogen-binding of nucleic acids (Butler, 1954) or -SH groups of proteins (Dustin, 1949). Extensive reviews on this aspect are available (Moutschen, 1967; Drake, 1969).

In explaining the effects of ultraviolet rays, it is suggested that the linkage between adenine and thymine undergoes a breakage and two adjacent thymines of a single strand of the nucleic acid undergo union, forming a thymine dimer, which causes difficulty in replication and chromosome breakage (*see* Setlow and Carrier, 1964; Hanawalt and Haynes, 1967). If the organisms are brought to light, there is photoreactivation of the repair enzyme complex and the frequency of the effects decreases (*see* Setlow, 1968; Witkin, 1969).

Research on this aspect, carried out in this laboratory, (Sen *et al.*, 1967; Sen, 1969), has shown that histone acts as repressor for the repair complex and the digestion of histone accelerates the u.v.-induced breaks.

SCHEDULES FOR TREATMENT WITH DIFFERENT TYPES OF AGENTS

X-RAYS

Apparatus—Consists in the simplest form of an x-ray tube with attached rectifier, transformer, control and dosimeter.

Dosage and period—Depends on the material, the stage of development, and the purpose of the treatment. For an entirely new material, the dosage and time should be fixed by a series of trials with different dosages combined with different periods of time.

Application on Plant Materials

(a) *Seeds*—Place dry seeds, spread in a single row, in a Petri dish. Expose to x-rays at 30 000 R. Germinate in moist sawdust. Remove root tips when about 1 cm long, treat in acetic–alcohol mixture (1:3) for 30 min and squash following orcein or Feulgen schedule. If soaked in water before irradiation, the effect is usually greater and may become drastic.

(b) *Seedlings and bulbs with root tips*—Place young seedlings on a Petri dish with their radicles pointing in the same direction. For bulbs, take healthy ones with a tuft of healthy roots about 2 cm long and place them with the root tips facing the source. Expose to x-rays for the desired period. Transfer them to sawdust or to nutrient medium, as the case may be. For immediate effect, remove root tips at regular intervals of an hour, fix for 30 min in acetic–alcohol (1:3), followed by staining by Feulgen or orcein squash methods. For prolonged effect, study the root tips similarly at intervals of 24 h.

(c) *Inflorescences*—Grow the plants in flower-pots. Expose the young inflorescences to x-rays by bending them to face the source. For immediate effect, select a flower bud of suitable size and observe the pollen mother cells after smearing in acetic-carmine solution. Allow the inflorescence to develop and observe meiotic stages at intervals of 24 h. Endosperm and pollen grain mitosis can be studied in a similar manner.

In tissue culture, different agents have been tried on tissues growing in culture.

Application on Animal Materials

For small animals—Place the animals to be irradiated inside a tube and expose to the x-rays. Transfer them to their normal conditions. At regular intervals, dissect out the gonads in a drop of Ringer's solution, stain with acetic-carmine and observe.

OTHER IONISING RAYS

Alpha Rays

The sources of α-rays are certain radioactive elements like radon and polonium. The rays are formed of particles having a low penetration and a high ion density. They may be used on both root tips and inflorescences of plants.

Beta Rays

The radiation emitted is formed of charged particles with low penetration. They can be obtained from special generators and applied in the same way as x-rays. Alternatively, radioisotopes like ^{3}H, ^{32}P, ^{14}C, etc., also emit β-particles (Cronkite and colleagues, 1959). The material can be treated by immersion in or injection with a known quantity of a radio-isotope solution of accepted concentration. The isotope can also be applied in known quantities in the food or tagged with certain essential constituents of nucleic acid and protein, thymidine, uridine (for RNA), different amino acids, etc.

Gamma Rays

Gamma rays are obtained as electromagnetic radiations either from radium itself or from the isotope ^{60}Co. The material can be exposed to these rays by placing it within a ^{60}Co field for known periods. The rays are more penetrating. The material can be fixed and stained as usual for observation.

Table 7 EFFECTS OF RADIATION RESULTING IN CHROMOSOME BREAKAGE IN SOME ANIMAL TISSUE

Radiation	Material used	Effective dosage	Author
1. Beta rays	*Drosophila*, larva	^{32}P in food at 1 mg per fly approx. 10 000 Å approx. 72, 144 and 216 h preceded and followed by 2000R x-rays	Bateman and Sinclair, 1950
2. Near infra-red	*Drosophila*, sperm	10 000 Å approx. before and after 3000R x-rays	Kaufmann and Gay, 1947
3. Neutrons	*Drosophila*, whole fly, male	Different dosages	Giles, 1943
4. X-rays	*Drosophila*, whole fly *Drosophila*, egg *Drosophila*, embryo *Locusta*, testes *Chortophaga*, embryo	1000–5000 R 500 R, 105 R 5000 R 500 R approx. 4–8000 R	Müller, 1940; Kaufmann, 1946 Ulrich, 1957 Geyer-Duszynzka, 1955 White, 1935 Carlson, Harrington and Gaulden, 1953; Carlson and Harrington, 1955

Other Radiations

Other radiations like fast neutrons can be obtained from special generators or from cyclotrons or atomic piles.

NON–IONISING RADIATIONS

Ultraviolet Rays

The apparatus consists, in simpler forms, of a mercury lamp or a quartz mercury arc in conjunction with a monochromator.

The method of treating the material, both plant and animal, is similar to that followed with x-rays. Bajer and Molé Bajer (1961) irradiated endosperm chromosomes during division with ultraviolet rays with microbeam apparatus constructed by Uretz and Perry (1955), the ultraviolet source being a mercury green lamp (see also chapter on u.v. microscopy).

Table 8   EFFECTS OF RADIATION RESULTING IN CHROMOSOME BREAKAGE IN SOME PLANT TISSUE

Radiation	*Material*	*Dosage*	*Author*
1. Alpha rays	*Tradescantia,* pollen tube	Different dosages	Catcheside and Lea, 1943
	Tradescantia, pollen grain	Inflorescence treated in radon soln.	Kotval and Gray, 1947
	Vicia faba, root tips	Treated in radon soln., 7–8 units	Thoday and Read, 1947
2. Beta rays	*Tradescantia,* pollen grain	^{32}P plaque, externally	Kirby-Smith and Daniels, 1953
	Tradescantia, anthers	(a) ^{14}C from ammonium carbonate at 0.9–8.2 μCi/cm^3 for 4–8 days	Kirby-Smith and Daniels, 1953
		(b) ^{32}P from sodium hydrogen phosphate at 1–10 μCi/cm^3 for 1–9 days	Giles and Bolomey, 1948
		(c) ^{3}H-thymidine 1 μCi/ml for 8–56 h	Wimber, 1959
3. Gamma rays	*Tradescantia,* pollen grain	(a) 5 per min for 100–2000 min	Koller, 1953
		(b) ^{60}Co. 100–400 R	Kirby-Smith and Daniels, 1953
		(c) ^{60}Co: 1.1–1.3 MeV in air and nitrogen	Swanson, 1955

Table 8—*continued*

Radiation	Material	Dosage	Author
4. Near infra-red	*Tradescantia,* pollen grain	(a) 10 000 Å approx. for 3 h before and after 107 R x-rays	Swanson, 1949
		(b) 10 000 Å approx. for 3 h before and after 90 and 350 R x-rays	Yost, 1951
	Tradescantia, meiosis	10 000 Å approx. for 3 h	Snoad, 1955
5. Neutrons	*Tradescantia,* pollen grains	Different dosages	Thoday, 1942
6. Ultraviolet	*Tradescantia,* pollen tube	(a) 2540 Å at 2×10^{-3} ergs/mm^2 for 60 s	Swanson, 1943
		(b) 2537 Å before and after x-rays	Swanson, 1944
	Zea mays, pollen grain	2537 Å at 546 000 ergs/cm^2	Fabergé, 1955
7. X-rays	*Hyacinthus,* root tip	150–1000 R	La Cour, 1953
	Lilium, p.m.c.	150 R	Sauerland, 1956
	Scilla, endosperm	50 R	La Cour and Rutishauser, 1954
	Tradescantia, p.m.c.	18 R and 360 R	Haque, 1953
	Tradescantia, anther	150 R	Sax, 1938
	Tradescantia, pollen grain early	(a) 360 R	Darlington and La Cour, 1945
		(b) 30–300 R	Lane, 1951; Sax, King and Luippold, 1955
	Tradescantia, pollen grain late	180 R. 250 R	Catcheside and Lea, 1943; Bishop, 1954
	Trillium, root tip	5–45 R	Darlington and La Cour, 1945
	Trillium, p.m.c.	50 R	Sparrow, Moses and Dubow, 1952
	Trillium, pollen grain	45–375 R	Darlington and La Cour, 1945
	Uvularia, p.m.c.	90 R	Darlington and La Cour, 1952
	Vicia, root tip	50–200 R	Thoday, 1953; Davidson, 1958
	Zea mays, pollen grain	800–1500 R	Catcheside, 1938

Infrared Rays

The apparatus generally consists of special types of tungsten lamps used for rapid drying. In some later models, arcs, for the production of both ultra-violet and infrared rays, have been built within one source with separate controls. The unwanted rays are screened off with suitable filters.

The method of application is similar to that of x-rays.

TREATMENT OF TISSUE WITH BOTH PHYSICAL AND CHEMICAL AGENTS

It may sometimes be necessary to treat the tissue with both physical and chemical agents, either singly or in a combined form. Radiation effects and the consequent cytological irregularities have often been found to be much minimised in the absence of oxygen or in the presence of certain chemicals (Riley, 1952, 1957; Riley, Giles and Beatty, 1951). As the effects of radiation on the organism as a whole are mostly deleterious, such chemicals are referred to as 'protective chemicals', against radiation damage. Their study is considered as one of the most important aspects of radiobiology, as it has opened up possibilities of affording protection against the destructive effects of radiation to human beings.

Protective chemicals can be applied before, after, as well as during, the time of radiation, the procedure varying according to the type of chemical and the nature of the radiation. From a cytological aspect, protection is afforded against chromosome breakage and inhibition of cell division. The mechanism through which protection is afforded is not clear in all cases, but in the absence of oxygen it can, however, be explained on the basis of the fact that lack of oxygen does not allow the formation of hydrogen peroxide. Similarly, some of the protective substances may act as *oxygen acceptors* removing the dissolved oxygen from the solution. Therefore, protective substances must be oxygen acceptors (Riley, 1954; Sharma and Sharma, 1964). On this basis, glucose, sodium nitrite, ferrous sulphate and stannous chloride are all to be considered as protective chemicals.

The principle of protection afforded by certain sulphydryl compounds such as BAL (2,3-dimercaptopropanol), cysteine and glutathione, is not yet fully clarified, but it has been suggested that they may affect the target itself, thus operating against the direct action of radiation. It is also possible that reaction following irradiation of water may be modified or the availability of oxygen necessary for the effect may also be reduced.

In addition to the compounds mentioned above, propene, magnesium chloride, diethyl ether (Nybom, Lundquist, Gustafsson and Ehrenberg, 1953), hydrogen sulphide, mercaptoacetic acid, cysteamine, i.e. β-mercaptoethylamine, histamine and other amines (Bacq and Herve, 1951), carbamyl choline, thiourea, cobalt, etc., may also be used. It is, however, very necessary to have protective chemicals with selective action, otherwise, along with other parts of the body, the radiation treated area may also develop radio-resistance, and so nullify the purpose of radiation treatment. A number of critical reviews have been published regarding chemical

Table 9 SOME OTHER CHEMICALS WHICH INDUCE CHROMOSOME BREAKS AND SHOW OTHER
MUTACHROMOSOMIC PROPERTIES (SHARMA AND SHARMA, 1960)

I. *Dyes*
   1. Brilliant cresyl blue
   2. Methyl blue
   3. Toluidine blue
   4. Orcein

II. *Coumarin and its derivatives*

III. *Plant pigments*

IV. *Vegetable oils, fats and essences*
   1. Vegetable oils and fats
   2. Essential oils like eugenol,
      lavender, eucalyptol, fennel,
      turpentine, etc.

V. *Drugs and bacterial products*
   1. Antibiotics
   2. BAL and arrhenal
   3. DDT insecticides
   4. Sulpha compounds
   5. Bacterial products

VI. *Alkaloids and related compounds*
   1. Putresin
   2. Podophyllotoxin
   3. Protoanemonine
   4. Veratrine
   5. Theobromine
   6. Theophylline
   7. Caffeine
   8. Berberine
   9. Extracts of *Alstonia scholaris*
      and *Holarrhena antidysentrica*
   10. Vincristine
   11. Vincaleucoblastine
   12. Griseofulvin

VII. *Hormones and growth-promoting*
      substances
   1. Hormones
   2. Nucleic acid
   3. Maleic hydrazide
   4. Vitamins
   5. Growth regulators like 4-chloro-
      2-methylacetic acid, 2, 4-
      dichlorophenoxyacetic acid

VIII. *Mustard and allied compounds*

IX. *Phenols*

X. *Hydrocarbons*

XI. *Azo compounds*

XII. *Heterocyclic bases*

XIII. *Purine derivatives*

XIV. *Miscellaneous*
   1. Acenaphthene
   2. Acetyl pyridine chloride
   3. Acridine derivatives
   4. Acriflavin
   5. Alkalis (several)
   6. Azine series
   7. Azotriprite
   8. Benzene vapour
   9. Benzpyrene
   10. Chloranil
   11. Chloroform
   12. Chromic acid
   13. Cyclohexylcarbamate
   14. *p*-Dichlorobenzene
   15. 2, 4-Dichlorophenoxyacetate
   16. Ethylenediaminetetraacetic
      acid
   17. Ethylene glycol
   18. Ethyl urethane
   19. Gallic acid
   20. Gammexane
   21. Guanidine nitrate
   22. Halogen derivatives
   23. Hydrogen peroxide
   24. Mercury compounds
   25. *m*-Nitroxylene
   26. Morphine
   27. Naphthalene derivatives
   28. *β*-Naphthoquinoline
   29. *o*-Isopropyl *n*-phenyl
      carbamate
   30. Oxygen
   31. Parasorbic acid
   32. Pentavalent arsenic
   33. Propiolactone
   34. Quinoline
   35. Rhamnose
   36. Phenanthrene derivatives
   37. Phenyl mercuric hydroxide
   38. Phenyl mercuric nitrate
   39. Phosphates
   40. Phosphomanganate
   41. Potassium dichromate
   42. Salicylic series
   43. Sodium *p*-aminosalicylic acid
   44. Sugars
   45. Trypaflavin
   46. Urethane
   47. Veronal
   48. Xyloquinone
   49. Lysergic acid diethylamide
   50. Ethyleimine
   51. *N*- nitrosomethylurea
   52. *N*-*N*-dinitroso-*N*-*N*-dimethyl-
      tetraphthalamide
   53. 1-Nitrosoimidazolidone
   54. Fluoro deoxyuridine
   55. Bromo deoxyuridine

Table 10 PROTECTION AGAINST RADIATION BY DIFFERENT CHEMICALS (RILEY, 1957). MATERIALS—BULB OF *Allium cepa* WITH ROOT TIPS IMMERSED IN DIFFERENT SOLUTIONS

Treatment	Dose	Percentages of protection against	
		interchanges	deletions
Water (Control)	114 R at 38 rev/min	—	—
Na-sulphydrate 4×10^{-3} M	,,	50	36
Na-thiosulphate 4×10^{-3} M	,,	59	48
Na-hydrosulphate 4×10^{-3} M	,,	55	35
Na-metabisulphite 4×10^{-3} M	,,	27	36
Na-pyrosulphate 4×10^{-3} M	,,	55	46
Na-peroxydisulphate 4×10^{-3} M	,,	0	0
Glucose $2 \cdot 8 \times 10^{-3}$ M	,,	7	0
Ethanol $1 \cdot 7$ M	,,	55	29
Water (Control)	300 R at 100 rev/min	—	—
Na-bisulphite 2×10^{-3} M	,,	5	43
Na-bisulphate 2×10^{-3} M	,,	1	0
Tryptophane 2×10^{-3} M	,,	28	20
Water (Control)	270 R at 45 rev/min	—	—
Na-peroxydisulphate 4×10^{-3} M	,,	0	0
Cysteine 4×10^{-3} M	,,	30	28
Cystine 4×10^{-3} M	,,	0	52
Uracil 4×10^{-3} M	,,	0	0
		Percentage of protection bridges/cell	
Water (Control)	112 R at 45 rev/min	—	
Cysteine 4×10^{-4} M	,,	100	
Cystine 4×10^{-4} M	,,	77	
Na-bisulphate 4×10^{-4} M	,,	27	
Na-bisulphite 4×10^{-4} M	,,	0	

protection as well as repair and recovery from chromosome damage (Riley, 1954; Evans, 1966, 1968).

Another purpose for which both the agents are often applied to the tissue is to secure a high number of metaphase stages for treatment with x-rays.

Schedule of Treatment

1. Place young denuded bulbs of *Allium cepa* in jars containing tap water and allow the roots to germinate till they are 1–3 cm long.

2. Place the bulbs horizontally in large glass Petri dishes filled with a solution of the chemical compound whose protective action is to be studied. For this purpose, 4×10^{-3} M sodium thiosulphate may be used. Treat for 30 min in this solution.

3. Irradiate the bulbs in the same solution of sodium thiosulphate with an x-ray dose of approx. 150 R at an output of 30–40 R/min. The x-ray machine may be operated at 200 kV peak and 15 mA.

4. After irradiation, keep the bulbs in the fluid for another 10 min. Transfer the bulbs to beakers containing tap water where they can be kept for 4–5 days for immediate observation and for observation of the root tips at regular intervals.

5. Cut healthy young root tips, treat in 0·5 per cent colchicine solution for 1 h to secure a large number of metaphase stages necessary for the study of chromosome interchanges and deletions.

6. Fix in acetic–alcohol (1:3) and follow the usual procedure for orcein or Feulgen staining of root tips and observe the interchanges and deletions at the metaphase stages.

7. To measure the protection afforded by the chemical, set up a control experiment in which all the above steps are followed, substituting distilled water in place of the chemical compound. The difference in the frequency of interchange and deletions between the two sets will give a measure of the protection afforded by the chemical.

8. For animal and plant tissues in culture, the tissue is irradiated *in vitro* and treated with chemicals added to the medium.

REFERENCES

Adams, R. L. P. and Lindsay, J. G. (1967). *J. Biol. Chem.* **242**, 1314
Adams, R. L. P., Abrams, R. and Lieberman, I. (1966). *J. Biol. Chem.* **241**, 903
Auerbach, C. and Robson, J. M. (1946). *Nature* **157**, 302
Bacq, Z. M. and Herve, A. (1951). *Arch. int. Physiol.* **59**, 348
Bajer, A. and Molé Bajer, J. (1961). · *Exp. Cell Res.* **25**, 251
Bateman, A. J. and Sinclair, W. K. (1950). *Nature* **65**, 117
Bell, S. and Wolff, S. (1964). *Proc. nat. Acad. Sci. U.S.* **51**, 195
Berger, C. L. (1938). *Publ. Carneg. Instn.* **496**, 209
Bhattacharyya, S. S. (1958). *Chromosoma* **9**, 305
Bishop, C. J. (1954). *J. Hered.* **45**, 99
Blakeslee, A. F. and Avery, A. (1937). *J. Hered.* **28**, 392
Brock, T. D. (1964). In *Experimental chemotherapy* **3**, 119, London; Academic Press
Butler, J. A. V. (1954). *Acta* **10**, 97

Carlson, J. G. and Harrington, N. G. (1955).   *Radiation Res.* **2,** 84
Carlson, J. G., Harrington, N. G. and Gaulden, M. E. (1953).   *Biol. Bull.* **104,** 313
Catcheside, D. G. (1938).   *J. Genet.* **36,** 321
Catcheside, D. E. and Lea, D. E. (1943).   *J. Genet.* **45,** 186
Christoff, M. and Christoff, M. A. (1948).   *Genetics* **33,** 36
Clever, U. (1965).   *Symp. Brookhaven Nat. Lab.* (BNL 931), 242
Collins, J. F. (1965).   *Brit. Med. Bull.* **21,** 223
Cronkite, E. P., Bond, V. P., Fleidner, T. M. and Rubissi, J. R. (1959).   *Lab. Invest.* **9,** 263
D'Amato, F. (1952).   *Caryologia* **4,** 311
Darlington, C. D. (1937).   *Recent advances in cytology.* London; Churchill
Darlington, C. D. and La Cour, L. F. (1945).   *J. Genet.* **46,** 180
Darlington, C. D. and La Cour, L. F. (1952).   *Heredity* **6,** 41
Darken, M. A. (1964).   *Pharmacol. Rev.* **16,** 223
Davidson, D. (1958).   *Exp. Cell Res.* **44,** 329
Davies, J., Gilbert, W. and Gorini, L. (1964).   *Proc. nat. Acad. Sci. Wash.* **51,** 883
Deysson, G. (1968).   *Int. Rev. Cytol.* **24,** 99
Drake, J. W. (1969).   *Ann. Rev. Genet.* **3,** 247
Dustin, P. (1949).   *Exp. Cell Res.* (Suppl.) **1,** 153
Eigsti, O. J. and Dustin, P. (1957).   *Colchicine in agriculture, medicine, biology and chemistry.* Ames;
   Iowa State College Press
Evans, H. G. (1962).   *Int. Rev. Cytol.* **13,** 221
Evans, H. G. and Bigger, T. R. L. (1961).   *Genetics* **46,** 277
Evans, H. J. (1966).   In *Genetical aspects of radiosensitivity. IAEA, Vienna* **31**
Evans, H. J. (1968).   *Symp. Brookhaven Nat. Lab.* (BNL) 50058 (C51)
Evans, H. J. and Sparrow, A. H. (1961).   *Symp. Brookhaven Nat. Lab.* (BNL) 675. 101
Fabergé, A. C. (1955).   *Z. KonstLehre* **87,** 392
Fox, D. P. (1966).   *Chromosoma* **20,** 173
Gale, E. F. (1963).   *Pharmacol. Rev.* **15,** 481
Gaul, H. (1967).   *Proc. Symp. on Induced mutation and their utilisation,* Berlin 269
Geitler, L. (1948).   *Öst. bot. Z.* **95,** 277
Geyer-Duszynzka, I. (1955).   *Zool. Polon.* **6,** 250
Ghosh, S. and Sharma, A. K. (1968).   *J. Cyt. Genet.* **3,** 54
Giles, H. and Bolomey, R. A. (1948).   *Symp. Quant. Biol.* **13,** 104
Giles, N. H. (1943).   *Genetics* **28,** 398
Glass, E. and Marquardt, H. (1967).   *Chromosoma* **21,** 1
Gregory, W. C. (1956).   *Proc. Int. Conf. Peaceful Uses Atomic Energy* **12,** 48
Gustafsson, Å. (1954).   *Acta. Agric. scand.* **4,** 601
Haque, A. (1953).   *Heredity* **6,** 57
Halkka, O. (1967).   *Hereditas* **58,** 248
Hanawalt, P. C. and Haynes, R. H. (1967).   *Sc. Amer.* **216,** 36
Haselkorn, R. (1964).   *Science* **143,** 682
Hata, T., Sano, Y., Sugawara, R., Matsumae, A., Kanamori, K., Shima, T. and Hoshi, T. (1956).   *J.*
   Antibiot. Tokyo **94,** 141
Heddle, J. A. (1965).   *Genetics* **52,** 1329
Hughes-Schrader, S. (1927).   *Z. Zellforsch.* **6,** 509
Huskins, C. L. (1947).   *Amer. Nat.* **81,** 401
Huskins, C. L. and Cheng, K. C. (1950).   *J. Hered.* **41,** 13
Huskins, C. L. and Steinitz, L. (1948).   *J. Hered.* **34,** 67
Ito, M., Hatta, Y. and Stern, H. (1967).   *Dev. Biol.* **16,** 54
Iyer, V. N. and Szybalski, W. (1967).   *Chromosoma* **21,** 463
Iyer, V. N. and Szybalski, W. (1963).   *Proc. nat. Acad. Sci. Wash.* **50,** 355
Jain, H. K. and Singh, U. (1967).   *Chromosoma* **21,** 463
Jardeztky, O. and Julian, G. R. (1964).   *Nature* **201,** 397
Kaufmann, B. P. (1946).   *J. exp. Zool.* **102,** 293
Kaufmann, B. P. and Gay, H. (1947).   *Proc. nat. Acad. Sci., Wash.* **33,** 366
Kaufmann, B. P., Gay, H. and McDonald, M. R. (1960).   *Int. Rev. Cytol.* **9,** 77
Keyl, H. G. (1956).   *Chromosoma* **9,** 441
Kihara, H. and Yamashita, K. (1938).   *Akemine Commem. Papers* **9**
Kihlman, B. A. (1964).   *Chromosome Conference* at Oxford University, England

Kihlman, B. A. (1966). *Action of chemicals on dividing cells.* Englewood, N. J.; Prentice-Hall
Kihlman, B. A. (1971). In *Advances in cell and molecular biology* **1**, New York: Academic Press
Kihlman, B. A. and Hartley, B. (1968). *Hereditas* **59**, 439
Kirk, J. M. (1960). *Biochim. biophys. Acta* **42**, 167
Kirby-Smith, J. S. and Daniels, D. S. (1953). *Genetics* **38**, 375
Kogut, M. and Lightbown, J. W. (1964). In *Experimental chemotherapy* **3**, 39, London; Academic Press
Koller, P. C. (1948). *Int. Congr. Cytol.* **85**
Koller, P. C. (1953). *Heredity* **6**, 5
Konzak, C. F., Nilan, R. A., Hurle, J. R. and Heiner, R. E. (1961). *Symp. Brookhaven Nat. Lab. BNL* 675
Kotval, J. P. and Gray, L. H. (1947). *J. Genet.* **48**, 135
Kucan, Z. and Lipmann, F. (1964). *J. biol. Chem.* **239**, 516
La Cour, L. F. (1953). *Heredity* **6**, 163
La Cour, L. F. and Rutishauser, A. (1954). *Chromosoma* **6**, 696
Lamprecht, H. (1956). *Agr. hort. Genet.* **14**, 161
Lane, G. R. (1951). *Heredity* **5**, 1
Lawley, P. D. and Brookes, P. (1965). *Nature* **206**, 480
Lea, D. E. (1955). *Actions of Radiation on Living Cells.* Cambridge; The University Press
Levan, A. (1949). *Proc. 8th Int. Congr. Genet., Stockholm, Hereditas,* 325
Lindahl, P. E. (1953). *Exp. Cell Res.* **5**, 416
Mackay, J. (1954). *Hereditas, Lund.* **40**, 65
Mager, J., Benedict, M. and Artman, M. (1962). *Biochim. biophys. Acta* **62**, 202
McLeish, J. (1953). *Heredity* **6**, 125
Metz, C. W. (1926). *Science* **63**, 190
Michaelis, A., Schoneich, J. and Rieger, R. (1965). *Chromosoma* **16**, 101
Mitra, J. and Steward, F. C. (1961). *Am. J. Bot.* **48**, 358
Moutschen, J. (1967). *Medi. Nucl. Radiobiol. Latina* **10**, 289
Moutschen, J., Jana, M. K. and Degraeve, N. (1966). *Carylogia* **19**, 531
Moutschen, J., Moutschen-Dahmen, M., Woodley, R. and Archambeau, J. (1968). *Rad. Res.* **34**, 488
Müller, H. J. (1927). *Science* **66**, 84
Müller, H. J. (1940). *J. Genet.* **40**, 1
Nemec, B. (1904). *Jb. wiss. Bot.* **39**, 645
Nilan, R. A. (1956). *Conf. Radioactive Isotopes in Agric.* U.S. Atomic Energy Commission, T.D. 7512, p. 151
Nandi, S. (1969). *Proc. Int. Seminar on chromosomes, Nucleus* suppl. 220
Nathans, D. (1964). *Proc. nat. Acad. Sci. Wash.* **51**, 585
Nescovic, B. A. (1968). *Int. Rev. Cytol.* **24**, 71
Nizam, J. (1969). *Proc. Int. Seminar on chromosomes, Nucleus,* 224
Nybom, N., Lundquist, U., Gustafsson, Å. and Ehrenberg, L. (1953). *Heriditas, Lund* **39**, 445
Oehlkers, F. (1943). *Z. insukt. Abstamm.-u. Vererblehre* **81**, 313
Padilla, G. M. and Cameron, I. L. (1968). *Int. Rev. Cytol.* **24**, 1
Pavan, C. (1965). *Symp. Brookhaven Nat. Lab.* (BNL 931), 222
Pavan, C. and Da Cunha, A. B. (1969). *Ann. Rev. Genet.* **3**, 425
Pricer, W. E. and Weissbach, A. (1964). *Biochem. biophys. Res. Commun.* **14**, 91
Rao, R. N. and Natarajan, A. T. (1965). *Cancer Res.* **25**, 1761
Rasmussen, R. E. and Painter, R. B. (1966). *J. Cell Biol.* **29**, 11
Reich, E. (1964). *Science* **143**, 684
Reich, E., Franklin, R. M., Shatkin, A. J. and Tatum, A. L. (1962). *Proc. nat. Acad. Sci. Wash.* **48**, 1238
Rendi, R. and Ochoa, S. (1962). *J. biol. Chem.* **237**, 3711
Riley, R. and Miller, T. E. (1966). *Mutation Res.* **3**, 355
Riley, H. P. (1952). *Genetics* **37**, 618
Riley, H. P. (1954). *8th Int. Congr. Bot., Paris* Sec. **9**, 17
Riley, H. P. (1957). *Genetics* **42**, 593
Riley, H. P., Giles, N. H. and Beatty, A. V. (1951). *Genetics* **36**, 572
Sauerland, H. (1956). *Chromosoma* **7**, 627
Sax, K. (1938). *Genetics* **23**, 494
Sax, K., King, E. D. and Luippold, H. (1955). *Rad. Res.* **2**, 171

Schultz, J. (1965).   *Symp. Brookhaven Nat. Lab.* (BNL 931), 116
Schwartz, H. S. (1962).   *J. Pharmacol. Exp. Ther.* **136,** 250
Sen, S. (1969).   *Proc. Internat. Seminar on chromosomes, Nucleus Suppl.* 213
Sen. S. (1970).   *Res. Bull.* **2**
Sen, S., Golechha, P. and Sharma, A. K. (1967).   *Proc. 54th Ind. Sci. Congr. III*
Setlow, R. B. (1968).   *Symp. Brookhaven Nat. Lab.* (BNL 50058 C 51)
Setlow, R. B. and Carrier, W. L. (1964).   *Proc. Nat. Acad. Sci.* **51,** 226
Sharma, A. K. and Sen, S. (1954*b*).   *Caryologia* **6,** 151
Sharma, A. K. and Mookerjea, A. (1954).   *Bull. bot. Soc. Beng.* **8,** 25
Sharma, A. K. and Roy, M. (1956).   *Chromosoma* **7,** 275
Sharma, A. K. and Sen, S. (1954*a*).   *Genet. iber.* **6,** 19
Sharma, A. K. and Bhattacharjee, D. (1954*b*).   *Caryologia* **6,** 151
Sharma, A. K. and Bhattacharyya, B. (1956).   *Caryologia* **9,** 38
Sharma, A. K. and Bhattacharyya, U. C. (1962). φ*yton* **18,** 39
Sharma, A. K. and Chatterji, A. K. (1962).   *Nucleus* **5,** 67
Sharma, A. K. and Chatterjee, T. (1963).   *Folia Biol.* **11,** 158
Sharma, A. K. and Chatterjee, T. (1964).   *Nucleus* **7,** 113
Sharma, A. K. and Chaudhuri, M. (1961).   *Nucleus* **4,** 157
Sharma, A. K. and Chaudhuri, R. K. (1963).   *Rev. Port. Biol. Zool. Geral.* **4,** 21
Sharma, A. K. and Bhattacharyya, G. N. (1967).   *Acta Biol. Hung.* **18,** 67
Sharma, A. K., Chaudhuri, M. and Chakraborti, D. P. (1963).   *Acta Biol. Med. Germ.* 11, 433
Sharma, A. K. and Datta, A. (1956). φ*yton* **6,** 71
Sharma, A. K. and Ghosh, S. (1965).   *Nucleus* **8,** 183
Sharma, A. K. and Ghosh, S. (1969).   *Acta Biol. Acad. Sci. Hung.* **20,** 11
Sharma, A. K. and Gupta, A. (1959).   *Nucleus* **2,** 131
Sharma, A. K. and Mukherjee, R. N. (1965).   *Genetica* **28,** 143
Sharma, A. K. and Sarkar, A. K. (1963).   *Rev. Port. Zool. Biol. Geral.* **4,** 29
Sharma, A. K. and Sarkar, A. K. (1967).   *Genet. Agrar.* **21,** 77
Sharma, A. K. and Sharma, A. (1960).   *Int. Rev. Cytol.* **10,** 101
Sharma, A. K. and Sharma, A. (1961).   *Histochemie* **2,** 260
Sharma, A. K. and Sharma, A. (1962).   *Nucleus* **5,** 127
Sharma, A. K. and Sharma, A. (1964).   *Proc. II Int. Congr. Histo and Cytochemistry* 218
Sharma, A. K. and Talukdar, C. (1964).   *Nucleus* **7,** 23
Sharma, A. K. and Talukdar, C. (1965).   *Biologia* **20,** 105
Singleton, U. R., Konzak, C. F., Shapiro, S. and Sparrow, A. H. (1956).   *Proc. Int. Conf. Peaceful Uses Atomic Energy* **12,** 25
Smith, H. H. (1958).   *Bot. Rev.* **24,** 1
Smith, L. (1946).   *J. agric. Res.* **73,** 291
Smith-Kielland, I. (1964).   *Biochim. biophys. Acta* **91,** 360
Snoad, B. (1955).   *Chromosoma* **7,** 451
Sparrow, A. H. (1951).   *Ann. N.Y. Acad. Sci.* **51,** 1508
Sparrow, A. H., Moses, M. J. and Dubow, R. J. (1952).   *Exp. Cell Res.* **2,** 245
Sparrow, A. H. and Singleton, W. R. (1953).   *Amer. Nat.* **87,** 29
Srivastava, B. I. S. (1967).   *Int. Rev. Cytol.* **22,** 360
Stadler, L. J. (1928).   *Anat. Rec.* **41,** 97
Stebbins, G. L. (1965).   *Symp. Brookhaven Nat. Lab.* (BNL 931), 204
Swaminathan, M. S. (1969).   *Proc. 12th Int. Congr. Genet. Tokyo* **3,** 309
Swaminathan, M. S. and Natarajan, A. T. (1957).   *Stain Tech.* **32,** 43
Swanson, C. P. (1943).   *J. Gen. Physiol.* **26,** 485
Swanson, C. P. (1944).   *Genetics* **29,** 61
Swanson, C. P. (1949).   *Proc. nat. Acad. Sci., Wash.* **35,** 237
Swanson, C. P. (1955).   *Genetics* **40,** 193
Taylor, J. H. (1962).   *Int. Rev. Cytol.* **13,** 39
Taylor, J. H. (1963).   *J. cell. comp. Physiol.* suppl. **62,** 73
Taylor, J. H., Haut, W. F. and Tung, J. (1962).   *Proc. Nat. Acad. Sci. U.S.* **48,** 190
Thoday, J. M. (1942).   *J. Genet.* **43,** 189
Thoday, J. M. (1953).   *Heredity* **6**
Thoday, J. M. and Read, J. (1947).   *Nature* **160,** 608
Tjio, J. H. and Hågberg, A. (1951).   *An. Estac. exp. Aula Dei* **2,** 149

Torrey, J. G. (1961). *Exp. Cell Res.* **23,** 281

Traut, R. R. and Monro, R. E. (1964). *J. mol. Biol.* **10,** 63

Turner, M. K., Abrams, R. and Lieberman, I. (1966). *J. biol. Chem.* **241,** 5777

Ulrich, H. (1957). *Zool. Anz.* **19**

Uretz, R. B. and Perry, R. P. (1955). *Rad. Res.* **3,** 355

Vazkuez, D. (1964). *Nature* **203,** 257

Wagner, T. E. (1969). *Nature* **222,** 1171

Wagner, J. H., Maher, M. N., Konzak, C. F. and Nilan, R. A. (1968). *Mutation Res.* **5,** 57

White, M. J. D. (1935). *Proc. Roy. Soc. B.* **119,** 61

Wimber, D. E. (1959). *Proc. nat. Acad. Sci. Wash.* **45,** 839

Witkin, E. M. (1969). *Ann. Rev. Genetics* **3,** 525

Wolff, S. (1967). *Ann. Rev. Genetics* **1,** 221

Yost, H. T. (1951). *Genetics* **36,** 176

Yarmolinsky, M. B. and Haba, G. L. de la (1959). *Proc. Nat. Acad. Sci. Wash.* **45,** 1721

Young, C. W., Schochetman, G. and Karnofsky, D. A. (1967). *Cancer Res.* **27,** 535

Zeuthen, E. (1964). *Synchrony in Cell Division and Growth.* New York; Intersciences

Zimmerman, F. K. and Schwaier, R. (1967). *Molec. Gen. Genetics* **100,** 63

Part II
Chemical Nature of Chromosomes

18 Introduction

The development of the present concept of the chemical make-up of chromosomes affords an excellent example of the extent to which refinements of methods aid in the advancement of a theoretical understanding. The earliest attempt towards analysing the chemical nature of the hereditary substance was made by Miescher (1874) by isolating 'nuclein' from the nuclei of various animals through a standard biochemical method of isolation. In fact, 'nuclein'—the invariable constituent of the cell nucleus—was later known to be nucleoprotein, on which vigorous research is at present being carried out in different centres. Even with considerable modifications and alterations in the outlook with regard to the chemical nature of chromosomes in recent years, the fundamental principle involved in Miescher's discovery still remains undisputed, although Miescher himself was unaware of the implications of his findings. It has been the work of research teams of the present century to correlate the findings of those engaged in different disciplines of biological science, namely, that of Strasburger on 'mitosis', of Mendel on 'hereditary factors' and of Miescher on 'nuclein'. Through this synthesis, the 'nuclein' of Miescher is established as the constituent of the gene substance, the behaviour of which in heredity Mendel analysed, and which takes part in the cell division that Strasburger followed.

From the time of Miescher up to 1924, there was a blank phase in the study of chromosome structure, during which period no significant improved method was devised. The year 1924, however, represents a landmark in the science of cytogenetics due to the outstanding work of Feulgen and Rossenbeck in devising a special technique, based on Schiff's reaction for aldehydes, for the detection of deoxyribonucleic acid *in situ* in chromosomes. The basis of the Feulgen staining and the validity of the test have already been discussed in Chapter 5.

The impetus given through the work of Feulgen and Rossenbeck led to the analytical methods of Kossel (1928), Levene and Bass (1931), Behrens (1938) and Gulick (1941), which established the polynucleotide structure of the DNA molecule in association with the protamines, histones, etc., of chromosomes. Gradually, the study of chromosome structure was shifted from a purely cytogenetical level to chemical and histochemical analysis, which has given an understanding of the pattern of organisation, even at the sub-microscopic level. In chemical terms, the chromosome is now visualised as a giant complex molecule made up of several but less complex molecules, the physicochemical units—the genes responsible for hereditary stability. Within the last 25 years tremendous advances have been made on the study of chromosome structure due to the cooperative efforts of workers engaged in different branches of science, namely genetics, cytology, biochemistry and biophysics. The critical methods evolved by them have demonstrated that the chromosome cannot be visualised as simply a deoxyribonucleic acid during functional phases as in procaryotes, but is to be considered in higher plants and animals as composed of at least two types of nucleic acid, namely, RNA and DNA, as well as two types of proteins—the histone and an acidic protein rich in tryptophane.

The clarification of the complex constitution of chromosomes has been aided through biochemical methods of isolation of the chromatin matter, cytochemical tests meant for the localisation of constituents *in situ*, ultra-structural study as well as methods for quantitation. For the latter purpose precise tests for nucleic acids and for proteins in general, and amino acids in particular, are available. One of the most important inventions is the method for the differential localisation of RNA and DNA by pyronin and methyl green (Brachet, 1950; see Talukder and Sharma, 1968). This work is aided by highly critical controlled tests, including digestion through specific enzymes like deoxyribonuclease and ribonuclease for the respective nucleic acids, and pepsin, trypsin and chymotrypsin for the digestion of different types of proteins. Similarly, methods for nucleic acid extraction through trichloracetic acid and perchloric acid have aided in finding out the precise seats of occurrence of the two acids, when applied in conjunction with specific tests. These techniques for localisation and control are not free from limitations. Nevertheless, in spite of these limitations, the conclusions arrived at from different aspects of study are nearly identical, indicating that the conclusions with regard to the chemical nature of chromosomes are not far from the truth. Moreover, spectrophotometric methods, based on ultra-violet absorption at 2650Å of purine components of the nucleic acid, devised originally by Caspersson and modified later by Pollister, Leuchtenberger and others (Leuchtenberger, 1958; Vendrely and colleagues, 1958, Pollister *et al.*, 1969), provide confirmatory evidence. These methods along with the study of ultrastructure, have also helped in ascertaining whether the chromosome undergoes any cyclic change in structure. Quantitative estimation of the different constituents has been aided through chromatography as well (Chargaff and Davidson, 1955). Studies on x-ray and electron diffraction patterns by Ambrose (1956) and several other workers have also aided in deducing the relationship between nucleic acids and proteins within the chromosomes (*see* chapter on x-ray microscopy).

The knowledge that has been gained with the aid of the above methods suggests that RNA and DNA lie in intimate connection with the proteins (Chargaff and Davidson, 1955; Nurnberger and Gordon, 1958), but uncontroversial evidence is lacking. Chargaff (1953) stated that lipoid is present in the chromosome in the form of a protein complex. Vendrely and colleagues (1958) suggested the presence of a pre-formed distinct nucleohistone part within the chromosome, merely on the basis of certain data relating to the constancy of chemical composition. That there is a distinct nucleoprotein complex within the chromosome in the living cell has been proved unequivocally by Kaufmann and his colleagues (Kaufmann, 1953; Bal and Kaufmann, 1959; Dutt and Kaufmann, 1959) in their works, involving the use of deoxyribonuclease on unfixed cells of *Tradescantia*, grasshopper, *Drosophila* and of trypsin on the fixed cells of *Drosophila*. Moreover, taking recourse to deoxyribonuclease and ribonuclease treatments, together with Feulgen and pyronin–methyl green staining, not only the presence of RNA in the chromosome has been confirmed but it has also been shown that the chromosome chain may be composed of both RNA and DNA (Kaufmann and colleagues, 1958; Kaufmann, McDonald and Gay, 1951; cf. Wolstenholme, 1965). The latter contention has, however been much debated (Kaufmann *et al.*, 1960; cf. Das *et al.*, 1964).

Even though RNA may be considered as a constituent of the chromosome, opinions still differ as to whether it is associated with the histone or the non-histone protein (Sharma and Roy, 1956; Sharma and Sharma, 1958). Mirsky and Ris (1947), on the basis of their technique for the isolation of chromosomes, suggested that RNA remains principally in association with the non-histone protein and DNA with the histone. On the other hand, other workers (Kaufmann, Gay and McDonald, 1960), through cytochemical tests involving ribonuclease treatment and specific acid extraction of different proteins followed by staining, have claimed that RNA remains in association with the histone. They demonstrated that dilute HCl (0·02 N) hydrolysis can remove histone but not acidic protein. DNA remains in combination with both histone and non-histone (Mirsky and Ris, 1951; Bernstein and Mazia, 1953; Bloch and Godman, 1955; Bloch, 1958; Kaufmann, Gay and McDonald, 1960; *see* Dupraw, 1965; Bonner, 1967; Lima-de-Faria, 1969). RNA of chromatin is bound both with DNA and histone. Walters (1968) noted the appearance of ribonucleoprotein structures during meiotic prophase, associated with the release of synthetic products. The importance of histones in gene regulation has been amply realised (Busch, 1965; Georgier, 1969). Chromosomes in the extended and active phase have been demonstrated to contain non-histones whereas the stage of inactivity and condensation has been marked by the presence of histones. Residual complexes of calf thymus chromatin have been shown to be made up of DNA and histone, and transcribing activity is associated with the gradual loss of histone (Georgier *et al.*, 1966; Paul and Gilmour, 1966; Clever, 1968; Sonnenbichler, 1969). Relevant to this observation is the finding made by Gall (1966) whose spreading technique of chromosome fibre enabled him to demonstrate that in *Triturus*, 250–300 Å thick inactive fibres are made of nucleohistone and owe their origin to the coiling of those thinner fibres which are actually involved in protein synthesis. This gradual deproteinisation associated with gene action is an index of regulation of genetic activity by histone.

Further complexity in chromosome structure has been introduced by the demonstration of the possible presence of lipids in the chromosomes (Brock, Stowell and Couch, 1952; Chayen, La Cour and Gahan, 1957; Serra, 1958; Idelman, 1958). Unfortunately, as lipid shares many of the chemical properties of other substances, it is very difficult to locate it in the chromosomes. Indications of its presence have been observed in root tips, anthers, salivary glands, etc. The modification of the phospholipid test is no doubt an important achievement (Serra, 1958), but the study of lipids has been much neglected (Deane, 1958). In all probability, lipids are present in the form of lipoprotein complexes, with specific sulphydryl proteins and, as pointed out by Kaufmann, Gay and McDonald (1960), they may have some correlation, as yet unexplored, with the nucleic acid components of the chromosomes.

Technical achievements have aided in working out the different constituents of chromosomes, and ample evidence has been gathered in support of the complex molecular pattern of DNA initially suggested by Watson and Crick (1953). According to this model, the DNA molecule is of the nature of a double stranded helix composed of two complementary polynucleotide strands connected with each other by hydrogen bonding between purine and pyrimidine bases. The helix may be compared with a ladder in which base pairs form the rungs and phosphoric acid forms the uprights. The arrangement of the four bases is not at random but distinct orderly pairing occurs between adenine and thymine on the one hand and guanine and cytosine on the other. These base pairs form the basis of the genetic code and, as such, the entire hereditary information is stored in this alphabet of four letters. It has been shown that the number of base pairs constituting a functional unit—the gene—varies and that they can provide immense diversities in form, for the control of diverse genetic characters, satisfying the requirements of a geneticist. Through the outstanding works of Nirenberg, Khorana and others, direct evidence has been gathered to show that three base sequences together, i.e. a triplet, control the synthesis of one amino acid and through this mechanism of *coding*, hereditary information inscribed in DNA is transcribed to RNA and finally translated to the synthesis of specific polypeptides of enzyme proteins on ribosomes through messenger RNA and amino-acyl transfer RNA. As 64 triplets are possible in four base sequences, the number of diversities required for the control of 20 amino acids is satisfied through these triplets. The universal, non-overlapping and degenerate nature of the code, implying that one triplet may code for several amino acids, has also been established. The implications of the works of Khorana and his colleagues on gene synthesis are well known. The synthesis of the different constituents needed for protein synthesis in cell free cultures has been achieved, and even the mechanisms of chain initiation through methionine, of chain termination, and recognition of initiation sites by ribosomes, have been clarified (Yanofsky, 1967; Geiduschek *et al.*, 1968; see Clark and Marcker, 1968; August, 1969; Roberts, 1969). As generally the protein molecule contains at least 200 amino acids, the length of a gene comes to about 600 base pairs, at least in the bacteriophage. The genes have also been further subdivided into cistron, recon and muton, by Benzer, the two latter being the units of recombination and mutation. Further the Watson–Crick model explains satisfactorily the self-duplicating mechanism which is an essential requirement of a chromosome. The two chains being complementary to each other, each of them can serve as a negative for the formation of positives or sister strands during the re-duplication after splitting. Mutation can also be explained plausibly on the basis of this model by assuming substitution, transposition, omission and duplication (Beadle, 1957).

Outstanding achievements have been made in the field of genetic control of differentiation through the concept of *operon*, the operating unit in a chromosome proposed by Jacob and Monod. According to this hypothesis, structural genes, responsible for enzyme protein synthesis, are controlled by an operator gene, which is regulated by a repressor (*see* Bretscher, 1968). The regulated mechanism of differentiation is controlled by the repression and derepression of an operon. The concept, as worked out in bacteria, can no doubt explain the mechanism of differentiation where the gene structure is a simple DNA molecule and differentiation involves only certain physiological properties, capacity of infection and patterning of the protein layer. But in the higher organisms, where growth and development are sequential and phasic, and where even intrachromosomal relationships are often manifested, especially in relation to euchromatin and heterochromatin, the functional operon may not be the same as in procaryotes. Even if the existence of operons in higher organisms is accepted, then it must be admitted that such operons are the constituents of much more complex operational units. It has been rightly suggested that in higher organisms, the chromosomes may constitute innumerable nucleotides, which function autonomously during development, as noted in puffing in Diptera, and each such chromomere is comparable to an entire phage chromosome (Beermann, 1967, and *see* Whitten, 1969a, b; Goldberg *et al.*, 1969). No doubt the operon concept provides a basic background to account for differentiation, but to explain the phasic pattern of development characteristic of higher plants and animals, a complicated mode of operation has to be visualised embracing translocation effects, interchromosomal relationship in the rhythmic functioning of the cell metabolism and triggering of the different developmental phases.

The technical achievements, as noted above, are no doubt principally responsible for bringing out the complex constitution of chromosomes, but even then no precise understanding has been obtained as to the pattern of their arrangement and the exact relationship between nucleic acids and protein. The difficulty encountered in obtaining this knowledge may be attributed to the complexity of the chromosome strand and DNA molecule. Ris (1957, 1967), Kaufmann, Gay and McDonald (1960) and others, through ultrastructural studies by electron microscopy, have demonstrated the multiple fibrillar constitution of the chromosomes, but there is still the question, what is the chemical basis which causes the fibrils to adhere to one another. This has not yet been precisely solved. Mazia (1954) suggested that bivalent cations, like calcium and magnesium, serve as bridges between the macromolecules of chromosomes, which was refuted by Kaufmann, Gay and McDonald (1960); Ambrose (1956) pointed out the necessity of hydrogen bonding in maintaining the structural integrity, whereas the importance of protein linkers has been suggested by several authors (*see* Ris, 1967).

Though DNA is considered to be the basic substance in the chromosome, it is yet to be found whether it exists in the form of a continuous strand throughout the length of the chromosome or is interrupted or attached at certain segments. This information is essential as it has a strong bearing on the DNA nature of the gene. Callan and MacGregor (1958) demonstrated that deoxyribonuclease treatment disintegrates isolated but unfixed lamp-brush chromosomes, suggesting that DNA forms a continuum throughout. In addition, 5'-fluorodeoxyuridine (FUDR), which is a pyrimidine base analogue capable of inhibiting synthesis of thymine in DNA replication, if applied in low concentrations before DNA replication, induces chromosome breakage (Taylor and colleagues, 1962). Ultra-violet radiation at 200 nm is also effective in causing chromosome breakage (Bloom and Keider, 1962 and *see* Moses, 1964). Even then the principal difficulty in proposing a typical model of the chromosome lies in the fact that it cannot be visualised as merely an aggregation of polynucleotide chains of DNA. Watson and Crick suggested a dextrose intertwining of complementary polynucleotide chains, but the chromonemata in chromosomes may have a sinistrose or irregular coiling and may remain in a paranemic association without being interlocked (Kaufmann, Gay and McDonald, 1960). Therefore, in spite of Watson and Crick's invention, refinements in cytochemical and ultra-structural methods are needed to effect a reconciliation between the double helix concept of DNA and the other constituents of chromosomes, as worked out cytochemically. The later models of chromosome structure, proposed by Taylor, Freese and others, appear to be quite reasonable. They suggested a side-chain model in which relationship between protein and nucleic acid can be explained (Freese, 1958; Taylor, 1959). A single protein core has been visualised, the two halves of which are placed on opposite sides with nucleohistone molecules in between forming the rungs of a ladder, the chain being attached to only one axis through a bond, which allows rotation. The complementary chain remains attached to the opposite axis. Further modification (Taylor, 1969) has assumed interruptions in the axis between alternate DNA double helices, on each side of the ladder, linked tandemly by linkers. The ladder model, when screwed and twisted in coiled chromosomes, may represent multi-strandedness (Kaufmann and De, 1956), but this theory still requires strong evidence to make it acceptable. Dupraw (1965) proposed a folded fibre model. De suggested a model, assigning a role to acidic proteins (1967). Several other models have also been proposed from time to time (*see* Hamilton, 1968). The importance in gene activity of RNA, now regarded as a possible constituent of the chromosome, is also not yet clear. McMaster-Kaye and Taylor (1958) and Kaufmann and colleagues (1959) suggested that crossing over and re-combination may be controlled by RNA . Marinozzi's observations (1964) further confirm the role of RNA in the formation of synaptonemal complex during synapsis (*see* also Menzel and Price, 1966). On the other hand, it has been suggested that the presence of RNA during this phase should await further ultrastructural and cytochemical evidence (Moses, 1968). The structure of the synaptonemal complex has been dealt with in detail by different workers in recent years (Moens, 1968; Smith and King, 1968; Gassner, 1969; Moses, 1969; Sen, 1969; Sheridan and Barnett, 1969; Stern and Hotta, 1969; Westergaard and Wettstein, 1970). Mazia (1961) considers that RNA may be responsible for chromosome movement or the chromosome may be their carrier. Sinsheimer (1957) went so far as to consider that RNA, DNA and protein may jointly contribute to the expression of hereditary characters (cf. Commoner, 1968) and it does seem likely that all the components of the chromosome are responsible for the maintenance of structural integrity and form an interconnected system. Kaufmann, Gay and McDonald (1960) stated that the pattern of this complex fibril changes with metabolism. The dynamicity in structural patterning of chromosomes is exhibited in its association with histones in the body cells, being replaced by protamines in the germinal line. Diversity in the nature of chromosomal histones has also been recognised. Lysine-rich histone has been demonstrated to combine with DNA, forming cross-linking with double helix, whereas histones rich in arginine are linked with the phosphoric acid groups along the double helix (Littau *et al.*, 1965). The heterogeneity of the DNA molecule and its consequent modifications, while combining with proteins, have been emphasised as well (Wake *et al.*, 1968). Sharma (1968) questioned the feasibility of proposing a universal model for chromosome structure in view of its dynamicity in different phases of development and growth. Chromosomes for transmission, as in spermatozoa, or meant for replacement of parts, as in crossing over, may have a structural difference from chromosomes of the somatic nucleus. The futility of proposing a universal model from a single phase of development is apparent.

The above résumé of our knowledge of the chemical nature of chromosomes, as aided through the different techniques, reveals the gaps still to be filled. However, with so much technical background already prepared, and refinements in methodology attained within the last few years, exact knowledge of the pattern of association of the nucleic acids and proteins, in maintaining the structural integrity of the chromosome, does not seem to be far off. The correlation of chromosome components, in quality and quantity, with phases of development and growth, would lead to a better understanding of the chemical basis and initiating mechanism of each genetically controlled reaction.

REFERENCES

August, J. T. (1969). *Nature* **222**, 121
Bal, A. K. and Kaufmann, B. P. (1959). *Nucleus* **2**, 51
Beermann, W. (1967). In *Heritage from Mendel*, Madison; University of Wisconsin
Beadle, G. W. (1957). *The Physical and Chemical Basis of Inheritance*. Condon Lecture Publication, Oregon; University of Oregon Press
Behrens, M. (1938). *Handb. biol. Arb. Meth.* **10**, 1363

Bernstein, M. H. and Mazia, D. (1953).   *Biochem. biophys. Acta* **10,** 600
Bloch, D. P. (1958).   *Frontiers of cytology,* p. 113. New Haven: Yale University Press
Bloch, D. P. and Godman, C. G. (1955).   *J. biophys. biochem. Cytol.* **1,** 17
Bloom, W. and Leider, R. J. (1962).   *J. Cell. Biol.* **13,** 269
Bonner, J. (1967).   In *Regulation of nucleic acid and protein synthesis.* Amsterdam; Elsevier
Brachet, J. (1940).   *C.R. Soc. Biol. Paris* **133,** 88
Bretscher, N. S. (1968). *Nature* **217,** 509
Brock, B., Stowell, R. E. and Couch, K. (1952).   *Lab. Invest.* **1,** 439
Busch, H. (1965).   *Histones and nuclear proteins.* New York; Academic Press
Callan, H. G. and MacGregor, H. C. (1958).   *Nature* **181,** 1479
Chargaff, E. (1953).   *Some conjugated proteins,* p. 36, New Brunswick; Rutgers University Press
Chargaff, E. and Davidson, J. N. (1955).   *The Nucleic acids.* New York; Academic Press
Chayen, J., La Cour, L. F. and Gahan, P. B. (1957).   *Nature* **180,** 652.
Clark, B. F. C. and Marcker, K. A. (1968).   *Sc. Amer.* **218,** 36
Clever, U. (1968).   *Ann. Rev. Genet.* **2,** 11
Commoner, B. (1968).   *Nature* **220,** 224
Das, N. K., Luykx, P. and Alfert, M. (1964).   *Proc. 2nd Int. Congr. Histochem.* Frankfurt, 234
De, D. N. (1964).   *Nature* **203,** 343
Deane, H. W. (1958).   *Frontiers of cytology,* p. 227. New Haven; Yale University Press
Dupraw, E. J. (1965).   *Nature* **296,** 338
Dutt, M. K. and Kaufmann, B. P. (1959).   *Nucleus* **2,** 85
Feulgen, R. and Rossenbeck, H. (1924).   *Hoppe-Seyl. Z.* **135,** 203
Freese, E. (1958).   *Cold. Spr. Harb. Symp. Quant. Biol.* **23,** 13
Gall, J. G. (1966).   *Chromosoma* **20,** 221
Gassner, G. (1969).   *Chromosoma* **26,** 22
Geiduschek, E. P., Brody, E. N. and Wilson, D. L. (1968).   In *Molecular association in biology.* New York; Academic Press
Georgier, G. (1969).   *Ann. Rev. Genet.* **3,** 155
Georgier, G., Ananieva, L. and Kozlov, J. (1966).   *J. Mol. Biol.* **22,** 365
Goldberg, Whitten, J. and Gilbert, (1969).   *J. Insect. Physiol.* **15,** 409
Gulick, A. (1941).   *Bot. Rev.* **7,** 433
Hamilton, L. D. (1968).   *Nature* **218**
Idelman, S. (1958).   *C.R. Acad. Sci. URSS* **246,** 1098 and 3282
Kaufmann, B. P. (1953).   *Exp. Cell Res.* **4,** 408
Kaufmann, B. P. and De, D. (1956).   *J. biophys. biochem. Cytol.* **2,** 49
Kaufmann, B. P., Gay, H., Fuscaldo, K. and Buchanan, J. (1958).   *Carnegie Inst. Wash. Year Book* **57,** 406
Kaufmann, B. P., Gay, H., Dutt, M. K., Bal, A. K. and Buchanan, J. (1959).   *Carnegie Inst. Wash. Year Book* **58,** 440
Kaufmann, B. P., Gay, H. and McDonald, M. R. (1960).   *Int. Rev. Cytol.* **9,** 77
Kaufmann, B. P., McDonald, M. R. and Gay, H. (1951).   *J. cell. comp. Physiol.* **38,** suppl. 1, 71
Kossel, A. (1928).   *The Protamines and histones.* London; Longman Green
Kurnick, N. B. (1952).   *Stain Tech.* **27,** 233
Leuchtenberger, C. (1958).   *General cytochemical methods,* p. 220, New York; Academic Press
Levene, P. A. and Bass, L. W. (1931).   *Am. Chem. Soc. Monogr. Sder.* **56,** New York; Chemical Catalogue Co.
Lima-de-Faria, A. (1969).   *Handbook of molecular cytology* Amsterdam; North-Holland
Littau, V. C., Burdick, C. J., Allfrey, V. G. and Mirsky, A. E. (1965).   *Proc. Nat. Acad. Sci. US* **54,** 1204
McMaster-Kaye, R. and Taylor, J. H. (1958).   *J. biophys. biochem. Cytol.* **4,** 5
Marinozzi, V. (1964).   *J. Ultrastruct. Res.* **10,** 433
Mazia, D. (1954).   *Proc. Nat. Acad. Sci. Wash.* **40,** 521
Mazia, D. (1961).   *Ann. Rev. Biochem.* **30,** 661
Menzel, M. and Price, J. M. (1966).   *Am. J. Bot.* **53,** 1079
Miescher, F. (1874).   *Ber. dtsch. Chem. Ges.* **7,** 376
Mirsky, A. E. and Ris, H. (1947).   *J. gen. Physiol.* **31,** 1
Mirsky, A. E. and Ris, H. (1951).   *J. Gen. Physiol.* **34,** 475
Moens, P. B. (1968).   *Chromosoma* **23,** 418
Moses, M. J. (1964).   *Cytology and cell physiology,* p. 423, New York; Academic Press
Moses, M. J. (1968).   *Ann. Rev. Genet.* **2,** 363

Moses, M. J. (1969). *Genetics* **61,** 41

Nurnberger, J. T. and Gordon, M. W. (1958). *Frontiers of cytology,* p. 167, New Haven; Yale University Press

Paul, J. and Gilmour, R. S. (1966). *J. Mol. Biol.* **16,** 242

Pollister, A. W., Swift, H. and Rasch, E. (1969). In *Physical techniques for biological research 3C,* 201, New York; Academic Press

Ris, H. (1957). *The Chemical Basis of Heredity,* p. 23, Baltimore; Johns Hopkins Press

Ris, H. (1967). In *Regulation of nucleic acid and protein biosynthesis.* Amsterdam; Elsevier

Roberts, J. W. (1969). *Nature* **224,** 1168

Sen, S. K. (1969). *Exp. Cell Res.* **55,** 123

Serra, J. A. (1959). *Rev. Port. zool. biol. geral.* **1,** 109

Sharma, A. K. (1968). *Sci. Cult.* **34,** suppl. 69

Sharma, A. K. and Roy, M. (1956). *La Cellule* **58,** 133

Sharma, A. K. and Sharma, A. (1958). *Bot. Rev.* **24,** 511

Sheridan, W. F. and Barnett, R. J. (1969). *J. Ultrastruct. Res.* **27,** 216

Sinsheimer, R. L. (1957). *Science* **125,** 1123

Sonnenbichler, J. (1969). *Nature* **223,** 205

Smith, P. A. and King, R. C. (1968). *Genetics* **60,** 335

Stern, H. and Hotta, Y. (1969). *Genetics* **61,** 27

Talukder, G. and Sharma, A. K. (1968). *Nucleus* **11,** 106

Taylor, J. H. (1959). *Oklahama Conf. on Radioisotopes in Agriculture* US Atomic Energy Commission, T.D. 7578, p. 123

Taylor, J. H. (1962). *Int. Rev. Cytol.* **13,** 39

Taylor, J. H. (1969). *Proc. 12th Int. Congr. Genet.* **3,** 177

Taylor, J. H., Haut, W. F. and Tung, J. (1962). *Proc. Nat. Acad. Sci. US* **43,** 821

Vendrely, R., Alfert, M., Matsudaira, H. and Knobloch, A. (1958). *Exp. Cell Res.* **14,** 295

Watson, J. D. and Crick, F. H. C. (1953). *Nature* **171,** 737 and 964

Walters, M. S. (1968). *Heredity* **23,** 39

Wake, K., Ochiai, H. and Tanifugi, S. (1968). *Jap. J. Genet.* **43,** 15

Westergaard, M. and Wettstein, D. von (1970). *Compt. Rend. Lab. Carlsberg* **37,** 239

Whitten, J. (1969*a*). *J. Morph.* **127,** 73

Whitten, J. (1969*b*). *Chromosoma* **26,** 215

Wolstenholme, D. R. (1965). *Chromosoma* **17,** 219

Yanofsky, C. (1967). *Ann. Rev. Genet.* **1,** 117

19 Nucleic Acid and its Components

Since nucleic acid is composed of three principal constituents—namely sugar, bases and phosphoric acid—the majority of chemical tests are based on the identification and localisation of any one of these constituents. The principal procedures, limitations and advantages of the different techniques are outlined in the following text.

TESTS FOR SUGAR

There are certain methods (Turchini, Castel and Kien, 1944) through which both the deoxyribose and the ribose sugars can be tested, but at the same time there are certain specific tests for the deoxy sugar, of which the Feulgen procedure is considered to be the most important. For the demonstration of aldehydes through fluoresence—Schiff's reagents, like acridine orange (Kasten, 1959; Joshi and Korgaonkar, 1959; Paolillo, 1964; Rigler, 1964; Roschlau, 1965), please see chapter on Fluorescence microscopy. Stoward (1963) employed salicyloyl hydrazide sequence instead of Schiff dyes.

FEULGEN REACTION

The Feulgen reaction is based on Schiff's reaction for aldehydes whereby, by acid hydrolysis, the liberated aldehydes of the deoxy sugar are allowed to react with fuchsin–sulphurous acid to yield a typical magenta colour reaction. For details of this reaction and its discussion, please refer to the chapter on staining.

The Feulgen test is not only specific for the localisation of DNA in chromosomes, but at the same time the intensity of the reaction may be considered to be an index of the amount of DNA present in the cell. Several authors (Pollister and Ris, 1947; Pollister, 1950; Pollister, Himes and Ornstein, 1951; Pollister, Swift and Alfert, 1951; Pollister, 1952; Mellors,

1955; Freed and Benner, 1964; Mendelsson, 1966; Wied, 1966; vide Freed, 1969) have employed spectrophotometric methods for measurement of the intensity of Feulgen colour for the quantitative estimation of DNA in the fixed nuclei. On the basis of the quantitative estimation, the DNA content of cells is inferred to be a constant factor and is proportional to the number of chromosomes present (Vendrely and Vendrely, 1948; Ris and Mirsky, 1949; Pasteels and Lison, 1950; Swift, 1950; and *see* Pearse, 1968). Stowell and Cooper (1945) modified the Feulgen reaction, and coupled with photometric recordings, measured mean amounts of DNA per unit volume per cell and correlation was brought about between morphological data and other results. Caspersson (1947) utilised the ultraviolet absorption technique and the Feulgen reaction and developed the concept that nuclear-associated chromatin plays a significant role in the nucleoprotein synthesis of malignant cells. For further details, please see chapter on photometry.

FEULGEN–NAPHTHOIC ACID HYDRAZIDE REACTION

This schedule is meant for testing the presence of deoxyribonucleic acid. Danielli (1947) utilised 2,4-dinitrophenylhydrazine for staining chromosomes yellow. This method avoided the difficulty of mild hydrolysis in revealing aldehydes and the use of Schiff's reagent. Pearse (1951) similarly secured good results by using 2-hydroxy-3-naphthoic acid hydrazide. This compound not only combines with aldehydes but with ketones as well, and if the resulting dark yellow compound is coupled with diazotised *o*-dianisidine in alkaline solution, a purple blue compound will be obtained. As the hydrazide reagent combines with tissue proteins as well, to produce protein hydrazides, the cytoplasmic details also are maintained; however, this reagent acts as an acidic dye as compared to the basicity of the basic fuchsin.

SCHEDULES

2,4-Dinitrophenylhydrazine

This was used by Danielli (1947) in three different methods which are briefly outlined below:
   1. Hydrolyse squashes of salivary glands of *Drosophila* with N HCl for 15 min. Treat with 2,4-dinitrophenylhydrazine. The chromosomes take up very little colour.
   2. Treat the squashed tissue at $0\,°C$ with a saturated solution of dinitrophenylhydrazine in 2N HCl or with a saturated solution, 0·5N in HCl, in 75 per cent alcohol, cover and observe. The bands take up colour.
   3. Hydrolyse a suspension of deoxyribosenucleic acid at $60\,°C$ by N HCl for 15 min. Allow it to react with 2,4-dinitrophenylhydrazine and the solution will slowly become cloudy. Keep for 30 min at room temperature, then cool to $0\,°C$ and place the squashes in the solution. The final

concentration of HCl is 0·02N. Remove the squashes after an hour. The chromosomes take up light stain.

The stain, in general, is fainter than that obtained after fuchsin staining.

Feulgen–Naphthoic Acid Hydrazide Test

Reagent a

2-Hydroxy-3-naphthoic acid hydrazide	0·1 g
50% ethyl alcohol	100 cm²
Glacial acetic acid	5 cm²

Reagent b

Tetrazotised *o*-dianisidine	0·1 g
Veronal buffer solution pH 7·4	100 cm²

Procedure

1. Bring the sections (originally fixed in alcoholic or formalin fixatives) down to water. Rinse in N HCl. Hydrolyse in N HCl at 60 °C for 6–15 min, depending upon the tissue and fixative.

2. Rinse successively in cold N HCl, distilled water and 50 per cent ethyl alcohol.

3. Immerse for 1–3 min in reagent *b* at 0 °C. Wash in distilled water.

4. Dehydrate through alcohol and xylol grades and mount in D.P.X.

5. DNA takes up bluish purple stain. Cytoplasm and other proteins may be stained red.

DISCHE REACTION (DISCHE, 1930, 1944)

The Dische reaction was considered for a long time to be a specific test for the deoxyribose sugar of the chromonucleic acid. Dische employed five reactions including the carbazole and the diphenylamine. The carbazole reaction is specific for pyrimidine-bound aldehyde DNA (Schneider, 1948) and is carried out in strong sulphuric acid solution, heating to 100 °C. In the diphenylamine reaction a blue coloration is obtained by the addition of diphenylamine to the tissue similarly treated with hot acid.

The Dische test is often employed for the quantitative estimation of DNA from isolated nuclei by subjecting them to the test and observing against a colorimeter. The matching of colour with the known samples of nucleic acid solutions gives an estimate of the amount of DNA present.

For cytochemical localisation of DNA *in situ*, this test cannot be applied; moreover, the specificity of the Dische reaction has been questioned (Stacey, 1950; Dische, 1955) on the ground that under strongly acid conditions 2-deoxypentoses become converted to laevulinic acid which gives no colour with the reagent, but an intermediate compound ω–hydroxylae-vulinic aldehyde yields colour with diphenylamine. As other deoxysugars, e.g. deoxyxyloses, also yield laevulinic aldehyde, the validity of the results remains questionable if not supplemented by other control experiments.

Several tests are in vogue for RNA. They involve the use of orcinol, phloroglucinol, cysteine and sulphuric acid. Dische (1955) suggested that the orcinol reaction of Bial (1903), Dische and Schwartz (1937) and Dische (1953) is comparatively more sensitive than the latter two, and the replace-ment of orcinol by phloroglucinol (Dische and Borenfreund, 1957) yields even better results. In this method, the minimum concentration of ribose and its esters can be much more precisely determined than that obtained through the use of orcinol. Specificity is accentuated by the fact that it is not affected by the presence of other sugars.

DISCHE'S SCHEDULES FOR COLOUR REACTIONS WITH DNA

The experiments have all been carried out on extracted nucleic acids. For extraction schedules, the reader is referred to the chapter on extraction.

Schedule of Reaction with Diphenylamine

Reagents

Extracted aq. DNA	50–500 µg/cm^3 of solution
Diphenylamine	1 g (twice crystallised from
	70% ethyl alcohol or ether)
Glacial acetic acid	100 cm^3
Conc. H$_2$SO$_4$	2·75 cm^3

For preparing the reaction mixture, dissolve the diphenylamine in the acetic acid and add sulphuric acid.

Procedure

1. Mix together one volume of DNA solution and two volumes of the reaction mixture.
2. Heat at 100 °C for at least 10 min.
3. For control, heat a quantity of water separately with the reaction mixture.

Observations

Observations show that the DNA solution takes up a blue colour, which persists for hours. An absorption curve drawn with a Beckman spectro-photometer shows a maximum at 595 nm. The colour is maximum after 10 min heating, but shorter periods up to 3 min may give enough coloration for qualitative purposes. The blue colour is produced by 2-deoxypentoses and not only 2-deoxyribose.

Schedule of Reaction of DNA with Cysteine and Concentrated H_2SO_4 (Dische, 1949)

Reagents

Extracted DNA solution with 50–500 μg/cm^3 of DNA	1 cm^3
Conc. H_2SO_4	4 cm^3

Procedure

1. Add together the DNA extract and H_2SO_4 under cooling. Keep for 1 h.
2. Add 1 cm^3 of 3 per cent solution of cysteine hydrochloride. Shake thoroughly. Keep the mixture for 24–48 h at room temperature.
3. Measure absorption with a Beckman spectrophotometer.

Observations

These show a sharp absorption maximum at 375 nm, which is reached after 48 h.

Modified Schedule of Reaction with Indole and HCl (Ceriotti, 1952)

Reagents

DNA solution containing 2·5–15 μg/cm^3 DNA	2 cm^3
0·04% indole C.P. solution in distilled water	1 cm^3
Conc. hydrochloric acid	1 cm^3
Chloroform—purified by extraction with conc. H_2SO_4, followed by water extraction and storage over $CaCl_2$ for 48 h	

Procedure

1. Add the other reagents to the DNA solution.
2. Immerse the test tube for 10 min in a boiling water bath.
3. Cool to room temperature under running water.
4. Extract three times with 4 cm^3 of chloroform. Separate water from chloroform by centrifuging.

Observations

Observations show that the water phase gives a yellow colour which persists for several hours. The chloroform phase shows a faint pink colour. The absorption curve, by a Beckman spectrophotometer, has a sharp peak at 490 nm and a smaller constant peak at 460 nm. The reaction is regarded as being produced by only the purine nucleotides of DNA (Dische, 1955).

Hydrazine–Benzaldehyde Schiff Reaction:

The principle of separating pyrimidines through anhydrous hydrazine at high temperature (Takemura, 1958) has been adapted by Smith and Anderson (1960) for nuclear DNA. Benzaldehyde is applied for complete removal of pyrimidine bases. This method may even be used for quantitation (*see* Pearse, 1968).

DISCHE'S SCHEDULE FOR COLOUR REACTIONS WITH RNA (DISCHE, 1955)

Schedule on Orcinol Reaction (Dische and Schwartz, 1937)

Reagents

Extracted solution of RNA	1.5 cm^3
Reaction mixture, prepared by dissolving 0.1 g of ferric chloride in 100 cm^3 of hydrochloric acid and adding 3.5 cm^3 of a 6% solution of orcinol in ethyl alcohol	3 cm^3

Procedure

1. Mix the RNA extract with reaction mixture.
2. Heat in a waterbath for 3 min.
3. Cool to room temperature under running water.
4. Measure the optical density with a Beckman spectrophotometer against a blank containing water and the reagent.

Observations

These show an optical density at 665 nm. The specificity of the reaction is rather low as not only RNA, but also 2-deoxryibose, DNA, methylpentose and hexuronic acids give a green colour with an absorption maximum around 670 nm.

Various modifications of this schedule are available.

Schedule on Phloroglucinol Reaction (Euler and Hahn, 1946)

Reagents
Extracted solution containing:

2 mg of RNA	1 cm^3

Reagent mixture containing:

0·1% solution of ferric chloride in a mixture of conc. HCl and glacial acetic acid (1:6)	8 cm^3
25% phloroglucinol solution in a mixture of conc. HCl, water and glacial acid (1:1:2)	1 cm^3

Procedure

  1. Add the reagent mixture to RNA extract and stir.
  2. Immerse the tube in a boiling water bath for 50 min.
  3. Cool to room temperature under running water.
  4. Add 1 cm^3 of the phloroglucinol solution and keep for 20 min at room temperature.
  5. Immerse the tube in a boiling water bath for 4 min.
  6. Cool down to room temperature and keep for 2–24 h.

Observations

Observations show that the maximum intensity of the colour appears after 10 h and the adsorption maximum is at 680 nm. DNA does not give any coloration, as prolonged heating might have destroyed the sugar of the purine nucleotides of DNA.

Schedule of Reaction with Cysteine and H_2SO_4

The schedule is similar to schedules II and III described under reactions for DNA (*see* page 414). The reading is taken 15 min after the addition of cysteine when the maximum absorption produced by pentoses is observed. The peak of absorption curve is observed at 390 nm.

Schedule for Observing Aldopentose in Presence of other Saccharides (Dische and Borenfreund, 1957)

Reagents
Extracted RNA solution containing:

4–40 µg/cm^3 of pentose	0·4 cm^3

Reaction mixture (freshly prepared) containing:

Glacial acetic acid	110 cm^3
Hydrochloric acid	2 cm^3
0·8% glucose	1 cm^3
5% phloroglucinol	5 cm^3

Procedure

1. Add 0·4 cm^3 of the RNA extract to 5 cm^3 of the reaction mixture.
2. Shake, immerse for 15 min in a vigorously and uniformly boiling water bath.
3. Cool to room temperature under tap water.

Observations

Observations show that all aldopentoses produce an intensely red colour with a sharp absorption maximum at 552 nm. The optical density of the colour produced by ribose or its phosphate esters is proportional to the concentration of the compounds. This reaction is not affected by the presence of other sugars.

TRYPTOPHANE–PERCHLORIC ACID CONDENSATION METHOD

Cohen's (1944) method involves freeing of deoxyribose sugar from the bases by treatment with hot perchloric acid. The liberated groups of the sugar are then allowed to condense with the secondary amine of tryptophane to give the colour reaction. The reaction is very slow and carried out in a test-tube. No *in situ* localisation is possible as perchloric acid quickly extracts DNA.

Schedule

Add 0·2 cm^3 of 1 per cent tryptophane solution and 1·2 cm^3 of 60 per cent perchloric acid to 1 cm^3 of a solution containing 100–500 μg of extracted DNA. Boil in a water bath for 10 min, cool in tap water and observe under a spectrophotometer. A purple colour is obtained and an absorption maximum is observed at 500 nm. An aldehyde intermediate probably causes the colour reaction.

TESTS FOR THE BASES

No specific chemical test is, in fact, at present available to test purine and pyrimidine components of nucleic acid. The only reliable method of their identification is measurement through ultra-violet absorption originally devised by Caspersson (1947). A strong absorption can be obtained at 2650 Å of the purine and pyrimidine components of the nucleic acid. However, this method takes for granted that all the purine and pyrimidine components are located in the nucleic acids. Moreover, it is also necessary to find out how far the absorption is affected by neighbouring compounds.

Danielli (1947) utilised tetrazotised benzidine for staining chromatin of the chromosome. Mitchell (1942) employed it for demonstrating cytoplasmic ribonucleoprotein. He claimed that tetrazotised benzidine reacts not only with tyrosine, tryptophane and histidine (*see* Chapter 15) but also with the purine and pyrimidine groups. For specific chromatin staining, the method involves the treatment of the sections in acetic anhydride (10 per cent solution in dry pyridine, treated for an hour at 100 °C) or in benzoyl chloride (10 per cent in dry pyridine, for 20 h at room temperature), which are termed blocking agents, prior to application of tetrazotised benzidine. The basic principle of this method is to allow the tissue to react with a reagent which will not allow a number of tissue components to undergo any reaction with tetrazotised benzidine. By using such blocking reagents as acetic anhydride or benzoyl chloride, tryptophane, tyrosine and histidine are apparently eliminated and, as such, the components resistant to benzoylation and reacting thereafter with tetrazotised benzidine are the purines and pyrimidines. However, Danielli stated that other substances may react with diazonium salts.

Pearse (1953, 1960, 1968) stated that tetrazonium reaction involves the protein part of the DNA protein as purified DNA, free from protein, does not give any reaction whatsoever. It has been claimed that even after benzoylation, the reaction is due to the amino acids of the protein which remain protected from benzoylation due to the physical state of nucleoprotein. Bernard and Danielli (1956) also suggested later that, after benzoylation, the reaction is due to histidine residues in the nucleoprotein fraction. As such, this test should not be used to demonstrate purine and pyrimidine components of nucleic acids.

SCHEDULE FOR THE COUPLED TETRAZONIUM REACTION

Main Schedule

Reagents
    Reagent a

1. Benzidine base		0·5 g
Conc. HCl		5 cm^3
Distilled water		20 cm^3
2. 2% aq. sodium nitrite solution		10 cm^3
3. 5% aq. ammonium sulphamate		10 cm^3
4. Anhydrous sodium carbonate·		28 g
Distilled water		95 cm^3

Prepare first reagents (1) and (4) separately. To 20 cm^3 of reagent (1) at 0–5 °C, add slowly, a few drops at a time, 7 cm^3 of 2 per cent NaNO$_2$

solution from a chilled pipette, for a period of 10–15 min. Agitate rapidly but do not allow the temperature to rise above 10 °C. Keep the mixture in cold to attain a temperature of about 5 °C or below. Add 3·5 cm^3 of 5 per cent aqueous ammonium sulphamate solution. Add 34 cm^3 of reagent (4) and stir. As effervescence ceases and the solution becomes alkaline, it changes to a clear dark yellow. Add enough distilled water to make the volume up to 150 cm^3.

Reagent b

Veronal acetate buffer pH 9·2

Reagent c

8-Aminonaphthol-3,6–disulphonic acid (H acid)	0·5 g
Veronal acetate buffer pH 9·2	25 cm^3

Procedure

Paraffin sections (fixed previously in formalin, alcohol or freeze dried) are used.

1. Bring down the sections to water.
2. Immerse them in fresh diazotised benzidine (reagent *a*) at 4 °C for 15 min.
3. Wash in water and three changes of veronal acetate buffer at pH 9·2 (reagent *b*) for 2 min in each change.
4. Immerse in a saturated solution of H acid in veronal acetate buffer at pH 9·2 (reagent *c*) for 15 min.
5. Wash in distilled water for 3 min.
6. Dehydrate in alcohol, clear in xylol and mount in D.P.X. or Canada balsam.

Alternative for step 2 is 0·2 per cent aqueous solution of Fast blue B salt in tris buffer at pH 9·2, for 5 min at room temperature.

Observations

Most of the tissue components are stained reddish brown, showing the presence of tyrosine, tryptophane and histidine. The colour produced is stable and remains for several months.

Precautions

The tetrazotised benzidine solution (reagent *a*) should be prepared freshly before use.

Schedule with Benzoylation

Procedure

1. Bring down sections to water and then gradually to absolute ethanol.
2. Immerse these sections or paraffin sections in petroleum ether for 3 min.
3. Remove and dry in air.
4. Heat the dried sections to 60 °C in an incubator or to 80 °C on a warm plate for 5–10 min. This step is optional, depending on the material.
5. Treat the dried sections in 10 per cent benzoyl chloride in dry pyridine for 10–16 h at room temperature. The period of treatment is shorter for alcohol-fixed tissues and longer for formalin-fixed ones.
6. Rinse in absolute acetone.
7. Immerse in absolute ethyl alcohol.
8. Bring down the sections, through decreasing alcohol grades, to water.
9. Next follow the coupled tetrazonium reaction, from step 2 in the previous schedule, to demonstrate the presence of bases.

Precautions

Keep the benzoyl chloride solution away from moisture, as it rapidly loses strength on contact with water. In controlled humidity, the solution lasts for about a month. Dry pyridine by distillation over barium sulphide. As mentioned already, this test is not specific for bases.

Schedule with Acetylation

Procedure

Steps 1, 2, 3 and 4 are the same as those of the previous schedule.
In step 5—Immerse the sections in 10 per cent acetic anhydride in dry pyridine and heat for 4–8 h at approx. 100 °C, under a reflux condenser.
The later steps are similar to the previous schedule.

TESTS BASED ON THE REACTION WITH PHOSPHORIC GROUPS

Tests for phosphoric groups of nucleic acids are not so precise as those meant for detecting the sugar components. Though it is generally considered that phosphate dye linkage is in the form of a salt, recent works indicate the role of hydrogen bonding in the reaction (*see* Pearse, 1968). However, one of the principal staining schedules, based on the use of methyl green and pyronin, involves reaction with phosphoric groups. It will be discussed under methods for differentiating DNA and RNA in tissues.

Serra and Lopes (1945) demonstrated phosphorus *in situ* in deoxyribonucleic acid in tissues, and photographs presented by them reveal clearly the presence of phosphorus in chromosomes. However, as phosphorus is an essential constituent of many compounds, the test should always be applied in two sets of tissue, one undergoing digestion with nuclease prior to the test and serving as the control. Without this control measure it is not possible to differentiate phosphorus of nucleic acid from that of other constituents.

This method involves, in principle, the treatment of the tissue in the hydrochloric acid–molybdic reagent for hydrolysis and precipitating phosphomolybdate for 2–3 weeks at a low temperature of 4–10 °C. The phosphomolybdate, when formed through ammonium molybdate, is demonstrated by means of acetic benzidine and sodium acetate, whereby it takes a blue colour. The method has, however, two serious limitations. In the first place, the sensitivity of phosphomolybdic reagent, which is often not very sharp, should be properly assessed. Secondly, all the shortcomings of precipitation tests are inherent in this method. Pearse (1960) has objected to the test on the basis that prolonged acid hydrolysis may lead to diffusion of phosphate ions causing artefacts.

METHODS FOR THE SIMULTANEOUS DETECTION OF DEOXYRIBOSE AND RIBOSE NUCLEIC ACIDS

Several methods are at present available by which both types of nucleic acids can be detected simultaneously. These techniques can be divided into two categories, those based on reaction with sugar moiety and those based on reactions with phosphoric groups.

METHOD DEVISED BY TURCHINI AND COLLEAGUES

The most important technique falling under the first category is that of Turchini and colleagues (1944). These authors first introduced fluorones for critical differentiation of DNA and RNA in the cell. The technique is based on the condensation of deoxyribose and ribose sugar with 9-phenyl-2,3,7-trihydroxy-6-fluorone, or more precisely xantheonone, having the structure:

This should be 9-phenyl 2,3,7-trihydroxy 6-oxoxanthene

OR 9-phenyl 2,6,7-trihydroxy 3-oxoxanthene

(OR ------- xanthen-6-one and xanthen-
3-one respectively)

Therefore, prior to this reaction, uncovering of sugars by acid hydrolysis is an essential step in the technique. It differs from the Feulgen reaction in the fact that both the sugars react, deoxyribose giving blue to violet and ribose yielding yellow to red coloration. As strong acid hydrolysis may cause extraction of RNA, the choice of a proper fixative is needed to prevent this extraction. The authors have recommended fixation with chromic or picric compounds. Moreover, for different types of organisms, different fixatives have been recommended and the period of hydrolysis has been varied.

Pearse (1960) used a variety of fixatives containing ethanol as well as formalin on human tissues, but in none of the cases could he secure any differentiation excepting a general pale rose to pink colour of the nuclei. Moreover, as the shades of colour often vary, even for DNA alone, it is not proper to utilise this method for quantitative estimation (Kurnick, 1955*a*, *b*).

Backler and Alexander (1952) used methyl substituted derivatives of fluorone and this modification of Turchini's method yielded successful results. The compound used by them is 9-methyl-2,3,7-trihydroxy-6-xantheonone having the chemical structure:

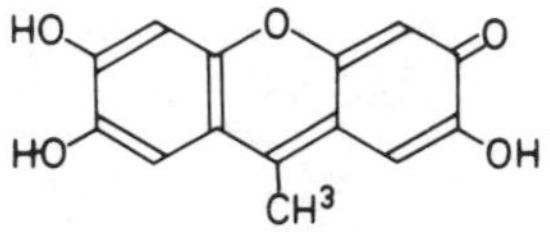

They obtained violet to blue black colour in DNA and yellow to red colour in RNA, but in this method also, similar to Turchini's, the fixative and period of hydrolysis are important factors which should be varied in order to secure reproducible results in a variety of tissues in different organisms. Because many factors control the reaction, the method is not considered as critical for quantitative and qualitative assessments for the nucleic acids as other methods in vogue, which are discussed later in this chapter. The only advantage of this method is that selective staining of the two nucleic acids can be obtained with the use of only one type of xantheonone compound.

SCHEDULES

Method of Turchini, Castel and Kien (1944)

   Reagents

Nucleic acid reagent	Dissolve 80 mg of 9-phenyl (or methyl)-2,6,7-trihydroxy-3-fluorone in 100 cm^3 of 95% ethyl alcohol containing 15 drops of conc. sulphuric acid
Hydrochloric acid	1 N (alternatively 25%) conc. HCl in 90% ethyl alcohol
Aq. sodium carbonate	1% solution

Procedure

1. Fix the tissue in Bouin's fluid.
2. Prepare and cut paraffin sections. Bring down to water.
3*a*. For treatment with methyltrihydroxyfluorone reagent: Hydrolyse the sections in 1N HCl at 60 °C for 5 min, wash with water, rinse in absolute ethyl alcohol and treat for 5–10 min with nucleic acid reagent. Wash with several drops of 90 per cent ethyl alcohol, then with 1 per cent sodium carbonate solution. Rinse in water. Dehydrate through ethanol and xylol grades and mount in balsam.
3*b*. For alternative treatment with phenyltrihydroxyfluorone reagent: The procedure is similar except that hydrolysis is carried out in cold temperature in alcoholic solution of 25 per cent HCl for 3–5 min.

Method of Backler and Alexander (1952)

This method has been used on human autonomic ganglia fixed in Bouin's fluid, rat tissues fixed in Bouin's, Zenker's, formol and formol–saline fixatives, and mouse liver, frozen and dried.

Reagents
(*a*) 9-methyl-2,3,7-trihydroxy-6-fluorone containing:

1,2,4-Triacetyltrioxybenzene	1 mole
Paraldehyde	1·25 mole
Ethyl alcohol	5 times the combined weight of the above reagents
Sulphuric acid	5–10% (v/v)

Mix the chemicals together and add sulphuric acid. Allow to stand for 18–24 h at room temperature. Add 30 volumes of distilled water. Allow to remain for 24 h when a reddish orange precipitate of the dye settles out. Filter and discard the filtrate. Dry the residue at 37–40 °C in an oven. Dissolve the precipitate in the minimum quantity of ethyl alcohol needed and filter. Add 30–60 volumes of water to the filtrate and allow it to stand for 24 h. Filter and dry again as before. The product will be a reddish orange powder, decomposing at 319 °C, moderately soluble in alcohol and practically insoluble in water.
(*b*) Sulphuric acid.
(*c*) Hydrochloric acid 1 N.
(*d*) 1 per cent solution of sodium carbonate in distilled water.

Procedure

1. Deparaffinise and bring down paraffin sections to water.
2. Immerse the slides in 1 N HCl for 2 min at 60 °C for partial hydrolysis.

3. Transfer immediately, without washing, to 80 per cent ethyl alcohol and keep for 15 s.

4. Transfer the slides directly to a mixture containing:

9-Methyl-2,3,7-trihydroxy-6-fluorone	1 g
Sulphuric acid	1 cm^3
95% ethyl alcohol	100 cm^3

Keep for 20 min in this solution.

5. Keep in 1 per cent aqueous sodium carbonate solution for 2 min.

6. Immerse the slides in distilled water for 2 min for rinsing.

7. Keep in acetone and distilled water mixture (1:1) for 3 min.

8. Transfer to pure acetone for 3 min.

9. Keep in acetone and xylol mixture (1:1) for 2 min.

10. Give two changes of pure xylol, $1\frac{1}{2}$ min in each.

11. Mount in piccolyte. RNA takes up yellow to red stain and DNA stains violet to blue black.

Note

1. Use the staining mixture for several weeks.

2. Use fresh sodium carbonate solution for each set of staining.

3. Use fresh distilled water for rinsing.

4. For dehydration, use acetone instead of ethyl alcohol, as the dye is extracted by ethyl alcohol treatment.

PYRONIN–METHYL GREEN TECHNIQUE

Of all the methods employed for the differential localisation of RNA and DNA, the universally accepted standard technique is that of Brachet (1940, 1942, 1958) based on Pappenheim (1849) and Unna's (1902) methyl green–pyronin G mixture. In this schedule, methyl green imparts a green colour to chromatin or, more precisely, DNA, whereas RNA present in the cytoplasm, nucleolus, etc., appears pinkish red with pyronin, in both cases the reaction involving the phosphoric groups of the nucleic acid moiety. The specificity of pyronin staining of RNA is also confirmed by staining a parallel slide originally digested with ribonuclease. The latter preparation, serving as the control, does not show the presence of RNA or pyronin staining, being exhausted of this nucleic acid by digestion through its specific enzyme. Another control set can be stained merely with 1 per cent toluidine blue (Kurnick, 1952) and the dye adheres only to RNA-containing areas.

Several modifications of the original technique have been adopted from time to time. In Brachet's schedule, the aqueous dye mixture is mixed with acetate buffer and acetone is used as the dehydrating agent. With low pH levels, pyronin is stongly positive, whereas at higher levels, methyl green yields the crisper colour. The optimum pH level for staining is considered to be 4·8. Kurnick (1955a, b), on the other hand, recommended pyronin

(05564, Gurr) specifically for the purpose, and used tertiary butyl alcohol for dehydration. Bhaduri and Mukherjee (1961) secured good results in plants by using toluol as the final clearing agent.

Methyl Green

It is correctly represented as 'Methyl Green OO', a basic dye of the triphenyl methane series. The structure can be represented as

It is slightly soluble in water and comparatively less soluble in alcohol.

   Balbiani (1881) first utilised this dye to secure green colour in salivary gland chromosomes of *Chironomus*, followed later by Pappenheim and Unna, who employed it in combination with pyronin. In Unna's modification, phenol was a constituent of the mixture which evidently adjusted the pH for optimum staining with methyl green (Kurnick, 1950*a*, *b*). The selective staining of DNA by methyl green was first demonstrated by Brachet (1942). Kurnick (1947) claimed that selectivity of DNA staining by methyl green is dependent on the polymerised nature of the DNA molecule, and he observed that depolymerised DNA could not be satisfactorily stained with methyl green, and also that histones compete with the dye for nucleic acid. The constant stoichiometry of the staining showed the formation of the chemical compound with the phosphoric group during staining. Kurnick and Mirsky (1950) asserted that one dye molecule combines with ten phosphoric groups of DNA, basing their observation on the stoichiometry of the reaction by dialysis, precipitation of stain–nucleic acid mixtures and staining of nuclei of known DNA content. This ratio of 1:10 was given by heptamethyl pararosaniline (C.I. 684) and the ratio 1:13 by hexamethyl pararosaniline (C.I. 685) (Pearse, 1960, 1968). Kurnick (1950*a*, *b*) and Errera (1951) claimed that stable binding of methyl green should involve at least two amino groups with two methyl groups of the dye. The stainability of methyl green is therefore controlled largely by different agents which under certain conditions may bring about depolymerisation (Kurnick, 1955*a*, *b*)). In this connection Godman and Deitch (1957) claimed that, even with all possible precautions, there is no certainty that DNA is not depolymerised. For quantitative analysis of DNA through methyl green staining, certain conditions must however be satisfied—such as, avoiding depolymerising fixing agents, optically homogeneous preparation of nuclei, elimination of histones, as well as the use of stain in aqueous solution. Kurnick (1955*a*, *b*) further suggested that proteins other than histones or protamines may possibly be bound to DNA at sites other than the phosphoric acid so that their presence does not stand against methyl green stainability.

Alfert (1952) suggested that the blocking of stainable groups of RNA by protein is principally responsible for specific staining of DNA by methyl green. There is, however, no fundamental difference between Alfert's claim and Kurnick's observations (Kurnick, 1955*a*, *b*; Pearse, 1960). Taft (1951) depolymerised DNA by varying pH and temperature without affecting methyl green stainability. Pearse (1960) pointed out that such depolymerisation may principally involve rupture of hydrogen bonding (Gulland and Jordan, 1947) which may be a reversible process.

It was suggested by Vercauteren (1950) that, in the nucleic acid molecule, the spacing of negatively charged phosphate residues corresponding to two positively charged sites on the methyl green molecule is the principal controlling factor in staining. Agents which cause breaking of weak bonds in the DNA molecule result in coiling of the molecule, thus causing alteration in spacing which is responsible for loss of methyl green staining. Rosenkranz and Bendich (1958) suggested that the double strand state of DNA is responsible for methyl green specificity.

Goldstein (1961) claimed that the explanation so far provided for methyl green staining of DNA is unnecessary, the chief factor being the weight of the dye cation, and he has suggested that dye cations, having a combined atomic weight between 350 and 500, can penetrate and adhere to unclean DNA whereas smaller cations adhere to denser molecules such as cytoplasmic RNA. Methyl green, having a cationic weight of 387, is specific for DNA. Heat or other agents may bring about alterations in the structure of DNA in such a way that it becomes denser and, as such, becomes stainable with pyronin. He confirmed this suggestion by studying the behaviour of a number of dyes with different cationic weights such as toluidine blue, malachite green, acridine red, celestine blue, etc. However, along with cationic weights, hydrogen ion concentration, mordant, solubility, etc., also control the stainability to a significant extent (Baker, 1958, 1966).

Van duijn (1962) criticised Goldstein's hypothesis of density difference being the basis of differential staining and suggested that more evidence was necessary to validate its cytochemical specificity. Cowden (1965) stained cytoplasmic RNA of oocytes with methyl green alone, while the same regions took up stain with pyronin Y in methyl green–pyronin mixture. Scott (1967), with the aid of his critical electrolyte concentration method, noted that the combination of both these dyes with polynucleotides is too strong to be explained only by electrostatic bonds. Planar dyes, such as pyronin, react with single-stranded RNA with accessible bases, whereas non-planar dyes like methyl green combine with double helical structures. The importance of the period of staining and dye concentration has been indicated; best results have been secured with staining for 16 h in 0·15 per cent methyl green and 0·25 per cent pyronin in 50 mM sodium acetate (pH 5·6) with 2M magnesium chloride.

One of the principal steps which has a profound influence on staining is the purification of the dye, which is essential as methyl violet remains as an impurity with methyl green. Kurnick (1950*a*, *b*) and Leuchtenberger (1950) suggested purification before use by chloroform extraction. Tatt (1951) performed purification by repeated chloroform extractions of a 0·5 per cent solution of methyl green in 0·1 M acetate buffer. Jordan and Baker (1955) recommended the extraction of freshly prepared 0·5 per cent aqueous methyl green with chloroform at least eight times, until all the impurities are removed and the chloroform becomes colourless.

The above discussion reveals that methyl green staining may yield erroneous results unless all the different factors affecting the process are strictly controlled; discrepant results often noted are due to these variable factors. However, under strictly limited conditions, with a special check on purification of the dye, fixation without any depolymerising effect, and pH during staining, the method can be safely applied for localisation and quantitation estimation of DNA.

Pyronin

Pyronin is a basic dye of the xanthene group and is available in three different forms, 'pyronin B', 'pyronin Y' and 'pyronin G'. Pyronin Y (Michrome No. 339, Gurr) is a tetramethyl whereas pyronin B (Michrome No. 44, Gurr) is a tetraethyl compound.

$$[(C_2H_5)_2N \quad\quad O \quad\quad N(C_2H_5)_2]^+ \quad Cl^- \qquad [(CH_3)_2N \quad\quad O \quad\quad N(CH_3)_2]^+ \quad Cl^-$$

Pyronin B   Pyronin Y

Both are, however, fairly soluble in water and sparingly soluble in alcohol.

Kurnick (1950*a*, *b*) first demonstrated that pyronin preferentially stains low polymers of nucleic acid, but stoichiometric studies did not reveal constancy in the binding of depolymerised DNA by pyronin (Kurnick and Mirsky, 1950). To secure selective staining, pyronin G or Y is suitable as pyronin B results in non-specific cytoplasmic staining. Evidently, methylation may have some connection with the staining property of pyronin.

The selectivity of pyronin staining of RNA, as already mentioned, is confirmed in Brachet's (1942) schedule through ribonuclease digestion. Kurnick (1955*b*), however, noted that if the dye is used alone, nuclei also stain red, indicating thereby that methyl green competes effectively with pyronin for polymerised DNA while used in a mixture. Kurnick (1952) noted that pyronin B or a number of pyronin Y preparations of American firms stain protein as well, and, as such, they are ineffective for quantitative analysis. On the other hand, pyronin Y of G. P. Gurr Ltd revealed specific RNA coloration, specially when used in an aqueous solution with methyl green, followed by rinsing in butyl alcohol. In any case, the pyronin Y of different firms can be used (Kurnick, 1955*a*; Kaufmann, Gay and Mc-Donald, 1960; Pearse, 1960, 1968) if extracted repeatedly with chloroform, like methyl green. Paolillo (1964) has discussed in detail the specificity of different pyronins for plant tissues.

The choice of a proper fixative, as mentioned for methyl green, is also important for pyronin staining, as the use of fixatives which cause depolymerisation of nucleic acid results in pyroninophilia of DNA. For the same reason, nuclei of degenerating cells also stain with pyronin (Harris and Harris, 1950). Kurnick (1955a) has however pointed out that depolymerisation may be of different degrees, ultimately resulting in mononucleotides, but even before this ultimate stage is reached, DNA may become pyronin positive and lose methyl green stainability. It has also been pointed out that a mere rise in temperature, such as from 80–100 °C, may cause depolymerisation to the extent of pyroninophilia but not a loss of protein precipitability (Swift, 1953). If the depolymerisation is so severe as to dissolve DNA, no pyronin staining can be observed. Laverack (1955) recommended the staining for the study of nucleic acid in unfixed frozen sections.

Importance of the Methyl Green–Pyronin Method

Considerable advances have been made in the study of the chemical make-up of chromosomes and related structures of various organisms with the aid of the differential staining schedule. In addition to its application to various plant and animal cells, this method is finding wide application in cancer cytology, cytological studies from tissues cultured in artificial medium and bone marrow, as well as in peripheral blood smears (Brachet, 1942; Stowell, 1946; White, 1947; Perry and Reynolds, 1956; Hoffman, 1956; Arakaki and Sparkes, 1963).

Kaufmann, McDonald and Gay (1951) and Kaufmann, Gay and McDonald (1960) carried out intensive investigations on the identification of chromosomal components through enzymic digestion of fixed tissues and various staining procedures, including especially pyronin and methyl green. They showed that chromosomes contain both RNA and DNA in the form of chains, in addition to basic and non-basic proteins, and that both histone and non-histone proteins remain associated with DNA, whereas RNA is associated principally with histones. This latter finding is contrary to that of Mirsky and Ris (1947) who assumed RNA to be associated principally with tryptophane-rich protein in 'residual chromosomes', responsible for maintaining the structural integrity. Kaufmann considered that all the components form an interconnected system and no single component should be considered as essential for structural integrity (*see* also Dupraw, 1965).

During the active phase, DNA is associated with non-histone protein and 'genetic inactivity' is correlated with DNA–histone. Spermatozoa of certain organisms which do not contain either histone or protamine may be exceptions (Bernstein and Mazia, 1953; Ris, 1958). The shifting pattern in different phases of metabolism has been emphasised (Bloch, 1958). Evidence from studies involving differential localisation of DNA and RNA amply indicates that RNA exerts a significant influence in changing the properties of chromosomes during the course of mitosis (Kaufmann and Das, 1954; Kaufmann, McDonald and Bernstein, 1955; Davidson, 1957).

'Mobilisable' RNA may play a part in crossing over and recombination, especially as ribonucleoprotein has been found in attachment plates holding the synapsing chromosomes in *Tradescantia* (Kaufmann and colleagues, 1959; Kaufmann, Gay and McDonald, 1960). RNA may play a role in chromosome movement, or chromosomes may simply be convenient carriers for RNA distribution (Mazia, 1961). In spite of the fact that DNA forms the essential genetic substance, the presence of RNA in chromosomes, as clearly evidenced through the methyl green-pyronin technique and ribonuclease digestion, has opened up new complexities in the study of chromosome behaviour and function, the unravelling of which is essential for understanding the genetic control of all vital activites.

The impact of Brachet's works has been amply felt in the ultimate analysis of the different types of RNA involved in transcription and translation—the essential steps in protein synthesis. All present concepts of gene action and regulation of differentiation may be at least traced back to the differential localisation of DNA and RNA in the cell through the method devised by Brachet.

SCHEDULES FOR PYRONIN–METHYL GREEN STAINING

Modified Schedule of Brachet (1942 and 1953)

The recommended fixation time is for 4–16 h at pH 7·0 in 10 per cent aqueous formalin.

Reagents

(*a*) *Methyl green*—Before use, wash methyl green repeatedly with chloroform or amyl alcohol to dissolve and remove the traces of methyl violet which are formed when methyl green is exposed to the atmosphere. Filter off the residual methyl green and dry it. Alternatively, shake an aqueous solution of methyl green with an excess of chloroform or amyl alcohol. The violet component is to be gradually removed. Allow to stand for 2–3 days. Remove the aqueous supernatant liquid for use.

(*b*) *Pyronin Y or pyronin G*—In general, the bluish shades are more satisfactory.

(*c*) *Ribonuclease*—Ribonuclease solution can be prepared by Brachet's method, as follows. Mince up finely 0·5–1 kg of ox pancreas by passing through mincer. Pound the minced meat into a smooth paste with mortar and pestle. Suspend the paste in an equal volume N/10 acetic acid for 24 h. Boil for 10 min, cool and filter. Bring the pH to 6·0. Filter again. Add a few thymol or camphor crystals as preservative and store in cold. The enzyme retains its activity for several months.

(d) *Preparation of the required reagents*

Pyronin–methyl green mixture

Methyl green (washed in chloroform and dried)	0·15 g
Pyronin Y	0·25 g
90% ethyl alcohol	2·5 cm^3
M/5 acetate buffer pH 4·7	97·5 cm^3

In an alternative method, modified from Trevan and Sharrock (1951, *see* Pearse, 1960, p. 825) the solutions used are:

Solution (i)

5% aq. pyronin solution	17·5 cm^3
2% aq. methyl green solution (chloroform-washed)	10 cm^3
Distilled water	250 cm^3

Solution (ii)

M/5 acetate buffer pH 4·8

Mix equal quantities of (*i*) and (*ii*) in a staining jar. Do not use mixture after keeping for a week.

Ribonuclease

0·1% solution in distilled water
adjusted to pH 6·0

This chemical has been used for control experiments in the original schedule by Brachet.

Procedure

Only a general schedule is given, which has to be modified according to requirements.

1. Fix sliced tissue in Carnoy's or Zenker's fluid. Wash. Dehydrate as usual. Embed in paraffin. Cut paraffin sections. Deparaffinise in toluene and bring the sections down through alcohol grades to distilled water.

2. Stain the sections in pyronin–methyl green mixture for 20 min.

3. Wash them rapidly with distilled water.

4. Differentiate in 95 per cent ethyl alcohol for 5–10 min.

5. Dehydrate in absolute ethyl alcohol, clear in toluene and mount in D.P.X.

6. For control experiments: (*a*) keep one set of slides, marked 'A', in distilled water at pH 6·0 in an oven at 37 °C for 1 h. Then stain, differentiate, dehydrate and mount as described before; (*b*) keep a second set of slides, marked 'B', in ribonuclease solution in an oven at 37 °C for 1 h and then follow the staining schedule described before.

Observations

1. In the normally stained tissue, RNA in nucleolus and cytoplasm takes up red colour, while DNA appears as green particles in the nuclear chromatin.

2. In control experiment (*a*) no green colour is observed, showing that DNA has been removed by warm water.

3. In control experiment (*b*), RNA is removed by ribonuclease, shown by the complete absence of red colour.

Methyl Green–Pyronin Y Schedule by Kurnick (1955b)

The recommended fixative is Carnoy's fluid or freeze–drying.

Reagents

Pyronin Y	2 g
Methyl green	2 g

Dissolve 2 g of pyronin Y in 100 cm^3 of distilled water. Add chloroform and shake the mixture in a separating funnel till the layer of chloroform becomes colourless. Separate the dissolved dye. Similarly prepare a 2 per cent solution of methyl green and extract it with chloroform. The solutions can be kept as stock. For use, mix together 12·5 cm^3 of pyronin Y solution and 7·5 cm^3 of methyl green solution and add 30 cm^3 of distilled water.

Procedure

1. Fix and embed the tissue. Bring down the paraffin sections to distilled water.

2. Immerse the sections in the staining mixture for 6 min. Freeze–dried sections can be stained directly.

3. Remove excess stain by blotting with filter paper.

4. Transfer to n-butyl alcohol, keeping for 5 min and then treat in a further change of n-butyl alcohol for 5 min.

5. Transfer to xylol and keep for 5 min.

6. Transfer to cedarwood oil and keep for 5 min.

7. Mount, preferably in Permount.

Observations

Chromatin is stained green while cytoplasm and nucleoli are bright red.

Alternatives

In an alternative schedule, increase the period of staining up to 10–30 min, rinse in distilled water, drain and blot, keep in 2 changes of n-butyl alcohol for 5 min each and mount directly in euparal (Darlington and La Cour, 1960).

Jordan and Baker's Modification of Brachet's Schedule (1955)

Reagents

1. Methyl green extract is freshly prepared by treating 0·5 per cent methyl green solution in distilled water with chloroform repeatedly till the chloroform layer is colourless.

2. Acetate buffer—prepared by mixing 119 cm^3 of 0·2 M sodium acetate solution and 81 cm^3 of 0·2 M acetic acid and maintained at pH 4·8.

3. 0·15 per cent aqueous pyronin G solution.

For the final mixture, mix together:

0·5% aq. pyronin G solution	37 cm^3
0·5% aq. extracted methyl green solution	13 cm^3
0·2 M acetate buffer	50 cm^3

This mixture retains its capacity for at least 4 months.

Procedure

1. Bring down the paraffin sections to distilled water and blot to remove moisture.

2. Immerse for 30 min in the buffered staining mixture.

3. Wash in distilled water for a few seconds and blot again.

4. Keep in pure acetone for 1 min.

5. Transfer to acetone–xylol mixture (1:1) and then to pure xylol, keeping for 1 min in each.

6. Mount in neutral balsam.

Observations

DNA takes up blue, blue green or green colour while RNA stains red.

Kay and Dounce's Modification (1953), Kay (1953):

The staining mixture is prepared by using 0·37 per cent methyl green and 0·11 per cent pyronin B dissolved in glycerol, 20 cm^3; 2 per cent aqueous phenol, 100 cm^3; and rectified spirit, 25 cm^3. It can be used for fresh suspensions or dried smears of tissue homogenates.

Methyl Green–Pyronin and Ribonuclease Method for RNA (according to Brachet (1942) and Trevan and Sharrock (1951))

1. Fix in 10 per cent formalin at pH 7 ± 0.2 for 4–16 h.
2. Since commercially prepared methyl green is a mixture of methyl green and methyl violet, for purification, shake the aqueous solution of the dye with excess of chloroform or amyl alcohol. After 2–3 days, remove aqueous supernatant for use. Though the methyl green thus purified slowly breaks down into methyl violet, the process is so gradual as to be negligible.
3. Prepare:

Solution A: 5 per cent aqueous pyronin 17.5 cm^3; 2 per cent purified aqueous methyl green 10 cm^3; distilled water 250 cm^3.

Solution B: 0.2M acetate buffer, pH 4.8. It may or may not contain 30 cm^3 of 1 per cent orange G. Alternatively, an acid citrate buffer (pH 5.0), prepared by adding 51.5 cm^3 of 0.2 disodium hydrogen phosphate to 48.5 cm^3 of 0.1 M citric acid can be used.

Mix equal volumes of A and B. This mixture should not be kept for longer than a week.
4. Bring down the sections to water.
5. Stain in methyl green–pyronin solution for 10 min to 24 h.
6. Rinse in distilled water for a few seconds.
7. Blot and dehydrate rapidly in absolute acetone.
8. Rinse briefly first in a mixture of acetone and xylene (1:1) and then in 10 per cent acetone in xylol.
9. Clear in two changes of xylol and mount in a synthetic resin.
10. The nuclear chromatin takes up green, bluish green or purplish green colour while the sites of RNA stain red.

Modified Schedule for Plant Tissues (Bhaduri and Mukherjee, 1961)

Reagents

1. Pyronin Y (No. 10779, Gurr) England—2 per cent solution in distilled water, purified by shaking with an equal quantity of chloroform in a separate funnel.
2. Methyl Green Chroma (No. 50.07) Germany—purify 2 g by shaking twice with 50 cm^3 of chloroform, filtering and drying the residue at room temperature in a desiccator. Prepare a 2 per cent solution in distilled water

Prepare staining mixture by adding together:

2% aq. pyronin solution	50 cm^3
2% aq. methyl green solution	30 cm^3
Distilled water	10 cm^3
Chloroform	10 cm^3

Procedure

1. Bring paraffin sections of tissues, previously fixed in Carnoy's fluid, down to distilled water.
2. Shake the staining mixture before use. Then immerse the slides in it for 15 min at 20 °C.
3. Blot excess stain with a filter paper and dip the slide in 50 per cent ethyl alcohol and pure n-butyl alcohol mixture (1:1) for 2 s.
4. Immerse in n-butyl alcohol for 10 min.
5. Keep in pure chloroform for 10 min.
6. Again immerse in n-butyl alcohol for 5 min.
7. Treat in toluol for 1 h.
8. Mount in neutral balsam.

Observations

Chromosomes and nuclear chromatin stain bright green while nucleolar and cytoplasmic RNA stain pink to deep rosy red.

Sulkin's Schedule (1951)

Fix tissues in 10 per cent formalin, Zenker's solution of 90 per cent alcohol, bring the paraffin sections to water and immerse, one lot in N/10 KOH for 45 min at room temperature and the other in water. Stain in toluidine blue, haematoxylin–eosin or phloxine–methylene blue. These dyes stain both DNA and RNA. Exposure to KOH inhibits the staining of RNA alone.

TRICHOME STAINING SCHEDULE

Another technique for the differential stain of the two types of nucleic acid was developed by Korson (1951). In this method, trichome staining combinations were employed—orange G for protein, methyl green for DNA and toluidine blue for RNA. All of them are used in a common mixture in aqueous solutions. Orange G (Michrome No. 411) is an acid dye of the azo group with the formula:

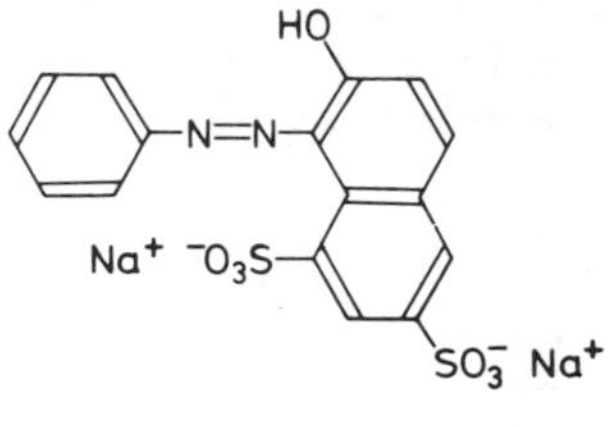

Toluidine blue (Michrome No. 404) is a basic dye of the thiazine group with the formula:

$$\left[(CH_3)_2N \underset{S}{\overset{N}{\bigcirc\bigcirc\bigcirc}} NH.CH_3 \right]^+ Cl^-$$

The fixative recommended is acetic–alcohol or acetone or even formalin-containing fixatives. Though normally methylene blue, toluidine blue, etc., stain RNA, it is difficult to distinguish the cytoplasmic RNA from the nuclear one. Following the trichrome technique, bright green coloration of chromatin is seen simultaneously with blue nucleolus and orange cytoplasm with blue-coloured structures embedded in it. The method involves staining to an 'end-point' and removing the excess dye through overnight treatment in butyl alcohol. Though the exact mechanism involved in the process is not clear, yet it is claimed to be a chemical reaction in which orange G combines with the proteins, leaving nucleic acids available for methyl green and toluidine blue staining. DNA combines with methyl green, and RNA is left over to react with toluidine blue. The specificity of this method has been claimed on the basis of control procedures. Extraction of RNA by cold perchloric acid removes those materials which stain in toluidine blue, whereas negative methyl green and toluidine blue reactions are obtained by extraction of both kinds of nucleic acids through hot trichloracetic acid. Similarly, the application of DNA and RNA before staining yields negative methyl green and toluidine blue reactions respectively.

This method, though recommended for the differential staining of DNA and RNA against a differentially stained cytoplasmic background, is not as widely applied as methyl green and pyronin, the latter method being particularly suitable for quantitative estimation. Moreover, the contrast is not very sharp when the green and blue coloration is used as the differentiating index of the two types of nucleic acids. One special advantage of this technique lies in its capacity to differentiate other cytoplasmic components from RNA positive bodies.

Several authors (Bradley and Wolf, 1959; Feder and Wolf, 1965; *see* Lamm *et al.*, 1965) have noted that metachromatic staining of DNA and RNA can be obtained with toluidine blue. Fixation in acrolein and embedding in ester wax have been recommended. Ghosh and Lettré (1969) worked out a schedule with toluidine blue and Feulgen reaction for double staining of DNA and RNA. The presence of DNA in nucleoli has also been demonstrated by Ebstein (1969), confirming the previous findings of Lettré and her colleagues.

Schedule for Differential Trichome Staining of DNA and RNA (Korson, 1951)

The fixatives recommended are Carnoy's fixative or cold acetone. The tissues stained were rat liver, spleen, thymus, intestine, spinal cord and pancreas, hepatoma and cholangioma and human thyroid and bone marrow. Marrow is smeared, air dried, fixed in absolute methyl alcohol for 3 min and dried again before use.

Staining solutions used

1. Orange G—4 per cent solution in distilled water.
2. Methyl green—purified by repeated washing in chloroform 0·15 per cent solution in distilled water.
3. Toluidine blue O—0·1 per cent solution in distilled water.

Procedure

1. Bring down the slides to water.
2. Treat in orange G solution for 2 min.
3. Dip in a jar containing distilled water to which a drop of orange G solution has been added.
4. Treat in methyl green solution for 15 min.
5. Immerse the slides in toluidine blue–methyl green mixture (1:1) for 5 min.
6. Rinse in absolute tertiary butyl alcohol.
7. Treat overnight in a change of absolute tertiary butyl alcohol.
8. Clear in xylol and mount.

Observations

The nuclear chromatin stains bright green (DNA), the nucleolus blue (RNA) and the cytoplasm orange with a variable amount of blue staining structure (RNA).

Several control experiments were set up to verify the results obtained. The schedules and their observations are given below.

Experiment 1—Bring down sections to water. Incubate with crystalline desoxyribonuclease to remove DNA selectively at room temperature for 7 h. The solution used contains 0·1 mg of desoxyribonuclease per 1 cm^3 of McIlvaine's buffer pH 6·5. Sufficient magnesium chloride is added to a 0·2 M solution. Stain as usual.

In the final preparation, chromatin is colourless, nucleoli are blue and cytoplasm is orange with blue granules.

Experiment 2—Incubate sections for 1 h at 45 °C with crystalline ribonuclease (0·01 mg/cm^3 solution) in McIlvaine's buffer pH 6·5 to remove RNA selectively. Stain as usual. Chromatin is found to be coloured green, nucleoli are colourless and cytoplasm is orange with colourless granules.

Experiment 3—Incubate the sections with desoxyribonuclease, followed by ribonuclease. Stain as usual. Both chromatin and nucleoli are colourless and cytoplasm is orange with colourless granules.

Experiment 4—Treat the sections with 10 per cent perchloric acid solution for 11 h at 4 °C for extracting RNA. Stain as usual. Chromatin stains green, nucleoli remain colourless and cytoplasm is orange with colourless granules.

Experiment 5—Treat sections with 5 per cent hot trichloracetic acid at 90 °C for 15 min or with 10 per cent hot perchloric acid at 80 °C for 15 min for extracting both nucleic acids. Stain as usual. Both chromatin and nucleoli are colourless and cytoplasm is orange with colourless granules.

(Acrolein–Toluidine Blue Method for DNA and RNA according to Feder and Wolf (1965))

1. Fix small bits of tissue in 10 per cent aqueous acrolein with 0·5 per cent calcium acetate.
2. Dehydrate in a mixture of methanol and methoxyethanol (1:1), pass through ethanol and *n*-propanol grades and embed in polyester wax following the usual schedules.
3. Cut sections 5 μm thick and mount on slides.
4. Bring the slides down to water and stain for 8 min in 0·1 per cent toluidine blue in phosphate buffer to pH 4·2 at 22 °C.
5. Dehydrate in tertiary butyl alcohol for 5 min, clear in xylol and mount. in a synthetic resin.
6. The sites of DNA take up a deep blue stain while those of RNA are pale purple.

GALLOCYANIN–CHROME ALUM FOR DIFFERENTIAL STAINING OF RNA AND DNA

Einarson (1935, 1949, 1951) suggested the use of a lake formed by gallocyanin and chrome alum for the differential staining of DNA and RNA. Gallocyanin (Michrome No. 143) is a dye of the oxazine group with the formula:

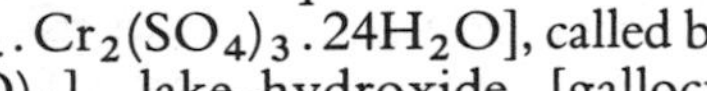

and acts as a weakly acid stain in aqueous solution. It forms three salts with chrome alum [$K_2SO_4 \cdot Cr_2(SO_4)_3 \cdot 24H_2O$], called by Einarson lake–cation [gallocyanin–$Cr(H_2O)_4$], lake–hydroxide [gallocyanin–$Cr(H_2O)_4OH$] and lake–sulphate [gallocyanin–$Cr(H_2O)_4SO_4$]. The lake–cation reacts with the phosphate groups of the nucleic acids to form combination dark-blue lake tissue salt. This reaction has been used for quantitative estimation of nucleic acid by Einarson. According to Sandritter, Diefenbach and Krantz (1954), 1 molecule of gallocyanin is bound by 15 phosphorus atoms in 'polymerised' RNA and by 23 atoms in heated RNA. Diefenbach and

Sandritter (1954) found the ratio of gallocyanin and DNA in their stoichiometric combination to be 1:3·7. Sandritter *et al.* (1963) noted a proportionate increase in red colour with heating up to over 30 min. Harms (1965) suggested that chromium lake of gallocyanin is formed by complexing of two hydroxyl groups lying adjacent to each other.

The advantages of gallocyanin–chrome alum technique for qualitative staining of nucleic acids are:

1. It is a progressive stain and is not washed out during dehydration and clearing.

2. Though the highest specificity of staining is at lower pH levels, between 1·5 and 1·75, yet staining can be done at almost any pH between 0·8 and 4·3.

However, the chief drawback of this method for quantitative staining is, according to Pearse (1968), that information about the proportion of combination with nucleic acids and stoichiometric data is still incomplete. Stenram (1954) found that even after ribonuclease extraction, staining with gallocyanin–chrome alum at a pH up to 4·0 was observed in Nissl substance in nerve cells. De Boer and Sarnaker (1956) extracted from gallocyanin solution, a compound which stained Nissl bodies bright blue while the rest of the cytoplasm remained colourless. Non-specific staining is held to be due to the attachment of the lake–sulphate to acid groups in the tissues through its dimethylamine grouping. Mayersbach (1956) regarded this method as more specific than pyronin–methyl green technique. Pearse (1968), however, regarded it to be a highly specific method, but not of the same calibre as pyronin–methyl green schedule. This method evidently does not require specific fixation and, though suitable for qualitative observations, is not very successful for the quantitative study of differentially stained substances.

Einarson's Schedule (Einarson, 1951)

Different fixatives have been recommended.

Reagents

Chrome alum	5 g
Gallocyanin	0·15 g
Distilled water	100 cm^3

Dissolve chrome alum in distilled water and add to it gallocyanin. Shake, and heat in a beaker on a flame and boil for 5 min. Remove the flame, allow the solution to cool to room temperature. Filter and collect the filtrate in a measuring cylinder. Add distilled water to the solution and filter until the filtrate measures 100 cm^3. This stock solution can be kept for about 1 month. The optimum pH is 1·64. If necessary, the pH of the solution can be adjusted by adding aqueous 1 N HCl or 1 M NaOH solution in the proportions shown in Table 11. If a precipitate is formed on adding NaOH, do not keep the solution for more than a week.

Table 11

Stock solution in cm³	pH *needed*	*Amount of* N HCl *to be added in* cm³	*Amount of* 1 M NaOH *to be added in* cm³
40	0·83	10	0
	0·90	9	0
	0·92	8	0
	0·94	7	0
	1·02	6	0
	1·10	5	0
	1·14	4	0
	1·18	3	0
	1·29	2	0
	1·44	1	0
	1·64	0	0
	1·84	0	1
	2·16	0	2
	2·90	0	3
	3·42	0	4
	3·76	0	5
	3·98	0	6
	4·07	0	7
	4·18	0	8
	4·27	0	9
	4·35	0	10

Procedure

1. Bring down the paraffin sections to water as usual.
2. Immerse and keep in dye–lake solution for 48 h at room temperature.
3. Rinse in distilled water.
4. Dehydrate through alcohol and xylol grades and mount in a suitable medium.

Observations

RNA and DNA take deep blue stain, the optimum pH with least non-specific staining being 0·83–0·94.

Modified Gallocyanin Schedule (De Boer and Sarnaker, 1956)

Various fixatives have been recommended for preparing paraffin blocks. Freeze-drying is employed after fixation in formalin.

Reagents

Gallocyanin	0·6 g
Chrome alum	10 g

Dissolve chrome alum in 200 cm^3 of distilled water. Shake the gallo-cyanin crystals in 200 cm^3 of distilled water for 1 min. Filter, and add the residue with the filter paper to chrome alum solution. Boil on a water bath for 30 min. Cool to room temperature. Filter, and add 1 per cent HCl to bring pH to 1·6.

Procedure

1. Bring paraffin sections down to distilled water as usual.
2. Immerse in the dye–lake for 24 h or more, the period being dependent on the tissue and the age of the stain.
3. Wash in acidulated distilled water, acidified with HCl at a pH 1·6, for 1 min. Change the solution and wash again until excess dye is removed.
4. Dehydrate as usual through alcohol and xylol grades and mount.

Observations

Deep blue colour is seen at the sites of nucleic acids.

Modified Gallocyanin Method for Nucleic Acids (according to de Boer and Sarnakar (1956))

1. Fix material in formalin and freeze-dry or fix in other fixatives and embed in paraffin.
2. For preparing the dye, shake 600 mg of gallocyanin in 200 cm^3 of distilled water for 1 min, filter and discard the filtrate. Transfer filter paper with residue to 200 cm^3 of 5 per cent chrome alum solution in distilled water. Boil in a water bath for 30 min. Cool, filter and adjust filtrate to pH 1·6 with 1 per cent HCl. Alternatively, following the method of Berube *et al.* (1966), mix 150 mg gallocyanin and 15 g chromalum in 100 cm^3 of distilled water, boil for 10–20 min, cool and filter. Bring filtrate to 100 cm^3 by washing the precipitate with water. The filtrate can be used directly for staining, or to separate the chelate, add dilute ammonia to bring filtrate to pH 8·0–8·5. Filter with suction through a medium-porosity fitted glass funnel. Wash the precipitate with anhydrous ethyl ether, dry and store. Before use, prepare a 3 per cent solution in normal sulphuric acid.
3. Bring down the sections to water.
4. Immerse in gallocyanin-chromalum solution for 24 h or more, depending on the maturity of the stain.
5. Wash for 1 min in distilled water acidulated with HCl at pH 1·6.
6. Wash repeatedly till no more dye comes out.
7. Dehydrate in ethanol, clear and mount in a synthetic resin.
8. The nucleic acids take up a deep blue stain.

STAINING OF RIBONUCLEIC ACID

Supravital staining of RNA by neutral red has been claimed by Dustin (1947). Neutral red is a basic dye of the azine group, having the formula:

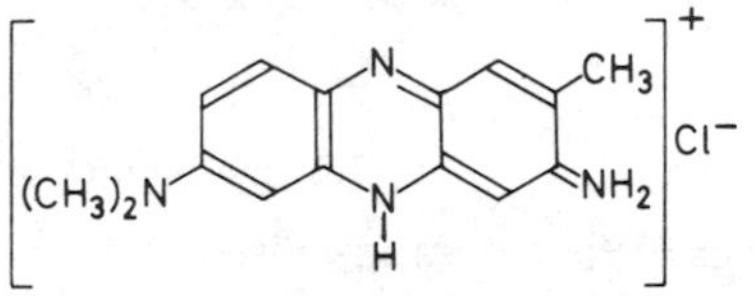

However, this stain is not widely applied as not all RNA in the cell is detected and chromosomal RNA cannot be studied through this method. Similar staining has also been noted with stilbamidine (Kurnick, Klein and Klein, 1950; Snapper and colleagues, 1951). No *in situ* localisation is possible in these methods as particles aggregating after diffusion may also take up the colour. Gram staining of bacteria is also dependent on its RNA (Stacey, 1947; Mitchell and Moyle, 1950).

In addition to toluidine blue (Hermann, Nicholas and Boricious, 1950), as mentioned above, methylene blue and several members of the thionin series have been considered (Stowell and Zorzoli, 1947) as stain for RNA. Methylene blue is a basic dye of the thiazine group, having the structural formula:

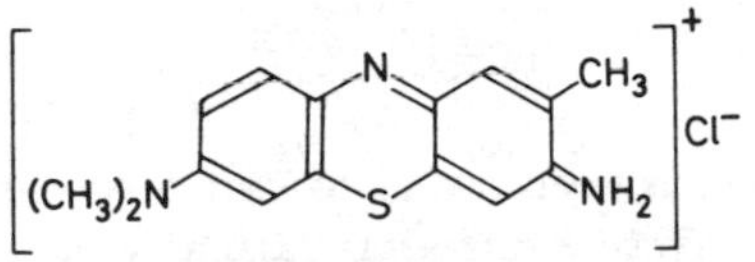

Pyronin is, however, more widely applied than any of these stains because of its superior specificity and capacity for yielding qualitative data.

(i) *Schiff-methylene blue staining* (Spicer, 1961): Bouin's fluid fixation preserves RNA. RNA stains with 0·02 per cent thiazine dye in aqueous McIlvaine phosphate–citrate buffer between pH 3 and 4; 24 h fixation in Bouin's fluid hydrolyses DNA, which stains in Schiff's reagent without further acid hydrolysis. Thus, after fixation in Bouin's fluid, direct Schiff staining followed by 0·02 per cent methylene blue in phosphate–citrate buffer at pH 3·0–3·5 colours DNA magenta and RNA blue.

(ii) *Pyronin staining for RNA* (Tepper and Gifford, 1962): Dehydrate fixed shoot or root-tips in tertiary butyl alcohol series, embed, cut at 7 μm, and mount with Haupt's adhesive. Place sections in a 2 per cent aqueous solution of pyronin Y for 6 min, blot, differentiate twice for 5 min each in tertiary butyl alcohol, clear twice for 5 min in xylol and mount in Harleco synthetic resin. The dye had previously been extracted 5 times with equal volumes of chloroform and kept at a pH of 3·5.

OTHER STAINS OF DNA

Similar to methyl green, stoichiometric data are claimed to have been obtained with crystal violet for DNA by Kurnick (1950b). Staining with aminoacridines too is stated to be due to the polymerised nature of DNA (Irvin and Irvin, 1952; Lawley, 1956).

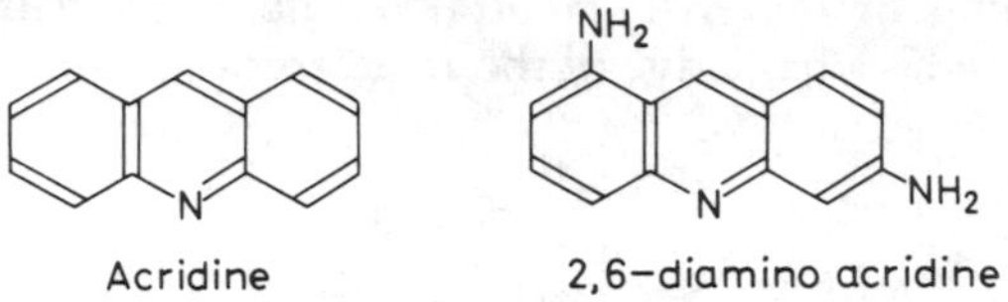

Acridine   2,6-diamino acridine

De Bruyn and colleagues (1953) demonstrated that the spacing of amino groups in 2,8-diaminoacridines is appropriate for forming hydrogen bonding with phosphate groups of DNA, being 7–8 Å apart. Due to its polymerised nature, DNA staining has also been reported with quinolines (Parker, 1949), methyl green, malachite green (Kurnick, 1950a, b) as well as rosaniline (Lawley, 1956). Malachite green, a basic dye of the triphenyl-methane group, has the structural formula:

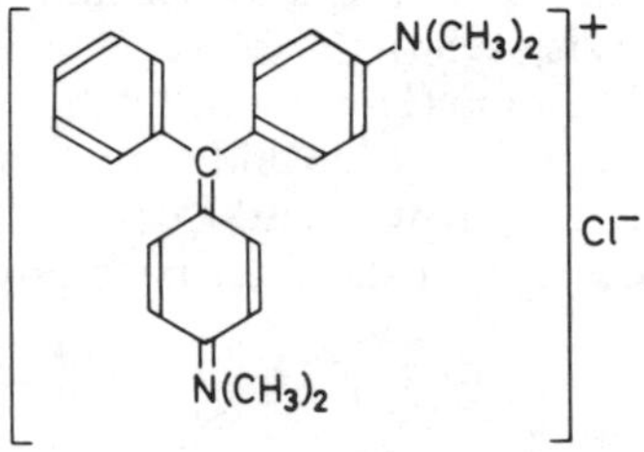

With rosaniline, the dye has been found to compete with metal cations for binding sites in DNA. Acriflavine is regarded as a supravital stain for chromatin (De Bruyn, Robertson and Farr, 1951; Oster and Grimson, 1949) but its specificity for DNA is not yet known. Brenner (1953) suggested that, even with haematoxylin, specific staining for DNA can be obtained with alum mordant. Celestine blue is also considered as a specific DNA stain by Sanders (1946) and Davidson (1948) who have utilised this dye with pyronin for the differential stain of two nucleic acids. However, its specificity is yet to be ascertained (Kurnick, 1955a). Celestine blue is a basic dye of the oxazine group, with the structural formula:

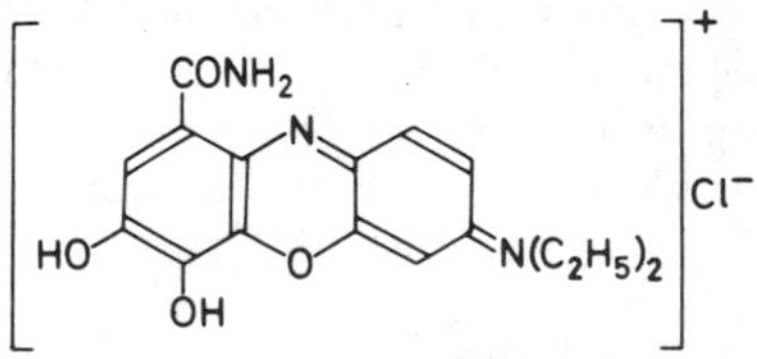

Sambucyanin has been considered as a stain for nucleic acids by Novelli (1954), which gives a red colour with both DNA and RNA.

Azure B staining has been so adjusted under appropriate conditions that DNA stains mainly orthochromatically and RNA metachromatically, so that DNA takes up a blue green colour, while cytoplasmic and nucleolar RNA are red purple (Flax and Himes, 1952).

Azure A can be used as a stain for nucleic acid if the dye is used in an aqueous solution and differentiation is carried out overnight in absolute ethyl alcohol. Its specificity has been confirmed by digestion with nucleases (Flax and Pollister, 1949).

Himes and Moriber (1956) stained paraffin sections successively in Feulgen reaction (azure A–Schiff reagent); periodic acid–Schiff (with basic fuchsin–Schiff reagent) and a 0·02 per cent solution of naphthol yellow S in 1 per cent acetic acid. It serves as a triple stain—blue green nuclei, red polysaccharides and yellow proteins.

REFERENCES

Alfert, M. (1952). *Biol. Bull., Wood's Hole* **103,** 145
Arakaki, D. T. and Sparkes, R. S. (1963). *Cytogenetics* **2,** 57
Arzac, J. P. (1950). *Stain Tech.* **25,** 187
Backler, B. S. and Alexander, W. F. (1952). *Stain Tech.* **27,** 147
Baker, J. R. (1958). *Principles of biological microtechnique.* London; Methuen
Baker, J. R. (1966). *Cytological techniques.* London; Methuen
Balbiani, E. G. (1881). *Zool. Anz.* **4,** 662
Benzer, S. (1967). *The Chemical basis of heredity,* p. 70. ed. by McElroy, W. D. and Glass, B. Baltimore; Johns Hopkins Press
Bernard, E. A. and Danielli, J. F. (1956). *Nature* **178,** 1450
Bernstein, M. H. and Mazia, D. (1953). *Biochem. biophys. Acta* **10,** 600
Berube, G. R., Powers, M. M., Kerkay, J. and Clark, G. (1966). *Stain Tech.* **41,** 73
Bhaduri, P. N. and Mukherjee, A. K. (1961). *Nucleus* **4,** 169
Bial, M. (1903). *Dtsch. med. Wschr.* **29,** 253 and 477
Bloch, D. P. (1958). *Frontiers of cytology,* p. 113, New Haven, Conn.; Yale University Press
Brachet, J. (1940). *Embryologie Chimique.* Paris; Masson
Brachet, J. (1942). *Arch. Biol. (Liège)* **53,** 207
Brachet, J. (1953). *Quart. J. micr. Sci.* **94,** 1
Brachet, J. (1958). *Biochemical cytology.* New York; Academic Press
Bradley, D. F. and Wolf, M. K. (1959). *Proc. Nat. Acad. Sci. Wash.* **45,** 944
Brenner, S. (1953). *Exp. Cell Res.* **5,** 257
Caspersson, T. (1947). *Symp. Soc. exp. Biol.* **1,** 127
Ceriotti, A. (1952). *J. biol. Chem.* **198,** 297
Cohen, S. J. (1944). *J. biol. Chem.* **156,** 691
Cowden, R. R. (1965). *Histochemie* **5,** 441
Danielli, J. F. (1947). *Symp. Soc. exp. Biol.* **1,** 101
Darlington, C. D. and La Cour, L. F. (1960). *Handling of chromosomes,* London; Allen and Unwin
Davidson, D. (1957). *Chromosoma* **9,** 39
Davidson, J. N. (1948). *Cold Spr. Harb. Symp. quant. Biol.* **12,** 50
De Boer, J. and Sarnaker, R. (1956). *Med. Proc. S.A.* **2,** 218
De Bruyn, P., Farr, R. S., Banks, H. and Morthland, F. W. (1953). *Exp. Cell Res.* **4,** 174
De Bruyn, P., Robertson, R. C. and Farr, P. S. (1951). *Anat. Rec.* **108,** 279
De Martino, C., Capanna, E., Civitelli, M. V. and Procicchiani, G. (1965). *Histochemie* **5,** 78
Diefenbach, H. (1944). *Proc. Soc. exp. Biol., N.Y.* **55,** 217
Diefenbach, H. (1949). *J. biol. Chem.* **181,** 379
Diefenbach, H. (1953). *J. biol. Chem.* **204,** 983
Diefenbach, H. (1955). *The Nucleic acids,* vol. 1, p. 285. New York; Academic Press
Diefenbach, H. and Borenfreund, E. (1957). *Biochem. Biophys. Acta* **23,** 639
Diefenbach, H. and Sandritter, W. (1954). *Acta Histochem.* **1,** 5
Diefenbach, H. and Schwartz, K. (1937). *Mikrochim. Acta* **2,** 13
Dupraw, E. J. (1965). *Nature* **206,** 338
Dustin, P. (1947). *Symp. Soc. exp. Biol., N.Y.* **1,** 114
Ebstein, B. S. (1969). *J. Cell Sci.* **5,** 27
Einarson, L. (1935). *J. comp. Neurol.* **61,** 101
Einarson, L. (1949). *Acta orthopaed. scand.* **19,** 27
Einarson, L. (1951). *Acta Path. Scand.* **28,** 82

Errera, M. (1951).   *Biochim. Biophys. Acta* **7,** 605
Euler, H. V. and Hahn, L. (1946).   *Svensk. Kem. Tidskr.* **58,** 251
Feder, N. and Wolf, M. K. (1965).   *J. Cell. Biol.* **27,** 327
Flax, M. H. and Himes, M. (1952).   *Physiol. Zool.* **25,** 297
Flax, M. H. and Pollister, A. W. (1949).   *Anat. Rec.* **99,** 56
Freed, J. J. (1969).   In *Physical techniques in biological research* 3C, 95, New York; Academic Press
Freed, J. J. and Benner, J. A. (1964).   *J. Roy. Microscop. Soc.* **83,** 74
Ghosh, S. and Lettré, R. (1969).   *Naturwiss.* **10,** 496
Godman, G. C. and Deitch, A. D. (1957).   *J. exp. Med.* **106,** 575
Goldstein, D. J. (1961).   *Nature, Lond.* **191,** 406
Gulland, J. H. and Jordan, D. R. (1947).   *Symp. Soc. exp. Biol.* **1,** 56
Harms, H. (1965).   *Handbuch der Farbstoffe* dür *die Mikroskopie*, Kamp-Lintfort: Staufen
Harris, S. and Harris, T. N. (1950).   *Proc. Soc. exp. Biol., N.Y.* **74,** 142
Hermann, H., Nicholas, J. S. and Boricious, J. K. (1950). *J. biol. Chem.* **184,** 321
Himes, M. and Moriber, L. (1956).   *Stain Techn.* **31,** 67
Hoffman, B. C. (1956).   *Mikroskopie* **10,** 251
Irvin, J. L. and Irvin, E. M. (1952).   *Fed. Proc.* **11,** 235
Jordan, B. M. and Baker, J. R. (1955).   *Quart. J. micr. Sci.* **96,** 177
Joshi, V. N. and Korgaonkar, K. S. (1959).   *Nature* **183,** 400
Kasten, F. H. (1959).   *Histochemie* **1,** 466
Kaufmann, B. P. and Das, N. K. (1954).   *Proc. nat. Acad. Sci., Wash.* **40,** 1052
Kaufmann, B. P., Gay, H., Dutt, M. K., Bal, A. K. and Buchanan, J. (1959).   *Carnegie Inst. Wash. Yearb.* **58,** 440
Kaufmann, B. P., Das, N. K. and McDonald, M. R. (1960).   *Int. Rev. Cytol.* **9,** 77
Kaufmann, B. P., McDonald, M. R. and Bernstein, M. H. (1955).   *Ann. N.Y. Acad. Sci.* **59,** 553
Kaufmann, B. P., McDonald, M. R. and Gay, H. (1951).   *J. cell. comp. Physiol.* **38,** Suppl. 1. 7
Kay, E. R. M. (1953).   *Stain Tech.* **28,** 41
Kay, E. R. M. and Dounce, A. L. (1953).   *J. Amer. Chem. Soc.* **75,** 4041
Korson, R. (1951).   *Stain Tech.* **26,** 265
Korson, R. (1964).   *J. Histochem. Cytochem.* **12,** 875
Kurnick, N. B. (1947).   *Cold. Spr. Harb. Symp. quant. Biol.* **12,** 141
Kurnick, N. B. (1950a).   *J. gen. Physiol.* **33,** 243
Kurnick, N. B. (1950b).   *Exp. Cell Res.* **1,** 151
Kurnick, N. B. (1952).   *Stain Tech.* **27,** 233
Kurnick, N. B. (1955a).   *Int. Rev. Cytol.* **4,** 221
Kurnick, N. B. (1955b).   *Stain Tech.* **30,** 213
Kurnick, N. B., Klein, E. and Klein, G. (1950).   *Experientia* **6,** 152
Kurnick, N. B. and Mirsky, A. E. (1950).   *J. gen. Physiol.* **33,** 265
Lamm, M. E., Childers, L. and Wolf, M. K. (1965).   *J. cell Biol.* **27,** 313
Laverack, J. O. (1955).   *Quart. J. micr. Sci.* **96,** 29
Lawley, P. D. (1956).   *Biochim. Biophys. Acta* **22,** 451
Leuchtenberger, C. (1950).   *Chromosoma* **3,** 449
Mayersbach, H. (1956).   *Acta Histochem.* **3,** 128
Mazia, D. (1961).   Biochemistry of the dividing cell. *Ann. Rev. Biochem.* **30,** 669
Mellors, R. C. (1955).   *Analytical cytology.* New York; McGraw-Hill
Mendelssohn, M. L. (1966).   In *Quantitative cytochemistry* New York; Academic Press
Mirsky, A. E. and Ris, H. (1947).   *J. gen. Physiol.* **31,** 1
Mitchell, J. S. (1942).   *Brit. J. exp. Path.* **23,** 296
Mitchell, P. and Moyle, J. (1950).   *Nature, Lond.* **166,** 218
Novelli, A. (1954).   *Nature* **173,** 691
Oster, G. and Grimson, H. (1949).   *Arch. Biochem.* **24,** 119
Paolillo, D. J. (1964).   *Acta Histochem.* **18,** 276, 283
Parker, F. S. (1949).   *Science* **110,** 426
Pasteels, J. and Lison, L. (1950).   *C. R. Soc. Biol. Paris* **230,** 780
Pearse, A. G. E. (1951).   *J. clin. Path.* **4,** 1
Pearse, A. G. E. (1953, 1960 and 1968).   *Histochemistry—theoretical and applied.* Boston, Maryland; Little, Brown
Perry, S. and Reynolds, J. (1956).   *Blood* **11,** 1132
Pollister, A. W. (1950).   *Rev. Hemat.* **5,** 527

Pollister, A. W. (1952). *Lab. Invest.* **1,** 106
Pollister, A. W., Himes, M. and Ornstein, L. (1951). *Fed. Proc.* **10,** 629
Pollister, A. W. and Ris, H. (1947). *Cold Spr. Harb. Symp. quant. Biol.* **12,** 147
Pollister, A. W., Swift, H. and Alfert, M. (1951). *J. cell. comp. Physiol.* **38,** 101
Rigler, R. (1964). *Proc. II Int. Cong. Histochemie,* 233 Heidelberg; Springer
Ris, H. (1958). *Colloq. Ges. physiol. chem.* **9,** 1
Ris, H. and Mirsky, A. E. (1949). *J. gen. Phys.* **33,** 125
Roschlau, G. (1965). *Histochemie* **5,** 396
Rosenkranz, H. S. and Bendich, A. (1958). *J. Biophys. Biochem. Cytol.* **4,** 663
Sanders, F. K. (1946). *Quart. J. micros. Sci.* **87,** 203
Sandritter, W., Diefenbach, H. and Krantz, F. (1954). *Experientia* **10,** 210
Sandritter, W., Kiefler, G. and Rick, W. (1963). *Histochemie* **3,** 318
Schneider, W. C. (1948). *Cold Spr. Harb. Symp. quant. Biol.* **12,** 169
Scott, J. E. (1967). *Histochemie* **9,** 30
Serra, J. A. and Lopes, A. Q. (1945). *Port. Acta Biol.* **1,** 111
Smith, S. W. and Anderson, P. N. (1960). *Anat. Rec.* **138,** 179
Snapper, I., Schneid, B., Lieben, F., Gerber, I. and Greenspan, E. (1951). *J. Lab. clin. Med.* **37,** 562
Spicer, S. S. (1961). *Stain Tech.* **36,** 337
Stacey, M. (1947). *Symp. Soc. Exp. Biol.* **1,** 86
Stacey, M. (1950). *Nature* **166,** 771
Stenram, U. (1954). *Acta Anat.* **20,** 36
Stoward, P. J. (1963). *D. Phil. thesis,* Oxford
Stowell, R. E. (1946). *Stain Tech.* **31,** 137
Stowell, R. E. and Cooper, Z. K. (1945). *Cancer Res.* **5,** 295
Stowell, R. E. and Zorzoli, A. (1947). *Stain Tech.* **22,** 51
Sulkin, H. M. (1951). *Proc. Soc. exp. Biol. N.Y.* **78,** 32
Swift, H. (1950). *Physiol. Zool.* **23,** 169
Swift, H. (1953). *Int. Rev. Cytol.* **2,** 1
Taft, E. B. (1951). *Stain Tech.* **26,** 205
Takemura, S. (1958). *Biochim. Biophys. Acta* **29,** 447
Tepper, H. B. and Gifford, E. M. (1962). *Stain Tech.* **37,** 52
Trevan, D. J. and Sharrock, A. (1951). *J. Path. Bact.* **63,** 326
Turchini, J., Castel, P. and Kien, K. V. (1944). *Bull Tech. Histol. Micr.* **21,** 124
Van duijn, C. (1962). *Nature* **193,** 999
Vendrely, R. and Vendrely, C. (1948). *Experientia* **4,** 434
Vercauteren, R. (1950). *Enzymologia* **14,** 134
White, J. C. (1947). *J. Path. Bact.* **59,** 223
Wied, G. (1966). Ed. *Introduction to quantitative cytochemistry.* New York; Academic Press

20 Proteins

Within the nucleus, proteins remain chiefly in combination with nucleic acids, forming nucleoproteins. It is universally accepted that the relationship between proteins and the nucleic acids, RNA or DNA, is very intimate (Chargaff, 1955; Kaufmann, Gay and McDonald, 1960). Nucleolipido-protein complexes are also possible (Serra, 1955, 1968). With the aid of special enzymic digestion methods, followed by tests for proteins and nucleic acids, evidence for the presence of two different types of proteins has been collected, one type being basic (e.g. protamine or histone), and the other acidic. The basic protein types are rich in arginine, lysine and occasionally histidine, and the acidic protein or the non-basic type is rich in tryptophane. However, several other kinds of proteins, and even non-individualised polypeptides, have also been found to exist. Histones may contain a number of amino acids whereas protamines may represent a special type of simplified protein of dibasic amino acids (Serra, 1942).

The two types of proteins can be analysed by ultra-violet spectrophoto-metry (Caspersson, 1940; Hyden, 1943) and the relative concentrations can be determined by chemical tests (Serra, 1946; Glenner and Lillie, 1957; Lillie, 1957). Isolated nuclei in 3 per cent aqueous sodium hydroxide solution can be fractionated by precipitation with mineral acids, permitting the basic proteins to remain in solution from which, in the presence of ammonium chloride, they can be precipitated out in 70 per cent alcohol (Mayer and Gulick, 1942; Serra, 1945).

Stedman and Stedman (1947, 1950) isolated from nuclei an acidic protein which they termed the 'chromosomin', an essential constituent of chromosomes. Mirsky and Ris (1947, 1951) separated nucleo-histones from residual chromosomes, in which tryptophane-containing protein was found with RNA. Later work revealed the existence of a well defined nucleo-histone complex in the chromosome (Vendrely and Vendrely, 1953; Vendrely and colleagues, 1958; Bal and Kaufmann, 1959; Dutt and Kaufmann, 1959), and direct relation between DNA and the arginine content of isolated nuclei has also been found (Vendrely and Vendrely,

446

1953). Dounce (1952, 1955) noted that DNA may also combine with non-histone proteins through its phosphate groups. By means of different cytochemical methods involving enzymic digestion, colour reaction, and extraction with dilute HCl for specifically removing histones, it has been concluded that DNA remains in the chromosomes in association with two types of proteins, histones and non-histones (Mirsky and Ris, 1951; Bernstein and Mazia, 1953; Bloch, 1958; Kaufmann, Gay and McDonald, 1960). Kaufmann and associates (1951) regard RNA as associated with histone, while Mirsky and Ris (1947) consider it to be connected with tryptophane-containing protein.

Though the existence of at least two different types of proteins in chromosomes of higher organisms has been unanimously accepted, their exact pattern of association and function are to some extent still obscure. Mirsky and Ris (1947), from their study on isolated chromosomes, inferred that the residual chromosome, containing both DNA and RNA with non-histone protein, is responsible for maintaining the structural integrity even when the nucleo-histone fraction is separated by sodium chloride treatment. Kaufmann, Gay and McDonald (1960) considered that all four components of the chromosome, that is, DNA, RNA, histone and non-histone proteins, form an interconnected system, and according to them, both nucleic acids and proteins are essential for retaining the chromosomal stability, thus no single component should be considered as an essential structural material.

Regarding the functional aspects of proteins at the chromosomal level, the direct correlation between the amount of DNA and the amount of histone is worth consideration. The function of histones as gene repressors is being well recognised (*see* Zubey, 1964; Busch, 1965). McLeish (1959) found that the nuclear arginine and nuclear DNA show positive correlation in photometric tests. During the interphase, prior to mitosis, doubling of DNA is correlated with histone synthesis (*see* Bloch, 1958). According to him DNA histone complex predominates at the time of DNA-replication, whereas non-histone proteins are associated with DNA at the time of physiological activity. However, as sperms of certain organisms contain neither histones nor protamines, this generalisation is debatable (Bernstein and Mazia, 1953; Ris, 1958; Kaufmann, Gay and McDonald, 1960). The amount of residual or non-histone protein corresponds with the amount of RNA in the cell, which is evidently related to cytoplasmic activity. The protein, associated with RNA, whether histone or non-histone (as in residual chromosome), may be concerned with anaphase movement (Ris and Kleinfeld, 1952) or synapsis, as in *Tradescantia* (Kaufmann and colleagues, 1959).

No universally acceptable model of chromosome structure, reconciling the double helix model of DNA and protein, has yet been proposed, but one has been suggested by Freese (1958) and Taylor (1963, 1969) in which double helix DNA molecules are situated in a linear arrangement, linked by protein blocks, possibly of histone type. The role of divalent cations as linkers has also been suggested. However, this suggested model does not accommodate the longitudinal variations in the chromosomes, especially of the euchromatic and heterochromatic segments (Lewis and John, 1963).

Disregarding the isolation procedures already considered, which do not allow the study of chromosomal proteins *in situ*, a number of chemical tests are at present available through which specific amino acids can be localised at the chromosomal level. They are the Sakaguchi reaction for arginine, tyrosine and tryptophane reactions, based mainly on the Millon's method, and sulphydryl tests for proteins by Barrnett and Seligman (1951).

ARGININE TEST

Of all the amino acid tests so far employed for the study of proteins in the chromosomes, that for arginine is the most important. The original Sakaguchi method (1925) has been modified by different workers (Serra, 1946; Baker, 1947; Thomas, 1950; Liebman, 1951), but is based on the principle (Baker, 1947) that a positive reaction can be obtained with compounds having the formula:

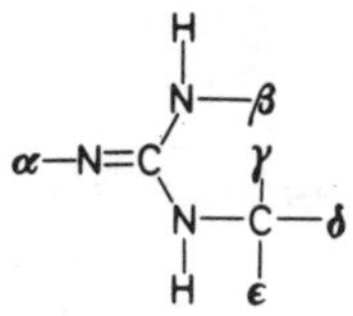

$$\alpha-N=C \begin{matrix} \overset{H}{N}-\beta \\ | \quad \gamma \\ N-C-\delta \\ | \quad \epsilon \\ H \end{matrix}$$

where $\alpha = \beta = $ H or CH_3; and $-\delta = \begin{matrix} -\gamma \\ \\ -\epsilon \end{matrix} = \equiv N$ and the positions of γ, δ and ϵ may be taken by various atoms or by a single nitrogen of dicyandiamide. The guanidine group is involved in the reaction. The compound formed by the reaction of arginine with α-naphthol may have the structure shown below (Gurr, 1958):

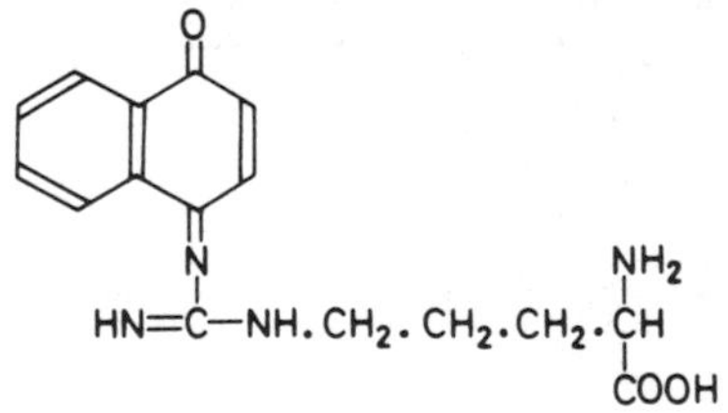

$$HN=C-NH.CH_2.CH_2.CH_2.\underset{\underset{COOH}{|}}{\overset{\overset{NH_2}{|}}{CH}}$$

Serra (1946) obtained positive arginine reaction in the meiotic chromosomes of both plants and animals. Sharma and Bhattacharyya (1957) noted an increase in the arginine content of the nucleus and cytoplasm following induced malignancy. The only limitation of this method is that the colour is unstable and fades rapidly. However, in Liebman's method the colour can be retained for about six months.

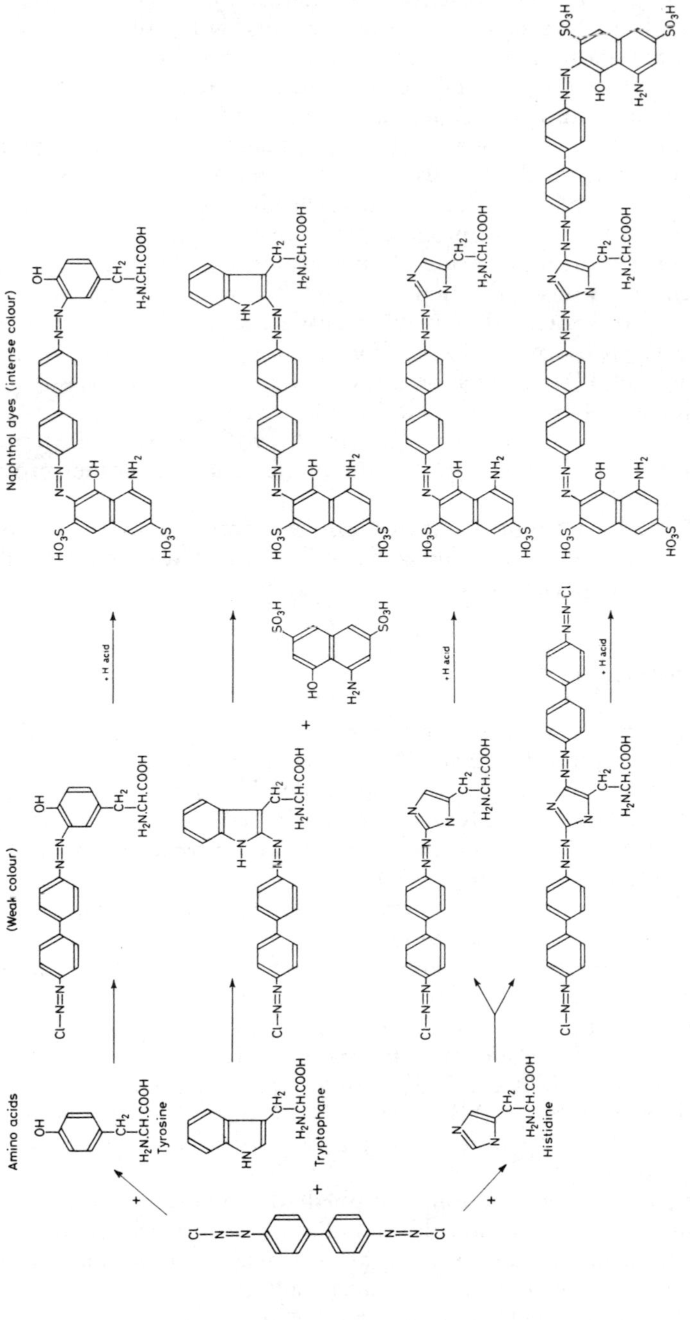
Amino acids
Tyrosine
Tryptophane
Histidine
(Weak colour)
Naphthol dyes (intense colour)
+ H acid
+ H acid
+ H acid

TYROSINE, TRYPTOPHANE AND HISTIDINE REACTIONS

Danielli's (1950) coupled tetrazonium reaction involving the use of tetrazotised benzidine gives positive results with all the above three protein-bound amino acid groups, but in order to locate each amino acid specifically, different blocking agents, namely benzoyl chloride, performic acid, 2,4-dinitrofluorobenzene, have been used prior to the addition of tetrazonium salts, these agents acting specifically on certain amino acids so that the reactions of these acids with tetrazonium salts are blocked. In this method, the application of a colour-producing agent is preceded by treatment with non-chromogenic reagents having a specific blocking capacity on certain amino acids. Of the various blocking agents, 2,4-dinitro-fluorobenzene blocks reaction of tyrosine, while performic acid acts against tryptophane, and benzoyl chloride against all three (Barnard and Danielli, 1956; Barnard, 1961). For coupling, 8-amino-1-naphthol-3,6-disulphonic acid (H acid) has been used. The reactions can be interpreted as given on pages 449 and 451 (Gurr, 1958).

In place of tetrazotised benzidine, michrome blue salt 250 (tetrazotised *o*-dianisidine) can be used which is quite stable, and the colour produced is blue instead of red.

In addition to coupled tetrazonium reactions, methods are available through which the three amino acids can be separately demonstrated *in situ*. These are now described.

Trytophane Reaction

This test is principally based on the indole reaction, which yields a blue or reddish violet colour (Ehrlich, 1901; Lison, 1936; Serra, 1946; Gurr, 1958). Gurr does not consider the test as specific unless confirmed through other methods, the process involving condensation of the reagent with phenols, amines or pyrrol derivatives, yielding coloured substance comparable with triphenyl-methane dyes (page 451).

In addition to the above principle, Glenner and Lillie (1957) have employed the benzylidine condensation reaction for localising indole derivatives. Here, *p*-dimethylaminobenzaldehyde is used for the reaction and azotate of 8-amino-1-naphthol-5-sulphonic acid (S) is used for post-coupling. The method is based on the principle that the indole form resonates to an α-indolalin and a negative centre is produced at the carbon atom in position 2 of the indole ring. Following condensation of the carbonyl group of *p* DMAB at the carbon 2 of the indole ring, a carbinol is formed. After dehydration, a purple violet coloured compound is produced (page 451).

The post-coupling reaction is applied to the condensation product to secure intensification of the colour, and in order to perform azo-coupling, under anhydrous conditions as far as possible, the post-coupling reaction with S acid is carried out in glacial acetic acid. A dark blue compound can be observed in the tissue. The reaction has been successfully employed by

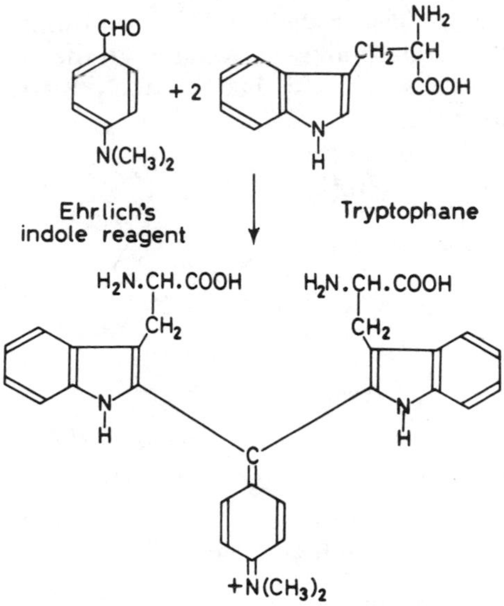

Ehrlich's
indole reagent

Tryptophane

Glenner and Lillie's
benzylidine condensation
method

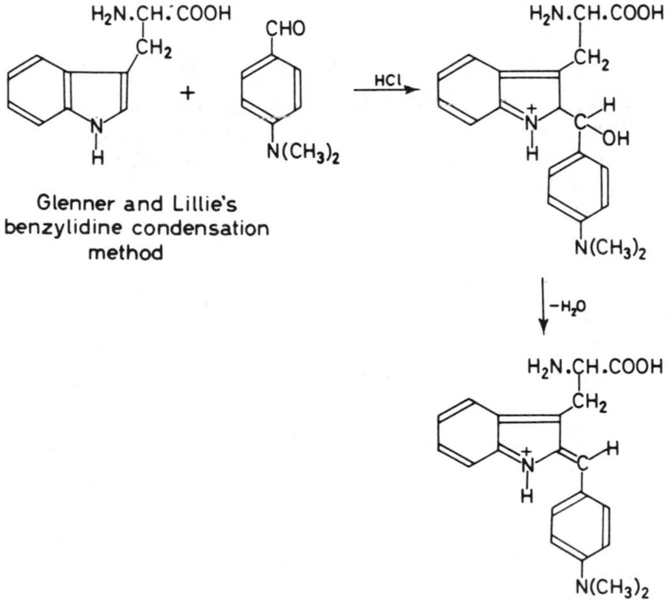

Sharma and Chatterji (1964) in the mitotic chromosomes of plants, where the mass of tissue, as such, is allowed to undergo the reaction before final smearing in 45 per cent acetic acid. The reaction is interpreted by Glenner and Lillie (1957) as:

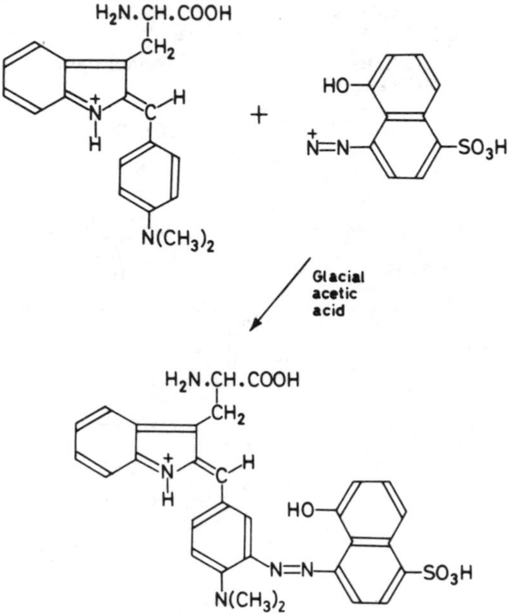

Tyrosine Reaction

As adopted by Bensley and Gersh (1933), Serra (1946) and Gurr (1958), this test is principally based on Millon's reaction for proteins. Mercuric nitrate is used as Millon's reagent and the hydroxyphenyl group of tyrosine reacts with mercuric nitrate to produce an unstable coloured compound (Lugg, 1937; Gurr, 1958).

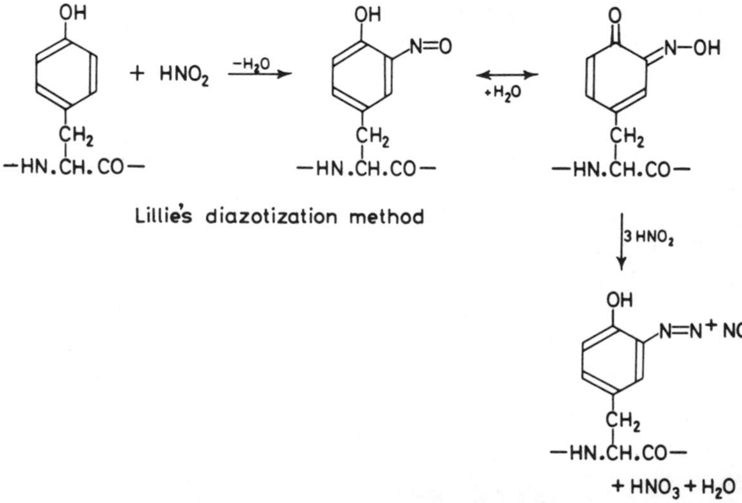

Lillie's diazotization method

Lillie (1957) applied protein diazotisation followed by coupling with a naphthol to localise protein bound tyrosine. The coupling reagent tried is S acid, 8-amino-1-naphthol-5-sulphonic acid. His method is based on Morel and Sisley's technique (1927), where the principle involves the formation of an o-C-nitrozation of tyrosine, followed by the formation of a quinone-oxime tautomer and, finally, the production of diazonium nitrate by reduction through more nitrous acid.

Histidine Reaction

Protein bound histidine is best demonstrated in Landing and Hall's technique (1956) which is based on the principle of allowing histidine to react with michrome blue salt 250 with the structure

The reaction takes place with the imidazole group of histidine, and two weakly coloured compounds are produced which, following coupling with H acid (8-amino-1-naphthol-3,6-disulphonic acid), form a mixture of intensely coloured azo dyes of brick red or reddish brown appearance. The reaction is almost identical to that of the coupled tetrazonium test, except that here, instead of tetrazotised benzidine, michrome blue salt 250 is used.

REACTIONS FOR S–S AND S–H GROUPS OF PROTEINS

Of all the tests meant for the S–S and S–H groups of proteins, Barrnett and Seligman's technique (1954) is widely applied and is based on the use of tetrazolium salts, whose oxidising property is utilised at a higher pH. At the site of reducing groups, strongly coloured insoluble formazan is produced from colourless tetrazolium salts. The formazans may possibly be tagged with other structures as well (Nineham, 1955).

Hyde and Paliwal (1958) employed with success a modified schedule for demonstrating chromosomal sulphydryl groups in onion root tips. In a separate method by Chevremont and Frederic (1943), the reaction involves the reduction of potassium ferricyanide by the sulphydryl groups within the tissue. The ferrocyanide thus produced, reacts with ferric sulphate to deposit insoluble prussian blue.

Test for Chromosomal Histones

In addition to the Sakaguchi reaction based on the arginine content of histones (McLeish, 1959), several other tests for nucleohistones are used. Of them, alkaline fast green (Alfert and Geschwind, 1953) is the one most widely employed, giving the best results with tissues fixed in formalin, or formaldehyde vapour, or freeze-dried (Cowden, 1966; Pearse, 1968). The removal of nucleic acid is desirable before applying the test (Davenport and Davenport, 1965). Naphthol yellow S (Deitch, 1955), amido black 10B (Geyer, 1960), alkaline eosin or bromophenol blue (Bloch and Hew, 1960), Biebrich scarlet (Spicer, 1962), etc. have also been utilised in securing colour reactions. Jobst and Sandritter (1964) used metaphosphoric acid treatment, followed by gallocyanin, whereas the ammoniacal silver nitrate method has been applied by others (Black and Ansley, 1964).

Formaldehyde fixatives have been regarded as suitable for the preservation of histones (McLeish, 1959) while Davies (1954) regarded freeze-substitution in methanol as a superior method of preservation. De (1961) recorded better preservation after freeze-substitution in methanol than in ethanol. He further noted that FAA fixation retains both basic and non-basic proteins and renders the histone resistant to extraction by 1/100 N HCl at 25 °C, for 4 h. This resistance has been attributed to cross-linkage formation in the protein side chains. There are different methods for the isolation of fractions, including chemical fractionation, chromatography and electrophoresis, which have been dealt with in detail by Busch (1965).

SCHEDULES

ARGININE REACTION

Liebman's Modification (1951) of Thomas's Method (1950)

Reagents

1. 1% α-naphthol solution in		
absolute ethyl alcohol	5 cm^3	
Distilled water	95 cm^3	

Mix together

2. 1 N sodium hypochlorite		
solution	15 cm^3	
1 N potassium hydroxide		
solution	5 cm^3	
Distilled water	80 cm^3	

Mix together

3. Urea	10 g	
Potassium hydroxide	5 cm^3	
Distilled water	15 cm^3	
Tertiary butyl alcohol	70 cm^3	

Mix the first three chemicals together. Then add butyl alcohol.

Procedure

1. Cut paraffin sections of materials (16 μm thick), previously fixed in Bouin's fluid, and mount on slides without the use of albumin. Deparaffinise by immersing successively in pure xylol and absolute alcohol. Dip in celloidin solution. Drain off excess celloidin and wipe the back of the slide. Before the celloidin solution has hardened completely, immerse successively in 90 and 70 per cent ethyl alcohol. Bring down to distilled water.

2. Treat the slides with cold α-naphthol solution (1) for 15 min at 0–4 °C.

3. Transfer immediately to cold hypochlorite–hydroxide mixture (2) and treat for $1\frac{1}{2}$ min at 0–4 °C.

4. Dip immediately in cold urea solution (3) and keep for 10 s with continuous stirring at 0–4 °C.

5. Change over to a fresh cold urea solution (3) and treat with stirring for 2 min at 0–4 °C.

6. Immerse in anhydrous tertiary butyl alcohol for 10 s, stirring the solution with the slides.

7. Treat in a change of tertiary butyl alcohol for $3\frac{1}{2}$ min.

8. Immerse in pure xylol with stirring for 10 s.

9. Treat with two more changes of pure xylol, keeping 1–2 min in the first and 2–3 min in the second.

10. Drain off xylol and mount in liquid paraffin seal.

Observations

Arginine or arginine-containing proteins take up an orange to reddish colour.

Precautions

1. Steps 2–5 should be carried out at 0–4 °C and the remaining ones at room temperature.

2. The reagents should be freshly prepared.

Serra's Test for Arginine (Serra, 1944)

Reagents

1. Reaction mixture containing:

1% α-naphthol solution in 96% alcohol, diluted 1:10 with 40% alcohol	0·5 cm^3
Normal sodium hydroxide solution	0·5 cm^3
40% aq. urea solution	0·2 cm^3

Keep at 0–5 °C.

2. 2 per cent sodium hypobromite solution, freshly prepared by pouring 0·7 cm^3 of liquid bromine in 100 cm^3 of 5 per cent NaOH, with stirring and cooling.

Procedure

1. Harden the pieces or sections in 10 per cent formalin solution for 12–24 h if the fixative does not contain formalin.

2. Immerse the pieces or sections for 15 min in the reaction mixture at 0–5 °C.

3. Add to the reaction mixture with tissue 0·2 cm^3 of the sodium hypobromite solution. Keep for 3 min at 0–5 °C with stirring.

4. Add another 0·2 cm^3 of 40 per cent urea solution to the mixture and stir.

5. Immediately add 0·2 cm^3 more of sodium hypobromite solution to the mixture with constant stirring.

6. Remove the material after 3 min and pass successively through four pure glycerin baths, keeping to 2–3 min in each.

Observations

Arginine-containing areas give positive reaction but the rest do not.

Alternative

Immerse the tissue in NaOBr solution for 3 min after step 5. Then transfer to glycerin baths.

Sakaguchi Dichloronaphthol Reaction for Arginine

According to McLeish *et al.*, (1957); Deitch, (1961); fixed material in acetic–ethanol, or by freeze-drying, followed by embedding in paraffin.

1. Prepare immediately before use: 4 per cent filtered barium hydroxide; 1 per cent sodium hypochlorite; 1·5 per cent 2,4-dichloronaphthol in tertiary butyl alcohol.

2. Bring down slides to water, wash twice in distilled water. Blot and transfer to an empty staining jar.

3. Add to a flask successively: barium hydroxide, 5 parts; sodium hypochlorite, 1 part; dichloronaphthol, 1 part, with shaking. Pour the mixture into the staining jar and keep the slide in it for 10 min at 22 °C.

4. Pass the slides through three changes of tertiary butyl alcohol, shaking vigorously in each change.

5. Keep in two changes of xylene containing 5 per cent tri-*N* butylamine for 30–60 s in each.

6. Drain and mount in Shillaber's Oil containing 10 per cent tri-*N*-butylamine. The presence of arginine is indicated by an orange-red colour.

Sakaguchi Oxine Reaction for Arginine

Modified according to Carver, Brown and Thomas (1953).

1. Bring down sections embedded in paraffin and fixed in Bouin, Carnoy's or formalin fixatives, to 70 per cent ethanol.

2. Treat for 15 min at room temperature in 0·3 per cent 8-hydroxy-quinoline in 30 per cent ethanol.

3. Transfer the slide immediately, without draining, to alkaline hypochlorite solution. (The solution of freshly prepared hypochlorite is standardised before use against 0·1N sodium thiosulphate, employing 1 cm^3 of the hypochlorite, 5 cm^3 of N-potassium iodine, 8 cm^3 of conc. HC1 and 50 cm^3 of water. The fresh hypochlorite solution is used as stock. Prepare a 0·15N stock solution of KOH. Before use, measure out suitable

quantities of the two stock solutions in a 100 cm^3 cylinder to make a final concentration of 0·15N chlorine and 0·015N KOH, taking the average chlorine content of fresh commercial hypochlorite to be 1·6N.) Keep slides for 60 s without moving.

4. Prepare alkaline urea solution by adding 15 g urea to 10 cm^3 of 0·15N KOH in a 100 cm^3 cylinder. Dilute with water to 25–30 cm^3 and mix until dissolved. Add 70 cm^3 tertiary butyl alcohol and mix. Transfer the slide immediately, without draining to the alkaline urea solution, and agitate gently for 10 s. Transfer and keep in a fresh alkaline urea bath for 2 min.

5. Treat slide in tertiary butyl alcohol for 4 min.

6. Keep in aniline oil for 3 min.

7. Wash in xylol for 10 s.

8. Mount in a synthetic resin containing 0·025 cm^3 aniline per 100 cm^3. Arginine sites are indicated by an orange colour.

TYROSINE REACTION

Bensley and Gersh's (1933) Modification of Millon's Reaction

Reagents

1. Aqueous nitric acid solution, prepared by mixing:

303 cm^3 of conc. nitric acid with 202 cm^3 of distilled water and keeping for 48 h	100 cm^3
Mercuric nitrate	180 g
Distilled water	900 cm^3

Shake at intervals for a period of several days. Filter.

2. Millon's reagent, prepared by mixing:

Sodium nitrite	1·4 g
66·6% aq. nitric acid solution (as described before)	3 cm^3
Mercuric nitrate solution in aq. nitric acid solution (solution 1 described before)	400 cm^3

3. Distilled nitric acid, prepared by mixing:

66·6% aq. nitric acid solution (as described before)	100 cm^3
Distilled water	390 cm^3

Procedure

1. Prepare and mount paraffin sections of formalin-fixed materials. Deparaffinise by passing through pure benzene and pure acetone grades. Dry by exposure to atmosphere. Frozen sections can be mounted directly on slides without water.

2. Treat several slides simultaneously in the Millon's modified reagent solution (2) at room temperature. Remove one slide after every 15 min or 30 min interval from the reagent and treat as follows.

3. Dip immediately in the distilled nitric acid solution (3).

4. Rinse thoroughly in distilled water.

5. Dehydrate quickly through 70 per cent and absolute ethanol grades, clear in xylol and mount in neutral balsam.

Observations

The areas which have tyrosine-containing proteins are coloured orange to brick red.

Precautions

1. The optimum period of treatment in Millon's reagent is determined by removing the slides at fixed intervals.

2. Incubation at 60 °C in Millon's reagent accelerates the reaction and it can be completed in 1 h instead of 3 h at room temperature.

Serra and Queiroz-Lopes's Modified Method (1945)

Reagents

1. Reaction mixture containing:

Magnesium sulphate	7·5 g
Magnesium chloride	5·5 g
Sodium sulphate	7 g

Dissolve in 85 cm^3 of distilled water, to which 12·5 g of concentrated sulphuric acid has been added. Dilute to 100 cm^3 with distilled water.

2. 1 M sodium nitrite solution, prepared by dissolving 6·9 g in 100 cm^3 of distilled water.

Procedure

1. Bring down the slides or pieces of tissue to water.

2. Incubate in a few cm^3 of the reaction mixture for 30 min at 60 °C in a stoppered jar.

3. Cool the container in running water and leave at room temperature for 10 min.

4. Add an equal volume of distilled water to the reaction mixture containing the tissue.

5. Add to the mixture a few drops of sodium nitrite solution.

6. After 3–4 min, place the tissues in free glycerin and either prepare squashes or mount as usual.

Observations

After 3 min treatment in $NaNO_2$ solution, the sites containing proteins with tyrosine take up colour, which lasts for several months.

Lillie's Modification (1957) of the Morel–Sisley Method for Tyrosine (1927)

The fixative recommended is 4 per cent formaldehyde solution for 3–48 h.

Reagents

1. Nitrosating mixture containing:

Sodium nitrite	6·9 g
Glacial acetic acid	5·8 cm³
Make up to 100 cm³ with distilled water	

2. Coupling mixture containing:

1-amino-8-naphthol-4-sulphonic acid (S acid)	1 g
Potassium hydroxide	1 g
Urea	2 g
70% ethanol	100 cm³

Procedure

1. Cut paraffin sections 5 μm thick. Mount and deparaffinise through xylol and alcohol grades and bring down to distilled water.
2. Nitrosate at 3 °C in the nitrosating mixture for 18 h.
3. Rinse in three changes of cold distilled water (0 °C), keeping 5 s in each.
4. Couple in the coupling mixture for 1 h at 3 °C.
5. Rinse in three changes of N/10 hydrochloric acid, keeping 5 min in each.
6. Wash in running water for 10 min, dehydrate in alcohol, clear in xylol and mount in neutral balsam.

Observations

Areas containing tyrosine take up a pinkish red colour.

Dinitrofluorobenzene (DNFB) Method for Tyrosine, SH and NH$_2$

According to Danielli (1950) and Burstone (1955); material is freeze-dried or fixed in ethanol, acetone, formalin, etc.

1. Bring down the sections to absolute acetone or ethanol, remove and dry in air.
2. Treat with a saturated solution of DNFB in 90 per cent ethanol saturated with sodium bicarbonate for 2–16 h at room temperature.
3. Wash in 3 changes of 90 per cent ethanol and finally in water.
4. Treat with 5 per cent sodium hydrosulphate for 30 min at 45 °C.
5. Wash in water.
6. Immerse in nitrous acid, prepared by adding 1 vol. of freshly prepared 5 per cent sodium nitrite to 4 vols. of 2N HCl at 0–4 °C for 30 min.
7. Wash in water.
8. Treat in a saturated solution of H acid in veronal acetate buffer at pH 9·4 for 15 min at 0·4 °C.
9. Wash in water, dehydrate through ethanol grades, clear in xylol and mount in balsam or DPX. The sites of DNA attachment in the tissues appear as reddish–purple.

TRYPTOPHANE REACTION

Indole Reaction

Reagent—Ehrlich's indole reagent or *p*-dimethylaminobenzaldehyde.

Procedure

1. Bring down frozen or paraffin sections to distilled water.
2. Keep for 1–2 h at 60 °C in a partly filled stoppered jar.
3. Bring down the jar and its contents to room temperature.
4. Wash the sections with absolute ethanol, clear in xylol and mount in a suitable medium.

Observations

The formation of a bluish or reddish violet coloration at the sites containing tryptophane or other substances containing the indole group.

Glenner and Lillie's Method for Tryptophane (1957)

Reagents

1. *p*-Dimethylaminobenzaldehyde   1 g
   Conc. hydrochloric acid (sp. gr. 1·19)   10 cm^3
   Glacial acetic acid   30 cm^3
2. Fresh diazotate of 8-amino-1-naphthol-5-sulphonic acid (S acid), prepared by adding 240 mg of S acid to 3 cm^3 of N HCl and 6 cm^3 of distilled water, cooling to 4 °C, adding 1 cm^3 of N NaNo$_2$ and stirring at 4 °C for 15 min   1 cm^3
   Glacial acetic acid   40 cm^3

Procedure

1. Fix tissues in 10 per cent calcium acetate formalin. Cut paraffin sections 5 μm thick, deparaffinise and bring down to absolute alcohol. Dry in air for 30 s.

2. Immerse for 5 min in *p*-dimethylaminobenzaldehyde mixture (1) at 25 °C.

3. Wash successively in three changes of glacial acetic acid for 30 min in the first change and 1 h in each of the other two.

4. Treat for 5 min at room temperature in S acid mixture (2).

5. Wash in two changes of glacial acetic acid, keeping for 30 s in each.

6. Treat for 5 min at room temperature in 40 cm^3 of glacial acetic acid containing 20 mg of new fuchsin.

7. Wash in two changes of glacial acetic acid, keeping for 1 min in each, pass through acetic acid–xylol and then pure xylol grades and mount.

Observations

When counter-stained with new fuchsin, sites of indole derivatives stain pale purple to deep blue. Otherwise they colour pale to dark blue.

Eosin—Light Green Method for Tryptophane (Hřsel, 1957):

Treat with 4 per cent CrO_3 for 30 min, wash in running water for 20–30 min, rinse in distilled water, treat in 1 per cent phosphomolybdic acid for 20 min, rinse, stain for 30 min in 1 per cent eosin, rinse, stain in 1 per cent light green for 10 min, dry with filter paper, dehydrate through isopropyl alcohol grades, clear and mount. Eosin stains tryptophane-containing proteins; light green is bound to amino groups.

Tryptophane Method for Formalin-Fixed Tissues

According to Adams (1960). The material is fixed in formalin and embedded in paraffin following the usual schedule.

1. Bring the sections down to absolute ethanol.

2. Immerse the slides in a solution containing: glycerol, 5 cm^3; 60 per cent ferric chloride, 1 cm^3; conc. sulphuric acid, 5 cm^3; methylated alcohol, 80 cm^3. The solution can be stored for several months.

3. Lift out slide with forceps, decant excess fluid, ignite in a small flame, holding the slide horizontal with the sections uppermost. Repeat 3 to 6 times.

4. Wash in absolute ethanol.

5. Rinse until clean in glacial acetic acid–ethanol mixture (1:1). Clear in xylol and mount.

6. The sites of tryptophane show a mauve coloration. Since the pigment is not stable, the slides must be examined within 24 h.

HISTIDINE REACTION

Landing and Hall's Method (1956)

Reagents

1. Gram's iodine	30 cm^3
Ammonia solution (sp. gr. 0·880)	2 cm^3
2. Diazo mixture containing:	
Naphthanil diazo blue B	
(Michrome blue salt 250)	0·05 g
Veronal acetate buffer pH 9·16	50 cm^3
3. H acid mixture containing:	
8-amino-1-naphthol-3,6-	
disulphonic acid (H acid)	1 g
Veronal acetate buffer pH 9·16	50 cm^3

Procedure

1. Fix in 10 per cent formalin. Prepare paraffin sections as usual and bring down to distilled water.

2. Treat in ammoniacal Gram's iodine solution (1) at room temperature for 24 h.

3. Wash in distilled water.

4. Wash in 95 per cent ethyl alcohol till the sections lose their yellow colour.

5. Wash successively in distilled water and veronal acetate buffer pH 9·16.

6. Treat in diazo mixture (2) for 15 min at 0–4 °C.

7. Rinse in distilled water.

8. Wash in three changes of veronal acetate buffer, keeping 2 min in each.

9. Treat in H acid mixture for 15 min at 0–4 °C, stirring.

10. Rinse thoroughly in water. Dehydrate through acetone or alcohol and xylol grades and mount.

Observations

The areas containing histidine stain brick red to reddish brown.

Metaphosphoric Acid–Gallocyanin for Basic Proteins

According to Jobst and Sandritter (1964). Air-dried smears may be used after alcohol or formalin fixation.

1. Keep smears for 15 min in 5 per cent aqueous trichloracetic acid at 95 °C to remove DNA.

2. Wash three times in 70 per cent ethanol, followed by water.

3. Dissolve completely 0·8 g of crystalline metaphosphoric acid (HPO$_3$) in 100 cm^3 of distilled water, by shaking at 20 °C.

4. Keep the slides in the above freshly prepared solution at 20 °C for 1 h.

5. Rinse in distilled water three times, 1 min each time.

6. Prepare gallocyanin–chrom alum solution by boiling 150 mg of gallocyanin for 10 min with 5 g chrome alum in 100 cm³ of distilled water and restore to a final volume of 100 cm³. Immerse the slides in the stain for 48 h at 20 °C.

7. Wash in running water for 5 min.

8. Dehydrate in ethanol, clear in xylol and mount in a synthetic resin.

9. The sites of basic protein stain deep blue due to the combination of metaphosphoric acid with NH_2 and guanidine groups.

Naphthol–Yellow S Method for Basic Proteins of Nuclei

According to Deitch (1955). The material can be post-fixed in ethanol after freeze-drying; or fixed in acetic-ethanol and then cut in the cold microtome; or embedded in paraffin after fixation in formalin or Carnoy's fluid.

1. Bring down the sections to water.

2. Immerse for staining 4–6 h in 0·5 per cent aqueous Naphthol yellow S at pH 2·7.

3. Wash in distilled water.

4. Dehydrate in 95 per cent ethanol, clear in xylol and mount in a synthetic resin.

5. The amount of yellow staining shows the number of available basic groups of protein. In the nuclei, if the nucleic acids have not previously been removed, only those basic groups not blocked by DNA will be shown.

COMBINED REACTION FOR TYROSINE, TRYPTOPHANE AND HISTIDINE

Danielli's Coupled Tetrazonium Reaction (1950)

Reagents

1. Benzidine solution containing:

Benzidine base	0·5 g
Conc. hydrochloric acid	5 cm³
Distilled water	20 cm³. Store at 0–5 °C

2. H acid soltuion containing:

8-amino-1-naphthol-3,6-disulphonic acid (H acid)	0·5 g
Veronal acetate buffer pH 9·0	25 cm³

Procedure

1. Prepare paraffin sections from alcohol or formalin-fixed tissues. Bring down to distilled water.
2. Add 7 cm^3 of cold 2 per cent aqueous sodium nitrite solution to 20 cm^3 of cold benzidine solution (1) drop by drop at 0–5 °C, not allowing the temperature to rise above 5 °C. Chill the solution to 0 °C.
3. Add to it 3·5 cm^3 of 5 per cent aqueous ammonium sulphamate solution, stirring, and then 34 cm^3 of aqueous sodium carbonate solution (28 g in 95 cm^3 of water). Stir well. After effervescence has stopped, make up to 150 cm^3 with distilled water. This solution is tetrazotised benzidine and it decomposes after an hour.
4. Treat the sections in tetrazotised benzidine solution for 15 min.
5. Rinse thoroughly in distilled water.
6. Wash in three changes of veronal acetate buffer (pH 9·0), keeping for 2 min in each.
7. Treat in H acid solution (2) for 15 min.
8. Rinse in running water for 3 min. Dehydrate through alcohol and xylol grades and mount.

Observations

The sites containing tyrosine, tryptophane and histidine stain reddish brown. The stain is stable. Blocking agents should be used prior to this test for specific localisation of each amino acid (*see* text).

SULPHYDRYL GROUPS

Modified Alkaline Tetrazolium Reaction

According to Deguchi (1964) for SS and SH groups.
1. Prepare a 5 per cent KCN solution and adjust the pH to 8·4 with 0·2M acetic acid. To 10 cm^3 add 10 cm^3 of 0·2 borate buffer (pH 8·4) and 15 mg nitro-BT. Keep in a covered glass jar. Store at −20 °C.
2. Bring down sections to absolute ethyl alcohol.
3. Cover with 1 per cent celloidin and allow to dry.
4. Immerse the slides in fresh 0·5M thioglycollic acid, adjusted to pH 8·0 with 1 per cent NaOH for 2–3 h at 50 °C.
5. Rinse successively in water, 1 per cent acetic acid, and again in water.
6. Immerse in the nitro-BT reagent for 30–60 min at 37 °C.
7. Wash in 1 per cent acetic acid and then in water.
8. Dehydrate, clear and mount in a synthetic medium.
Protein-bound SS and SH give dark blue to red colour in the section.

Tetrazolium Method for Both S–S and S–H Groups (Barnett and Seligman, 1954)

Reagents

1.	Trichloracetic acid	1 g
	80% ethyl alcohol	100 cm^3
2.	Incubation mixture containing:	
	10% aq. potassium cyanide	
	solution	48 cm^3
	1 N sodium hydroxide	2 cm^3
	Blue tetrazolium chloride	0·025 g

Procedure

1. Fix tissues in alcoholic trichloracetic acid solution (1) for 24 h.
2. Prepare paraffin blocks. Cut sections 5–10 μm thick. Fix to slides with minimum quantity of albumin. Dry at 37 °C. Deparaffinise with xylol and bring down to absolute alcohol.
3. Dip in 0·5 per cent celloidin dissolved in absolute alcohol and ether mixture (1:1). Drain and dry.
4. Treat in 90 and 70 per cent alcohol, keeping 2 min in each. Bring down to distilled water.
5. Treat in incubation mixture (2) at 37 °C for 8–12 h, till the sections take up a blue colour.
6. Rinse thoroughly in water. Mount in glycerin jelly.

Observations

The sites of sulphydryl groups are indicated by the blue colour.

Alternative—Neotetrazolium chloride may be used in place of blue tetrazolium. The incubation period is 2–3 h and the colour developed is purplish red.

Test of Chromosomal Histone (Black and Ansley, 1964)

Squash salivary glands in 45 per cent acetic acid, make air dry preparations, fix in 10 per cent acetate buffered neutral formalin (15 m), rinse in water, place in A–S solution (fresh aqueous 10 per cent AqNo$_3$ added to conc. NH$_4$OH till turbid), agitate for 5–6 s, wash in water, develop in 3 per cent formalin (2 m) wash in water, dehydrate and mount, A–S reacts with DNA-bound histone.

REFERENCES

Adams, C. W. M. (1960).   *J. Path. Bact.* **80,** 442
Alfert, M. and Geschwind, I. I. (1953).   *Proc. Nat. Acad. Sci. Wash.* **39,** 991
Baker, J. R. (1947).   *Quart. J. Sci.* **88,** 115
Bal, A. K. and Kaufmann, B. P. (1959).   *Nucleus* **2,** 51
Barnard, E. A. (1961).   In *Gen. cytochemical methods* **2,** 203
Barnard, E. A. and Danielli, J. F. (1956).   *Nature* **178,** 1450
Barrnett, J. R. and Seligman, A. M. (1951).   *Science* **114,** 579
Barrnett, J. R. and Seligman, A. M. (1954).   *J. nat. Cancer Inst.* **14,** 769
Bensley, R. R. and Gersh, I. (1933).   *Anat. Rec.* **57,** 217
Bernstein, M. H. and Mazia, D. (1953).   *Biochim. Biophys. Acta* **10,** 600
Black, M. and Ansley, H. R. (1964) *Science* **143,** 693
Bloch, D. P. (1958).   In *Frontiers of cytology,* chap. 6, ed. by Palay, S. L. New Haven, Conn.; Yale
    University Press
Bloch, D. P. and Hew, H. C. Y. (1960).   *J. Biophys. Biochem. Cytol.* **8,** 169
Burstone, M. S. (1955).   *J. Histochem. Cytochem.* **3,** 32
Busch, H. (1965).   *Histones and other nuclear proteins.* New York; Academic Press
Carver, M. J., Brown, F. C. and Thomas, L. E. (1953).   *Stain Tech.* **28,** 89
Caspersson, T. (1940).   *J. Roy. micr. Soc.* **60,** 8
Chargaff, E. (1955).   *The nucleic acids,* chap. 10, ed. by Chargaff, E. and Davidson, J. N. New York;
    Academic Press
Chevremont, M. and Frederic, J. (1943).   *J. Arch. Biol.* **54,** 589
Cowden, R. R. (1966).   *Histochemie* **6,**
Danielli, J. F. (1950).   *Cold Spr. Harb. Symp. quant. Biol.* **14,** 32
Davenport, R. and Davenport, J. C. (1965).   *J. Cell Biol.* **25,** 319
Davies, H. G. (1954).   *Quart. J. Micr. Sci.* **95,** 433
De, D. N. (1961).   *Nucleus* **4,** 1
Deguchi, Y. (1964).   *J. Histochem. Cytochem.* **12,** 261
Deitch, A. D. (1955).   *Lab. Invest.* **4,** 324
Deitch, A. D. (1961).   *J. Histochem. Cytochem.* **9,** 477
Dounce, A. L. (1952).   *Exp. Cell Res. Suppl.* **2,** 103
Dounce, A. L. (1955).   *The nucleic acids,* chap. 18, ed. by Chargaff, E. and Davidson, J. N. New York;
    Academic Press
Dutt, M. K. and Kaufmann, B. P. (1959).   *Nucleus* **2,** 85
Ehrlich, P. (1901).   *Dt. med. Wschr.* **27,** 865
Freese, E. (1958).   *Cold Spr. Harb. Symp. quant. Biol.* **23,** 13
Geyer, G. (1960).   *Acta Histochem.* **10,** 286
Glenner, G. G. and Lillie, R. D. (1957).   *J. Histochem. Cytochem.* **5,** 279
Gurr, E. (1958).   *Methods of analytical histology and histochemistry.* London; Leonard Hill
Hŕsel, I. (1957).   *Acta Histochem.* **4,** 47
Hyde, B. B. and Paliwal, R. L. (1958).   *Amer. J. Bot.* **45,** 433
Hyden, H. (1943).   *Acta physiol. scand.* **6,** Suppl. 17, 136
Jobst, K. and Sandritter, W. (1964).   *Histochemie* **4,** 277
Kaufmann, B. P., McDonald, M. R., Gay, H., Rowan, M. E. and Moore, E. C. (1951).   *Carnegie
    Inst. Wash. Yearb.* **50,** 203
Kaufmann, B. P., Gay, H., Dutt, M. K., Bal, A. K. and Buchanan, J. (1959).   *Carnegie Inst. Wash.
    Yearb.* **58,** 440
Kaufmann, B. P., Gay, H. and McDonald, M. R. (1960).   *Int. Rev. Cytol.* **9,** 77
Landing, B. H. and Hall, H. E. (1956).   *Stain Tech.* **31,** 97
Lewis, K. R. and John, B. (1963).   *Chromosome marker.* London; Churchill
Liebman, E. (1951).   *Stain Tech.* **26,** 261
Lillie, R. D. (1957).   *J. Histochem. Cytochem.* **5,** 528
Lison, L. (1936).   *Histochimie Animale.* Paris; Gautier Villars
Lugg, J. W. H. (1937).   *Biochem. J.* **31,** 1422
McLeish, J. (1959).   *Chromosoma* **10,** 686
McLeish, J., Bell, L. G. E., La Cour, L. F. and Chayen, J. (1957).   *Exp. Cell Res.* **12,** 120
Mayer, D. T. and Gulick, A. (1942).   *J. biol. Chem.* **146,** 433
Mirsky, A. E. and Ris, H. (1947).   *J. gen. Physiol.* **31,** 1

Mirsky, A. E. and Ris, H. (1951). *J. gen. Physiol.* **34,** 475
Morel, A. and Sisley, P. (1927). *Bull. Soc. chim. Fr.* **41,** 1217
Nineham, N. (1955). *Chem. Rev.* **55,** 2
Pearse, A. G. E. (1960 and 1968). *Histochemistry.* Boston; Little, Brown
Ris. H. (1958). *Colloq. Ges. physiol. Chem.* **9,** 1
Ris, H. and Kleinfeld, R. (1952). *Chromosoma* **5,** 363
Sakaguchi, S. (1925). *J. Biochem., Tokyo* **5,** 25
Serra, J. A. (1942). *Bol. Soc. broteriana* **16,** 83
Serra, J. A. (1944). *Portug. Acta biol.* **1,** 1
Serra, J. A. (1945). *Acta I Reun. Biol.* **1,** 47
Serra, J. A. (1946). *Stain Tech.* **21,** 5
Serra, J. A. (1955). *Handbuch der Pflanzenphysiologie* **1,** 413
Serra, J. A. and Queiroz-Lopes, A. (1945). *Portug. Acta biol.* **1,** 51
Serra, J. A. (1968). *Modern Genetics* 3, New York; Academic Press
Sharma, A. K. and Bhattacharyya, B. (1957). *Bull. bot. Soc. Beng.* **11,** 34
Sharma, A. K. and Chatterji, A. K. (1964). *J. Histochem. Cytochem.* **12,** 266
Spicer, S. S. (1962). *J. Histochem. Cytochem.* **10,** 691
Stedman, E. and Stedman, E. (1947). *Cold Spr. Harb. Symp. quant. Biol.* **12,** 224
Stedman, E. and Stedman E. (1950). *Biochem. J.* **47,** 508
Taylor, J. H. (1969). *Proc. 12th Int. Congr. Genet.* **3,** 177
Taylor, J. H. (1963). In *Molecular genetics,* part I, p. 65, ed. by Taylor, J. H. New York; Academic Press
Thomas, L. E. (1950). *Stain Tech.* **25,** 143
Vendrely, R. and Vendrely, C. (1953). *Nature* **172,** 30
Vendrely, R., Alfert, M., Matsudaira, H. and Knobloch, A. (1958). *Exp. Cell Res.* **14,** 295
Zubey, G. (1964). In *The Nucleohistones,* p. 95, ed. by Bonner, J. and Tsio, P. San Francisco; Holden-Day

21 Enzymes

ALKALINE PHOSPHATASE

Of all the tests employed for the *in situ* localisation of enzymes in chromosomes, the technique for the demonstration of alkaline phosphatase activity is the most important, the term 'alkaline phosphatase' being loosely used for histochemical purposes, which is precisely 'phosphomonoesterase I'. This enzyme can act at an alkaline pH of 9·2 to 9·6 and is specific for monoesters of *ortho*-phosphoric acid.

$$R-O-\overset{\overset{O}{\|}}{\underset{\underset{OH}{|}}{P}}-OH \quad + \quad H.OH \quad \longrightarrow \quad R.OH \quad + \quad HO-\overset{\overset{O}{\|}}{\underset{\underset{OH}{|}}{P}}-OH$$

Phosphate ester   *ortho*Phosphoric acid

The demonstration of the enzyme activity in the cell, and especially in the chromosome, can be carried out principally through two different methods: (*a*) calcium phosphate deposition; and (*b*) azo dye methods.

Calcium Phosphate Method

The calcium phosphate precipitation technique was first developed independently by Gomori (1939) and Takamatsu (1939), and was based on the principle that if a tissue containing the enzyme is incubated at 37 °C in a medium containing phosphate salt as the principal constituent at an alkaline pH (9·4), the liberated phosphoric acid can be deposited at the site of the enzyme as an insoluble calcium phosphate precipitate if a calcium salt is present as one of the components. The visualisation of calcium phosphate can be done by converting it into silver sulphate and then to metallic silver or into cobalt phosphate through a cobalt salt and finally to a black precipitate of cobalt sulphide through ammonium sulphide. The method as modified at present (Danielli, 1950, Gomori, 1946, 1952) contains in the medium magnesium sulphate or chloride as activator (Kabat and Furth, 1941). A barbiturate solution is used to buffer the medium between 9·0 and 9·4 and a citrate buffer is often used to remove preformed calcium salt. The way through which magnesium activates the enzymes is not fully explained. Three alternative modes of action are possible (Pearse, 1960, 1968): (a) the metal may form an essential part of the active centre of the enzyme; (b) it may bind the enzyme with the substrate; or (c) the activation may be caused by changing the surface charge of the protein and, consequently, its electrokinetic potential.

Several phosphate salts have been used as substrates, such as α-, β-glycerophosphate, fructose diphosphate, adenosine triphosphate, sodium dihydrogen phosphate, etc., but the most commonly used one is sodium-β-glycerophosphate. Substrate specificity has been observed in a number of cases and alkaline phosphatase is found to represent a large number of enzymes (Gould, 1944; Glick and Fischer, 1946; Zorzoli and Stowell, 1947; Dempsey, 1949; Sharma, Mookerjea and Ghosh, 1953; Dixon and Webb, 1964). Moss (1964) recorded the presence of alkaline phosphatases in different human tissues, differing in electrophoretic and enzymatic properties. Eaton and Moss (1966) demonstrated that alkaline phosphatase and pyrophosphatase are identical.

A number of factors influence the activity of the enzyme. Fixation has a profound effect, and the most effective fixation is freezing, but as there are a number of disadvantages in this procedure, chilled alcohol or acetone (4 °C) has been found to be quite suitable as it does not cause any loss of activity. Cold neutral formalin-fixation also gives successful results (Seligman, Chauncey Nachlas, 1951). However, even with a metallic fixative, Sharma and Roy (1956a) secured satisfactory results in plant chromosomes.

In addition to fixation, the temperature for embedding in paraffin and for mounting or storage also affects the activity to a significant extent. Embedding in a mixture of paraffin waxes of m.p. 42° and 55 °C yields excellent results (Dalgaard, 1956). In no case should a temperature higher than 37 °C be used for drying slides (Ruyter and Newmann, 1949; Pearse, 1960).

As the enzyme is effective in an alkaline medium, the maintenance of proper pH (9·2–9·8) is absolutely necessary. Moreover, the calcium phosphate becomes soluble at a low pH and, as such, the problem of diffusion may result; Robbins (1950), Cacioppo and colleagues (1953) and Sonnenschein and Kopac (1953) dealt with this aspect in detail. The problem of activation is also important. In addition to magnesium, ascorbic acid, salts of bile acids, zinc, etc., have been found to act as activators (Cloetens, 1941; Newman, Kabat and Wolf, 1950; Sadasivan, 1950; Hoare and Delroy, 1955). The choice of proper substrates is a major factor, and though sodium glycerophosphate is the most commonly used substrate, phenyl phosphate is more easily hydrolysed at an alkaline pH (Hill, 1956; Pearse, 1960).

Activity of alkaline phosphatase in nuclei and chromosomes has been demonstrated by several workers (Krugelis, 1942; Danielli and Catcheside,

1945; Brachet and Jenner, 1948; Sulkin and Gardner, 1948; De Nicola, 1949; Bhattacharjee and Sharma, 1951; Chevremont and Firket, 1953; Sharma, Mookerjea and Ghosh, 1953; Sharma and Roy, 1956a, b). Danielli and Catcheside (1945) observed phosphatase-positive areas in the bands of the salivary gland chromosomes of *Drosophila* and compared them with the Feulgen-positive areas. Bhattacharjee and Sharma (1951) demonstrated greater activity in meiotic cells of plants than in the mitotic ones. Sharma, Mookerjea and Ghosh (1953) correlated the phosphatase activity in chromosomes, throughout the divisional cycle, with the Feulgen stainability. Sharma and Roy (1956a) obtained positive phosphatase reaction in chromosomes and nucleoli, even after nucleic acid extraction, and attributed this residual activity to phospholipoid complexes. These authors (1956c) also noted changes in phosphatase activity after treatment with x-rays.

The validity of the test has, however, been questioned on two grounds:

1. The possibility of the diffusion of the enzyme from the site of its occurrence and its adsorption at a different site (Feigen, Wolf and Kabat, 1950; Newman and colleagues, 1950).

2. Diffusion of calcium phosphate and its adsorption at a different site (Cleland, 1950; Novikoff, 1955).

Several authors claimed that a positive phosphatase reaction may not indicate high activity at that particular site (Martin and Jacoby, 1949; Novikoff, 1952, 1955). Danielli (1947) suggested the reliability of this technique for *in situ* localisation of the enzyme. Lison (1948) also agreed, to a major extent, with Danielli's observations. Sharma, Mookerjea and Ghosh (1953) performed a number of control experiments, including inactivation of the enzyme through different methods, and claimed the validity of the test for *in situ* localisation (*see* Serra, 1955).

In any case, in view of the problem of diffusion and adsorption raised by certain authors (Pearse, 1968), several modifications have been suggested to eliminate these limitations. Deimling (1964) suggested the use of lead ions to combine with phosphoric acid, avoiding calcium and cobalt, but the membrane effect of lead is a limitation. Gomori (1951), in his schedule, increased the concentration of calcium ions in the substrate mixture. Martin and Jacoby (1949) suggested the elimination of magnesium salt from the medium, and also stated that if the slides are coated with celloidin and dried in air for a few minutes, reproducible results may be obtained. Maengwyn-Davies and Friedenwald (1950) prepared the incubation medium saturated with calcium phosphate. Even incubation in the substrate medium before the removal of paraffin (Goetsch, Reynolds and Bunting, 1952) was suggested. Novikoff (1955) considered that such procedures may all lead to the reduction of enzyme activity (*see* Osawa, 1951). According to him, if these measures can be adopted without reducing enzyme activity, the Gomori–Takamatsu technique can then be applied for *in situ* localisation. Fishman *et al.* (1963) and Watanabe and Fishman (1964) noted the inhibition of intestinal alkaline phosphatase activity by L-phenylalanine. Thomas and Aldridge (1966) recorded a similar inhibition by Be ions.

Gomori (1950) suggested that the alkaline phosphatase technique can be applied for quantitative estimation. Danielli (1950) stated that quantitative results can be estimated on the principle that the sites of highest activity appear before those of less activity. Several methods for the quantitative estimation of alkaline phosphatase activity, with the aid of spectrophotometry, radioactivity as well as interferometry, have been developed (Doyle, Omoto and Doyle, 1951; Barka and colleagues, 1952; Barter, Danielli and Davies, 1955; Shugar, Szenberg and Sierakowska, 1957; cf Pearse, 1968).

Azo Dye Method

Coupling Methods

The azo dye technique is based on the principle of precipitating the alcoholic part of the phosphate ester instead of the phosphoric acid. The method was originally developed by Menten, Junge and Green (1944). They used β-naphthyl phosphate as the substrate and the liberated β-naphthol, after hydrolysis, is reacted upon *in situ* by diazotised α-naphthylamine at pH 9·4, whereby a red precipitate is obtained (*see* reaction as outlined). It can be applied to both fresh and formalin-fixed tissues.

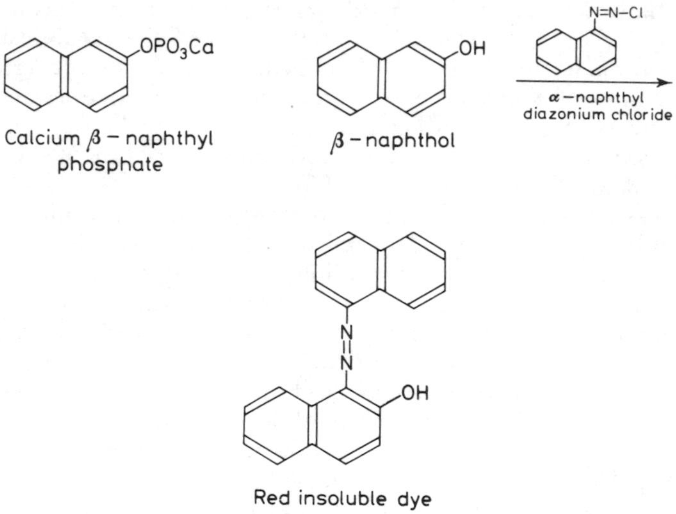

Calcium β – naphthyl phosphate

β – naphthol

α – naphthyl diazonium chloride

Red insoluble dye

Danielli (1946) suggested the use of phenyl phosphate and β-naphthyl phosphate to obtain the reaction. Mannheimer and Seligman (1948) used magnesium ion as activator and modified the procedure and used α-naphthyldiazoniumnaphthalene-1,5-disulphonate for coupling. The purplish red dye, which precipitates out, has been found to be insoluble.

This method has the added advantage over the original technique of Menten, Junge and Green (1944), in that the coupling reagent used here is stable as compared to the diazonium salt used by them. In fact, in acetone-fixed paraffin sections, the phosphatase-rich areas can be observed within 1 min. The low temperature (usually 10 °C) and short period of incubation reduce the background staining to a significant extent. In Mannheimer–Seligman's method (1948), development of the pigment can be controlled, as desired. In Gomori's modification of the original method (1951), sodium α-naphthyl phosphate has been used instead of calcium β-naphthyl phosphate, as the sodium salt is relatively less soluble (Friedman and Seligman, 1950). It does not cause any haziness or false adsorption. In order to avoid blurring,

Calcium β–naphthyl   β–naphthol
phosphate

Azodye

he has suggested that a non–optimal pH and temperature should be used, which would cause the naphthol to be liberated at a slower rate. That α-naphthyl salts are suitable as a substrate has also been confirmed by King (1952), Novikoff (1955) and Goessner (1958).

However, diazonium salts have certain disadvantages, namely, (a) several of them cause inhibition of enzyme activity, which may be eliminated by using a very low concentration of the solution, and (b) non-specific staining of tissue may result from their breakdown products (Pearse, 1968). When diazotised o-dianisidine is used as the diazonium compound, it has been found that the precipitation of colour varies with the pH. At a higher pH, both ends of the molecule are available for reaction, whereas at a lower pH, only one end undergoes coupling (Lojda et al., 1964). Loustalot (1955) noted that, following storage, enzyme activity was observed in residual substrate, but these storage artefacts can be eliminated by treatment through Lugol's iodine treatment. However, Gomori's modification has so far been found to be quite reliable and is much simpler than other techniques (Gurr, 1958). The most important advantage of Gomori's schedule is that the process can be carried out in higher temperatures, which results in good enzyme activity. Pearse (1968) made a detailed study of the diffusion, if any, of reaction products.

Another application of the coupling azo dye technique involves the use of substituted naphthol phosphate (Burstone, 1958a, b, 1962; Rutenberg and colleagues, 1958, etc.). Phosphate esters of several complex arylides of 2-hydroxy-3-naphthoic acid were used by Burstone. They are all highly stable esters and their reaction products are sparingly soluble. Several esters while used in 2 per cent dimethyl formamide at an alkaline pH have been found to be suitable, such as

(a) Naphthol – AS – MX

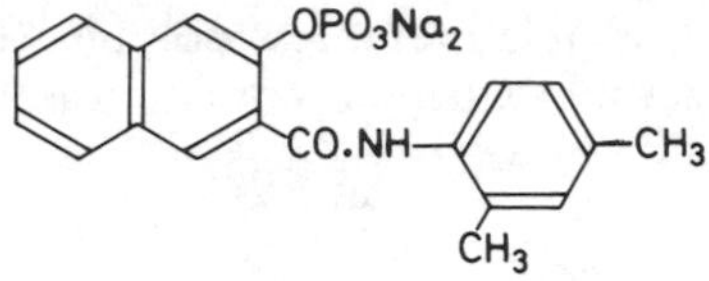

(b) Naphthol – AS – TR

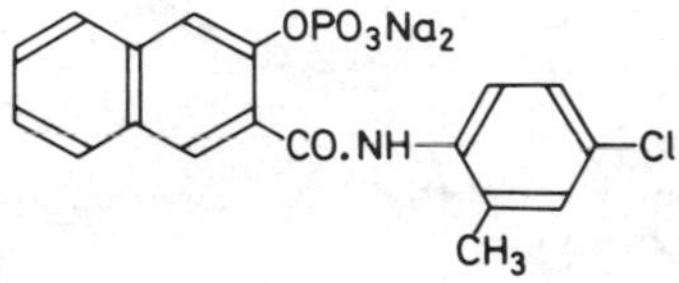

(c) Naphthol – AS – BI

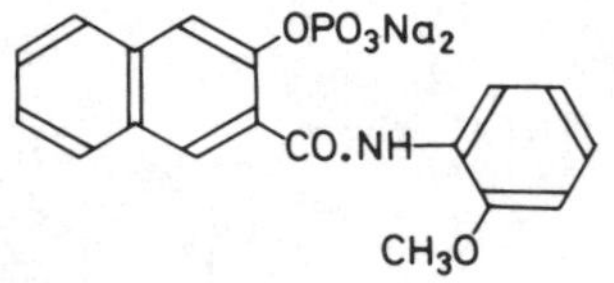

For coupling, fast blue RR, fast red RC, fast violet, etc., have been applied. The method has been widely used on freeze–dried tissues as well as to study leucocyte phosphatases in blood smears, bone marrow, the low solubility of the reaction products making precise localisation possible (Ackerman, 1962; Kaplow, 1963; Wetzel *et al.*, 1963, Wulff, 1967). Hexazotised bases were introduced by Davis and Ornstein (1959), using hexazonium para rosaniline (HPR) with naphthol AS substrate (Barka and Anderson, 1962). Similarly, Lojda *et al.* (1964) used tetrabromofuchsin and Stutte (1967) applied New Magenta (triaminotritolylmethane chloride). Moreover, being stable esters, their stock solutions can be stored for a long time. The limitation of this method lies in the facts that: (*a*) being sparingly soluble in water, the required concentration is scarcely obtained, (*b*) being composed of large molecules, access to the enzyme is not adequate, and (*c*) the low coupling capacity of hydroxy naphthoic arilides often hampers visualisation of the precipitate (Defendi and Pearse, 1955; Pearse, 1968).

Non-coupling Methods

In order to avoid the instability of the diazonium salts, their inhibitory effects and the use of low temperature, Loveless and Danielli (1949) used the phosphorylated azodye technique which involves the application of a coloured phosphate ester which yields an insoluble coloured base on hydrolysis. *p*-Nitrobenzeneazo-4-naphtholphosphate

O₂N—⟨benzene⟩—N–N—⟨naphthalene⟩—OPO₃H₂

has been used for the purpose. It does not inhibit the activity of the enzyme and undergoes ready hydrolysis. According to Pearse (1968), the free base is slightly soluble in water and, as such, diffusion occurs. This method has yet to be standardised for application in the study of chromosomes.

In addition to the above methods, post-coupling azodye techniques (Danielli, 1946) with sodium phenolphthalein phosphate have been used. Indoxyl phosphate has also been used as a substrate for the demonstration of enzyme activity (Holt, 1954; Seligman, Heymann and Barnett, 1954). For criticisms, the reader is referred to Defendi's work (1957) and Pearse (1968).

ACID PHOSPHATASE

This enzyme, which is actually 'phosphomonoesterase', acts on mono-esters of phosphoric acid at a pH of 5·0 to 5·3. Gomori (1941) originally devised the techniques, based on the same principle as alkaline phosphatase. As in acid pH, calcium salts are soluble in water, lead nitrate is used here instead of calcium salts and finally visualisation is brought about through the deposition of brown lead sulphide precipitate. It does not require magnesium as an activator and fluoride salts inhibit its action (Moog, 1943).

Several workers have, however, criticised the technique, claiming non-specific deposition of the precipitate (Moog, 1943; Wolf, Kabat and Newman, 1943; Hard and Lassek, 1946; Lassek, 1947). Newman, Kabat and Wolf (1950), Lovelock (1954) and Clarkson and Kench (1958) have discussed the details of the procedure and the different factors controlling it. Newman, Kabat and Wolf suggested that this method, with glycerophosphate as the substrate, yields reproducible results, provided control experimental sets are maintained. Novikoff (1955) stated that any procedure which minimises diffusion with alkaline phosphatase holds good for acid phosphatase as well.

Several modifications of the procedure were later published, and of these the more important ones are those of Gomori (1950), Goetsch and Reynolds (1951) and Tandler (1953). Gomori suggested that the precipitation may not be homogeneous due to patchy fixation as well as due to other causes. In his modified schedule, the substrate concentration has been raised and phosphates, which undergo more rapid hydrolysis than glycerophosphate, have been tried. The role of the buffer concentration has been emphasised. Moreover, rinsing in 1–2 per cent acetic acid, as suggested by Gomori (1952), is not recommended at present (Barka and Anderson, 1963; Bitensky, 1963; Goldfischer *et al.*, 1964; Ruyter, 1964; Janigan, 1965; Lake, 1965; Pearse, 1968), since Desmet (1962) demonstrated that acid rinsing may result in complete loss of precipitate.

Goetsch and Reynolds (1951) emphasised the procedure for mounting and deparaffinisation as important factors controlling the process, and according to them, the temperature during the entire procedure should not be raised above 38–40 °C (*see* Woodward, 1951). Egg albumin was regarded as having an inhibitive influence. Best results were obtained when the slide was not dewaxed at all. Ruyter (1964) did not record any such inhibition but recommended chilled 80 per cent acetone fixation.

Tandler (1953) used cobalt ions in place of lead, but at a lower pH; the precipitation of phosphate with this ion is incomplete. For better and non-diffusible precipitation through acid phosphatase activity, other metallic salts should be tried (Pearse, 1960). In another schedule, suggested by Takeuchi and Tanoué (1951), the pH level has been raised together with other modifications. For a comparative discussion of the different schedules, the reader is referred to Grogg and Pearse (1952).

With the aid of Gomori's method, the activity of acid phosphatase in nuclei has been reported by several authors (Wachstein, 1945; Rabinovitch, 1949). In plant chromosomes, the entire phosphatase cycle, during nuclear division, has been studied by Sharma and Roy (1956b). Sharma and Bhattacharyya (1957) analysed acid phosphatase activity in the chromosomes of malignant cells of plants, induced through different agents. Several authors (Avers, 1961; Beneš *et al.*, 1961; Harrington and Altschul, 1963; Poux, 1963; Beneš and Opatrna, 1964; Gahan, 1965) have demonstrated acid phosphatases in different plant organs. In animals, it is distributed profusely in spleen and liver. Its presence has been shown in fungi and protozoa as well (Atkinson and Shaw, 1955; Klamer and Fennell, 1963), and also in lysosomes (Novikoff, 1961; Wolman, 1965; Herveg *et al.* 1966; Wolman and Bubis, 1966; Gaham, 1965, 1967). More than one acid phosphatase has also been recorded (Connor and MacDonald, 1964; Elliot and Bak, 1964; Arsenis and Touster, 1967; Goodlad and Mills, 1967).

Seligman and Mannheimer (1949) used coupling azodye technique for the demonstration of acid phosphatase activity. As in acid pH, β-naphthol does not couple satisfactorily with diazonium salts, α-naphthyl phosphate has been used as the substrate. Anthraquinone-1-diazoniumchloride, being a stable diazonium compound, has been employed and sodium chloride is used to check diffusion. Similar to alkaline phosphatase, a number of modifications have been proposed, using naphthol AS phosphate and hexazotised fuchsin as diazonium compound, and for details the reader is referred to Burton (1954), Rutenberg and Seligman (1955), Gomori (1956); Goldberg and Barka (1962); Lojda (1963); Lojda *et al.* (1964); Meany *et al.* (1967); Pearse (1968).

CYTOCHROME OXIDASE

Another enzyme, the activity of which needs analysis, in connection with processes related to chromosome metabolism, is cytochrome oxidase. The method for its demonstration is outlined in the schedules.

SCHEDULES

Lead Nitrate Method for Acid Phosphatase (Gomori, 1950)

Reagents

1. Incubation mixture containing:
3% aq. sodium-β-glycero-phosphate solution	50 cm^3
M/20 acetate buffer, pH 5·0	500 cm^3
Lead nitrate	0·6 g

2. Concentrated ammonium
sulphide solution	10 cm^3
Distilled water	40 cm^3

Mix together

3. 2% aq. eosin solution   50 cm^3

Procedure

The fixative recommended is chilled acetone for 24 h for paraffin or celloidin sections, 10 per cent cold formalin for frozen sections and post-fixation in Wolman's fluid for 1–2 h.

1. Cut frozen sections 10–15 μm thick and mount on slides. Dry thoroughly at room temperature for adherence to the slides. For materials fixed for paraffin sectioning, cut paraffin sections 4–8 μm thick at 4 °C. Mount and post-fix in cold acetone for 1–2 h.

2. Bring down to water.

3. Incubate in the incubating mixture for 30 min to 16 h at 37 °C. The optimum period is usually 4 h.

4. Bring down to room temperature, wash first in distilled water, then in 1–2 per cent acetic acid and then again in distilled water, giving a few dips in each.

5. Treat with yellow ammonium sulphide solution for 1–2 min.

6. Rinse in water.

7. Immerse in eosin solution for 2–5 min.

8. Wash in water and mount in glycerin jelly.

9. For permanent preparations, omit step 8, dehydrate through alcohol and alcohol–xylol grades and mount in D.P.X.

Observations

A deposition of black precipitate at the sites of phosphatase activity.
Alternative—In step 4, rinsing with acetic acid can be omitted.

Modified Lead Nitrate Method for Acid Phosphatase (Takeuchi and Tanoué, 1951)

Reagents

1. Incubation mixture containing:

2% aq. sodium-β-glycero-phosphate solution	20 cm^3
0·1 M acetate buffer (pH 5·0–6·0)	10 cm^3
2% aq. lead acetate solution	10 cm^3
1–5% aq. magnesium chloride solution	3 cm^3

2. Ammoniacal silver nitrate solution, prepared by adding 28 per cent ammonia water drop by drop to 5 per cent aqueous silver nitrate solution, with shaking, until the precipitate just dissolves.

3. 5 per cent aqueous sodium thiosulphate solution.

Procedure

The earlier steps are similar to the previous schedule.

1. Bring down the sections to distilled water.

2. Incubate the sections at 37 °C in the incubating mixture for 30 min to 2h.

3. Wash in distilled water.

4. Immerse in ammoniacal silver nitrate solution for 30 min.

5. Treat in 5 per cent sodium thiosulphate solution for 5 min.

6: Mount directly in glycerin jelly, or for permanent preparations dehydrate through alcohol and alcohol–xylol grades and mount in D.P.X. or a suitable medium.

Observations

A deposition of brown precipitate at the sites of acid phosphatase activity.

Coupling Azodye Method for Acid Phosphatase

Reagents

1. Incubation mixture: Dissolve 10–20 mg of sodium-α-naphthyl-phosphate in 20 cm^3 of 0·1 M veronal acetate buffer at pH 5·0. Dissolve in it 1·5 g of polyvinyl pyrrolidone with stirring. Add 20 mg of the stable diazotate of *o*-aminoazotoluene or diethyl-sulphamino-*o*-anisidine. Stir and filter.

2. Mayer's aqueous haemalum solution.

Procedure

The recommended fixative is 10 per cent formalin at 4 °C for 10–16 h for frozen sections, later cut 10–15 μm thick, mounted and dried thoroughly at room temperature. Alternatively, cold microtome sections can be used after mounting on cover slips and post-fixing in cold acetone or Wolman's fixative.

1. Bring the sections down to distilled water.
2. Incubate at 37 °C in the incubation mixture. The period of treatment ranges from 30 s to 1 min for dog prostate, to 30–60 min for rat liver.
3. Rinse thoroughly in water.
4. Immerse in Mayer's haemalum solution for 4–6 min.
5. Rinse thoroughly in running water.
6. Mount directly in glycerin jelly or after dehydration in neutral balsam.

Observations

Sites of acid phosphatase activity are reddish brown while the nuclei are deep blue.

Alternative—0·1 M acetate buffer can be used instead of veronal acetate buffer in the incubation mixture.

Burstone's Naphthol AS–BI Phosphate Method (1958a, b)

The initial steps are similar to the previous schedules.

Reagents

1. Incubation mixture containing, 4 mg of naphthol AS–BI phosphate dissolved in 0·25 cm^3 of dimethyl formamide and added to 25 cm^3 of 0·2 M acetate buffer (pH 5·2–5·6). Add to it 35 mg of red violet LB diazonium salt and 2 drops of 10 per cent aqueous manganese chloride solution. Shake and filter.
2. Mayer's aqueous haemalum solution.

Procedure

1. Bring down the sections to water.
2. Incubate in the incubation mixture for 30 min to 6 h at 37 °C. 3, 4, 5 and 6 are similar to the same steps in the previous schedule.

Observations

The sites of acid phosphatase activity take up various shades of red, while nuclei are blue.

Alternative—AS–TR and AS–MS phosphates can be used instead of AS–BI phosphate. For frozen sections, the amount of dimethyl formamide used is 0.1 cm^3.

Rutenberg and Seligman's Post-coupling Method for Acid Phosphatase (1955)

Recommended fixatives are the same as in the previous schedules. Free-floating frozen sections and cold microtome sections, mounted and unfixed, can also be used.

Reagents

1. Incubation mixture containing:
Sodium-6-benzoyl-2-naphthylphosphate—Prepare by dissolving 14 g of 6-benzoyl-2-naphthol and 8.6 g of phosphorus oxychloride in 100 cm^3 of dry benzene and heating under reflux. Add 5 cm^3 of pyridine and heat for 30 min. Allow to cool. Filter. Distil the filtrate on a water pump at 70–80 °C till a thick syrupy liquid is formed. Keep in a desiccator in low vacuum over saturated aqueous KOH solution for a few days. Separate out the crystals of acid phosphate. Dissolve in a mixture of methyl alcohol and sodium methoxide in methyl alcohol (1:1). Suspend the crystals in water, filter, wash with methyl alcohol and ether and dry.
Dissolve 25 g of sodium-6-benzoyl-2-naphthylphosphate in 80 cm^3 of distilled water. Add 20 cm^3 of 0.5 *m*-acetate buffer (pH 5.0). Add 2 per cent sodium chloride solution.
2. Aqueous solution of diazonium salt (preferably fast blue β-*o*-dianisidine), made alkaline with sodium bicarbonate with the formula:

$$\left[\text{N}\equiv\text{N}-\overset{\displaystyle \text{CH}_3\text{O}}{\underset{}{\bigcirc}}-\overset{\displaystyle \text{OCH}_3}{\underset{}{\bigcirc}}-\text{N}\equiv\text{N} \right]^{2+} \text{ZnCl}_4^{2-}$$

Procedure

1. Bring down sections to water and wash successively in 0.8, 1.0 and 2 per cent aqueous sodium chloride solutions.
2. Incubate at 37 °C for 30 min to 2 h in case of fixed sections or for 10–60 min for fresh sections in the incubation mixture.
3. Rinse in several changes of distilled water. Use cold saline solution for fresh sections.
4. Treat for 3–5 min in cold aqueous diazonium salt solution.
5. Rinse in three changes of cold saline.
6. Mount in glycerin jelly or dehydrate as usual, clear and mount in balsam. In case of fresh unfixed sections, fix in 10 per cent cold formalin for 2 h before mounting.

Observations

The acid phosphatase sites are marked by a blue or reddish blue precipitate.

Mathers and Norman's Method (1956)

This method has been applied for the study of acid phosphatase activity in malignant and benign tumours of the prostate glands.

Reagents

Incubation mixture containing M/10 aqueous sodium glycerophosphate solution, buffered to pH 4·9 with acetate buffer containing 0·12 per cent lead nitrate solution.

3% aq. acetic acid solution
2% aq. ammonium sulphide solution
1% aq. light green solution

Procedure

1. Prepare 10 μm thick frozen sections and bring to water.
2. Incubate in the incubation mixture at 37 °C for 5–15 min.
3. Wash in 3 per cent acetic acid solution.
4. Treat with fresh ammonium sulphide solution for 2 min.
5. Rinse thoroughly in water.
6. Stain with light green solution.
7. Rinse again in distilled water.
8. Mount in glycerin jelly or dehydrate and mount in neutral balsam.

Observations

Dark brown or black precipitate is formed at the sites of acid phosphatase activity.

DETECTION OF ALKALINE PHOSPHATASE

Gomori's Calcium Cobalt Method for Alkaline Phosphatase (1946)

Reagents

1. Incubation mixture containing:

3% aq. sodium-β-glycerophosphate solution	10 cm^3
2% aq. sodium diethylbarbiturate	10 cm^3
Distilled water	5 cm^3
2% aq. calcium chloride solution	20 cm^3
5% aq. magnesium sulphate solution	1 cm^3

2. 2 per cent aqueous cobalt nitrate or acetate solution.

3. Aqueous dilute yellow ammonium sulphide solution (about 50 drops of concentrated liquid in 20 cm^3 of water).

Procedure

1. Preparation of paraffin sections—Fix in cold acetone at 4 °C with three changes for 24 h. Transfer through progressive alcohol grades to absolute ethyl alcohol, keeping 30 min in each grade. Treat with ethyl alcohol–ether mixture (1:1) for 1 h and transfer to 1 per cent celloidin. Decant excess celloidin and treat successively in chloroform and benzene, keeping 1 h in each. Embed in paraffin. Cut sections 5 μm thick and mount on albuminised slides.

Dry at 37 °C and store at 4 °C.

2. Bring down the slides in distilled water through immersion successively in petroleum ether and absolute acetone.

3. Incubate in the incubating mixture at 37 °C for 30 min to 16 h, 4 h being the optimum period.

4. Wash thoroughly in distilled water after bringing down the slides to room temperature.

5. Immerse in 2 per cent cobalt nitrate solution for 3–5 min. Wash in distilled water.

6. Immerse in yellow ammonium sulphide solution for 1–2 min.

7. Rinse in distilled water. Stain with 1 per cent eosin solution for 5 min if required.

8. Dehydrate and clear through alcohol and xylol grades and mount in neutral balsam.

Observations

A deposition of a black precipitate at the sites of alkaline phosphatase activity.

Alternatives

1. For frozen sections, cut very thin sections and mount on slides without albumin. Dry at room temperature for 1–2 h. The remaining steps are similar to paraffin sections.

2. In a later schedule by Gomori (1952), the incubation mixture contains:

3% aq. sodium glycerophosphate solution	20 cm^3
2% aq. sodium diethylbarbiturate solution	30 cm^3
2% aq. calcium chloride solution	4 cm^3
2% aq. magnesium sulphate solution	2 cm^3
Distilled water	30 cm^3

Cobalt chloride can be used instead of cobalt nitrate or acetate.

3. Non–metallic fixatives give quite good results. The preparation of paraffin blocks can be done through the usual alcohol–chloroform grades.

Caution

1. In no step, from fixation to mounting, should the temperature exceed 56 °C.
2. The incubation mixture should be freshly prepared before use.

Fredericsson's Modification of Gomori's Method (1952, 1956)

The method is recommended for paraffin sections of materials fixed in alcohol.

Reagents

1. Incubation mixture containing:

2% aq. sodium-β-glycerophosphate solution	25 cm^3
2% aq. sodium veronal solution	25 cm^3
2% aq. calcium nitrate·solution	5 cm^3
0·8% aq. magnesium chloride solution	5 cm^3
Acetone	40 cm^3

2. 2 per cent cobalt nitrate solution in 40 per cent acetone.
3. Dilute yellow ammonium sulphide solution in 40 per cent acetone.

Procedure

1. Fix tissues in 90 per cent ethyl alcohol at 22 °C for 24 h. Dehydrate through 96 per cent and absolute alcohol grades and then 2 changes of benzene, keeping 30 min in each. Embed in paraffin, cut sections 3–10 μm thick and mount without stretching in water. Deparaffinise through xylol and 40 per cent acetone grades. Bring down to distilled water.
2. Incubate in the incubation mixture at 37 °C for 10 min to 1 h.
3. Bring to room temperature and wash in 40 per cent acetone.
4. Immerse in cobalt nitrate solution in acetone for 5 min.
5. Wash again in 40 per cent acetone.
6. Immerse in ammonium sulphide solution in acetone for 3 min.
7. Wash in pure acetone.
8. Dehydrate through alcohol and xylol grades and mount in neutral balsam.

Observations

The formation of black precipitate at the sites of alkaline phosphatase activity.

α-Naphthyl Phosphate Method for Alkaline Phosphatase (Gomori, 1951)

Reagents

1. Incubation mixture containing:

Sodium-α-naphthyl phosphate	0·05 g
5% aq. borax solution	10 cm³
10% aq. magnesium chloride or	
sulphate solution	0·5 cm³
Cold distilled water (20 °C)	100 cm³
A stabilized diazonium salt	0·25 g

Either tetrazotised-*o*-dianisidine (michrome blue salt 250), 3-nitroanisole-4-diazonium chloride (michrome red salt 606), 4-chloroanisole-2-diazonium chloride (michrome red salt 612) or 3-nitrotoluene-4-diazoniumnaphthalene-1, 5-disulphonate (michrome scarlet salt 618).

2. Haematoxylin solution in water.

3.

70% ethyl alcohol	99 cm³
Glacial acetic acid	1 cm³

Mix together.

Procedure

1. For paraffin block preparations, fix thin cold slices of tissue at 4 °C for 24 h in three or four changes of acetone. Clear in xylol. Embed in paraffin rapidly. Cut sections and attach to albuminised slides. Deparaffinise as usual.

2. Rinse three times in pure acetone and then in three changes of distilled water.

3. Incubate in the incubation mixture at 37 °C for 10–30 min or more. Remove a slide at intervals and observe under the microscope until the correct brightness of colour is attained. Stir the mixture mechanically during incubation.

4. Rinse thoroughly in distilled water after bringing the slides to room temperature.

5. Immerse in haematoxylin solution for counterstaining.

6. Treat with acetic–alcohol mixture for 5–10 min.

7. Rinse thoroughly in water.

8. Mount in glycerin jelly or dehydrate, clear and mount in neutral balsam.

Observations

The sites of alkaline phosphatase activity stain purplish black with michrome blue salt 250, purplish brown with michrome red salt 606 and reddish brown with michrome red salts 612 and 618.

Alternative—Sodium β-naphthylphosphate may be used instead of sodium-α-naphthylphosphate but α-salts yield a more specific and non-diffusible precipitate.

A Modified Coupling Azo Dye Method for Alkaline Phosphatase (see Pearse, 1968)

For Frozen Sections

1. Fix thin slices of tissue in 10 per cent neutral formalin in the cold for 10–16 h, or use fresh frozen cold microtome sections, mounted on cover slips.
2. Cut frozen sections 10–15 μm and mount on slides without adhesive. Dry in air for 1–3 h.
3. Dissolve 10–20 mg sodium–naphthyl phosphate in 20 cm^3 0·1M stock tris buffer (pH 10). Add to it, with stirring, 20 mg of the stable diazotate of 5-chloro-*o*-toluidine. Cover the sections on the slides with the filtered solution and incubate at room temperature for 15–60 min.
4. Rinse in running water for 1–3 min.
5. Counterstain for 1–2 min in Mayer's haemalum.
6. Rinse in running water for 30–60 min and mount in glycerine jelly.
7. The sites of alkaline phosphatase activity appear brown with Fast Red TR and Fast Violet B or black with Fast Black B. The nuclei are dark blue.

For paraffin sections of material fixed in cold acetone.

1. Bring down the sections to water after passing successively through light petroleum and absolute acetone.
2. Cover with freshly prepared filtrate of substrate-diazonium salt mixture as in previous schedule.
3. Incubate for 30 min to 4 h for salt Fast Blue RR or for up to 2 h for salt Fast Red RC or for up to 12 h for salt Fast Red TR.
4. Rinse in water, counterstain with haemalum as given in previous schedule, wash in running water and mount in glycerine jelly.
5. Salt Fast Red TR gives the best results, the sites of alkaline phosphatase activity appearing reddish brown and the nuclei blue.

DETECTION OF CYTOCHROME OXIDASE

G *Nadi Reaction*

Reagents

1. α-naphthol solution—Add 1 g of α-naphthol to 100 cm³ of distilled water and boil till α-naphthol begins to melt. Add to it 40 per cent aqueous potassium hydroxide solution drop by drop till the mixture turns yellowish blue.

2. Dimethyl-*p*-phenylenediamine base solution—Add 0·5 g of dimethyl-*p*-phenylenediamine from a sealed tube to 100 cm³ of distilled water, breaking the tube inside the water. Keep for 24 h. Aid dissolution by occasional shaking.

3. Gram's iodine solution.

4. Carmalum solution.

5. 0·5 per cent aqueous lithium carbonate solution—2 drops added to 100 cm³ of distilled water.

Procedure

1. Fix thin tissues in formol-saline for 3–5 h. Cut frozen sections and bring them down to water.

2. Mix together equal volumes of α-naphthol and dimethyl-*p*-phenylenediamine base solutions. Filter. Treat the slides in this mixture until the sections turn blue.

3. Wash in distilled water.

4. Treat with Gram's iodine solution until the sections turn brown.

5. Keep in lithium carbonate solution for 15 min to 24 h until the sections again become blue.

6. Rinse in distilled water.

7. Immerse in carmalum for 2–5 min.

8. Mount in glycerin jelly.

Observations

The nuclei take up pink and the oxidase granules take up blue colour.

REFERENCES

Ackerman, G. A. (1962). *Lab. Invest.* **11**, 563
Arsenis, C. and Touster, O. (1967). *J. biol. Chem.* **242**, 3399
Atkinson, T. G. and Shaw, M. (1955). *Nature* **175**, 993
Avers, C. J. (1961). *Am. J. Bot.* **48**, 137
Barka, T. and Anderson, P. J. (1962). *J. Histochem. Cytochem.* **10**, 741
Barka, T. and Anderson, P. J. (1963). *Histochemistry, theory, practice and bibliography.* New York; Hoeber
Barka, T., Szalay, S., Posalaky, Z. and Kertesz, L. (1952). *Acta Anat.* **16**, 45

Barter, R., Danielli, J. F. and Davies, H. G. (1955).　*Proc. Roy. Soc. B.* **144,** 442

Beneš, K. and Opatrná, J. (1964).　*Biol. Plant. Praha* **6,** 8

Beneš, K., Lojda, A. and Horavka, B. (1961).　*Histochemie* **2,** 313

Bhattacharjee, D. and Sharma, A. K. (1951).　*Sci. & Cult.* **17,** 268

Bitensky, L. (1963).　*Quart. J. micr. Sci.* **104,** 193

Brachet, J. and Jeener, R. (1948).　*Biochim. Biophys. Acta* **2,** 423

Burstone, M. S. (1958a).　*J. Histochem. Cytochem.* **6,** 87

Burstone, M. S. (1958b).　*J. nat. Cancer Inst.* **20,** 601

Burstone, M. S. (1962).　*Enzyme histochemistry.* New York; Academic Press

Burton, J. F. (1954).　*J. Histochem. Cytochem.* **2,** 88

Cacioppo, F., Quagliariello, C., Coltorti, M. and Della Pietra, G. (1953).　*Arch Sci. biol., St. Petersb.* **37,** 563

Chevremont, M. and Firket, H. (1953).　*Int. Rev. Cytol.* **2,** 261

Clarkson, T. B. and Kench, J. E. (1958).　*Biochem. J.* **69,** 432

Cleland, K. M. (1950).　*Proc. Linn. Soc. N.S.W.* **75,** 35

Cloetens, R. (1941).　*Biochem. Z.* **310,** 42

Connor, R. L. and MacDonald, L. A. (1964).　*J. Cell. Comp. Physiol.* **64,** 257

Dalgaard, J. B. (1956).　*J. Histochem. Cytochem.* **4,** 14

Danielli, J. F. (1946).　*J. exp. Biol.* **22,** 110

Danielli, J. F. (1947).　*Symp. Soc. exp. Biol.* **1,** 101

Danielli, J. F. (1950).　*Nature* **165,** 762

Danielli, J. F. and Catcheside, D. G. (1945).　*Nature* **156,** 294

Davis, B. J. and Ornstein, L. (1959).　*J. Histochem. Cytochem.* **7,** 297

Defendi, V. (1957).　*J. Histochem. Cytochem.* **5,** 1

Defendi, V. and Pearse, A. G. E. (1955).　*J. Histochem. Cytochem.* **3,** 203

Deimling, O. H. (1964).　*Histochemie* **4,** 48

Dempsey, E. W. (1949).　*Ann. N.Y. Acad. Sci.* **50,** 336

De Nicola, M. (1949).　*Quart. J. micr. Sci.* **90,** 391

Desmet, V. J. (1962).　*Stain Tech.* **37,** 373

Dixon, M. and Webb, E. C. (1964).　*Enzymes,* 2nd ed. London; Longmans

Doyle, W. L., Omoto, J. and Doyle, M. E. (1951). *Exp. Cell. Res.* **2,** 20

Eaton, R. H. and Moss, D. W. (1966).　*Proc. Biochem. Soc.* **100,** 45

Elliot, A. M. and Bak, I. J. (1964).　*J. cell. Biol.* **20,** 113

Feigen, I., Wolf, A. and Kabat, E. A. (1950).　*Amer. J. Path.* **26,** 647

Fishman, W. H., Green, S. and Inglis, N. I. (1963).　*Nature* **198,** 685

Friedman, O. M. and Seligman, A. M. (1950).　*J. Amer. Chem. Soc.* **72,** 624

Fredricsson, B. (1952).　*Anat. Anz.* **99,** 97

Fredricsson, B. (1956).　*Acta Anat.* **26,** 246

Gahan, P. B. (1965).　*J. exp. Bot.* **16,** 350

Gahan, P. B. (1967).　*Int. Rev. Cytol.* **21,** 2

Glick, D. and Fischer, E. E. (1946).　*Arch. Biochem.* **11,** 65

Goessner, W. (1958).　*Histochemie* **1,** 48

Goetsch, J. B. and Reynolds, P. M. (1951).　*Stain Tech.* **26,** 145

Goetsch, J. B., Reynolds, P. M. and Bunting, H. (1952).　*Proc. Soc. exp. Biol., N.Y.* **80,** 71

Goldberg, A. F. and Barka, T. (1962).　*Nature* **195,** 297

Goldfischer, S., Essner, E. and Novikoff, A. B. (1964).　*J. Histochem. Cytochem.* **12,** 72

Gomori, G. (1939).　*Proc. Soc. exp. Biol., N.Y.* **42,** 23

Gomori, G. (1941).　*J. cell. comp. Physiol.* **17,** 71

Gomori, G. (1946).　*Amer. J. clin. Path.* **16,** 177

Gomori, G. (1950).　*Stain Tech.* **25,** 81

Gomori, G. (1951).　*J. Lab. clin. Med.* **37,** 526

Gomori, G. (1952).　*Microscopic histochemistry.* Chicago; Chicago University Press

Gomori, G. (1956).　*J. Histochim.* **4,** 453

Goodlad, G. A. J. and Mills, G. T. (1957).　*Biochem. J.* **66,** 346

Gould, B. S. (1944).　*J. biol. Chem.* **156,** 365

Grogg, E. and Pearse, A. G. E. (1952).　*Nature* **170,** 578

Gurr, E. (1958).　*Methods of analytical histology and histochemistry.* London; Leonard Hill

Hard, W. L. and Lassek, A. M. (1946).　*J. Neurophysiol.* **9,** 121

Harrington, J. F. and Altschul, A. M. (1963).　*Fed. Proc.* **22,** 475

488   *Enzymes*

Herveg, J. P., Beckers, C. and De Visscher, M. (1966).   *Biochem. J.* **100,** 540
Hill, M. (1956).   *Čs. Morphol.* **4,** 1
Hoare, R. and Delory, G. E. (1955).   *Arch. Biochem. Biophys.* **59,** 465
Holt, S. J. (1954).   *Proc. Roy. Soc.* B142, 160
Janigan, D. T. (1965).   *J. Histochem. Cytochem.* **13,** 476
Kabat, E. A. and Furth, J. (1941).   *Amer. J. Path.* **17,** 303
Kaplow, L. S. (1963).   *Am. J. Clin. Path.* **39,** 439
King, E. J. (1952).   Personal communication, referred to in Pearse, 1960
Klamer, B. and Fennell, R. A. (1963). *Exp. Cell Res.* **29,** 166
Krugelis, E. J. (1942).   *Biol. Bull.* **90,** 220
Lake, B. D. (1965).   *J. Roy. Micr. Soc.* **85,** 73
Lassek, A. M. (1947).   *Stain Tech.* **22,** 133
Lison, L. (1948).   *Bull. Histol. Tech. micr.* **25,** 23
Loustalot, P. (1955).   Personal communication, referred to in Pearse, 1960
Lojda, Z. (1962).   *Čs. Morfol.* **10,** 46
Lojda, Z., Vecerek, B. and Pelichova, H. (1964).   *Histochemie* **3,** 428
Loveless, A. and Danielli, J. F. (1949).   *Quart. J. micr. Sci.* **90,** 57
Lovelock, J. E. (1954).   *Biochem. J.* **60,** 692
Maengwyn-Davies, G. D. and Friedenwald, J. S. (1950).   *J. cell. comp. Physiol.* **36,** 421
Mannheimer, L. H. and Seligman, A. M. (1948).   *J. nat. Cancer Inst.* **9,** 181
Martin, B. F. and Jacoby, F. (1949).   *J. Anat.* **83,** 351
Mathers, G. L. and Norman, T. D. (1956).   *Lab. Invest.* **5,** 276
Meany, A., Gahan, P. B. and Maggi, V. (1967).   *Histochemie,* **11**
Menten, M. L., Junge, J. and Green, M. H. (1944).   *J. biol. Chem.* **153,** 471
Moog, F. (1943).   *J. cell. comp. Physiol.* **22,** 95
Moss, D. W. (1964).   *Scientific basis of Med. Ann. Rev.,* 334
Newman, W., Feigin, I., Wolf, A. and Kabat, E. A. (1950).   *Amer. J. Path.* **26,** 257
Newman, W., Kabat, E. A. and Wolf, A. (1950).   *Amer. J. Path.* **26,** 489
Novikoff, A. B. (1952).   *Exp. Cell Res.* suppl. **2,** 123
Novikoff, A. B. (1955).   In *Analytical cytology,* chap. 2, ed. by Mellors, R. C. New York; McGraw-Hill Hill
Novikoff, A. B. (1961).   In *The Cell* **2,** New York; Academic Press
Osawa, S. (1951).   *Embryologia* **2,** 1
Pearse, A. G. E. (1960) and (1968).   *Histochemistry.* Boston; Little, Brown
Poux, N. (1963).   *J. Microscopie* **2,** 485
Rabinovitch, M. (1949).   *Nature,* **164,** 878
Robbins, S. L. (1950).   *Amer. J. med. Sci.* **219,** 376
Rutenberg, A. M., Barrnett, R. J., Tsou, K. C., Monis, M. and Teague, R. (1958).   *J. Histochem. Cytochem.* **6,** 90
Rutenberg, A. M. and Seligman, A. M. (1955).   *J. Histochem. Cytochem.* **3,** 455
Ruyter, J. H. C. (1964).   *Histochemie* **3,** 521
Ruyter, J. H. C. and Newmann, H. (1949).   *Biochem. Biophys. Acta* **3,** 125
Sadasivan, V. (1952).   *Nature,* **169,** 418
Seligman, A. M., Chauncey, H. H. and Nachlas, M. M. (1951).   *Stain Tech.* **26,** 14
Seligman, A. M., Heymann, H. and Barrnett, R. J. (1954).   *J. Histochem. Cytochem.* **2,** 441
Seligman, A. M. and Mannheimer, L. H. (1949).   *J. nat. Cancer Inst.* **9,** 427
Serra, J. A. (1955).   *Handbuch der Pflanzenphys.* **1,** 413
Sharma, A. K. and Roy, M. (1956a).   *Cellule* **58,** 109
Sharma, A. K. and Roy, M. (1956b).   *Фyton* **7,** 23
Sharma, A. K. and Roy, M. (1956c).   *Cellule* **57,** 337
Sharma, A. K. and Bhattacharyya, B. (1957).   *Bull. bot. Soc. Beng.* **11,** 34
Sharma, A. K., Mookerjea, A. and Ghosh, C. (1953).   *Portug. acta biol.* **3,** 341
Shugar, D., Szenberg, A. and Sierakowska, H. (1957).   *Exp. Cell Res.* **13,** 424
Sonneschein, N. and Kopac, M. J. (1953).   *Anat. Rec.* **117,** 611
Stutte, H. J. (1967).   *Histochemie* **8,** 327
Sulkin, W. M. and Gardner, J. H. (1948).   *Anat. Rec.* **100,** 143
Takamatsu, H. (1939).   *Trans. Soc. Japan* **29,** 429
Takeuchi, T. and Tanoué, M. (1951).   *Kumamoto med. J.* **4,** 41
Tandler, C. J. (1953).   *J. Histochem. Cytochem.* **1,** 151

Thomas, M. and Aldridge, W. N. (1966).   *Biochem. J.* **98,** 94
Wachstein, M. (1945).   *Arch. Path.* **40,** 51
Watanabe, K. and Fishman, W. H. (1964).   *J. Histochem. Cytochem.* **12,** 252
Wetzel, B. K., Horn, R. G. and Spicer, S. S. (1963).   *J. Histochem. Cytochem.* **11,** 812
Wolf, A., Kabat, E. A. and Newman, W. (1943).   *Amer. J. Path.* **19,** 423
Wolman, M. (1965).   *Z. Zellforsch.* **65,** 1
Wolman, M. and Bubis, J. J. (1966).   *Histochemie* **7,** 105
Woodard, H. Q. (1951).   *J. Urol.* **65,** 688
Wulff, H. R. (1967).   *Med. dansk. Munksgard, Copenhagen*
Zorzoli, A. and Stowell, R. E. (1947).   *Anat. Rec.* **97,** 495

22 Micrurgy and Isolation

The isolation procedure is often adopted, not only for the chemical analysis of chromosomes but also for culturing the nuclei and chromosomes in natural or synthetic media. The latter procedure requires micrurgical operation of the cell without the use of any chemical. It allows a study of the initial steps in chromosome metabolism and of the mechanism of genetic regulation of differentiation. On the other hand, the chemical method of chromosome isolation, in which only non-injurious compounds are employed, is useful for the study of the chemical make-up of the chromosomes at different stages of development.

Micrurgical Method of Isolation of Nuclei and Chromosomes and Their Culture

The technique for micrurgical isolation has mainly been developed with respect to the salivary gland chromosomes of diptera, where puffing at different segments in different phases of development provides adequate proof of the genetic control of differentiation and the change of pattern following treatment with different agents. Short term *in vitro* culture can be carried out following the hanging drop method (as mentioned in the chapter on tissue culture), involving culturing in a drop of medium on the cover slip inverted over a depression slide, and sealed with oil, or on a slide covered with oil. Blowing through a pipette is recommended for adequate oxygen supply. The best medium for hanging drop culture is no doubt haemolymph but several other media, including TC 199 and Jones and Cunningham's medium (1961) for sciarids and chironomids, respectively, can also be used. Hadorn *et al.* (1963) devised a method for *in vivo* culture, in which glands, after explantation, were transferred to younger hosts by injection. One of the serious limitations of polytene chromosome culture is the fact that salivary gland cells have a strong ionic barrier against haemolymph, whereas owing to the presence of a special membrane, there is no such barrier between adjacent cells (Kanno and Loewenstein, 1964;

Kroeger and Lezzi, 1966). Consequently, in the case of any cell leakage in haemolymph or lumen, there is a complete loss of ionic equilibrium, because all the cells are simultaneously affected. Therefore, when excising the gland, extreme care must be taken not to rupture, or shear, the ligaments adjacent to it. Moreover, because of the low regeneration rate of glands after injury, this method allows a study of DNA synthesis and transcription only, not of the entire process of mitosis.

The Isolation and Culturing of Nuclei in Medium Under Oil

The essential requisites are, (*a*) a suitable medium, (*b*) a suitable oil and (*c*) siliconised slides. A good dissecting microscope serves the purpose for observation. Of the media used, the egg contents of *Drosophila* have been found to be very satisfactory. They have to be diluted at the later stages and culture can be prolonged even up to 4 h (Kroeger, 1963, 1966; Lezzi, 1961, 1965; *see* Von Borstel, 1959). Sugar medium with synthetic compounds, having the following composition, gives very satisfactory results (Frenster *et al.*, 1960; Sirlin and Schor, 1962; Rey, 1963; Lezzi, 1965; Kroeger, 1966):

Saccharose	64·1805 g
Glucose	3·3741 g
$MgCl_2$	1·7866 g
NaCl	1·6659 g

Tris buffer—0·5 litre (3·025 g tris + 20·7 cm^3 1/n HCl),

supplemented with polyvinylpyrrolidone or Luviskol-K90. The oil needed to cover the culture for checking against desiccation is hydrofluorocarbon oil (Kal-F No. 10) (Kopac, 1955), the viscosity of which can be adjusted by mixing with paraffin. Another oil, often employed, is a mixture of heavy mineral oil and Oronite Polybutane 128 (2:1).

The slide can be siliconised by dipping in a mixture of a few drops of silicone oil in 250 cm^3 of acetone and drying for 24 h at 20–25 °C. If required, the time, temperature and concentration can be varied.

The schedule followed for isolation and culturing is given below (Kroeger, 1966):
(a) Cover the donor tissue with a drop of oil on a siliconised slide.
(b) From a second oil drop on the slide containing the culture medium, transfer two spheres of medium to the first oil drop containing the donor tissue. Of the two spheres, sphere A should be about $\frac{1}{5}$ to $\frac{1}{10}$th the volume of the other sphere, and B should be about 20 times larger in volume than the donor tissue. The transfer is carried out under a dissecting microscope, after bringing the medium in sphere A in contact with the donor tissue so that the medium in sphere A forms just a rim surrounding the donor tissue. The method is as follows:
  (i) For semi-isolation, gradually cut off and separate the cytoplasm and other extranuclear components from the nucleus in A, with the aid of a bent tungsten needle, leaving the nucleus surrounded by a very thin layer of cytoplasm. The tungsten needle must be sharpened previously by immersion in a hot mixture of potassium

nitrate and sodium nitrite. Glass needles, drawn out in a gas micro-burner, can also be used. Draw the large droplet B and join it with A. Push the nucleus into the larger sphere B and finally cut off the connection between the two spheres.

(ii) Alternatively, for complete isolation, puncture the cell with a glass needle and squeeze out the nucleus and follow the procedure given in (i).

(c) For staining, after the incubation period, place a large drop of acetic-orcein solution on the material, avoiding conglomeration of cells, and stain for 20 min or more. Pass through acetone to remove the oil, rinse in water, blot off excess fluid, add a drop of acetic-orcein solution, wait for a few minutes and mount under a cover slip. Lactic-orcein solution, used alternatively, may be prepared according to Beermann's schedule, that is, boiling 2 g of orcein in a mixture of 50 cm^3 each of acetic acid and lactic acid, shaking and filtering. If necessary, a contrasting stain may also be applied, prepared by mixing light green (FS) 0·1 per cent in 96 per cent ethanol, and orange G, 0·2 per cent in 70 per cent ethanol, in the proportion of 55:45, followed by adding 1–2 drops of acetic acid to bring to pH 5·0 (Clever, 1961).

Isolation of Chromosomes

To secure fixed chromosome preparations, the method is quite a simple one. The chromosomes are fixed as usual in 45 per cent acetic acid and then removed from the cell with a needle.

To obtain unfixed chromosome preparations, the most convenient method is to puncture the salivary gland with a needle and squeeze out the chromosomes in sugar medium, as mentioned above (Lezzi, 1965). There are two other methods, one involving treatment of the gland with pronase followed by homogenisation and differential centrifugation (Karlson and Löffler, 1962) whereas in the other, the glands are dipped in 0·25 per cent solution of dried eggwhite for 2–3 h and then the stiffened chromosomes are isolated with needles (Buck and Malland, 1942). For studying ribonucleic acid metabolism, the required precursors and other factors have to be supplied to the medium (*see* Kroeger and Lezzi, 1966).

In *transplantation experiments*, a foreign cytoplasm is inserted within the host cell by piercing the cell without touching the nucleus, pressing the cytoplasm through the oil and then the slit with the help of a needle. Similarly, chromosomes can be donated to the host cytoplasm by taking out a chromosome, rolling it into a bundle and pushing it through the slit of the host cell. It is even possible to make a slit in the nucleus and transplant the chromosome inside the nucleus (Kroeger, 1966). The degree of perfection achieved in these procedures depends on experience, skill and steadiness of operation.

The main disadvantage of the micrurgical method is the difficulty in securing a long term culture and in performing an operation without injury to the tissue.

Isolation of Mammalian Metaphase Chromosomes by Extraction from Cell Culture

As the majority of the techniques of cell isolation by chemical means result in a dissociation of nucleohistones (Busch, 1965) and chromosome aggregation, a very simple method has been suggested by Maio and Schildkraut (1966*a*). The isolation is carried out at neutral pH utilising the high specific gravity of chromosomes, without the use of acid, alkali or enzyme. The method, claimed to yield stable and intact chromosomes, is outlined below:

1. *Cell culture*—Grow suspension cultures of HeLa strain S3 cells, mouse strain L-cells, Chinese hamster cells, strain V-79-379A from the lung of a female and a strain of Syrian hamster cells transformed by SV-40 virus, in Eagle's medium supplemented with non-essential amino acids and 5 per cent foetal calf serum. The generation time for the Chinese hamster cell line was 12 h while that for the other cells lines varied from 22–24 h at 37 °C.

2. *Metaphase arrest*—To these logarithmically growing cultures add vinblastine sulphate (Eli Lilly and Co.) to obtain a final concentration of 0·01 $\mu g/cm^3$ for HeLa and Chinese hamster cells, and 0·5 $\mu g/cm^3$ for Syrian hamster and L-cells, incubate at 37 °C for 15 h and 8 h, respectively, for the two former strains and 11 h for the two latter strains. The HeLa cells give the highest frequency of metaphase (often as high as 90 per cent) and do not form any micronuclei, as against a frequency of 20–40 per cent in the other strains. However, the only disadvantage of HeLa cells is the uniformity of their chromosomes, which are difficult to identify. The selection of the optimum incubation period should be checked against the formation of micronuclei, because, once formed they are very difficult to separate from the chromosomes and the chromosome yield is thus lessened.

Chromosome Extraction

Hypotonic treatment and homogenisation: In each experiment for extraction, 4–12 litres of cell culture containing 2×10^9 to 6×10^9 cells are harvested. Centrifuge at 500 g for 15 min in the cold at 10 °C and wash the sedimented cells twice in Earle's balanced salt solution. Re-suspend in a mixture (TM) containing 0·0001 M of each of the chlorides of calcium, magnesium and zinc in 0·02 M tris, maintained at pH of 7·0, in a proportion of cell suspension to medium (1:10) and keep for 20 min. Add 5 per cent filtered saponin solution to make a final concentration of 0·05 per cent and keep for 5 min. Transfer aliquots of the suspension to a 40 cm^3 capacity Dounce homogeniser with pestle, and break up the cells with a few strokes to release the chromosomes.

2. To *prepare crude chromosome suspension*—add twice the volume of TM, containing 0·05 per cent saponin, to the homogenate and transfer to centrifuge tubes. Fill about 3 cm of each tube and centrifuge for 5 min at 120 g. The resulting sediment contains most of the nuclei and unbroken cells. The suspension liquid, containing the chromosomes, is decanted and

stored. Re-suspend the sediment in each tube by pipetting in TM solution to the original volume, centrifuge and store the supernatant to extract the chromosomes left in the sediment. Repeat the process to extract most of the chromosomes; observing through a phase contrast microscope.

3. To obtain a purified extract of chromosomes, centrifuge the collected suspension in an anglehead centrifuge at 2500 g for 10 min. Wash the sediment in only 0·02 M tris containing 0·1 per cent saponin (pH 7·0), because the divalent metallic salts in the TM solution prevent the liberation of chromosomes from the debris and solution of the amorphous material. Re-suspend the sediment in 2·2 M sucrose in 0·02 M tris (pH 7·0) and 0·1 per cent saponin. Layer the suspension in cellulose nitrate ultracentrifuge tubes over 10 cm³ of the dence sucrose solution (sp. gr. 1·28). Stir with a glass rod to mix the two solutions, leaving about 1 cm of the sucrose solution undisturbed at the bottom of each tube. Centrifuge at 50 000 g for 1 h. The sediment contains the chromosomes (sp. gr. approx. 1·35) plus the few nuclei that are left as contaminant, while the supernatant, with the fine cytoplasmic debris, is discarded. Re-suspend the sediment with purified chromosomes in TM and observe under phase contrast. If the chromosomes are still mixed with impurities, repeat the process of washing in tris-saponin buffer and centrifuging in dense sucrose. Remove the sucrose by suspending the chromosomes in TM and centrifuge at 2500 g for 10 min. The procedure is repeated three times.

From the HeLa metaphase cells, about 30–40 per cent of the chromosomes are recovered, as counted in a phase contrast haemocytometer. The average DNA content of isolated chromosomes from HeLa metaphase cells has been estimated to be 0·5 µµg per chromosome and therefore less than 10 per cent of DNA is lost from the chromosomes during the isolation procedure. The suspension containing chromosomes may be stored in the TM solution at 0 °C in ice buckets. Longer storage is recommended at -20 °C, after adding glycerol, to obtain a final concentration of 20 per cent. For experimental purposes, the chromosomes can be maintained both in a contracted state and in monodisperse suspension in a buffer containing 0·005 M calcium chloride and 0·05 per cent saponin in 0·02 M tris at a pH of 7·0. Dispersion may be achieved by aspirating repeatedly through a syringe with a No. 22 spinal tap needle.

In the method followed for permanent mounts, an equal volume of acetic-methanol (1:3) fixative is added to the suspension, mixed thoroughly by aspirating through a pipette, and centrifuged at 2500 g for 3 min. The sediment is re-suspended in the fixative and centrifuged, the process being repeated several times. The final pellet is redispersed in a small amount of fixative. Air-dried slides are prepared by spreading a drop of the concentrated solution over a grease-free slide and allowing it to dry. The preparations may be stained using the usual stains for chromosomes (*see* chapter on human chromosomes). The advantages of having isolated chromosomes for experimental work, as seen from Chinese hamster and HeLa cells are:

1. The karyotype is usually maintained; there is little evidence of loss through breakage and the chromosomes are strongly Feulgen-positive.

2. Phase contrast photographs show the chromosomes to be plump,

refractive and rigid, possibly due to the adherence of histones which are generally lost in normal fixation procedure.

3. The chromosomes can be studied with regard to changes in ionic environment and enzyme treatment.

4. Isolated chromosomes of Chinese hamster cells can be separated into three classes by low-speed centrifugation at a steep sucrose gradient (Maio and Schildkraut, 1966*b*). The chemical composition of the morphologically distinguishable chromosomes can then be studied differentially.

The inherent disadvantages of the method include:

1. Selective loss or the possibility of loss, of small chromosomes during centrifugation and,

2. Absorption of the ribosomes on the chromosomes during extraction.

The future possibilities of the method and refinements suggested are:

1. In addition to the divalent metal cations needed for maintaining the chromosome structure at a neutral pH during isolation, several polyamines and even histones may be used for the same purpose.

2. The purity of the preparations depends on phase contrast observations, to exclude the presence of nuclei, membrane fragments and cytoplasmic debris, but extraneous materials may be adsorbed on the chromosomes or the chromosome constituents may get lost. Since they cannot be detected through phase contrast, refinements in this aspect are needed.

Isolation of Nuclei in S3 HeLa Cells (Prescott *et al*., 1966)

1. Grow monolayers of S3 HeLa cells in F-10 medium supplemented with 10 per cent calf serum, penicillin (50 units/cm^3) and streptomycin (50 µg/cm^3).

2. Add 5 cm^3 0·25 per cent trypsin in saline D-2 at pH 7·0–7·4 (Ham, 1963). Incubate for 5–10 min at 36·7 °C and suspend in an additional quantity of 5 cm^3 saline D$_2$.

3. Centrifuge at 1000 g at 0 to 2 °C for 5 min and remove the trypsin. Wash thrice in 50 cm^3 of 0·154 M KC1 and centrifuge at 1000 g for 5 min.

4. For lysing the cells and isolating the nuclei, re-suspend the cells (0·1 cm^3) in 4 cm^3 of 0·1 per cent (v/v) Titron X-100 isolation medium (Rohm and Haas, Philadelphia) containing 0·001 per cent (w/v) spermidine phosphate trihydrate (spermine) dissolved in redistilled water. Agitate continuously by blowing through a micro-pipette to rupture cell membrane. After 2–4 min, fix a drop in Carnoy's fluid, stain with toluidine blue and observe under a microscope to find if the nuclei are completely freed from cytoplasmic contamination.

5. After complete lysis, add 0·25 M sucrose with 0·001 per cent spermine to a final volume of 50 cm^3 and centrifuge at 700 g at 0–2 °C for 20 min. Re-suspend the sediment in 0·25 M sucrose with spermine and centrifuge. Repeat twice. Spermine prevents excess rupture of the nuclear membrane and saline D$_2$ with spermine is more satisfactory than sucrose.

REFERENCES

Buck, J. B. and Malland, A. M. (1942).   *J. Hered.* **33,** 173

Busch, H. (1965).   *Histones and other nuclear proteins.* New York; Academic Press

Clever, U. (1961).   *Chromosoma* **12,** 607

Frenster, J. H., Allfrey, V. G. and Mirsky, A. E. (1960).   *Proc. Natl. Acad. Sci. US* **46,** 432

Hadorn, E., Gehring, W. and Staub, M. (1963).   *Experientia* **19,** 530

Ham, R. G. (1963).   *Expt. Cell Res.* **29,** 515

Jones, B. M. and Cunningham, I. (1961).   *Expt. Cell Res.* **23,** 368

Kanno, Y. and Loewenstein, W. R. (1964).   *Science* **143,** 959

Karlson, P. and Löffler, U. (1962).   *Z. Physiol. Chem.* **327,** 286

Kopac, M. J. (1955).   *Trans. N.Y. Acad. Sci.* **17,** 257 and **18,** 22

Kroeger, H. (1963).   *Nature* **200,** 1234

Kroeger, H. (1966).   In *Methods in Cell Physiology* **2,** 61, New York; Academic Press

Kroeger, H. and Lezzi, M. (1966).   *Ann. Rev. Entomol.* **11,** 1

Lezzi, M. (1961).   Diploma thesis, E.T.H. Zürich

Lezzi, M. (1965).   *Expt. Cell Res.* **39,** 289

Maio, J. J. and Schildkraut, C. L. (1966a).   In *Methods in Cell Physiology* **2,** 133, New York; Academic Press

Maio, J. J. and Schildkraut, C. L. (1966b).   *Federation Proc.* **25,** 707

Prescott, D. M., Rao, M. V. N., Evenson, D. P., Stone, G. E. and Thrasher, J. D. (1966).   In *Methods in Cell Physiology* **2,** 131, New York; Academic Press

Rey, V. (1963).   Diploma thesis, E.T.H., Zürich

Sirlin, J. L. and Schor, N. A. (1962).   *Expt. Cell Res.* **27,** 165 and 363

Von Borstel, R. C. (1959).   *Federation Proc.* **18,** 164

23 Extraction

The importance of extraction techniques in the study of chromosomes cannot be over-estimated and, through refinements, not only can nucleic acids and nuclei be extracted from the tissue, but even the chromatin threads can be isolated. For the extraction of nucleic acids, several techniques are employed and extraction through chemicals or digestion through specific enzymes are considered to be suitable for the purpose.

EXTRACTION THROUGH CHEMICALS

Schneider (1945), following the method of Cohen (1944), first demonstrated that hot trichloracetic acid ($CC1_3COOH$) treatment (5 per cent treatment at 40 °C for 15 min) removes nucleic acids from the tissue, the method enabling the complete removal of both DNA and RNA from the cell. Several authors (White, 1950; Kaufmann and colleagues, 1951c; Sharma and Bhattacharjee, 1952) have shown that this method of extraction leaves a protein part showing positive staining with acid dyes. Distinct protein staining has been recorded, both in chromosomes and nucleoli. Atkinson (1952) effectively removed RNA without affecting DNA by using trichloracetic acid at 60 °C.

Sharma (1951) observed that heterochromatic regions of chromosomes can be sharply stained if acetic–alcohol fixed root tips are treated with 0·25 M trichloracetic acid at 60 °C for 40 min followed by hydrolysis at N HC1 for 20 min, prior to Feulgen staining. Following this procedure, the metaphase chromosomes lose DNA from their segments, whereas heterochromatic segments adjacent to the centromere show positive Feulgen reaction. This procedure can, therefore, be applied for staining heterochromatic segments of chromosomes.

Boivin, Vendrely and Vendrely (1949) demonstrated that hydrolysis with 1 N HCl results in complete removal of RNA and depolymerisation of DNA. The same concentration can remove both types of nucleic acids if applied at 37 °C for 3 h (Dempsey, Singer and Wislocki, 1950).

Perchloric acid was employed initially by Ogur and Rosen (1949) for extracting nucleic acids from root tips of onion. Erickson, Sax and Ogur (1949) as well as Sulkin and Kuntz (1950) suggested that specific extraction of RNA as performed through ribonuclease can be effectively substituted by perchloric acid. Cold perchloric acid treatment (10 per cent at 4 °C) for a period of 4 h or over has been used to remove RNA, and the application of 10 per cent acid at 70 °C for 20 min results in the complete removal of both DNA and RNA. Seshachar and Flick (1949) pointed out that the extraction of RNA by cold perchloric acid can be completed only under strictly limited conditions. Moreover, this procedure is associated with depolymerisation of DNA, though not apparently losing any Feulgen stainability. Following such a procedure, pyroninophilia of chromosomes increases rapidly, which evidently is due to the depolymerised nature of DNA. Pearse (1960, 1968) agreed with Seshachar and Flick and considered that digestion with ribonuclease cannot yield results parallel to perchloric acid because in addition to depolymerisation, various proteins, glyco- and lipo-proteins, are also removed by the latter procedure. Kurnick (1955) has shown that in unfixed and fixed tissue homogenates, 10 per cent perchloric acid treatment cannot be employed for quantitative separation of nucleic acids, though principally RNA is extracted at low temperatures. It has been suggested (Pearse, 1960) that perchloric acid can be employed for the specific extraction of RNA under limited conditions of fixation and staining and also depending on the type of tissue used (Koenig and Stahlecker, 1951, 1952; Atkinson, 1952; Wenderoth, 1953; Franz, Warden and Mayer-Arendt, 1954; Goessner, 1954). Aldridge and Watson (1963) noted that no protein is extracted following cold perchloric acid treatment in acrolein-fixed tissues, whereas after acetic-ethanol fixation, there was very little protein extraction but Kasten (1965) observed that post-washing in water causes displacement of protein.

In addition to acids, even alkali treatment has been employed to remove nucleic acids from cells. Levene (1901) employed sodium or ammonium chloride solution for the purpose. Due to the depolymerising effect of salt solutions on nucleohistones, Mirsky and Pollister (1946) could separate ribonucleoprotein from deoxyribonucleoprotein from liver through 0·15 M and 1 M sodium chloride treatment. Complete removal of RNA from fixed materials was secured by Opie and Lavin (1946) by treatment with 0·17 M sodium chloride solution at 56 °C for 2 h. Chargaff, Crampton and Lipshitz (1953), following the analysis of bases after extraction by NaCl, demonstrated that DNA has a constant mean composition within each species but varies from species to species. In addition to these acid and alkali extraction methods, a few others have also been employed by some authors. Caspersson, Hammersten and Hammersten (1935) employed malonic and formic acids for the extraction of nucleic acids. Brachet (1940) observed that treatment in water at 70 °C gradually removed RNA from tissues. A number of buffer solutions on the alkaline side have also been found to remove RNA selectively (Stowell and Zorzoli, 1947). It is necessary to prepare and use depolymerising enzyme solutions in glass distilled water and at 37 °C so that the action of electrolytes and the effect of water can be eliminated.

EXTRACTION THROUGH ENZYMATIC DIGESTION

The principle of this method is based on the application of a specific enzyme to the cell and, on the basis of its digestive property, location of the particular substance in the tissue or the cell. Enzymes that are employed for the digestion of nucleoproteins from chromosomes belong to two categories, namely nucleases and proteases. The former, which bring about digestion of nucleic acids, are available in the form of deoxyribonuclease and ribonuclease, the name being derived from the medium on which they act. For the complete breakdown of nucleic acid and its components, at least three different types of enzymes are necessary, which would be capable of breaking: (a) the polymeric linkage into component nucleotides; (b) the ester linkage maintaining the nucleotide structure, and (c) the glycosidic linkage for maintaining nucleosidic structure, respectively. Nucleases, also known as nucleopolymerases, responsible for the first action are primarily applied in chromosome studies.

Proteases may be trypsin or chymotrypsin, which digest basic proteins, and pepsin, which degrades both histone and more acidic proteins (Daly, Mirsky and Ris, 1951; Kaufmann and colleagues, 1951c).

Ribonuclease is generally extracted from ox pancreas, spleen, or liver in crystalline form (Kunitz, 1940; Brachet, 1942; Ledoux and Brachet, 1955; Kaplan and Heppel, 1956). Its molecular weight is 15 000 and the enzyme, on purification, is a soluble protein. It can resist very high temperatures (Pearse, 1961). Ribonuclease preparation has often been found to be contaminated with proteolytic enzymes (Mazia, 1941; Schneider, 1946; Kaufmann, 1950) so that it is always necessary to check the effect of the enzyme on protein structures. Brachet (1940) first employed ribonuclease for the critical detection of RNA in the tissue in his pyronin–methyl green schedule. In all the modern staining techniques for the detection of RNA and DNA, application of ribonuclease is always used as a control measure to check the site of RNA. Stowell and Zorzoli (1947) pointed out that electrolytes in solution and distilled water at 60 °C may remove RNA and, as such, the enzyme should be dissolved in glass distilled water and used at 37 °C (Kaufmann, 1950). The general procedure for destroying the proteolytic activity of ribonuclease is to boil the extract for 3 min in acid solution, prior to buffering in veronal acetate buffer at a pH of 6·75 (White, 1947). A similar purpose is served by heating at 80 °C for 10 min. The thermostable nature of ribonuclease does not harm its potency under such severe treatment.

Fixatives affect the activity of the enzyme to a significant degree. Chromic acid itself, and fixatives containing it, prevent ribonuclease action. Results are also not satisfactory with Bouin or certain alcoholic fixatives (*see* Pearse, 1960). In general, formalin fixation is recommended for effective digestion with ribonuclease.

A considerable amount of fundamental work on chromosomes has been aided through the application of ribonuclease. Several authors (Kaufmann, Gay and McDonald, 1950; Pollister and colleagues, 1950; McDonald and Kaufmann, 1954; *see* Kaufmann, Gay and McDonald, 1960) have utilised specific staining procedures, combined with the use of ribonuclease. These experiments reveal that RNA forms one of the principal components of the chromosomes and, in all likelihood, is linked with an acidic protein rich in tryptophane.

Similar to ribonuclease, deoxyribonuclease has been employed for causing digestion of DNA. It is a protein of the albumin type, containing both tyrosine and tryptophane. Methods have been evolved for the preparation of an extremely pure form of deoxyribonuclease, free from proteolytic contaminants or ribonuclease (Kunitz, 1948, 1950; Kaufmann and colleagues, 1951c). The crystalline enzyme shows activation by magnesium ions and becomes inactivated by heat. If the enzyme preparation is available in a pure form, the complete loss of methyl green and Feulgen stainability is immediate (Brachet and Shaver, 1948; *see* Vercauteren, 1950; Kurnick, 1955; *see* Pearse, 1960).

In addition to the nucleic acid enzymes, as mentioned already, trypsin and chymotrypsin are applied for the digestion of the basic protein while pepsin removes both histone and more acidic proteins (Daly, Mirsky and Ris, 1951; Kaufmann and colleagues, 1951c; *see* Kaufmann, Gay and McDonald, 1950). In fact, the application of these two enzymes, combined with nucleases, has helped in the interpretation of chromosome structure as an interconnected system of two types of nucleic acids and proteins.

Even in view of the applicability of the enzyme digestion procedures, the limitations of these techniques cannot be ignored. Pure forms of enzymes, free from any contaminants, should be used but it is difficult to obtain such pure forms. Danielli (1947) rightly stressed the difficulty of finding out if an enzyme preparation acts on only one substrate or on a family of substrates. It is also necessary to know how far an enzyme can penetrate through different barriers to the substrate, as the acidity of the substrate, concentration of the electrolysing solution, temperature, inhibiting or activating factors, etc., should all be taken into consideration. Danielli (1953) emphasized the necessity of finding out the extent to which isolated enzymes represent their state *in vivo*, as they may undergo certain modifications during extraction. However, in spite of the limitations, the methods based on enzymatic hydrolysis have helped the cytologists to clarify the problems of chromosome structure, growth and differentiation to a significant extent.

In addition to the above methods for the extraction of DNA and RNA, techniques have been devised to extract intact nuclei, and even chromatin can be isolated from the tissue (Dounce, 1952). Meischer in 1871 first isolated 'nuclein' by digesting pus cells with HCl and pepsin. Later, the method was followed by Kossel (1928) and his collaborators.

Isolation procedures may be direct or indirect. In the former, the nucleus is isolated directly from the cytoplasm with the aid of a micromanipulator and observed under the microscope (Edström, 1953); however, the schedule is beset with several limitations and cannot be applied conveniently to all tissues. For chemical analysis, therefore, the indirect method of isolating in masses is adopted.

In mass isolation technique, the general principle involves, firstly, the disintegration of the cytoplasm through mechanical or chemical means, keeping the nucleus intact, followed by filtration through mesh or cheese cloth, differential centrifugation in suitable liquids and sedimentation. The general specific gravity of the liquid lies between the range of 1·35–1·45 and centrifugation at 1000–6000 rev/min is needed for separation.

For chromatin isolation, sodium chloride solutions of different strengths have been applied. They are found to dissociate nucleohistones from non-basic proteins.

The isolation procedure, however, has its disadvantages. The principal limitation is the use of methods for separation and the liquids needed for suspension. It is claimed that even with weak solvents there is the possibility of decomposition of sensitive compounds, adsorption of cytoplasmic contents by the nucleus and elimination or diffusion of some of the nuclear components (Danielli, 1953; Serra, 1955). However, as the entire process is carried out at 0 °C, the chances of autolysis are very low, but even then the solvents that are employed, either aqueous or non-aqueous, may lead to diffusion. In the former case, diffusion is very heavy (Daly, Allfrey and Mirsky, 1952) but, even in the latter, lipoid components may diffuse out (Serra, 1955). Dounce (1952) noted that even physiological saline solution results in the extraction of nuclear components.

Isolation techniques no doubt enable a worker to handle the chromosomes directly in the laboratory and thus a correct analysis is possible but, in view of the limitations mentioned above, they must be applied with caution, especially with a proper check on the solvent used for suspension and separation. Further, isolation can be carried out on chromosomes at a particular phase of division, such as on metaphase (Chorhazy and colleagues, 1963). With the aid of isolation procedures coupled with autoradiography, it has been possible to demonstrate that RNA synthesis in isolated thymus nuclei is dependent on DNA (Allfrey and Mirsky, 1962). Moreover, nuclear RNA resembling DNA in base composition has been analysed (Sibatani and colleagues, 1962). The relationship between DNA and RNA seems to be best clarified through these procedures.

SCHEDULES

ISOLATION OF NUCLEI

FROM ERYTHROCYTE CELLS OF BIRDS

Laskowski's Method (1942)

Reagents

Isotonic salt solution, 0·1 M aqueous potassium dihydrogen phosphate solution and lysolecithin solution (prepared by mashing poison glands of

100 bees with an emulsion of 5 g of lecithin in 30 cm^3 of phosphate buffer at pH 7·0, digesting at 37 °C for 24 h and filtering through a Birkfield filter).

Procedure

1. Draw out fresh citrated blood, collect the erythrocytes by centrifuging and wash them repeatedly in fresh changes of isotonic salt solution by alternately suspending them in the fresh solution and then centrifuging.

2. Take about 40 cm^3 of the saline solution containing the final suspension of cells and add about 5–10 cm^3 of lysolecithin solution. Treat for 30–40 min at room temperature, observing at intervals.

3. After complete haemolysis, wash the cells again in salt solution by alternate suspension and centrifugation several times in fresh solution.

4. Store in cold as a suspension in normal saline solution or in 0·1 M KH$_2$PO$_4$ solution. The nuclei take blue colour with aqueous methylene blue solution.

Modification

Bensley (*see* Glick, 1949) used a few drops of ether instead of lysolecithin solution.

Dounce and Lan's Method (1943)

Reagents

0·9 per cent aqueous sodium chloride solution, 0·11 M phosphate buffer (pH 6·8–7·0) containing 0·3 g of Merck's purified saponin.

Procedure

1. Similar to the previous schedule, wash the erythrocytes twice in sodium chloride solution.

2. Suspend the cells in a quantity of saline solution equal to the amount of blood used and add one-tenth its volume of phosphate buffer. Treat for 5 min.

3. Wash the nuclei repeatedly in the saline solution and add each time 2–3 cm^3 of phosphate buffer to the nuclei just before adding the saline solution.

STONEBURG'S METHOD OF RAT TUMOUR CELLS (1939)

Reagents

5 per cent aqueous citric acid solution, 1 per cent hydrochloric acid, 0·8 per cent sodium chloride solution, pepsin (1/10 000).

Procedure

1. Chop up the cleaned tissue to small bits.
2. Keep overnight in five times its volume of citric acid. Stir occasionally.
3. Add an equal quantity of distilled water and filter through 8 layers of cheesecloth.
4. Add an equal quantity of water to the filtrate and keep for 3 h.
5. Centrifuge in a cylinder lined with cellophane. Wash the nuclei repeatedly with water by centrifuging and suspending alternately.
6. Treat the residue in equal volume of a mixture of HCl and NaCl with pepsin (1:1) for 4 h at 37 °C. Stir at intervals.
7. Remove the nuclei settled at bottom and wash.

DOUNCE'S METHOD FOR RAT LIVER NUCLEI (1952)

1. Freeze freshly dissected rat liver.
2. Add 100 g of it to 500 cm^3 of citrated water containing ice chips (1·05 cm^3 of 1 M citric acid in 100 cm^3 of water, pH 6·0–6·2) in a Waring blendor while stirring continuously for 10–15 min.
3. Filter once through 2 layers and then again through 4 layers of cheesecloth.
4. Centrifuge for 20 min at about 2000 rev/min. Separate the sediment and make it up with water to 400 cm^3. Add a drop of caprylic alcohol to break up the foam.
5. Centrifuge for 15 min. Suspend the residue in 400 cm^3 of distilled water and centrifuge for 10 min.
6. Repeat washing the tissue and centrifugation 4–5 times, adding 200 cm^3 of water each time, centrifuging for periods decreasing from 10–3 min at slower revolutions and decanting the supernatant each time.
7. Filter through four layers of cheesecloth.
8. Keep the nuclei for 45 min in 100 cm^3 of distilled water. Decant. Centrifuge and re-suspend nuclei in a few cm^3 of distilled water.
9. To prevent agglutination, add a few drops of 1 M citric acid to the washing water each time.
10. This method can also be applied to tumour tissue, varying the amount of citric acid to get a final pH of 3·0.

HOERR'S MODIFICATION OF LAZAROW'S METHOD FOR LIVER NUCLEI (1943)

1. Occlude the inferior vena cava of the liver for 10–20 s/min, thus changing the hydrostatic pressure from 2–4 ft. Pass cold physiological salt solution simultaneously through the liver *in situ*, replacing the blood completely.

2. Dissect out the liver and chill to 0 °C without freezing.

3. Grind in 0·85 per cent sodium chloride solution and squeeze through bolting silk.

4. Wash out in 0·5–0·7 per cent sodium chloride solution at a pH of 6·0–6·2.

5. Separate and wash by successive centrifugation and suspension at a temperature of nearly 0 °C.

GULICK'S MODIFICATION OF BEHRENS' (1939) METHOD FOR THYMUS AND LYMPH NUCLEI

1. Dissect out the organ. Chop into small bits and freeze in liquid air.

2. Dehydrate in changes of dry acetone at 20 °C followed by extraction 'in ether. Remove excess ether in a desiccator over conc. H_2SO_4.

3. Mash in power mill, suspend in benzene–carbon tetrachloride mixture (sp. gr. 1·25).

4. Grind again in a ball mill with glass beads.

5. Centrifuge at slow speed (30–50 rev/min) for 1–2 months.

6. Suspend successively in pure benzene and benzene–carbon tetra-chloride mixture, with increasing proportions of carbon tetrachloride, centrifuging after each change till no more powder settles.

7. Then add benzene till the denser nuclear fraction comes down under centrifugation.

8. Separate and store.

SCHNEIDER AND PETERMANN'S SCHEDULE (1950, REFERRED TO BY SERRA, 1955)

This has been used for animal tissues. The experiment is carried out at 20 °C.

1. Chop 50 g of fresh calf thymus into small bits.

2. Mix in a Waring blendor with 50 cm^3 of 0·5 M aqueous sucrose solution and 400 cm^3 of a mixture of 0·25 M sucrose and 0·0018 M calcium chloride solutions for 4 min at 350 rev/min.

3. Squeeze through a double layer of gauze and a single layer of flannel-lette.

4. Centrifuge at 2000 rev/min for 10 min.

5. Suspend the sediment in 90 cm^3 of the mixed sucrose–calcium chloride solution and allow to settle for 10 min.

6. Filter through a double layer of gauze.

7. Centrifuge the filtrate at 2000 rev/min for 10 min.

8. Collect the sediment containing the nuclei. Purify by washing in sucrose–calcium chloride solution and re-suspension.

STERN AND MIRSKY'S SCHEDULE (1952)

This is suitable for plant materials.

1. Extract 6 g of wheat flour with petrol ether.
2. Suspend in 300 cm^3 of fresh ether and grind for 46 h in a ball mill.
3. Suspend in 500 cm^3 of a mixture of cyclohexane and carbon tetra-chloride, adjusted at a specific gravity of 1·395. Centrifuge.
4. Separate the supernatant. Add to it one-third its volume of ether and centrifuge. The sediment gives some nuclei with debris.
5. Suspend the sediment from step 3 in cyclohexane and carbon tetra-chloride mixture (sp. gr. 1·447), centrifuge at 6000 rev/min for 6 min.
6. Collect the supernatant, suspend in petrol ether and again centrifuge. The supernatant contains now largely nuclei.
7. The supernatant can be purified by repeating step 5 with 6000 rev/min and a mixture of sp. gr. 1·416 and then by repeating step 6.

VENDRELY'S SCHEDULE (1952)

This is suitable for animal material.

1. Cool the fresh tissue to −25 °C.
2. Chop into very small pieces and suspend in about 10 cm^3 of M/3 citric acid at 2 °C.
3. Shake thoroughly in a shaker and filter.
4. Centrifuge at 2 °C for 5 min at 3500 rev/min. Suspend the sediment in M/18 citric acid.
5. Repeat step 4 four times, each time reducing the speed of centrifuga-tion, and discarding the supernatant.
6. A final centrifugation for 5 min at 1000–1200 rev/min gives a homo-geneous suspension of nuclei.

ISOLATION OF CHROMOSOMES

CLAUDE AND POTTER'S METHOD (1943)

The tissues used were spleen cells of leukaemic mice and liver cells of normal rats and guinea-pigs.

1. Chill the tissue to 0 °C. Grind with an equal quantity of dry sand in a mortar for 3 min.
2. Slowly add 6 times its quantity of distilled water or 0·9 per cent sodium chloride solution (pH 7·4).
3. Centrifuge for 1 min at 1500 rev/min. Discard the residue.
4. Centrifuge again the supernatant liquid for 10 min. Collect the sedi-ment containing chromatin threads.

5. Suspend the sediment in one and a half times its volume of saline solution.

6. Centrifuge by alternately allowing the centrifuge to attain 1500 rev/min and then switching it off. Finally centrifuge at 1500 rev/min for 10 min.

7. Repeat steps 5 and 6 again with supernatant fluid. The sediment contains white threads of chromatin.

MIRSKY AND POLLISTER'S METHOD FOR EXTRACTING NUCLEOPROTEIN (1946)

This is suitable for plant tissues.

1. Extract 50 g of wheat flour with petrol ether, dry at room temperature and wash with 0·14 M sodium chloride solution.

2. Extract with 1 M sodium chloride solution and mix thoroughly.

3. Centrifuge at 10 000 rev/min.

4. Pour out into 10 volumes of distilled water. A fibrous mass of nucleoprotein precipitates out, which is found to be a mixture containing chiefly thymonucleohistone.

EXTRACTION OF NUCLEIC ACIDS

NON-ENZYMATIC METHODS

Removal of Both DNA and RNA

Dempsey's method (Dempsey and colleagues, 1947)

1. Bring paraffin sections of fixed materials down to water.

2. Treat with 1 N hydrochloric acid at 37 °C for 3 h.

3. Rinse thoroughly in distilled water.

4. Stain following pyronin–methyl green or some other staining schedule described for RNA and DNA and observe.

5. On comparison with control preparations, the extracted tissue shows absence of staining or less intense staining with basic dyes and more with acid dyes.

Erickson's method (see Gurr, 1958)

1. Bring paraffin sections down to water.

2. Treat in 5 per cent aqueous perchloric acid solution for 20–30 min at 60 °C.

3. Then treat in 1 per cent aqueous sodium carbonate solution for 1–5 min.

4. Rinse thoroughly in distilled water.

5. Stain as usual in pyronin–methyl green or some other double stain and observe.

6. The results are similar to those obtained with Dempsey's method.

Schneider's method (1945)

1. Bring paraffin sections down to water.
2. Treat in 4 per cent aqueous trichloracetic acid solution for 15 min at 90 °C.
3. Rinse carefully in distilled water to remove the acid completely.
4. Stain and observe as in the previous schedules in comparison with control preparations.
5. The results are similar to those observed in the previous schedules.

Removal of RNA alone

Erickson's method

1. Bring down paraffin sections of materials fixed in formalin to water.
2. Treat with 10 per cent aqueous perchloric acid solution at 4 °C for 12–18 h.
3. Treat with 1 per cent aqueous sodium carbonate solution for 1–5 min.
4. Rinse thoroughly in distilled water.
5. Stain following one of the schedules described earlier for staining nucleic acids.
6. When compared with control sections, the extracted tissue does not show any staining for RNA. If, after extraction, the tissue stains less intensely with basic dyes and more intensely with acid ones, the effect is due to the removal of RNA.

Foster and Wilson's method (see *Gurr, 1958*)

1. Bring down paraffin sections as usual to water.
2. Treat the slides with 2 per cent aqueous sodium tauroglycocholate solution at 60 °C for 24–48 h. Pass a continuous stream of oxygen through the reagent during this period.
3. Rinse carefully in distilled water.
4. Stain and observe as in the previous schedule.
5. Similar results are obtained.

Removal of DNA from Euchromatic Segments only (*Sharma, 1951*)

Fixation recommended is acetic ethanol (1:1).
1. Bring down paraffin sections to water as usual, or for squash preparations wash the fresh tissue, such as cut root tips, in water.

2. Treat with 0·25 M aqueous trichloracetic acid solution at 60 °C for 40 min.

3. Hydrolyse in N hydrochloric acid at 60 °C for 20 min.

4. Wash thoroughly in distilled water and stain in leuco-basic fuchsin solution.

5. For entire tissues, squash in 45 per cent acetic acid solution. In case of paraffin sections, run through acetic–alcohol and alcohol–xylol grades and finally mount in balsam.

6. The heterochromatic regions, particularly of the centromere and telomere, stain sharply in Feulgen, the rest of the chromosome remaining unstained. Secondary constriction regions do not take up stain.

ENZYMATIC METHODS

Extraction of RNA

The mode of preparation and use of ribonuclease for the extraction of RNA has been described under Brachet's schedule for pyronin–methyl green staining (*see* chapter on Nucleic Acids).

Extraction of DNA with Deoxyribonuclease

Reagents

Dissolve 3·6 mg of magnesium sulphate in 10 cm^3 of glass distilled water. Add 0·1 N sodium hydroxide solution to bring the pH to 6·0. Dissolve 0·25–0·5 mg of crystalline deoxyribonuclease in 1 cm^3 of the magnesium sulphate solution; 0·1 per cent gelatin can be added as stabiliser. Store at 4 °C.

Procedure

1. Bring down paraffin sections or squashes fixed in Carnoy's fluid or acetic–alcohol to distilled water.

2. Immerse a set of slides in the deoxyribonuclease solution and another set in magnesium sulphate solution for control. Treat at 37 °C for 1–2 h.

3. Hydrolyse in N HCl at 60 °C for 12 min and stain following Feulgen schedule as usual. Mount in 45 per cent acetic acid.

4. Chromosomes in the slides treated with deoxyribonuclease are colourless while those in the control slides stain bright magenta.

EXTRACTION OF PROTEINS

Extraction with Pepsin for the Removal of Tryptophane-containing Proteins

Recommended fixative is Carnoy's fluid.

1. Bring down paraffin sections or frozen sections to water as usual. Rinse in distilled water.

2. Incubate one set of slides in a covered Petri dish, lined with filter paper, and containing 0·005 g of crystalline pepsin dissolved in 5 cm^3 of 0·5 N hydrochloric acid, for 2–3 h at 37 °C.

3. For control, incubate a second set of slides in 5 cm^3 of 0·5 N hydrochloric acid under the same conditions.

4. Rinse both sets thoroughly in distilled water.

5. Stain both sets according to the schedule for staining proteins (*see* chapter on Proteins).

6. In the slides digested in pepsin, tryptophane-containing proteins do not take up stain as compared to the stained control preparations.

Extraction with Trypsin

The recommended fixative is Carnoy's fluid.

1. Bring down paraffin-embedded sections to distilled water.

2. Incubate a set of slides, in a covered pair of Petri dishes lined with filter paper, containing buffer solution (pH 6·0) for control at 37 °C for 15 min to 1 h.

3. Incubate a second set of slides in an identical pair of Petri dishes containing 0·025 g of crystalline trypsin dissolved in 25 cm^3 of the buffer solution (pH 6·0) at 37 °C for 15 min to 1 h.

4. Rinse thoroughly in distilled water and stain both sets following staining schedule for proteins.

5. In the control set, proteins take up stain according to the method employed. In the set digested with trypsin, proteins (mainly basic) are removed and so the sites do not take up stain.

REFERENCES

Aldridge, W. G. and Watson, M. L. (1963). *J. Histochem. Cytol.* **11**, 773
Allfrey, V. G. and Mirsky, A. B. (1962). *Proc. nat. Acad. Sci., Wash.* **48**, 1590
Atkinson, W. B. (1952). *Stain Tech.* **27**, 153
Behrens, M. (1939). *Hoppe-Seyl. Z.* **258**, 27
Boivin, A., Vendrely, R. and Vendrely, C. (1949). *C.N.R.S. Paris,* 67
Brachet, J. (1940). *C. R. Soc. biol. Paris* **153**, 88 and 90
Brachet, J. (1942). *Arch. Biol., Paris* **53**, 207
Brachet, J. and Shaver, J. R. (1948). *Stain Tech.* **23**, 177
Caspersson, T., Hammarsten, E. and Hammarsten, H. (1935). *Trans. Faraday Soc.* **31**, 367
Chargaff, E., Crampton, C. F. and Lipshitz, R. (1953). *Nature,* **172**, 289
Chorhazy, M., Bendich, A., Berenfreund, E. and Hutchinson, D. J. (1963). *J. Cell Biol.* **19**, 59
Claude, A. and Potter, J. S. (1943). *J. exp. Med.* **77**, 345
Cohen, S. J. (1944). *J. biol. Chem.* **156**, 691
Daly, M. M., Allfrey, V. G. and Mirsky, A. E. (1952). *J. gen. Physiol.* **36**, 173
Daly, M. M., Mirsky, A. E. and Ris, H. (1951). *J. gen. Physiol.* **34**, 439

510   *Extraction*

Danielli, J. F. (1947).   *Symp. Soc. exp. Biol.* **1,** 101
Danielli, J. F. (1953).   *Cytochemistry—A critical approach.* New York; Wiley
Dempsey, E. W., Bunting, H., Singer, M. and Wislocki, G. B. (1947).   *Anat. Rec.* **98,** 417
Dempsey, E. W., Singer, M. and Wislocki, G. B. (1950).   *Stain Tech.* **25,** 73
Dounce, A. L. (1952).   *Exp. Cell Res.* Suppl. **2,** 103
Dounce, A. L. and Lan, T. H. (1943).   *Science* **97,** 584
Edström, J. E. (1953).   *Biochim. biophys. Acta* **12,** 361
Erickson, R. O., Sax, K. O. and Ogur, M. (1949).   *Science* **110,** 472
Franz, F., Warden, I. and Mayer-Arendt, J. (1954).   *Naturwissenschaften* **7,** 165
Glick, D. (1949).   *Techniques of histo- and cyto-chemistry.* New York and London; Interscience Publishers
Goessner, W. (1954).   *Z. wiss. Mikr.* **61,** 377
Gurr, E. (1958).   *Methods of analytical histology and histochemistry.* London; Leonard Hill
Hoerr, N. L. (1943).   *Biol. Symp.* **10,** 185
Kaplan, H. S. and Heppel, L. A. (1956).   *J. biol. Chem.* **222,** 907
Kasten, F. H. (1965).   *J. Histochem. Cytochem.* **13,** 13
Kaufmann, B. P. (1950).   *Portug. Acta Biol. A.,* Goldschmidt memorial vol. 813
Kaufmann, B. P., Gay, H. and McDonald, M. R. (1950).   *Cold Spr. Harb. Symp. quant. Biol.* **14,** 85
Kaufmann, B. P., Gay, H. and McDonald, M. R. (1951a).   *Amer. J. Bot.* **38,** 268
Kaufmann, B. P., Gay, H. and McDonald, M. R. (1951b).   *J. cell. comp. Physiol.* **38** Suppl. 1, 71
Kaufmann, B. P., Gay, H., McDonald, M. R., Rowan, M. E. and Moore, E. C. (1951c).   *Carnegie Inst. Wash. Yearb.* **50,** 203
Kaufmann, B. P., Gay, H. and McDonald, M. R. (1960).   *Int. Rev. Cytol.* **9,** 77
Koenig, H. and Stahlecker, H. (1951).   *J. nat. Cancer Inst.* **12,** 237
Koenig, H. and Stahlecker, H. (1952).   *Proc. Soc. exp. Biol., N.Y.* **79,** 159
Kossel, A. (1928).   *The Protamines and histones.* London; Longmans
Kunitz, M. (1940). *J. gen. Physiol.* **24,** 15
Kunitz, M. (1948).   *Science* **108,** 19
Kunitz, M. (1950).   *J. gen. Physiol.* **33,** 349 and 363
Kurnick, N. B. (1955).   *Int. Rev. Cytol.* **4,** 221
Laskowski, M. (1942).   *Proc. Soc. exp. Biol., N.Y.* **44,** 354
Ledoux, L. and Brachet, J. (1955).   *Biochim. biophys. Acta* **16,** 290
Levene, D. (1901).   *J. med. Res.* **6,** 135
Mazia, D. (1941).   *Cold Spr. Harb. Symp. quant. Biol.* **9,** 40
McDonald, M. R. and Kaufmann, B. P. (1954).   *J. Histochem. Cytochem.* **2,** 387
Meischer, F. (1871).   *Medicinisch-Chemische Untersuchungen,* vol. 4, p. 441, ed. by Hoppe-Seyler, F. Berlin; Hirschwald
Mirsky, A. E. and Pollister, A. W. (1946).   *J. gen. Physiol.* **30,** 117
Ogur, H. and Rosen, G. (1949).   *Fed. Proc.* **8,** 234
Opie, E. L. and Lavin, G. I. (1946).   *J. exp. Med.* **84,** 107
Pearse, A. G. E. (1968).   *Histochemistry—Theoretical and Applied.* Boston; Little, Brown
Pollister, A. W., Flax, M., Himes, M. and Leuchtenberger, C. (1950).   *J. nat. Cancer Inst.* **10,** 1349
Schneider, W. C. (1945).   *J. biol. Chem.* **161,** 293
Schneider, W. C. (1946).   *J. biol. Chem.* **164,** 241
Serra, J. A. (1955).   *Handbuch der Pflanzen physiologie,* vol. 1, p. 413
Seshachar, B. R. and Flick, E. W. (1949).   *Science* **110,** 659
Sharma, A. K. (1951).   *Nature,* **167,** 441
Sharma, A. K. and Bhattacharjee, D. (1952).   *Nature,* **169,** 117
Sibatani, A., de Kloet, S. R., Allfrey, V. G. and Mirsky, A. E. (1962).   *Proc. Nat. Acad. Sci. Wash.* **48,** 471
Stern, H. and Mirsky, A. E. (1952).   *J. gen. Physiol.* **36,** 181
Stoneburg, C. A. (1939).   *J. biol. Chem.* **129,** 189
Stowell, R. E. and Zorzoli, A. (1947).   *Stain Tech.* **22,** 51
Sulkin, N. M. and Kuntz, A. (1950).   *Proc. Soc. exp. Biol., N.Y.* **73,** 413
Vendrely, C. (1952).   *Bull. biol.* **86,** 1
Vercauteren, R. (1950).   *Enzymologia* **14,** 134
Wenderoth, M. (1953).   *Acta haemat.* **9,** 47
White, J. C. (1947).   *J. Path. Bact.* **59,** 223
White, J. C. (1950).   *Proc. biochem. Soc., Biochem. J.* **47,** 16

Appendix I

CULTURE MEDIUM FOR DROSOPHILA
(According to Bridge's formula, 1932)

Materials

Corn meal	10 g
Treacle .	13·5 cm^3
Agar-agar	1·5 g
Water	75 cm^3

A trace of a preservative, such as Nipagin and Moldex.

Preparation

Add the agar to the water and keep overnight, then boil to dissolve it completely. Add the remaining ingredients and mix thoroughly. Add a few drops of yeast suspended in water. Cool. Transfer a small quantity of the food to a bottle 3 in. × 1 in. and mate the flies at 25 °C in this bottle. After 2 days, transfer the flies to bottles prepared by placing a quantity of food about 1 in. deep in each, together with a piece of folded sterilised crêpe paper. Stopper the bottles with cotton wool. After laying eggs for a day in a bottle, the parents must be removed to a fresh bottle. Keep at room temperature. When the larvae come out, give the cultures another yeasting. Keep the bottles at 18 °C in trays containing running water to slow down growth till the larvae are needed.

Appendix II

BUFFER SOLUTIONS

M/5 BUFFERS

STOCK SOLUTIONS

For M/5 *Acetate Buffer*

Solution A:

Glacial acetic acid	12·01 cm^3
Distilled water	987·09 cm^3

Solution B:

Sodium acetate trihydrate	27·21 g
Distilled water—make up to	1000 cm^3

For M/5 *Borate Buffer*

Solution A:

Aq. boric acid solution	1·24%

Solution B:

Aq. sodium biborate solution	1·9%

For M/5 *Citrate Buffer*

Solution A:

Aq. citric acid solution	4·2%

Solution B:

Aq. sodium citrate solution	5·9%

512

For M/5 Maleate Buffer

Solution A:

Maleic acid	46·4 g	
1 N sodium hydroxide	400	cm^3
Distilled water—make up to	1000	cm^3

Solution B:

1 N sodium hydroxide solution

Table A.1 CONCENTRATIONS NEEDED FOR DIFFERENT pH

pH	M/5 Acetate buffer		M/5 Borate buffer		M/5 Citrate buffer		M/5 Maleate buffer		Distilled water
	A cm^3	B cm^3	A cm^3	B cm^3	A cm^3	B cm^3	A cm^3	B cm^3	cm^3
2·7	20·0	0							
2·8	19·0	0·1							
2·9	19·8	0·2							
3·0	19·7	0·3							
3·1	19·5	0·4							
3·2	19·4	0·6							
3·3	19·2	0·8							
3·4	19·0	1·0	—	—	80	20	—	—	—
3·6	18·6	1·5	—	—	76	24	—	—	—
3·7	18·0	2·0	—	—	—	—	—	—	—
3·8	—	—	—	—	70	30	—	—	—
3·9	17·0	3·0	—	—	—	—	—	—	—
4·0	16·5	3·5	—	—	6·5	3·5	—	—	—
4·1	16·0	4·0	—	—	—	—	—	—	—
4·2	15·0	5·0	—	—	61	39	—	—	—
4·3	14·0	6·0	—	—	—	—	—	—	—
4·4	13·0	7·0	—	—	—	—	—	—	—
4·5	12·5	7·5	—	—	55	45	—	—	—
4·6	10·0	10·0	—	—	—	—	50	0·5	49·5
4·7	9·0	11·0	—	—	—	—	—	—	—
4·8	8·0	12·0	—	—	46	54	50	1	49
4·9	7·0	13·0	—	—	—	—	—	—	—
5·0	5·5	14·5	—	—	40	60	50	1·8	48·2
5·1	5·0	15·0	—	—	—	—	—	—	—
5·2	4·0	16·0	—	—	—	—	50	2·8	47·2
5·3	3·5	16·5	—	—	35	65	—	—	—
5·4	3·0	17·0	—	—	—	—	50	4·0	46
5·5	2·5	17·5	—	—	30	70	—	—	—
5·6	2·0	18·0	—	—	—	—	50	5·8	44·2
5·7	1·5	18·5	—	—	—	—	—	—	—
5·8	—	—	—	—	—	—	50	7·6	42·4
5·9	1·0	19·0	—	—	—	—	—	—	—
6·0	—	—	—	—	—	—	50	10	40
6·1	—	—	—	—	—	—	—	—	—
6·2	0·5	19·5	—	—	—	—	50	12·5	37·5
6·3	—	—	—	—	—	—	—	—	—
6·4	—	—	—	—	—	—	50	14·5	35·5
6·5	0	20·0	—	—	—	—	—	—	—
7·4	—	—	90	10	—	—	—	—	—
7·6	—	—	85	15	—	—	—	—	—
7·8	—	—	80	20	—	—	—	—	—
8·0	—	—	70	30	—	—	—	—	—
8·2	—	—	65	35	—	—	—	—	—
8·4	—	—	55	45	—	—	—	—	—
8·6	—	—	45	55	—	—	—	—	—
8·8	—	—	30	70	—	—	—	—	—
9·0	—	—	20	80	—	—	—	—	—
9·2	—	—	10	90	—	—	—	—	—

M/10 BUFFERS

STOCK SOLUTIONS

M/10 *Phosphate Buffers*

Solution A: M/10 disodium hydrogen phosphate containing:

Disodium hydrogen phosphate, anhydrous 10% aqueous	141·98 cm^3
Distilled water	859·02 cm^3

Solution B: M/10 sodium dihydrogen phosphate containing:

10% aq. sodium dihydrogen phosphate solution	138·05 cm^3
Distilled water	861·95 cm^3

M/10 *Veronal Buffers*

Solution A:

N/10 hydrochloric acid

Solution B:

Sodium diethylbarbiturate powder, dried in an oven and dissolved in distilled water to give M/10 solution.

Table A.2 CONCENTRATIONS NEEDED FOR DIFFERENT pH

pH	M/10 Phosphate buffer		M/10 Veronal buffer	
	A cm^3	B cm^3	A cm^3	B cm^3
5·3	2·6	97·4		
5·4	3·2	96·8		
5·5	4·0	96·0		
5·6	5·1	94·9		
5·7	6·4	93·6		
5·8	8·0	92·0		
5·9	9·9	90·1		
6·0	12·3	87·7		
6·1	15·1	84·9		
6·2	18·6	81·4		
6·3	22·5	77·5		
6·4	26·7	73·3	49	51
6·5	31·7	68·3	—	—
6·6	37·5	62·5	48·6	51·4
6·7	43·3	56·7	—	—
6·8	49·1	50·9	47·8	52·2
6·9	55·1	44·9	—	—
7·0	61·1	38·9	46·4	53·6
7·1	66·6	33·4	—	—
7·2	72·0	28·0	44·6	55·4
7·3	76·8	23·2	—	—
7·4	80·8	19·2	41·9	58·1
7·5	84·1	15·9	—	—
7·6	87·0	13·0	38·5	61·5
7·7	89·4	10·6	—	—
7·8	91·5	8·5	33·8	66·2
7·9	93·2	6·8	—	—
8·0	94·7	5·3	28·4	71·6
8·2	—	—	23·1	76·9
8·4	—	—	17·7	82·3
8·6	—	—	12·9	87·1
8·8	—	—	9·2	90·8
9·0	—	—	6·4	93·6
9·2	—	—	4·8	95·2
9·4	—	—	2·6	97·4
9·6	—	—	1·5	98·5
9·8	—	—	0·7	99·3

VERONAL ACETATE BUFFERS

STOCK SOLUTIONS

Solution A:

Sodium acetate	9·714 g
Sodium diethylbarbiturate	14·714 g
Distilled water	500 cm³, boiled and then cooled

Solution B:

8·5% aq. sodium chloride solution

Solution C:

N/10 hydrochloric acid

Table A.3

pH	Solution A cm³	Solution B cm³	Solution C cm³	Distilled water cm³
2·62	5	2	16·0	2·0
3·20	5	2	15·0	3·0
3·62	5	2	14·0	4·0
3·88	5	2	13·0	5·0
4·13	5	2	12·0	6·0
4·33	5	2	11·0	7·0
4·66	5	2	10·0	8·0
4·93	5	2	9·0	9·0
5·32	5	2	8·0	10·0
6·12	5	2	7·0	11·0
6·75	5	2	6·5	11·5
6·99	5	2	6·0	12·0
7·25	5	2	5·5	12·5
7·42	5	2	5·0	13·0
7·66	5	2	4·0	14·0
7·90	5	2	3·0	15·0
8·18	5	2	2·0	16·0
8·55	5	2	1·0	17·0
8·68	5	2	0·75	17·25
8·90	5	2	0·5	17·50
9·16	5	2	0·25	17·75

pH 5·2 TO 8·6

TRIS (HYDROXYMETHYL) AMINOMETHANE-MALEATE BUFFER. Stock solutions: 0·2M-tris-maleate (24·2 g tris+23·2 g nucleic acid, or 19·6 maleic anhydride, in 1 litre of distilled water).
Stock 50 cm^3+x cm^3 0·2M NaOH, made up to 200 cm^3.

Table A.4

pH	x	pH	x
5·2	7·0	7·0	48·2
5·4	10·8	7·2	51·0
5·6	15·5	7·4	54·0
5·8	20·5	7·6	58·0
6·0	26·0	7·8	63·5
6·2	31·5	8·0	69·0
6·4	37·0	8·2	75·0
6·6	42·5	8·4	81·0
6·8	45·0	8·6	86·5

pH 5·0 TO 7·4

0·2 M CACODYLATE BUFFER. For stock solution, dissolve 42·8 g of Na(CH$_3$)$_2$.AsO$_2$.3H$_2$O in 1 litre of distilled water. (Sabatini *et al.*, 1963)
Take 50 cm^3 of cacodylate, add x cm^3 of 0·2 M-HCl and make up to 200 cm^3.

Table A.5

pH	x	pH	x
5·0	47·0	6·4	18·3
5·2	45·0	6·6	13·3
5·4	43·2	6·8	9·3
5·6	39·2	7·0	6·3
5·8	34·8	7·2	4·2
6·0	29·6	7·4	2·7
6·2	23·8		

pH 2·7 TO 5·7

Stock solutions (Lewis, 1962):

(i) 1·0 M-sodium acetate, 27·22 g in 200 cm^3 water;
(ii) 0·2 M-formic acid: 14·2 cm^3 90 per cent formic acid, 11·69 g NaCl in 1 l.

Table A.6   pH VALUES AT TWO CONCENTRATIONS

Molar proportion of acetate	pH values at		Molar proportion of acetate	pH values at	
	0·02M	0·1M		0·02M	0·1M
0·00	2·72	2·38	0·48	4·11	4·07
0·04	2·85	2·55	0·52	4·25	4·21
0·08	2·96	2·72	0·56	4·39	4·35
0·12	3·06	2·87	0·60	4·52	4·48
0·16	3·15	3·02	0·64	4·65	4·61
0·20	3·25	3·17	0·68	4·78	4·74
0·24	3·36	3·30	0·72	4·91	4·87
0·28	3·48	3·42	0·76	5·04	5·00
0·32	3·59	3·54	0·80	5·18	5·14
0·36	3·71	3·66	0·84	5·32	5·28
0·40	3·84	3·79	0·88	5·48	5·45
0·44	3·97	3·93	0·92	5·71	5·68

Plates

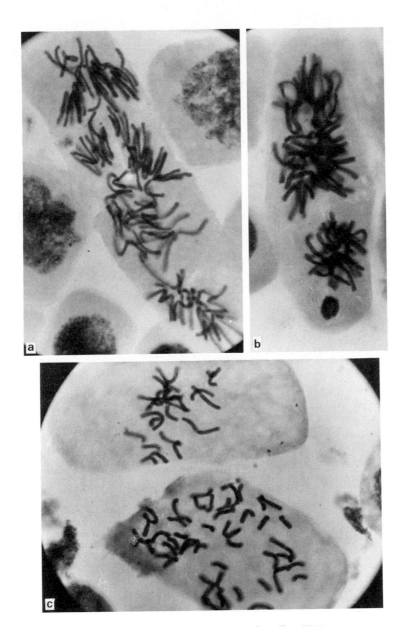

Plate 1:
(a) and (b)—*Multispindles in polyploid cells of* Allium cepa *root-tip, induced by treatment in sat. gammexane for 24 h, acetic-orcein stain;* (c)—*Polysomaty in somatic cells of* Zephyranthes mesochloa, *clarified through pre-treatment in 8-hydroxyquinolene, acetic-orcein stain* (courtesy of Drs. N. K. Bhattacharyya and M. Sarma, Cytogenetics Laboratory, Department of Botany, University of Calcutta)

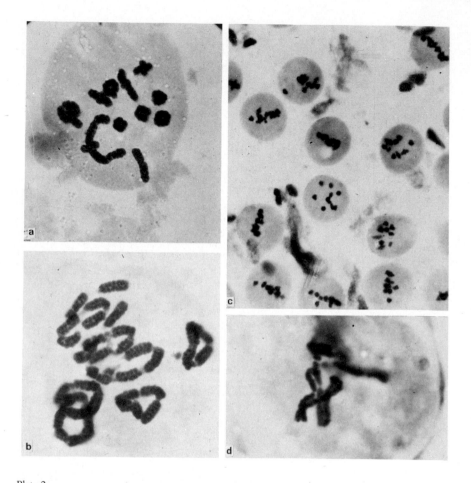

Plate 2:
(a) to (c)—*Acetic-carmine pollen mother cell smears:* (a)—*of* Commelina zebrina, *showing one quadrivalent and ten bivalents,*
(b)—*of* Tradescantia virginiana *showing quadrivalents and,*
(c)—*of* Cyanotis axillaris *showing secondary association of* 10 *bivalents.* (a), (b), (c) (courtesy of Drs. U. C. Bhattacharyya and S. Sen, Cytogenetics Laboratory, Department of Botany, University of Calcutta)
(d)—*Somatic crossing-over in* Haplopappus gracilis (2n = 4), *from a free floating cell in tissue culture* (courtesy of Drs. J. Mitra and F. C. Steward and the Editor, *American Journal of Botany*)

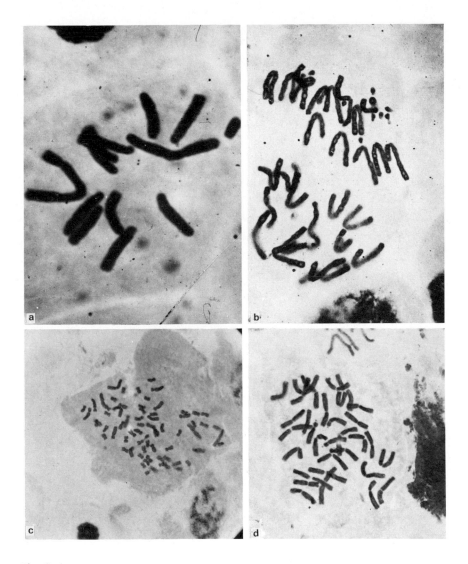

Plate 3:
(a)—*Somatic chromosomes of* Vicia faba *root, showing labile nature of secondary constriction region following treatment with 1000 R of x-rays.*
(b)—*Somatic anaphase stage in* Allium stracheyii *root, showing B-chromosomes, Feulgen stain*
(c)—*Somatic metaphase in* Ophiopogon intermedium *root, following pre-treatment with sat. isopsoralene solution, acetic-orcein stain*
(d)—*Somatic metaphase in* Allium tuberosum *root (4n), showing four satellited chromosomes after pre-treatment with aesculine and acetic-orcein stain* (courtesy of Drs. S. Sen, M. Sarma and Ashoke Chatterjee, Cytogenetics Laboratory, Department of Botany, University of Calcutta)

Plate 4:
Lactic-propionic-orcein squashes (courtesy of Prof. B. John)
(a)—*First metaphase of male meiosis of (upper)* Omocestus viridulus $(2n = 17)$ *and (lower)* Perapleurus alliaceous $(2n = 23)$ *with a heteropycnotic X univalent respectively*
(b)—*Diplotene of male in* Stauroderus scalaris $(2n = 17)$, *showing blocks of heterochromatin in autosomes and secondary constriction in X*

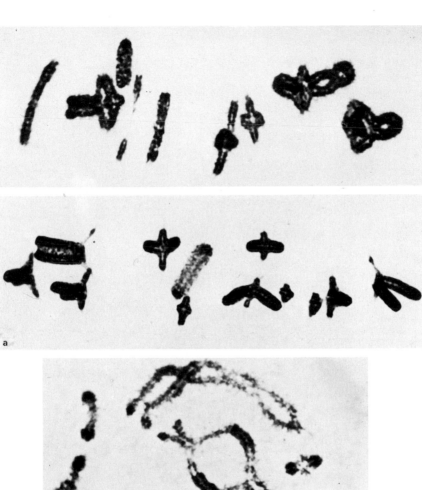

a

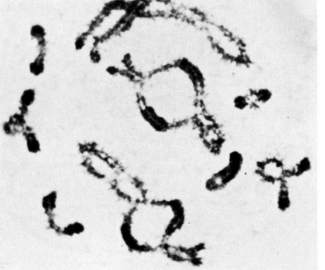

b

Plate 5:
Lactic-propionic-orcein squashes (courtesy of Prof. B. John)
(a)—*Neuroblast mitosis from a male embryo of* Myrmeleotettix maculatus (2n = 16 + X)
(b)—*First anaphase of meiosis in the male* Pezotettix giornii (2n = 23), *showing heteropycnotic X*
(c)—*Early mitotic anaphase in a neuroblast cell of a female embryo of* Schistocera gregaria (2n = 24)

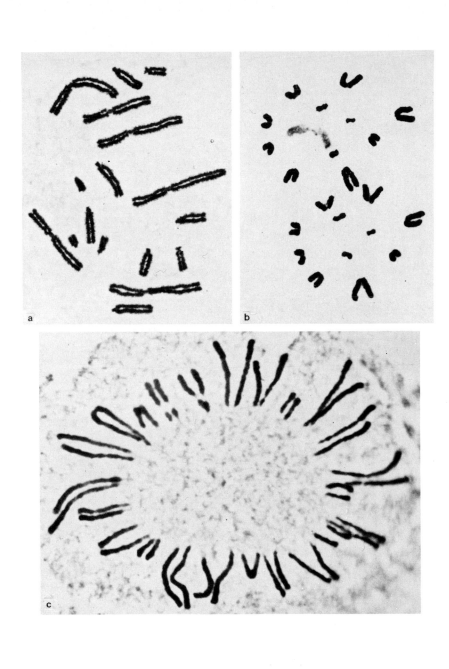

Plate 6:
Demonstration of ultrastructure of submicroscopic bands in a certain region of salivary gland chromosome X of Drosophila melanogaster, *following acetic–methanol fixation and uranyl acetate staining*
(a)—*Photomicrograph from the chromosome end taken from Darcupan embedded squash by means of phase contrast* (X 2500, reduced by four-fifths on reproduction)
(b) *and* (c)—*Electron micrographs of the same* (X 17000 *and* X 80000 *respectively*). (courtesy of Dr. V. Sorsa)

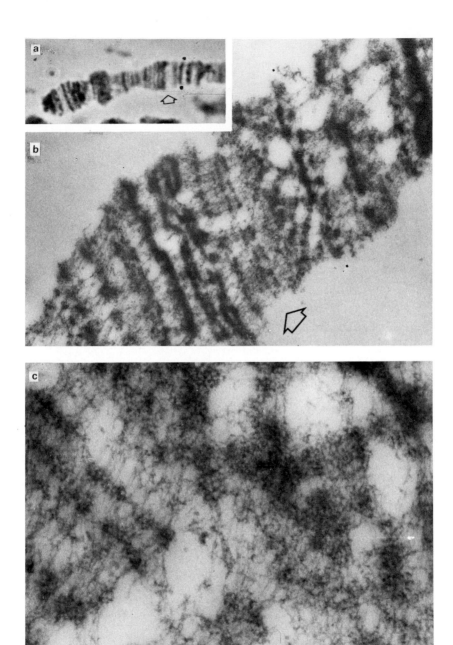

Plate 7:
(a)—*Man–mouse hybrid cell showing human metacentric and mouse acrocentric chromosomes, prepared following Sendai–virus mediated cell fusion method of Harris and Watkins* (courtesy of Prof. H. Harris and the Editors of the *Journal of Cell Science*)
(b)—**YACIR** *(Moloney virus induced lymphome in A/Sn mice having 40 acrocentric chromosomes) and MSWB (methyl cholanthrene induced sarcoma in ASW mice having 29 chromosomes) hybrid cell showing 69 or 70 chromosomes, prepared following the above-mentioned method* (courtesy of Dr. F. Weiner)

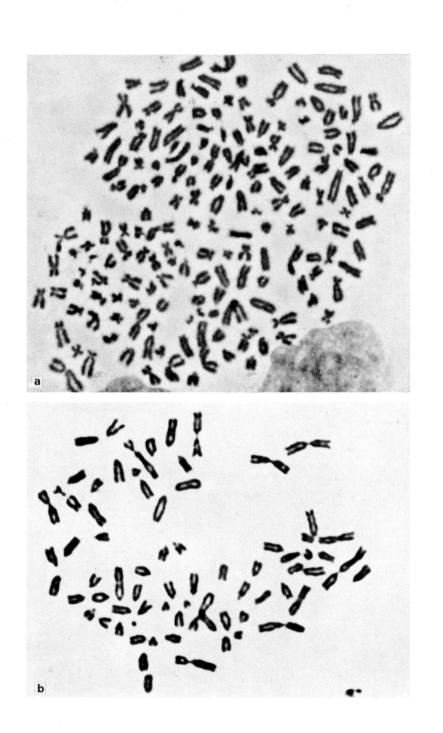

Plate 8:

(a)—*Acetic-orcein squash following several weeks' fixation in Nawashin's fluid of testis of grasshopper* Thericles whitei *Dersh, showing metaphase and anaphase stages* (courtesy of Prof. M. J. D. White)

(b)—*Tritiated thymidine autoradiograph of an ovarian follicle of the morabine grasshopper species* 'P45b' *(neo-XY race) showing heavy labelling in the proximal region of neo-X. (Method followed: thymidine injection followed after 11 h by colcemid injection and squashing after 5 h; preparations stained in Feulgen, dipped in K2 emulsion and exposed for* 3 weeks (courtesy of Mr. G. C. Webb, Dept. of Genetics, University of Melbourne)

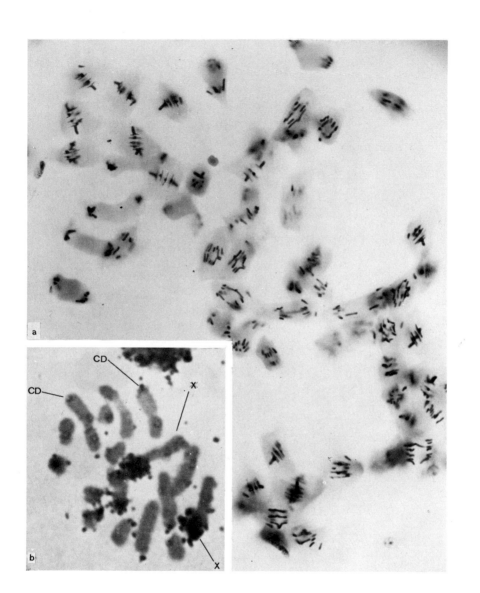

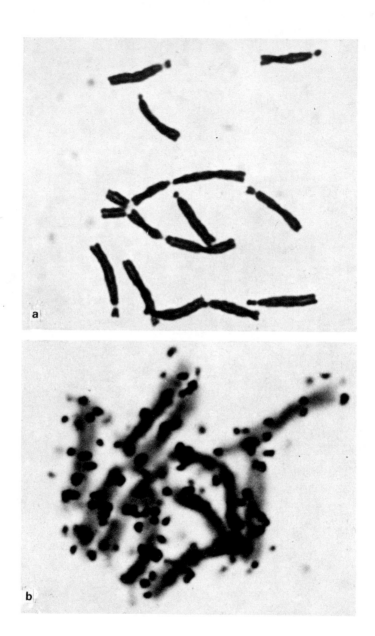

Plate 9:
(a)—*Localised chromosome exchange induced by hydroxyurea in* Vicia faba (courtesy of Prof. B. A. Kihlman)
(b)—*Autoradiograph of a root-tip cell of* Vicia faba, 7 h *after removal from* 9·5 h *labelling of the chromosomal protein with tritiated arginine* (courtesy of Drs. H. H. Smith and W. Prensky and Academic Press)

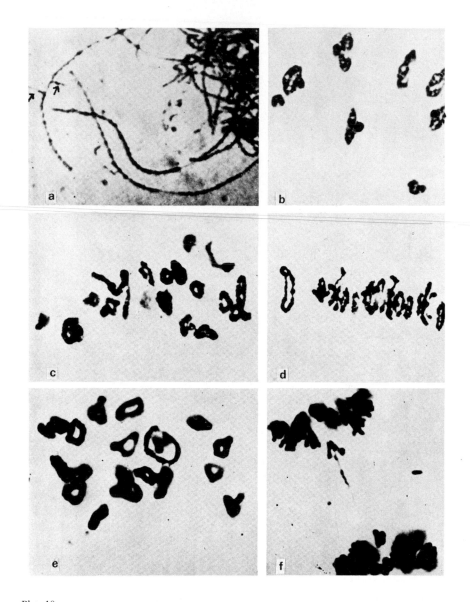

Plate 10:
Effect of oil treatment on meiosis in Triticum *species*
(a)—T. aestivum, 24 h *treatment in peanut oil, showing deficiency-duplication in pachytene*
(b)—T. monococcum, 6 h *treatment in mustard oil*
(c), (d), (e)—T. aestivum, *treatment in mustard, castor and peanut oil respectively showing different configurations* (courtesy of Dr. M. S. Swaminathan and the Editor, *Stain Technology*)

Plate 11:

(a)—*Hybridisation of mouse chromosome DNA with c-RNA synthesised from mouse satellite DNA—only centromeric heterochromatin labelled. (Method—Fix in acetic-ethanol* (1:3) *for* 10 min, *mount and mince the tissue in* 5 μl 45 *per cent acetic acid on a siliconised cover slip after draining off the excess fixative; invert cover slip on subbed slide (thin layered with gelatin and chrome alum); warm on a hot plate* (50 °C) *for* 3–5 min *till the liquid evaporates, flattening out the tissue; blot off excess acetic acid with filter paper; freeze-dry cover slip and slide in dry ice, separate and place slide in ethanol, later dry and store. For removal of RNA and residual proteins, keep slide in* 0·2 N HCl *at* 26 °C *for* 30 min *to remove basic protein, wash in water and air-dry. Put* 200 μl *Ribonuclease* (100 μg/ml *of pancreatic ribonuclease in* 2X SSC, *containing* 0·15 M NaCl, 0·015 M Na *citrate, brought to pH* 7·0 *with HCl) on the tissue, cover with a cover slip and keep in moist chamber (petriplate lined with filter paper) at* 37 °C *for* 1 h. *Remove cover slip in a jar filled with* 2X SSC, *rinse twice in the same mixture, pass through ethanol grades and dry in air. For denaturation, keep slide in* 0·07 N NaOH *at* 26–28 °C *for* 2 min, *transfer to* 70 *per cent ethanol and keep for* 10 min; *change twice; rinse thrice in* 95 *per cent ethanol for* 10 min *each and air-dry.*

For hybridisation, place 50 μl *of tritiated c-RNA on the tissue, cover with a square* (22 mm) *cover glass and keep slide in moist chamber at* 65 °C *overnight. Place slide in* 2X SSC *at room temperature to remove the cover slip and excess tritiated RNA; change twice in* 2X SSC, *transfer to ribonuclease solution* (20 μg/ml *in* 2X SSC) *at* 37 °C *for* 1 h. *Rinse several times in* 2X SSC *at room temperature, pass through ethanol grades, air dry and store.*

For autoradiography, follow the usual method of liquid emulsion autoradiography, using Kodak NTB-2 liquid emulsion, develop in Kodak D-19, stain with Giemsa and mount in permount.

For preparation of tritiated c-RNA by in vitro *transcription, use* E. coli *RNA polymerase (prepared from frozen cells of* E. coli *strain A-19—Burgess, R. et al., Nature, 221, 43, 1969). To* 0·25 ml *of buffer* (0·04 M *Tris pH* 7·9; 0·15 M KCl; 0·0046 M MgCl$_2$; 0·002 M MnCl$_2$; 7 × 10^{-5} M EDTA; 0·0058 M β*-mercaptoethanol), add* 4–5 *units of the enzyme,* 5–10 μg *of DNA* 50 nm mole GTP, 100 μc *of tritiated ATP, CTP, UTP (sp. activity—*10–30 c/m mole). *Incubate for* 90 min *at* 37 °C, *add DNAase I and keep at room temperature for* 20 min. *Add as carrier non-radioactive* E coli *RNA, extract mixture with* 200 μl 5 *per cent sodium dodecyl sulphate and* 2 ml *phenol; place aq. layer on Sephadex G-50 column; elute with water; pool TCA-insoluble count fraction; heat to* 85 °C (3 min); *cool and filter through nitrocellulose* (0·45 μm *pore size). The strength of the complementary polynucleotide is* 10 c/m mole = 0·7 × 10^8 dpm/μg *(courtesy of* Prof. J. G. Gall *and* Dr. M. L. Pardue)

(b)—*Landschütz ascites in a* ♀♀ CBA *mouse with* 44 *chromosomes, including* 5 *markers, one metacentric, one with a secondary constriction and three minutes from preparations made by intraperitoneal injection of colcemid soln followed by the usual orcein-air drying schedule (courtesy of* Dr. P. C. Koller *and* Dr. C. Talukdar)

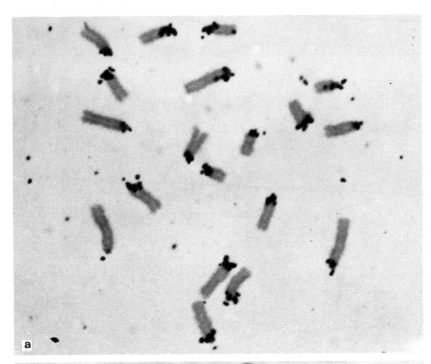

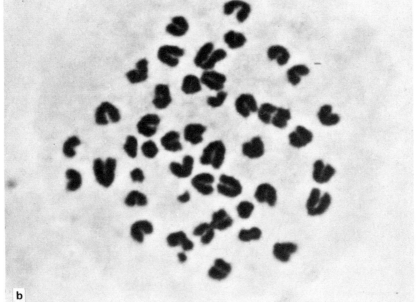

Plate 12:

(a)—*Venereal tumour of dog showing* 59 *chromosomes, acetic-dahlia stain and squash* (courtesy of Prof. S. Makino and Dr. M. S. Sasaki)

(b)—*Human leucocyte culture after irradiating blood with* 300 R *x-rays, showing a dicentric chromosome and an acentric fragment*

(c)—*Chromosomes of* Bandicota bengalensis bengalensis (*Indian mole rat* ♀, 2n = 42) *following bone marrow air drying method*

(d)—*Chromosomes from spleen culture of* Suncus murinus (*Indian shrew* ♂, 2n = 40). ((b), (c) and (d)—courtesy of Prof. S. P. Ray Chaudhuri)

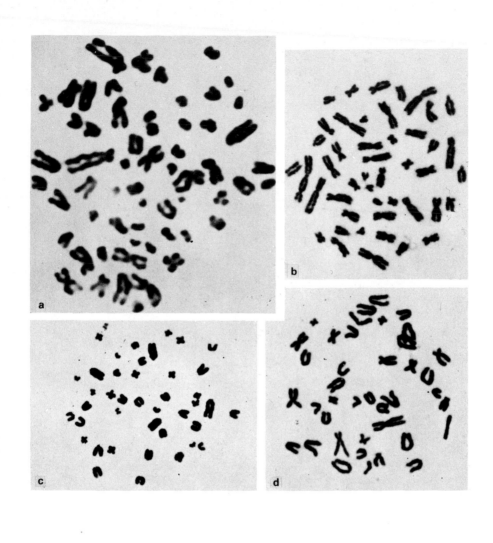

Plate 13:

(a)—*Salivary gland chromosome IV of* Chironomus tentans *showing Balbiani rings, fixed in acetic-alcohol* (1 : 3), *squashed in* 50 per cent *acetic acid, stained in toluidine blue (after Pelling) showing red RNA and blue DNA staining, photographed in blue-green light*

(b)—*Tritiated uridine autoradiograph of a similar chromosome following* 10 min *pulse labelling prior to fixation, acetic-carmine stain.* ((a) and (b)—courtesy of Prof. W. Beermann, Max-Planck-Institüt, Tübingen)

(c)—*Air-dried preparations of spermatogonical metaphase from testis of* Lycodon aulicus (*snake,* 2n = 36) *and,*

(d)—*from* Ptyas mucosus (*snake,* 2n = 36) *showing diakinesis.* ((c) and (d)—courtesy of Prof. S. P. Ray Chaudhuri)

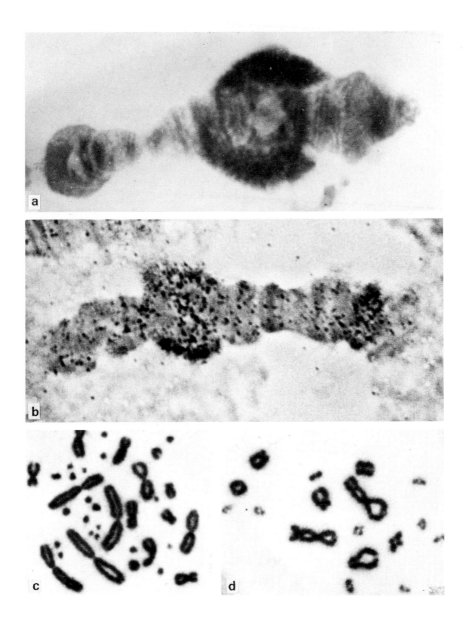

Plate 14:

Chromosomes of a gonosomic mosaic in creeping vole (Nicrotus oregoni), *showing*

(a)—*Somatic cell* (2n = 17, XO)

(b)—*Oogonium* (2n = 18, XX) *of female*

(c)—*Somatic cell* (2n = 18, XY)

(d)—*Spermatogonium* (2n = 17, OY) *of male*

Hypotonic treatment in distilled water (pH 7·0) *for* 30 min; *fixation in* 50 per cent *acetic-alcohol for* 40 min, *smear under cover glass; cover slip removed by freezing in methanol and dry ice; immersion in* 1 N HCl *for* 15 min *at* 60 °C; *Giemsa stain* (courtesy of Dr. S. Ohno)

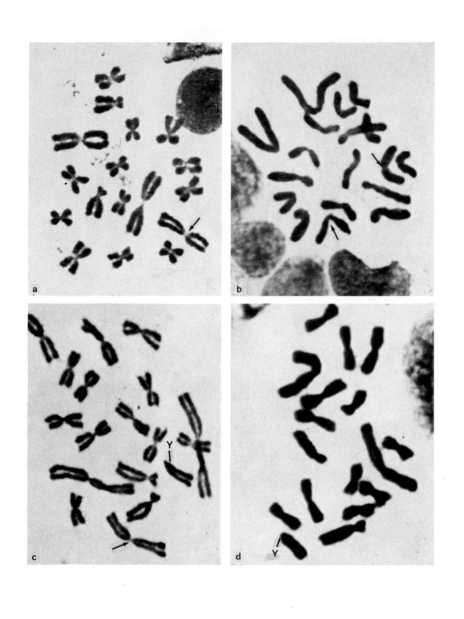

Plate 15:
(a)—*Meiosis in* Crinum latifolium (2n = 22), *a ring of 6 and 8 bivalents*
(b)—*Meiosis in* Verbena tenuisecta, 4 *bivalents and* 1 *trivalent.* ((a) *and* (b)—*acetic-ethanol fixation and acetic-carmine stain;* courtesy of Drs. T. N. Khoshoo, S. N. Raina and O. P. Arora)
(c), (d) and (e)—*Different meiotic configurations in* Triticum *species, showing multivalent formation as well; acetic-ethanol fixation and acetic-carmine stain* (courtesy of Prof. R. P. Roy)

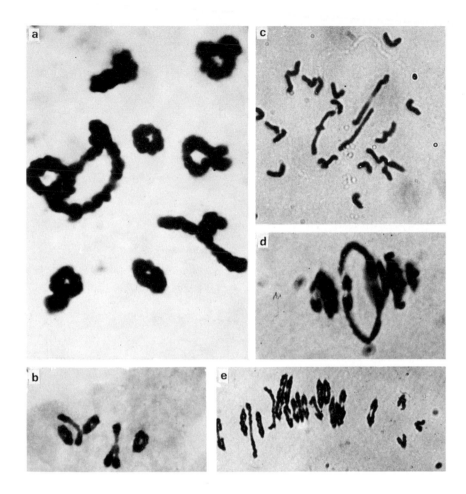

Plate 16:
(a) (b) and (c)—*Pachytene and diakinesis stages in* Neurospora crassa (n = 7). *Method followed: Dip strips of agar containing perithecia after 4 days of crossing in fixative (ethanol: acetic acid: 85 per cent lactic acid: 6:1:1) in closed vials and store in deep freeze. To prepare stain, mix a stock solution of 47 ml acetic acid and 20 ml lactic acid solution (1 ml 85 per cent lactic acid in 24 ml distilled water) with 5 ml N HCl and 28 ml distilled water and store at room temperature. Mix 5 ml of this fluid with 100 mg natural green (Gurr) and reflux for 4 min after boiling over a low flame, in a small beaker covered by a petri plate bottom containing two ice cubes. Tease out the asci in a drop of stain, mount after separation in another drop of stain under a cover glass, heat slightly and seal with dental wax* (courtesy of Prof. E. G. Barry)
(d)—*The left end of lampbrush bivalent VIII of* Triturus cristatus carnifex *showing in particular the telomeres and a sphere fusion* (courtesy of Prof. H. G. Callan and the Royal Society)

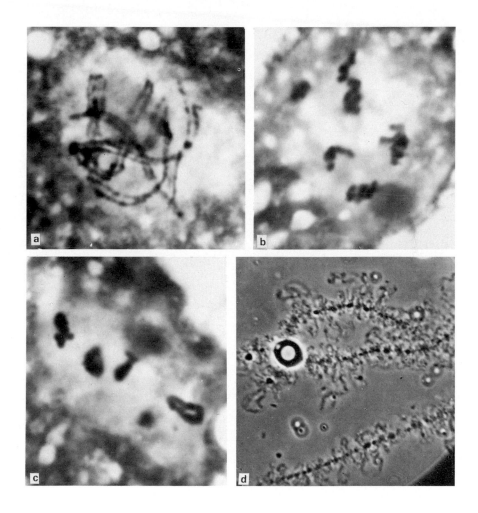

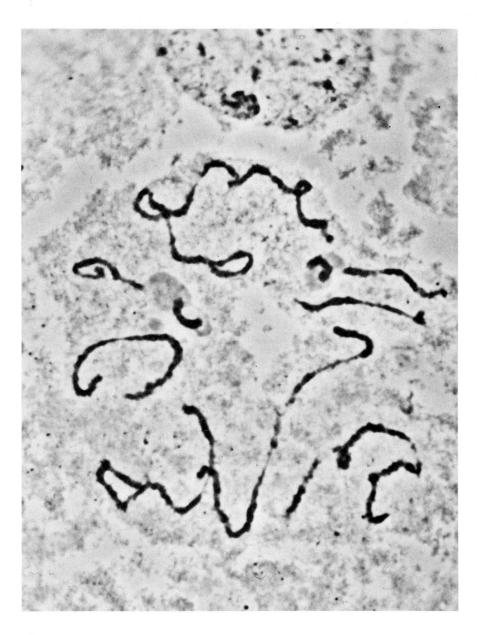

Plate 17:
Secondary constriction of the X chromosome attached to the nucleolus in the cell line $Pt-K_1$ *of female rat kangaroo,* Potorous tridactylus (2n = 12), *acetic-orcein squash without pre-treatment* (courtesy of Dr. T. C. Hsu)

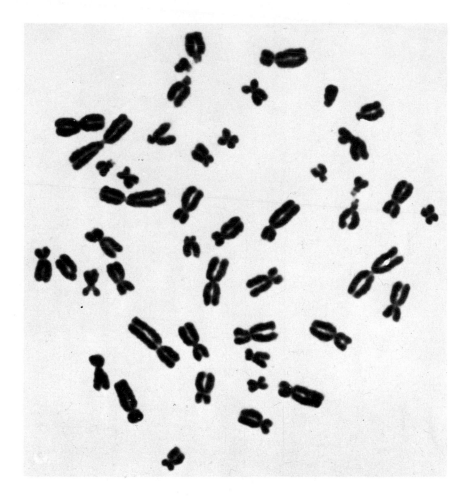

Plate 18:
Human leucocyte chromosomes (♂), following colchicine-hypotonic-acetic-orcein squash (courtesy of Drs. A. Levan and W. W. Nichols)

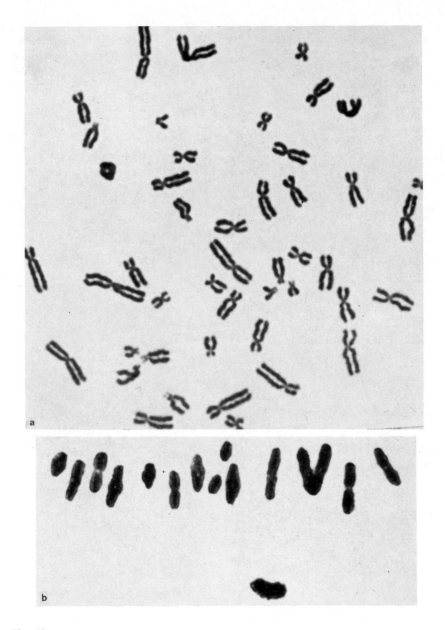

Plate 19:
(a)—*Human chromosomes from peripheral leucocyte culture (♀), with a ring chromosome* 18 (courtesy of Dr. M. Ray)
(b)—*First meiotic metaphase of* Ameles heldreichi *(Insecta, Mantodea), showing* $1_{III}+12_{II}+X$, *acetic-orcein squash* (courtesy of Prof. J. Wahrmann, Laboratory of Genetics, The Hebrew University, Jerusalem)

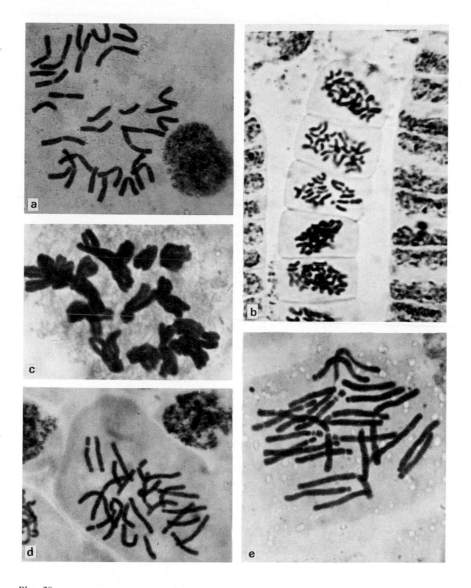

Plate 20:

(a)—*Endosperm of* Nothoscordum fragrans, *pre-treatment in aesculine and acetic-orcein stain*

(b)—*Antheridial filament of* Chara socotrensis *(green alga) showing polysomaty and synchronous division, acetic-orcein stain*

(c)—*Somatic chromosomes of* Vicia faba *following IBA treatment, inducing excess endo replication, acetic-orcein stain*

(d) and (e)—*Somatic chromosomes of* Lilium giganteum (2n = 24) *and* L. longiflorum (2n = 24) *respectively, showing pronounced secondary constrictions after treatment in colchicine and p–dichlorobenzene mixture, acetic-orcein stain* (courtesy of Drs. S. Sen, and P. Chatterjee, Cytogenetics Laboratory, Department of Botany, University of Calcutta)

Author Index

Page numbers in italics indicate complete references.

553

Subject Index